Ichthyology and Aquatic Biology

Ichthyology and Aquatic Biology

Edited by Rory Curtis

SYRAWOOD
PUBLISHING HOUSE
New York

Published by Syrawood Publishing House,
750 Third Avenue, 9th Floor,
New York, NY 10017, USA
www.syrawoodpublishinghouse.com

Ichthyology and Aquatic Biology
Edited by Rory Curtis

International Standard Book Number: 978-1-68286-467-8 (Hardback)

Cataloging-in-Publication Data

Ichthyology and aquatic biology / edited by Rory Curtis.
p. cm.
Includes bibliographical references and index.
ISBN 978-1-68286-467-8
1. Ichthyology. 2. Fishes. 3. Aquatic biology. 4. Aquatic organisms. 5. Marine biology. I. Curtis, Rory.
QL614.5 .I34 2017
597--dc23

Printed in China.

TABLE OF CONTENTS

PREFACE

This book includes some of the vital pieces of work being conducted across the world, on various topics related to ichthyology or the study of marine fish and its varied branches such as aquatic biology. The text also includes modern techniques of ichthyology that are practiced across the globe. While understanding the long-term perspectives of the topics, the book makes an effort in highlighting their impact as a modern tool for the growth of the discipline. This book is appropriate for students seeking detailed information in this area as well as for experts. It includes contributions of international scientists which will provide innovative insights into this field.

This book is the end result of constructive efforts and intensive research done by experts in this field. The aim of this book is to enlighten the readers with recent information in this area of research. The information provided in this profound book would serve as a valuable reference to students and researchers in this field.

At the end, I would like to thank all the authors for devoting their precious time and providing their valuable contributions to this book. I would also like to express my gratitude to my fellow colleagues who encouraged me throughout the process.

Editor

1

Are We Predicting the Actual or Apparent Distribution of Temperate Marine Fishes?

Jacquomo Monk[1]*, Daniel Ierodiaconou[1], Euan Harvey[2], Alex Rattray[1], Vincent L. Versace[3]

1 School of Life and Environmental Sciences, Faculty of Science and Technology, Deakin University, Warrnambool, Victoria, Australia, **2** School of Plant Biology (Oceans Institute M470), The University of Western Australia, Crawley, Western Australia, Australia, **3** School of Information Systems, Faculty of Business and Law, Deakin University, Warrnambool, Victoria, Australia

Abstract

Planning for resilience is the focus of many marine conservation programs and initiatives. These efforts aim to inform conservation strategies for marine regions to ensure they have inbuilt capacity to retain biological diversity and ecological function in the face of global environmental change – particularly changes in climate and resource exploitation. In the absence of direct biological and ecological information for many marine species, scientists are increasingly using spatially-explicit, predictive-modeling approaches. Through the improved access to multibeam sonar and underwater video technology these models provide spatial predictions of the most suitable regions for an organism at resolutions previously not possible. However, sensible-looking, well-performing models can provide very different predictions of distribution depending on which occurrence dataset is used. To examine this, we construct species distribution models for nine temperate marine sedentary fishes for a 25.7 km^2 study region off the coast of southeastern Australia. We use generalized linear model (GLM), generalized additive model (GAM) and maximum entropy (MAXENT) to build models based on co-located occurrence datasets derived from two underwater video methods (i.e. baited and towed video) and fine-scale multibeam sonar based seafloor habitat variables. Overall, this study found that the choice of modeling approach did not considerably influence the prediction of distributions based on the same occurrence dataset. However, greater dissimilarity between model predictions was observed across the nine fish taxa when the two occurrence datasets were compared (relative to models based on the same dataset). Based on these results it is difficult to draw any general trends in regards to which video method provides more reliable occurrence datasets. Nonetheless, we suggest predictions reflecting the species apparent distribution (i.e. a combination of species distribution and the probability of detecting it). Consequently, we also encourage researchers and marine managers to carefully interpret model predictions.

Editor: James Bell, Victoria University Wellington, New Zealand

Funding: This work was supported by the National Heritage Trust/Caring for our Country and Parks Victoria as part of the Victorian Marine Habitat Mapping Project. The funders had no role in data collection and analysis, decision to publish, or preparation of the manuscript.

Competing Interests: The authors have declared that no competing interests exist.

* E-mail: jacquomo.monk@deakin.edu.au

Introduction

Worldwide human activity is having adverse impacts on the structure and function of marine ecosystems [1]. In response, many initiatives are underway to identify, prioritize and ultimately preserve areas of importance [2,3,4,5,6]. An initial step in this process often involves delineating the distribution of species, assemblages or habitats [7]. This allows areas that support high diversity to be given the highest priority, which is particularly important when the maintenance and enhancement of biodiversity is the central goal of a management initiative [8]. To support such strategies, management agencies are increasingly seeking the provision of accurate, quantitative and spatially-explicit information on patterns of species distributions at scales relevant to the assessment and management process [3,9].

In this context, predictive modeling of species' distribution has become a fundamental tool [7]. These models have provided a popular analytical framework for relating geo-located observations of occurrence to environmental variables that contribute to a species distribution [7]. This relationship is based on statistically or theoretically derived response functions that characterize the environmental conditions associated with the ecological niche of a given organism [10]. Presence/absence models are frequently used to predict species distributions, but there is a common problem related to uncertainty in determining absences [11]; especially where the species is difficult to survey and does not appear to occupy all available suitable habitats [12]. In such cases, researchers have two options; (1) model presence/pseudo-absence (or background) data (e.g. [13]), or (2) model presence-only data (e.g. [14]). However, the use of a random sample from the background population to supply pseudo-absences may have unexpected consequences on results when true absences are expected [15,16]. For example, Wisz and Guisan [15] suggested that models built using random pseudo-absence datasets are expected to have lower predictive performance than models built with actual absences. In fact, it may be argued that, on a theoretical basis at least, a presence-only approach may be preferable because there is no requirement for truly exhaustive and exclusive absences; a requirement that is not met by most biodiversity data.

Often in the marine environment species distribution models (SDMs) are based on occurrence data collected by the researcher.

This often results in predictions that are reasonable depictions of the focal species distribution. While there are many different methods available to provide occurrence datasets for demersal fishes in the marine environment, baited and towed systems are being used as they overcome many issues associated with traditional survey methods (for a review of these issues see [17]). Moore et al. [18], for example, modeled temperate marine fish distributions based on baited-video-derived occurrence and fine-scale multibeam sonar datasets using classification trees and generalized additive models (GAMs). Similarly, Stoner et al. [19] used towed-video observations to model the relationships between *Lepidopsetta polyxystra* (northern rock sole) abundance and environmental variables (e.g. depth, sediment qualities, macroalgae) in five near-shore nursery grounds. The value of these data is not questioned; however the representativeness of these occurrence localities is dependent on which survey technique is used. For example, a comparison between baited and unbaited video, found that a greater number of individuals and species were recorded by the baited system; especially carnivorous fishes [20]. While research suggests that deploying a combination of survey methods used concurrently provides a better assessment of fish assemblages (e.g. [21]), the logistical or financial constraints of most studies limit fish biodiversity assessments to only one method. Consequently, understanding the influence of a survey method used to collect occurrence datasets for modeling of fine-scale habitat suitability is especially crucial. If an observation of zero individuals has arisen because it was present but not detected then any statistical inference based on such data are likely to be incomplete or wrong [22]. Consider a fish species that was observed in 10% of a study region. This fish may actually occur throughout the entire area but was only detected 10% of the time. Alternatively, it may also be found in only 10% of the area and has been detected perfectly.

Consequently, the aim of this paper is to highlight the potential differences in model predictions caused by the choice of survey method used to collect fish occurrence datasets. In this paper towed and baited underwater video methods are used to provide occurrence datasets for nine temperate marine sedentary fishes. Further, given the potential for presence/absence and presence-only models to produce considerably different predictions of habitat suitability, three commonly applied SDMs (i.e. generalized linear models; GLMs, GAMs and maximum entropy; MAXENT) are applied to the baited and towed video occurrences datasets for the nine fish taxa. This will provide a comparison of survey method as well as determine how these datasets potentially influence both types of SDMs (i.e. presence/absence and presence-only). With increasing application of SDMs in the marine environment, this paper will bring into focus the need for careful interpretation of predictions.

Results

Model Evaluation

The Area Under the receiver-operating characteristic Curves (AUC) was used to evaluate the models of habitat suitability for the nine demersal fish taxa. All of the 54 models of habitat suitability returned AUC values >0.5 (Table 1). Of these, MAXENT provided the top 12 highest performing models as measured by AUC. On 15 occasions GLMs and GAMs produced the same performing models as each other (as measured by AUC); with four of these being the equal highest performing model. In isolation, GLMs and GAMs only provided one model each that performed highest.

Similarity in Predictions of Habitat Suitability

The similarity between predictions of habitat suitability for the nine demersal fish taxa from the two observation techniques was assessed using the I statistic [23]. These I values were grouped by thresholds adapted from Roubicek et al. [24], which indicated I values <0.7 as low, 0.7–0.8 moderate, and >0.8 as highly degree of similarity between predictions of habitat suitability.

Similarity in habitat suitability models derived from baited-video. Generally, all three modeling approaches provided predictions with a high degree of similarity for the nine focal fish taxa based on the baited-video dataset (i.e. $I > 0.80$; Table 2; Figure 1; Figure 2). The GAMs and GLMs provided very similar model predictions (i.e. $I \sim 1$) for all nine fish taxa. Only slight differences were observed between MAXENT and the other two modeling approaches; with the greatest difference being observed for predictions of habitat suitability for Pempheris multiradiata (common bullseye; Table 2; Figure 2).

Similarity in habitat suitability models derived from towed-video. When compared to the baited-video datasets, more variation in similarity between the three modeling approaches was observed across the nine fish taxa based on the towed-video datasets (Table 2; Figure 1; Figure 2). Similar to the baited-video-derived models, GAM and GLM provided identical or very similar predictions for eight of the nine taxa; with only moderate differences being observed for *Upeneichthys vlamingii* (southern goatfish; $I = 0.78$). Similarly, comparison between MAXENT and the other two modeling approaches showed a high degree of similarity for five of the nine fish taxa. For *U. vlamingii* GAM was moderately different to both GLM and MAXENT, but no difference was observed between GLM and MAXENT ($I = 1$). Similarly, both GAM and GLM were moderately different to MAXENT for *Parequula melbournensis* (silverbelly), but no difference was observed between GAM and GLM (Table 2). *Meuschenia scaber* (cosmopolitan leatherjacket) showed the same trend as *P. melbournensis*, albeit to a greater degree (i.e. low similarity; Table 2). The GLM and MAXENT predictions showed only moderate similarity for *Odax cyanomelas* (herring cale).

Similarity in habitat suitability models between observation techniques. In contrast to models derived from the same dataset (i.e. baited or towed video), greater dissimilarity were observed when the predictions of habitat suitability based on the two video observation techniques were compared (Table 2). *Caesioperca* spp. (perch) was the only taxa to exhibit a high degree similarity for all modeling approaches between the two observation datasets (Table 2; Figure 1). With exception of the baited-video-derived MAXENT model for *Cheilodactylus nigripes* (magpie morwong) that showed a high degree of similarity with towed-video-derived MAXENT, models for this species had a moderate degree of similarity between observation datasets irrespective of modeling approach (Table 2). Identical predictions of habitat suitability were observed for GAM and GLM between observation datasets for *M. scaber* (Table 2). A high degree of similarity was observed for towed-video-derived MAXENT and baited-video-derived GAM and GLM (Table 2). Towed-video-derived GLM also showed a high degree of similarity with baited-video-derived MAXENT. A moderate degree of similarity was also observed between the towed-video-derived GLM and the baited-video-derived MAXENT. However, the towed-video-derived MAXENT showed a low degree of similarity with baited-video-derived MAXENT. For *Notolabrus tetricus* (blue-throated wrasse) a high degree of similarity was observed between the two observation datasets for GAM, GLM and MAXENT. Towed-video-derived MAXENT, however, showed a moderate degree of similarity with baited-video-

Table 1. Summary of model performances as measured by AUC for the baited and towed video datasets.

Taxon	Observation technique	GAM AUC	GLM AUC	MAXENT AUC
Caesioperca spp.	Baited video	0.73	0.80	0.63
	Towed video	0.57	0.57	0.61
Cheilodactylus nigripes	Baited video	0.66	0.66	0.65
	Towed video	0.77	0.77	0.84
Meuschenia scaber	Baited video	0.96	0.96	0.92
	Towed video	0.69	0.59	0.68
Notolabrus tetricus	Baited video	0.89	0.89	0.90
	Towed video	0.75	0.75	0.84
Odax cyanomelas	Baited video	0.87	0.87	0.92
	Towed video	0.81	0.75	0.90
Parequula melbournensis	Baited video	0.70	0.70	0.82
	Towed video	0.67	0.67	0.59
Pempheris multiradiata	Baited video	0.82	0.82	0.89
	Towed video	0.70	0.70	0.73
Pseudolabrus psittaculus	Baited video	0.90	0.90	0.87
	Towed video	0.60	0.60	0.68
Upeneichthys vlamingii	Baited video	0.72	0.72	0.74
	Towed video	0.54	0.54	0.63

derived GAM, GLM and MAXENT (Table 2). Predictions of habitat suitability using GAM, GLM and MAXENT based on either observation technique showed a high degree of similarity for *O. cyanomelas*. Towed-video-derived GLM, however, showed a moderate degree of similarity with all three models based on baited-video data. Towed-video-derived MAXENT for *P. melbournensis* showed a high degree of similarity with all three baited-video-derived models. Both towed-video-derived GAM and GLM showed the same trend; with a low degree of similarity between GAM and GLM between observation datasets (Table 2). Baited-video-derived MAXENT showed a moderate degree of similarity (i.e. I = 0.70). A moderate degree of similarity between the two observation datasets was observed for all models for *Pseudolabrus psittaculus* (rosy wrasse; Table 2). Towed-video-derived MAXENT for *U. vlamingii* showed a high degree of similarity with all three baited-video-derived models. In contrast, GAM and GLM showed a moderate degree of similarity with all three baited-video-derived models (Table 2). A low degree of similarity between the two observation datasets was observed for all models for *P. multiradiata* (Table 2; Figure 2).

Discussion

This study explored two commonly used underwater video techniques to provide occurrence data to develop and compare presence/pseudo-absence and presence-only fine-scale, habitat suitability models for nine species of temperate marine sedentary fishes. The habitat suitability models built in this study performed considerably better than random when assessed by AUC. The AUC values recorded in this study are similar to those observed in previous marine and terrestrial habitat suitability modeling studies (e.g. [14,25,26,27]). Despite the fact that AUC has recently been criticized (see [28,29]), it does provide a preliminary indication of the usefulness of a model for the identification of suitable habitat for a particular species [14]. This study has also demonstrated this to be the case, and further, that baited and towed video survey techniques are capable of providing models of similar quality (AUC); a conclusion supported by other recent studies [18,25,30,31,32,33]. For example, Moore et al. [18] compared the ability of presence/absence methods (GAM and classification and regression trees) to predict fine-scale habitat suitability for demersal fishes based on baited-video and multibeam sonar datasets. They found that baited-video and multibeam sonar datasets were useful in providing a detailed understanding of demersal fish-habitat associations, as well as accurately predicting species distributions across unsurveyed locations where continuous spatial seafloor data were available [18]. Similarly, Monk et al. [25] used towed-video and multibeam sonar derived datasets to compare commonly used presence-only methods (i.e. BIOCLIM, DOMAIN, Ecological-Niche Factor Analysis, MAXENT). They concluded that towed-video-based occurrence data provided well-performing, fine-scale models and encouraged the ongoing use of presence-only approaches, particularly MAXENT, in modeling suitable habitat for demersal marine fishes [25]. Despite these studies supporting the idea that underwater-video-based occurrence and multibeam sonar-derived datasets are capable of providing well-performing models, this study is one of the first to contrast these two video observation techniques for generating occurrence datasets for predictions of fine-scale habitat suitability for temperate marine fishes (using both presence/pseudo-absence and presence-only modeling techniques).

While numerous studies have compared modeling approaches in terms of model performance (i.e. via AUC or kappa; [14,25,26,27,34,35]), the main purpose of this study was to highlight how sensible-looking, well-performing (based on AUC) models can provide very different predictions of habitat suitability depending on which video observation dataset was used. Overall, greater dissimilarity between the three modeling approaches was observed across the nine fish taxa when models based on the two occurrence datasets were compared (relative to models based on the same survey method). This finding suggests

Table 2. Summaries of the similarity between habitat suitability predictions using the *I*-statistic (*I* ≈ 1: identical, *I* ≈ 0: completely different).

			Baited			Towed	
			GAM	**GLM**	**MAXENT**	**GAM**	**GLM**
Caesioperca spp.	Baited	GLM	0.96				
		MAXENT	0.87	0.88			
	Towed	GAM	0.92	0.93	0.88		
		GLM	0.92	0.93	0.88	1.00	
		MAXENT	0.87	0.87	0.83	0.87	0.87
Cheilodactylus nigripes	Baited	GLM	0.99				
		MAXENT	0.82	0.82			
	Towed	GAM	0.76	0.76	0.83		
		GLM	0.76	0.76	0.83	1.00	
		MAXENT	0.76	0.76	0.82	0.90	0.90
Meuschenia scaber	Baited	GLM	0.98				
		MAXENT	0.90	0.90			
	Towed	GAM	1.00	1.00	0.75		
		GLM	0.75	0.75	0.82	0.80	
		MAXENT	0.82	0.82	0.64	0.66	0.60
Notolabrus tetricus	Baited	GLM	0.95				
		MAXENT	0.90	0.90			
	Towed	GAM	0.87	0.87	0.84		
		GLM	0.87	0.87	0.84	1.00	
		MAXENT	0.79	0.79	0.77	0.86	0.86
Odax cyanomelas	Baited	GLM	0.99				
		MAXENT	0.89	0.89			
	Towed	GAM	0.81	0.81	0.80		
		GLM	0.76	0.76	0.75	0.81	
		MAXENT	0.82	0.82	0.81	0.83	0.76
Parequula melbournensis	Baited	GLM	0.98				
		MAXENT	0.86	0.86			
	Towed	GAM	0.68	0.68	0.70		
		GLM	0.68	0.68	0.70	1.00	
		MAXENT	0.82	0.82	0.87	0.72	0.72
Pempheris multiradiata	Baited	GLM	0.95				
		MAXENT	0.77	0.77			
	Towed	GAM	0.63	0.63	0.67		
		GLM	0.63	0.63	0.67	0.97	
		MAXENT	0.60	0.60	0.70	0.83	0.83
Pseudolabrus psittaculus	Baited	GLM	0.97				
		MAXENT	0.87	0.88			
	Towed	GAM	0.75	0.75	0.75		
		GLM	0.75	0.75	0.75	0.89	
		MAXENT	0.71	0.71	0.70	0.89	0.89
Upeneichthys vlamingii	Baited	GLM	0.99				
		MAXENT	0.87	0.87			
	Towed	GAM	0.78	0.78	0.79		
		GLM	0.78	0.78	0.79	0.78	
		MAXENT	0.80	0.80	0.82	0.78	1.00

that the characteristics of the occurrence data are important. This concept is supported by Kadmon et al. [36] who suggested that models are influenced by the reliability of occurrence data and distribution characteristics of the modeled species. The latter has been thoroughly discussed in previous marine and terrestrial studies (e.g. [25,26,36,37]), and suggest that narrowly distributed species that exhibit minimal niche variation provide more reliable models.

There are many factors that potentially influence the reliability of occurrence data of fishes, including traits such as; body size, crypticity, schooling behavior, habitat, camera avoidance and observer biases [38,39,40]. These issues are all inherently linked to the choice of survey method. For example, on numerous occasions, males of *N. tetricus* were observed aggressively guarding the bait against conspecific males and, in some cases, other species. Guarding behavior from territorial species could potentially lead to underestimates of species and densities on deployments where it is occurring [41]. Similarly, after reviewing some of the baited-video footage where the deployment vessel is heard approaching (to retrieve the unit after the 60-min deployment), it was noticed that on several occasions the fishes in the field of view rapidly departed. Avoidance behavior towards boat noise could potentially bias the towed-video dataset and has been reported by other researchers (e.g. [42,43]). Sarà et al. [42] found that agonistic behavior of *Thunnus thynnus* (northern bluefin tuna) was more evident when exposed to sounds from outboard motors. Further, Stoner et al. [44] found that the presence of underwater camera systems (e.g. towed-video system or remotely operated vehicles) could potentially bias the fish species observed; albeit difficult to quantify as most avoidance (or attraction) occurs outside the field of view.

Another possible explanation for the disparity between predictions of habitat suitability may be attributed to the deployment differences between the surveying methods (i.e. stationary *v.* moving). Consider habitat patches with comparable fish population density but varying in natural shelters such as crevices or macroalgae. Sampling that relies on the use of a moving platform (e.g. an obliquely angled towed-video camera that is flown 2 m above the seafloor) to provide visual observations could result in incomplete detection in habitat patches with more natural cavities or canopy forming macroalgae (e.g. kelp). By contrast, sampling that relies on a stationary platform (e.g. baited-video) to provide a visual observation dataset may allow time for species that are hiding amongst the crevices or canopy forming macroalgae to be observed. However, the use of bait in these systems is well documented to increase the number of pelagic or epibenthic carnivores in the vicinity of the camera deployment [45,46,47], which may result in reduced observations of cryptic or prey species recorded by these systems. The deployment of unbaited stationary cameras may decouple the effects of bait and seafloor structure, and warrants further investigation.

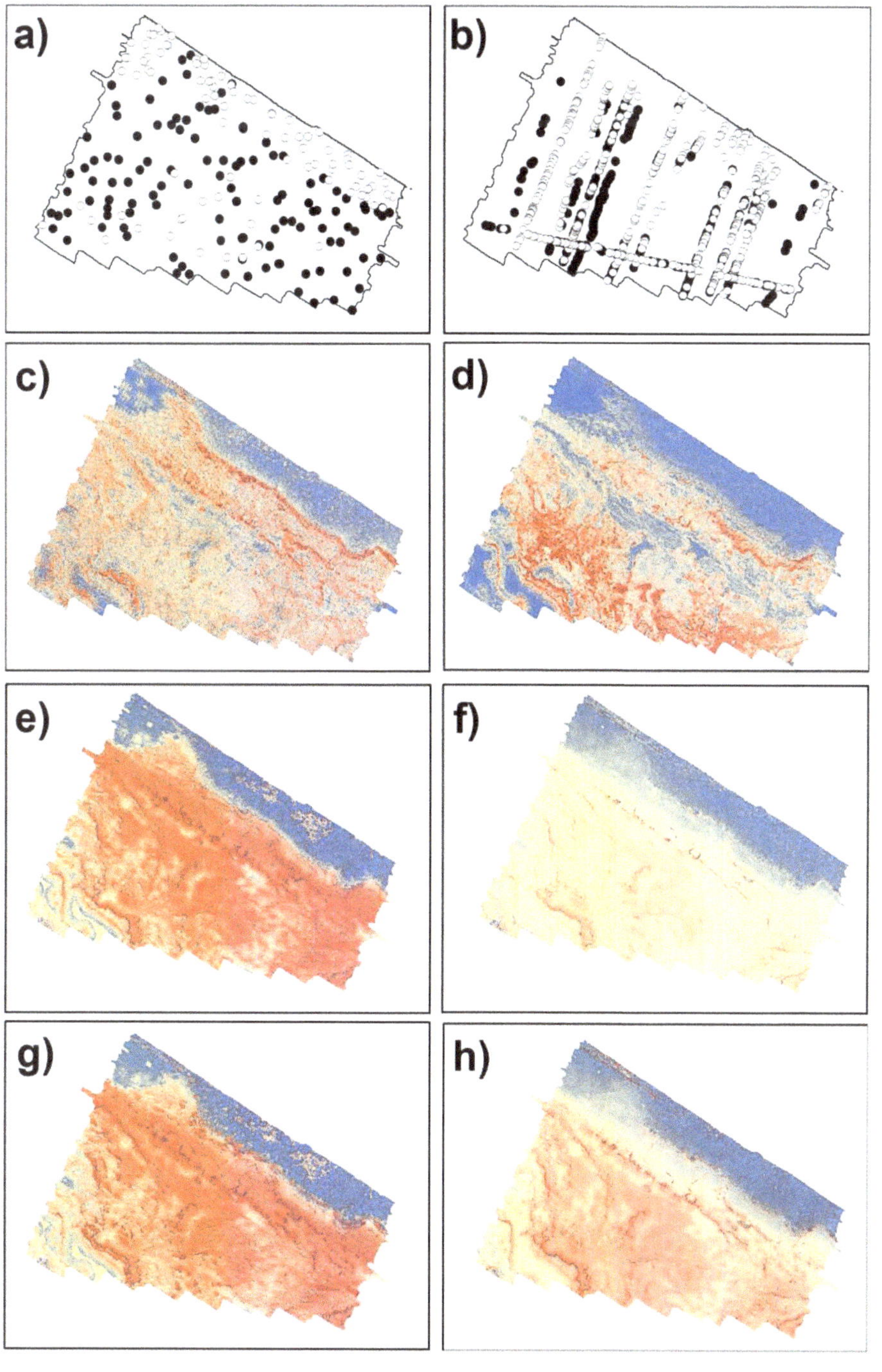

Figure 1. Example of similar habitat suitability predictions. Example of predicted habitat suitability for *Caesioperca* spp. showing very similar predictions based on the baited and towed video datasets. Left column: baited video. Right column: towed video. (a–b) presence/pseudo-absence localities (presence: black; pseudo-absence: white). (c–d) MAXENT predictions. (e–f) GLM predictions (g–h) GAM predictions. Red shading indicates high suitability, while blue highlights low suitability.

While the use of presence-only methods (such as MAXENT in this study) potentially reduces the issue of non-detections, the fact still remains that if a species is less detectable in a subset of its niche by a particular survey method, then this may influence model predictions. As highlighted by the differences between MAXENT model predictions observed in this study, the two video survey methods detected presences in slightly different environmental niches. For example, the fish taxon that provided predictions that were most similar was *Caesioperca* spp. (Figure 1). These are conspicuous aggregating fish species that are commonly observed in cloud-like schools feeding above reef crests [48]. Consequently, both video methods detected this conspicuous species in the same ecological niche, and are thus reflected in the similar predictions of suitable habitat. Conversely, *P. multiradiata* showed the lowest similarity in model predictions between survey methods (Figure 2). This timid species inhabits caves ranging from shallow (~10 m)

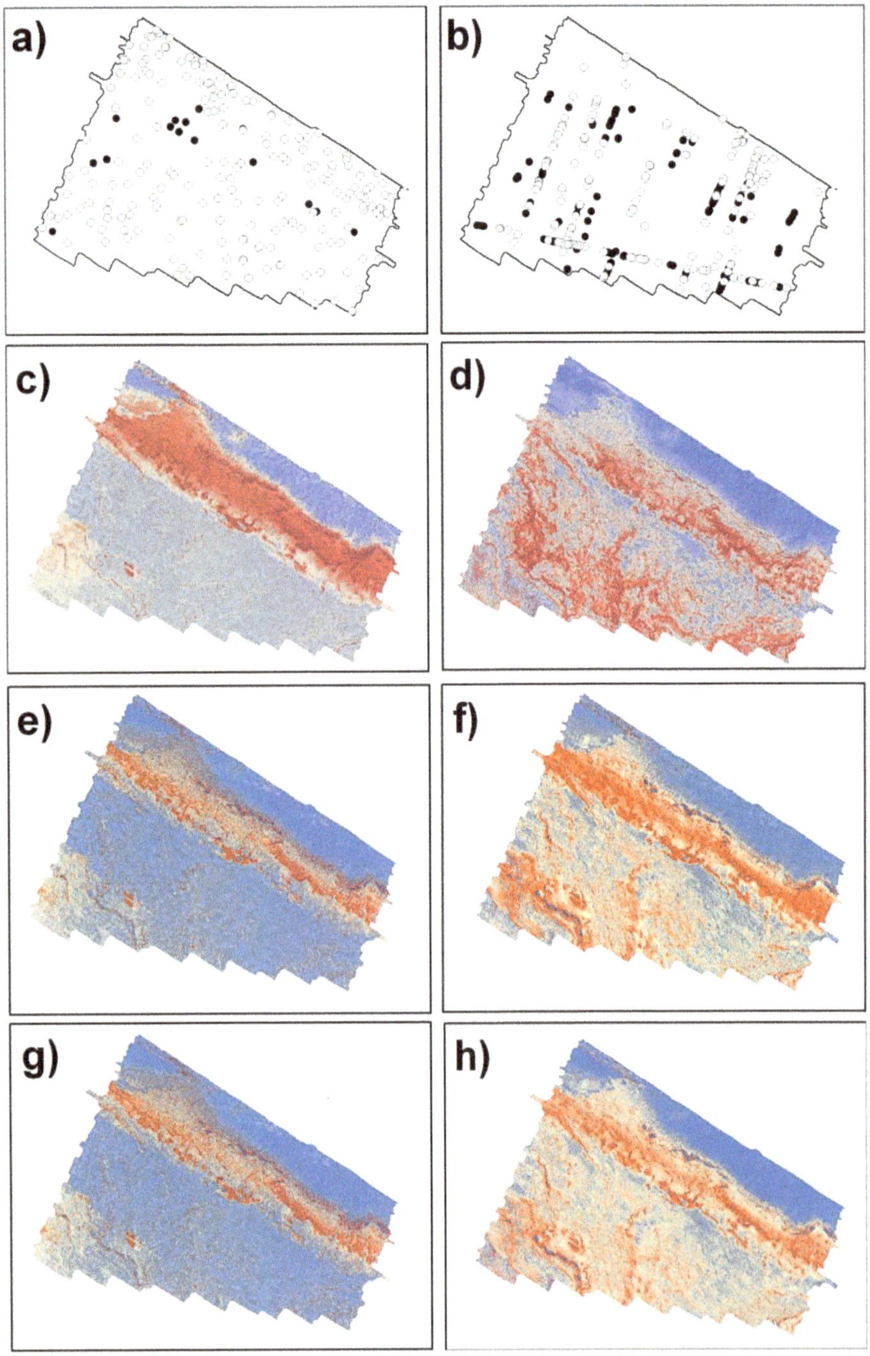

Figure 2. Example of dissimilar habitat suitability predictions. Example of predicted habitat suitability for *Pempheris multiradiata* showing dissimilar predictions based on the baited and towed video datasets. Left column: baited video. Right column: towed video. (a–b) presence/pseudo-absence localities (presence: black; pseudo-absence: white). (c–d) MAXENT predictions. (e–f) GLM predictions (g–h) GAM predictions. Red shading indicates high suitability, while blue highlights low suitability.

macroalgal-dominated reefs to deeper (~50 m) invertebrate colonized systems [49]. For this species, the two video methods detected individuals in different environmental niches. For example, the stationary characteristic of the baited-video method enabled individuals hiding among the shallow, complex reef systems to be recorded. However, the baited system recorded fewer occurrences in the deeper regions, which can possibly be attributed to the higher number of pelagic and epibenthic carnivores that were attracted to (and recorded by) the baited-video throughout these areas. This may result in avoidance by *P. multiradiata* from the baited-video on deployments were these predatory fishes were present in high numbers. By contrast, the towed video did not attract these predatory individuals, and recorded more *P. multiradiata* in the deeper invertebrate dominated regions of the study site.

Whilst all model predictions for the nine fish taxa reflect aspects of their known ecology, the results from this study suggest that

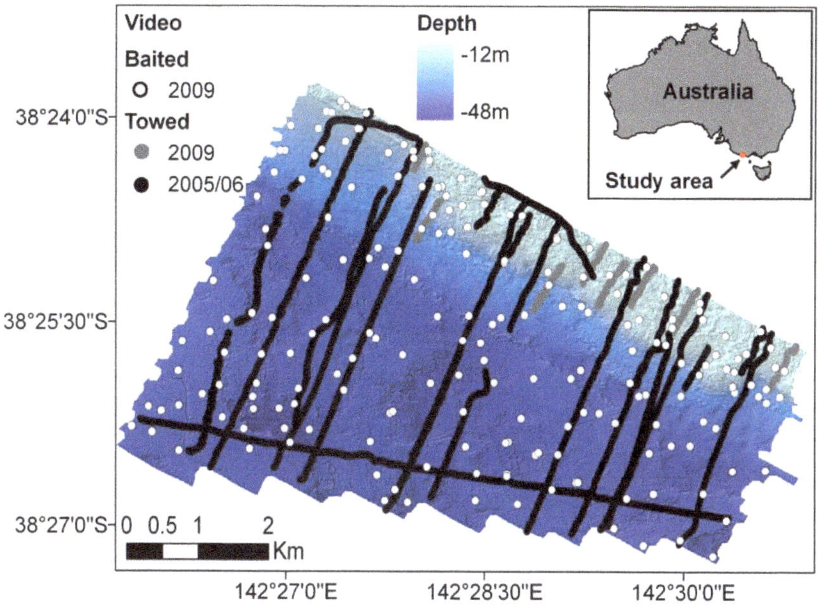

Figure 3. Study area. The location of the Warrnambool study area off the south-eastern coast of Australia. Shading indicates water depth. Black lines indicate towed video transects. White dots indicate baited video deployments. Red line delineates the southern extent of the Hopkins Bank.

differences between model predictions are actually reflecting the apparent species distribution (i.e. a combination of the habitat suitability of a fish species and the probability of detecting it; [50]). This bias cannot be resolved without consideration of variations in detectability that may arise from differences between survey methods, habitats and species [51]. Additionally, the strength of a habitat variable relationship in a SDM may be underestimated whenever imperfect detection is not accounted for, even with constant detectability [52]. Although some conventional SDMs allow for the problem of missing non-detection data to be partially addressed (i.e. missing zeros; [53]) or permit very general functional forms of covariates to be fitted, such as regression trees [54] and boosted regression trees [55], site-occupancy models may provide a useful alternative [50]. Site-occupancy models use the presence/absence (or more correctly termed detection/non-detection) patterns at sites surveyed multiple times (i.e. at least twice) to separate the sampling method from the ecological process and thus obtain estimates of the true species distribution along with unbiased estimates of variable importance [50,52,56]. However, temperate marine fish studies have rarely addressed the issue of detectability in video-derived occurrence datasets as surveying a site (especially for the purpose of building localized, fine-scale SDMs) more than once is often impractical due to limited weather windows (for more multiple surveys in a single field season) and deployment costs (both in a single season and between seasons). While underwater visual census methods have developed distance sampling techniques (e.g. [57]) that enable detectability to be accounted for, further research is needed to determine the relative detectability of fishes using towed or baited video systems.

Conclusions

This study has demonstrated that the characteristics of the video-derived occurrence data are potentially more important than the chosen modeling technique in developing fine-scale models of habitat suitability for temperate marine fishes. However, based on the results in the present study it is difficult to draw any general trends in regards to which circumstances what survey method provides more reliable occurrence datasets. Nonetheless, the main objective here was not to directly compare model performance, or even emphasize which of the two video methods compared is better for building models of habitat suitability for marine fishes. Instead, the purpose of this paper was to raise awareness that interpretation of habitat suitability models needs to account for the potential influence that the choice of survey method used to provide occurrence datasets may have. Whilst limitations within the datasets used in the present study precluded the use of site-occupancy models, which incorporate measures of detectability, it is suggested that these models may provide a practical alternative to conventional SDMs to predict the distribution of suitable habitat for demersal fishes. However, in the absence of repeat surveys (which would enable the use of enable site-occupancy models), researchers should select the method that is most likely to best detect their focal species; given the known behavior and ecology of the species (e.g. will the species react to bait? Or will the species respond to a moving camera?). Alternatively, conventional SDMs could be built utilizing occurrence datasets derived from different survey methods deployed at the same study site in a single field season. The similarity between these predictions can then be assessed (e.g. using *I*-statistic) to ensure that model outputs reflect as close to the actual distribution of suitable habitat for marine fishes as possible.

Materials and Methods

Study Site

The study site encompassed an area 25.7 km^2 that was situated offshore from the city of Warrnambool (38° 43′ S, 142° 43′ E), south-eastern Australia (Figure 3). The site ranged in depth from 12 to 50 m (calculated from multibeam sonar coverage for the study area which is based on lowest astronomical tide datum). The deeper regions consisted of a mixture of low (<1 m) profile reef and sandy sediments dominated by mixed red algae, sponges, ascidians, bryozoans and gorgonian corals [58]. The shallow reefs along the northern section (Hopkins Bank) of the study site were dominated by canopy-forming kelp (*Ecklonia radiata* and *Phyllospora comosa*). A large sandy area in the north-western region of the study site also supported a sparse cover of seagrass (Zosteraceae).

Fish Surveying Techniques

Ethics statement. This study was undertaken in strict accordance with the recommendations and procedures determined by the Prevention of Cruelty to Animals Act 1986 and its Regulations, and the Australian Code of Practice for the Care and Use of Animals for Scientific Purposes. This protocol was approved by the Deakin University Animal Welfare Committee (permit number A9-2009).

Video deployments. Two video methods were deployed; (1) baited and (2) towed. The sampling strategy for each method was designed to provide adequate spatial coverage of the study site and to be representative of the dominant substrata types and benthic biological habitats, whilst also being appropriate for SDM analysis.

Baited-video deployments. The sampling strategy for the baited-video drops was a stratified random design. Stratified deployments were allocated utilizing the multibeam sonar bathymetry and accurately predicted biotic habitat map available for the study area. The validation test for this biota map returned overall accuracy of 83%. Detailed descriptions of the method used to validate the biota map are available in Ierodiaconou et al. [58]. Ten replicate drops were preformed across three multibeam sonar derived variables; (1) bathymetry was grouped into 10 m depth strata (i.e. 10–19, 20–29, 30–39 and 40–49 m), (2) rugosity was reclassified into high, medium and low strata and (3) Benthic Position Index (BPI) was classed into trough, flat and peak. For predicted biotic habitat, 10 replicate drops were performed in each of the six predicted habitats (e.g. mixed brown algae, for details see; [58]). To ensure adequate spatial coverage 59 additional drops were randomly allocated throughout the deeper (i.e. >20 m) regions of the study site.

The baited-video systems used comprised two Sony HC 15E video cameras mounted 0.7 m apart on a base bar inwardly converged at 8° to gain an optimized field of view with visibility of ~7 m distance (water clarity dependent; [59]). Detailed information on the design and photogrammetric specifics can be viewed in Harvey and Shortis [60]. Up to five baited-video systems were deployed at any one time to increase sampling efficiency. Each baited-video system was deployed by boat and left to film on the seafloor for a period of 1 h. At least 36 min of filming time is recommended to obtain measures for the majority of fish species, though 60 min is advisable to obtain measures of numerous targeted fish species [61]. Each camera system was equipped with a synchronizing Light Emitting Diode (LED) that was visible in the fields of view of both video cameras. The LED was used to check synchronization of the video footage, thereby eliminating systematic error of motion parallax [59]. The LED emitted minimal light and was standard across all drops. Each system was also baited with ~800 grams of crushed *Sardinops sagax* (pilchard) in a closed plastic-coated wire mesh basket, suspended 1.2 m in front of the two cameras. Adjacent concurrent drops were separated by at least 250 m to avoid overlap of bait plumes and reduce the likelihood of fish moving between sites within the 1 h sampling period [62]. This distance is well accepted by baited camera operators to reduce the effects of bait plume between concurrent deployments [20,62,63,64]. All drops were deployed between 08:00 and 18:00 to minimize the effects of diurnal changes in fish behavior [65].

Fish were sampled from February to March 2009, with a total of 219 60-min baited-video deployments precisely positioned with a differential GPS to ensure accurate spatial location (Figure 3). However, 15 deployments were excluded from the analysis due poor visibility or being smothered in kelp. Footage from each baited-video deployment was interrogated to obtain the maximum number of fish belonging to each species present in the field of view of the cameras at one time (MaxN; [66]). These observations were obtained using the program EventMeasure (SeaGIS Pty Ltd; www.seagis.com.au) and converted to presence/pseudo-absence and presence-only datasets (see '*Data Treatment*').

Towed-video transects. Twenty nine towed-video transects were used to provide fish occurrence data. The 29 transects were selected to encompass the main physical gradients of the study site (e.g. depth, topographic variation, exposure). Additional transects were undertaken throughout the shallow heterogeneous regions of the site to ensure adequate representation of habitats throughout the study site. The 29 transects covered 68 linear km of the study area (Figure 3). Over eight days (December-March 2005/06 and February-March 2009) a micro remotely-operated vehicle (VideoRay Pro 3) was towed along the 29 transects at $0.5-1\ ms^{-1}$ (1–2 knots). The oblique angled camera was maintained approximately 2 m from the bottom with a field of view of ~7 m distance (water clarity dependent). The distance from the seafloor was monitored and maintained using live-feed video and a vessel-mounted winch system. A text overlay containing a time stamp and transect ID were recorded with the video using a Sony MiniDV recorder. The video footage was interrogated to identify fish species and the spatial position of each occurrence locality was then determined by matching the time stamp of the video with the corresponding survey positional data. The towed-video system was geo-located by integrating vessel location (Omnistar satellite dGPS), motion sensor (KVH) and acoustic camera positioning (Tracklink Ultra Short Baseline). The total propagated error at the maximum depth of the study site was ±5 m accuracy.

Species Distribution Models

The SDMs were built for the nine most common fish taxa (Table 3) across the two video survey techniques. For each of the fish taxon, GLM, GAM and MAXENT models were built using the same training and evaluation data derived from either the baited and towed video datasets. By using the same training and evaluation dataset derived from baited and towed video model performance can be directly compared.

Generalized linear models. The GLM is often used in ecological studies, and therefore serves as a benchmark for the other model types [67]. The GLMs were built in the Marine Geospatial Ecology Tool kit (MGET; Duke University), which interfaces between statistical software '*R*' (and its contributing packages; [68]) and ArcGIS 9.3. Each fish taxon was individually modeled using a logit link and a binomial error term. All models were fitted with the predictor variables listed in Table 4 using a backward stepwise procedure. The Akaike Information Criterion (AIC) was used to determine variable contribution as predictor variables were sequentially added and then dropped from the model.

Generalized additive models. The GAM is an extension of GLMs, allowing several transformations to be applied to individual independent variables before addition to the model. This improves the ability of the model to deal with nonlinear data. The GAMs were implemented using the *R* 'gam' package within MGET. Where necessary local spline smoothers equivalent to two degrees of freedom were used [69]. Backward stepwise procedure was again used to determine variable importance based on the AIC.

Table 3. Summary of the number of occurrences used in model building for each taxon based on the two video methods.

Taxon	Video method	Presence	Pseudo-absence
Caesioperca spp.	Baited	115	87
	Towed	431	431
Cheilodactylus nigripes	Baited	38	164
	Towed	32	32
Meuschenia scaber	Baited	106	96
	Towed	37	37
Notolabrus tetricus	Baited	114	88
	Towed	50	50
Odax cyanomelas	Baited	39	163
	Towed	56	56
Parequula melbournensis	Baited	29	173
	Towed	15	15
Pempheris multiradiata	Baited	15	187
	Towed	154	154
Pseudolabrus psittaculus	Baited	61	141
	Towed	90	90
Upeneichthys vlamingii	Baited	37	165
	Towed	31	31

Maximum entropy. The MAXENT approach has emerged as a powerful and flexible alternative to GLM and GAM for assessing species habitat suitability (see [14]). This general-purpose machine learning approach is designed for modeling species distributions based on presence-only data to determine the largest spread (i.e. maximum entropy) in a geographic dataset of species presences in relation to a set of background environmental variables [70]. This is essentially the same as maximizing the log likelihood of the data associated with the presence sites minus a penalty term [70]. This is conceptually similar to other commonly used information criteria (e.g. AIC). The penalty term regulates each environmental variable (known as a feature) by weighting it according to how much it adds complexity to the model; the sum of these weightings (including a regularization parameter, which is determined empirically) determines how much the likelihood should be penalized to prevent over-fitting [70]. Models used default settings; convergence threshold (0.00001), maximum iterations (1000), auto features, regularization multiplier ($r = 1$) and background points (10000).

Data Treatment

The GLMs and GAMs were fitted using presence/pseudo-absence, while MAXENT used only the presence datasets from each survey method. For the towed-video-derived models a 1:1 ratio of presence/pseudo-absence points were used (i.e. if there were 50 occurrences, then 50 pseudo-absence points were randomly generated; Table 3). For these towed-video datasets, pseudo-absence points were randomly generated along transects where no fish taxa were observed. For the baited-video-derived models pseudo-absences were generated from every deployment where the particular fish taxon was not observed (Table 3). All models used a set of relatively uncorrelated (i.e. spearman $rho<0.5$) multibeam sonar derived seafloor habitat variables as predictors (Table 4), and the fish occurrences as response variables (Table 3). For more detail on the 5 m^2 cell resolution multibeam sonar and habitat variables see Monk et al. [31]. Habitat variables were log- or root-transformed as necessary to prevent extreme frequency distributions within GLM and GAM.

Model Evaluation

Using the occurrence datasets that were set aside for model testing, model performance was evaluated using the threshold-

Table 4. Description of the nine seafloor variables retained to model building.

Variables	Variable description	Software
Aspect-Eastness Northness	Aspect (azimuthal bearing of steepest slope) has a inherent circularity built in, to overcome this, two trigonometric transformations [60] were applied; northness (sin(aspect)) and eastness (cos(aspect)). These two variables represent proxies for exposure.	Spatial Analyst- ArcGIS 9.3
Bathymetry	Bathymetry provides a measure of water depth based on lowest astronomical tide datum.	Fugro Starfix suite 9.1
Benthic position index	Measure of a location relative to the overall landscape. Calculated by comparing the elevation of a cell with the mean elevation of surrounding cells by the three analysis extents. Regions with positive values are higher than their surroundings, whereas areas negative values are lower. Flat areas have values closer to zero [61]	Benthic Terrain Modeler Tool for ArcGIS
Euclidean distance to bank	Hopkins bank is a major reef feature along the north section of the study region. This bank feature was extracted from a predicted reef class from a substratum map that was generated using a decision tree classifier [45]. The Euclidean distance to this feature was calculated in meters.	Spatial Analyst- ArcGIS 9.3
Euclidean distance to nearest reef	A predicted reef class from a substratum map, generated using a decision tree classifier [45], was used to calculate Euclidean distance (in meters) to nearest reef.	Spatial Analyst- ArcGIS 9.3
HSI-b	Hue-saturation-intensity (HSI) is a transformation of backscatter (proxy for seafloor hardness/softness), initially developed to decrease noise in radar reflectance [62]. Since backscatter represents seafloor reflectance, a HSI transformation may improve the separation of high and low frequency signal scattering properties of the substratum.	ENVI 4.2
Maximum Curvature	Maximum Curvature provides the greatest curve of either the profile or plan convexity relative to the analysis window [63].	ENVI 4.2
Rugosity	Rugosity provides the ratio of surface area to planar area within the analysis window and is to represent a measure of structural complexity [64].	Benthic Terrain Modeler Tool for ArcGIS

independent AUC of the ROC (receiver operating characteristic) [71]. The ROC plots sensitivity (the fraction of occurrence records that are classified as presence) against 1 – specificity (the portion of absence points that are classified as absent) for all possible thresholds. A curve that maximizes sensitivity for low values of the false positive fraction is considered a good model and is quantified by calculating the AUC. An AUC value of 0.5 implies the model predicts species occurrence no better than random, and a value of 1.0 implies perfect prediction [71]. The ROC curves and the AUC values were calculated in DIVA-GIS.

Potential Spatial and Temporal Confounding Factors

Semi-variograms and Moran's *I* statistics were built using SAM (Spatial Analysis in Macroecology) to check all model residuals for spatial autocorrelation. Only very weak spatial auto-correlation (i.e. all taxa<0.1) was found and corrections were not needed [72].

It is accepted that the collection of the two datasets three years apart is not ideal. Research suggests that many of the nine species used were highly territorial and maintain strict territories year round (e.g. labrids, monacanthids and pempherids). Barrett [73], for example, studied the short- and long-term patterns of six temperate marine fishes (from labrid and monacanthid families) and found that these species appeared to be permanent residents of the reef. Accordingly, these nine fish taxa should exhibit similar niche characteristics over time. However, to statistically address the time discrepancies within the towed-video dataset a permutation analysis of variance (PERMANOVA; [74]) was used to test differences in the nine fish taxa between 2005/6 and 2009. The Bray-Curtis coefficient was used to create resemblance matrix. To account for undefined values caused by joint absences a dummy species value of 1 was added in all samples. The PERMANOVA was run with 9999 permutations to obtain P(perm) values under unrestricted permutation of raw data using time as a factor. No significant difference was detected between 2005/6 and 2009 for towed video (pseudo-$F_{1,\ 161} = 2.28$, $P(perm) > 0.05$). This evidence, combined with aforementioned research, indicates that time within and between video datasets is not a confounding factor.

Similarity between Distribution Predictions

As proposed by Warren et al. [23], a modified Hellinger distance was used in order to compare between model predictions derived from the two observation techniques. This statistic (*I*) allows quantitative similarity assessments between distribution predictions (i.e. GIS grid layers) by computing the differences between them cell by cell. The *I*-values range from 0, indicating that the two predictions are completely different, to 1, suggesting that both are equal. The *I*-statistic is independent of sample size and predicted range sizes, making it superior to other metrics that have been proposed earlier [23]. This study adapted thresholds used by Roubicek et al. [24], and considered *I* values >0.8 are indicative of high degree of prediction overlap, values between 0.7 and 0.8 are moderate and values <0.7 indicate low similarity.

Acknowledgments

Towed-video and multibeam bathymetry datasets were collected as part of the Victorian Marine Habitat Mapping Project. Thank you to the crews aboard *Courageous II* and *Tanamerah* for undertaking the towed and baited video surveys, and also to the crew aboard the Australian Maritime College vessel *Bluefin* and Fugro Survey for the collection of multibeam sonar data. We would also like to thank Dr Mark Norman (Museum Victoria) and Dr Matt Edmunds (Australian Marine Ecology Pty Ltd) for their help with fish species identification. Also, we would like to thank Dr Tanya Compton (New Zealand Institute of Water and Atmospheric Research) for her constructive comments on an earlier version of the paper. The constructive comments from the two anonymous reviewers and the editor also greatly improved this paper. Analyses were undertaken at Deakin University, Warrnambool, Victoria, GIS Laboratory facility.

Author Contributions

Conceived and designed the experiments: JM DI VV. Performed the experiments: JM. Analyzed the data: JM. Wrote the paper: JM DI VV EH AR.

References

1. Jackson JBC, Kirby MX, Berger WH, Bjorndal KA, Botsford LW, et al. (2001) Historical overfishing and the recent collapse of coastal ecosystems. Science 293: 629–637.
2. Aswani S, Lauer M (2006) Benthic mapping using local aerial photo interpretation and resident taxa inventories for designing marine protected areas. Environmental Conservation 33: 263–273.
3. Harris PT, Whiteway T (2009) High seas marine protected areas: benthic environmental conservation priorities from a GIS analysis of global ocean biophysical data. Ocean and Coastal Management 52: 22–38.
4. Jordan A, Lawler M, Halley V, Barrett N (2005) Seabed habitat mapping in the Kent Group of islands and its role in marine protected area planning. Aquatic Conservation-Marine and Freshwater Ecosystems 15: 51–70.
5. Ward TJ, Vanderklift MA, Nicholls AO, Kenchington RA (1999) Selecting marine reserves using habitats and species assemblages as surrogates for biological diversity. Ecological Applications 9: 691–698.
6. Zacharias MA, Gregr EJ (2005) Sensitivity and vulnerability in marine environments: an approach to identifying vulnerable marine areas. Conservation Biology 19: 86–97.
7. Guisan A, Zimmermann NE (2000) Predictive habitat distribution models in ecology. Ecological Modelling 135: 147–186.
8. Myers N, Mittermeier RA, Mittermeier CG, da Fonseca GAB, Kent J (2000) Biodiversity hotspots for conservation priorities. Nature 403: 853–858.
9. Vanderklift MA, Ward TJ (2000) Using biological survey data when selecting Marine Protected Areas: an operational framework and associated risks. Pacific Conservation Biology 6: 152–161.
10. Austin M (2007) Species distribution models and ecological theory: a critical assessment and some possible new approaches. Ecological Modelling 200: 1–19.
11. Hirzel AH, Helfer W, Metral F (2001) Assessing habitat-suitability models with a virtual species. Ecological Modelling 145: 111–121.
12. Gibson L, Barrett B, Burbidge AH (2007) Dealing with uncertain absences in habitat modelling: a case study of a rare ground-dwelling parrot. Diversity and Distributions 13: 704–713.
13. Engler R, Guisan A, Rechsteiner L (2004) An improved approach for predicting the distribution of rare and endangered species from occurrence and pseudo-absence data. Journal of Applied Ecology 41: 263–274.
14. Elith J, Graham CH, Anderson RP, Dudik M, Ferrier S, et al. (2006) Novel methods improve prediction of species' distributions from occurrence data. Ecography 29: 129–151.
15. Wisz M, Guisan A (2009) Do pseudo-absence selection strategies influence species distribution models and their predictions? An information-theoretic approach based on simulated data. BMC Ecology 9: 8.
16. VanDerWal J, Shoo LP, Graham C, Williams SE (2009) Selecting pseudo-absence data for presence-only distribution modeling: how far should you stray from what you know? Ecological Modelling 220: 589–594.
17. Murphy HM, Jenkins GP (2010) Observational methods used in marine spatial monitoring of fishes and associated habitats: a review. Marine and Freshwater Research 61: 236–252.
18. Moore CH, Harvey ES, Van Niel KP (2009) Spatial prediction of demersal fish distributions: enhancing our understanding of species-environment relationships. ICES Journal of Marine Science 66: 2068–2075.
19. Stoner AW, Spencer ML, Ryer CH (2007) Flatfish-habitat associations in Alaska nursery grounds: Use of continuous video records for multi-scale spatial analysis. Journal of Sea Research 57: 137–150.
20. Harvey ES, Cappo M, Butler JJ, Hall N, Kendrick GA (2007) Bait attraction affects the performance of remote underwater video stations in assessment of demersal fish community structure. Marine Ecology-Progress Series 350: 245–254.
21. Colton MA, Swearer SE (2010) A comparison of two survey methods: differences between underwater visual census and baited remote underwater video. Marine Ecology-Progress Series 400: 19–36.
22. Wintle BA, Kavanagh RP, McCarthy MA, Burgman MA (2005) Estimating and dealing with detectability in occupancy surveys for forest owls and arboreal marsupials. Journal of Wildlife Management 69: 905–917.
23. Warren DL, Glor RE, Turelli M (2008) Environmental niche equivalency versus conservatism: quantitative approaches to niche evolution. Evolution 62: 2868–2883.

24. Roubicek AJ, VanDerWal J, Beaumont LJ, Pitman AJ, Wilson P, et al. (2010) Does the choice of climate baseline matter in ecological niche modelling? Ecological Modelling 221: 2280–2286.
25. Monk J, Ierodiaconou D, Versace V, Bellgrove A, Harvey E, et al. (2010) Habitat suitability for marine fishes using presence-only modelling and multibeam sonar. Marine Ecology-Progress Series 420: 157–174.
26. Segurado P, Araújo MB (2004) An evaluation of methods for modelling species distributions. Journal of Biogeography 31: 1555–1568.
27. Tsoar A, Allouche O, Steinitz O, Rotem D, Kadmon R (2007) A comparative evaluation of presence-only methods for modelling species distribution. Diversity and Distributions 13: 397–405.
28. Lobo JM, Jiménez-Valverde A, Real R (2008) AUC: a misleading measure of the performance of predictive distribution models. Global Ecology and Biogeography 17: 145–151.
29. Peterson AT, Papes M, Soberón J (2008) Rethinking receiver operating characteristic analysis applications in ecological niche modeling. Ecological Modelling 213: 63–72.
30. Moore CH, Harvey ES, Van Niel K (2010) The application of predicted habitat models to investigate the spatial ecology of demersal fish assemblages. Marine Biology 157: 2717–2729.
31. Monk J, Ierodiaconou D, Bellgrove A, Harvey E, Laurenson L (2011) Remotely sensed hydroacoustics and observation data for predicting fish habitat suitability. Continental Shelf Research 31: S17–S27.
32. Wilson MFJ, O'Connell B, Brown C, Guinan JC, Grehan AJ (2007) Multiscale terrain analysis of multibeam bathymetry data for habitat mapping on the continental slope. Marine Geodesy 30: 3–35.
33. Pittman SJ, Costa BM, Battista TA (2009) Using LiDAR bathymetry and boosted regression trees to predict the diversity and abundance of fish and corals Journal of Coastal Research 53: 27–38.
34. Leathwick JR, Elith J, Hastie T (2006) Comparative performance of generalized additive models and multivariate adaptive regression splines for statistical modelling of species distributions. Ecological Modelling 199: 188–196.
35. Ready J, Kaschner K, South AB, Eastwood PD, Rees T, et al. (2010) Predicting the distributions of marine organisms at the global scale. Ecological Modelling 221: 467–478.
36. Kadmon R, Farber O, Danin A (2003) A systematic analysis of factors affecting the performance of climatic envelope models. Ecological Applications 13: 853–867.
37. Sérgio C, Figueira R, Draper D, Menezes R, Sousa AJ (2007) Modelling bryophyte distribution based on ecological information for extent of occurrence assessment. Biological Conservation 135: 341–351.
38. MacNeill MA, Tyler EHM, Fonnesbeck CJ, Rushton SP, Polunin NVC, et al. (2008) Accounting for detectability in reef-fish biodiversity estimates. Marine Ecology-Progress Series 367: 249–260.
39. Bozec Y-M, Kulbicki M, Laloë F, Mou-Tham G, Gascuel D (2011) Factors affecting the detection distances of reef fish: implications for visual counts. Marine Biology. pp 1–13.
40. Godø OR, Huse I, Michalsen K (1997) Bait defence behaviour of wolffish and its impact on long-line catch rates. ICES Journal of Marine Science 54: 273–275.
41. Thomsen B, Humborstad O-B, Furevik DM (2010) Fish Pots: Fish Behavior, Capture Processes, and Conservation Issues. Behavior of Marine Fishes: Wiley-Blackwell. pp 143–158.
42. Sarà G, Dean JM, Amato DD, Buscaino G, Oliveri A, et al. (2007) Effect of boat noise on the behaviour of bluefin tuna *Thunnus thynnus* in the Mediterranean Sea. Marine Ecology-Progress Series 331: 243–253.
43. Popper A, Hastings M (2009) The effects of anthropogenic sources of sound on fishes. Journal of Fish Biology 75: 455–489.
44. Stoner AW, Ryer CH, Parker SJ, Auster PJ, Wakefield WW (2008) Evaluating the role of fish behavior in surveys conducted with underwater vehicles. Canadian Journal of Fisheries and Aquatic Sciences 65: 1230–1243.
45. Watson DL, Harvey ES, Anderson MJ, Kendrick GA (2005) A comparison of temperate reef fish assemblages recorded by three underwater stereo-video techniques. Marine Biology 148: 1189–1189.
46. Willis TJ, Babcock RC (2000) A baited underwater video system for the determination of relative density of carnivorous reef fish. Marine and Freshwater Research 51: 755–763.
47. Willis TJ, Millar RB, Babcock RC (2000) Detection of spatial variability in relative density of fishes: comparison of visual census, angling, and baited underwater video. Marine Ecology-Progress Series 198: 249–260.
48. Edgar G (2000) Australian marine life: the plants and animals of temperate waters. Sydney: Reed Books. 544 p.
49. Gomon MF, Bray D, Kuiter RH (2008) The fishes of Australia's south coast; Gomon MF, Bray D, Kuiter RH, editors. Sydney: New Holand. 928 p.
50. Kéry M, Gardner B, Monnerat C (2010) Predicting species distributions from checklist data using site-occupancy models. Journal of Biogeography 37: 1851–1862.
51. Kéry M, Schmidt BR (2008) Imperfect detection and its consequences for monitoring for conservation. Community Ecology 9: 207–216.
52. Tyre AJ, Tenhumberg B, Field SA, Niejalke D, Parris K, et al. (2003) Improving precision and reducing bias in biological surveys: estimating false-negative error rates. Ecological Applications 13: 1790–1801.
53. Ward G, Hastie T, Barry S, Elith J, Leathwick JR (2009) Presence-only data and the EM algorithm. Biometrics 65: 554–563.
54. Breiman L (2001) Statistical modeling: the two cultures. Statistical Science 16: 199–215.
55. Elith J, Leathwick JR, Hastie T (2008) A working guide to boosted regression trees. Journal of Animal Ecology 77: 802–813.
56. MacKenzie DI, Nichols JD, Lachman GB, Droege S, Royle JA, et al. (2002) Estimating site occupancy rates when detection probabilities are less than one. Ecology 83: 2248–2255.
57. Kulbicki M, Sarramégna S (1999) Comparison of density estimates derived from strip transect and distance sampling for underwater visual censuses: a case study of Chaetodontidae and Pomacanthidae. Aquatic Living Resources 12: 315–325.
58. Ierodiaconou D, Laurenson L, Burq S, Reston M (2007) Marine benthic habitat mapping using multibeam data, georeferenced video and image classification techniques in Victoria, Australia. Journal of Spatial Sciences 52: 93–104.
59. Harvey E, Shortis M (1996) A system for stereo-video measurement of subtidal organisms. Marine Technology Society Journal 29: 10–22.
60. Harvey ES, Shortis MR (1998) Calibration stability of an underwater stereo-video system: Implications for measurement accuracy and precision. Marine Technology Society Journal 32: 3–17.
61. Watson DL (2006) Use of underwater stereo-video to measure fish assemblage structure, spatial distribution of fishes and change in assemblages with protection from fishing [PhD Thesis]: The University of Western Australia.
62. Cappo M, Speare P, Wassenberg TJ, Harvey E, Rees M, et al. (2001) Use of baited remote underwater video stations (BRUVS) to survey demersal fish-how deep and meaningful? : Fisheries Research and Development Corporation, Australia. pp 63–71.
63. Watson D, Harvey E, Fitzpatrick B, Langlois T, Shedrawi G (2010) Assessing reef fish assemblage structure: how do different stereo-video techniques compare? Marine Biology 157: 1237–1250.
64. Langlois TJ, Radford BT, Van Niel KP, Meeuwig JJ, Pearce AF, et al. (2011) Consistent abundance distributions of marine fishes in an old, climatically buffered, infertile seascape. Global Ecology and Biogeography doi: 10.1111/j.1466-8238.2011.00734.x.
65. Willis TJ, Badalamenti F, Milazzo M (2006) Diel variability in counts of reef fishes and its implications for monitoring. Journal of Experimental Marine Biology and Ecology 331: 108–120.
66. Cappo M, Speare P, De'ath G (2004) Comparison of baited remote underwater video stations (BRUVS) and prawn (shrimp) trawls for assessments of fish biodiversity in inter-reefal areas of the Great Barrier Reef Marine Park. Journal of Experimental Marine Biology and Ecology 302: 123–152.
67. Moisen GG, Frescino TS (2002) Comparing five modelling techniques for predicting forest characteristics. Ecological Modelling 157: 209–225.
68. R Core Development Team (2008) R: a language and environment for statistical computing.
69. Hastie T (2008) GAM: Generalized Additive Models. R Package.
70. Phillips SJ, Anderson RP, Schapire RE (2006) Maximum entropy modelling of species geographic distributions. Ecological Modelling 190: 231–259.
71. Fielding AH, Bell JF (1997) A review of methods for the assessment of prediction errors in conservation presence/absence models. Environmental Conservation 24: 38–49.
72. Dormann CF, McPherson JM, Araujo MB, Bivand R, Bolliger J, et al. (2007) Methods to account for spatial autocorrelation in the analysis of species distributional data: a review. Ecography 30: 609–628.
73. Barrett NS (1995) Short- and long-term movement patterns of six temperate reef fishes (Families Labridae and Monacanthidae). Marine and Freshwater Research 46: 853–860.
74. Anderson MJ (2001) A new method for non-parametric multivariate analysis of variance. Austral Ecology 26: 32–46.

2

A Small-Scale Comparison of Iceland Scallop Size Distributions Obtained from a Camera Based Autonomous Underwater Vehicle and Dredge Survey

Warsha Singh[1]*, Erla B. Örnólfsdóttir[2], Gunnar Stefansson[3]

1 Science Institute, University of Iceland, Reykjavik, Iceland, **2** Holar University College, Holar, Iceland, **3** Science Institute, University of Iceland, Reykjavik, Iceland

Abstract

An approach is developed to estimate size of Iceland scallop shells from AUV photos. A small-scale camera based AUV survey of Iceland scallops was conducted at a defined site off West Iceland. Prior to height estimation of the identified shells, the distortions introduced by the vehicle orientation and the camera lens were corrected. The average AUV pitch and roll was 1.3 and 2.3 deg that resulted in $<2\%$ error in ground distance rendering these effects negligible. A quadratic polynomial model was identified for lens distortion correction. This model successfully predicted a theoretical grid from a frame photographed underwater, representing the inherent lens distortion. The predicted shell heights were scaled for the distance from the bottom at which the photos were taken. This approach was validated by height estimation of scallops of known sizes. An underestimation of approximately 0.5 cm was seen, which could be attributed to pixel error, where each pixel represented 0.24×0.27 cm. After correcting for this difference the estimated heights ranged from $3.8-9.3$ cm. A comparison of the height-distribution from a small-scale dredge survey carried out in the vicinity showed non-overlapping peaks in size distribution, with scallops of a broader size range visible in the AUV survey. Further investigations are necessary to evaluate any underlying bias and to validate how representative these surveys are of the true population. The low resolution images made identification of smaller scallops difficult. Overall, the observations of very few small scallops in both surveys could be attributed to low recruitment levels in the recent years due to the known scallop parasite outbreak in the region.

Editor: Shi Wang, Ocean University of China, China

Funding: This research work was funded by the Icelandic Research Fund, Rannis (grant number 110209023 http://www.rannis.is/) and the United Nations University - Fisheries Training Programme (http://www.unuftp.is/). Grants were received by WS, EBÖ and GS. The funders had no role in study design, data collection and analysis, decision to publish, or preparation of the manuscript.

Competing Interests: The authors have declared that no competing interests exist.

* Email: singh.warsha@gmail.com

Introduction

Length frequency distributions form the most basic data source for investigating the population dynamics of a species. In its simplest form a size distribution may provide a description of a recruitment signal and a few years of such data may indicate whether recruitment is variable or stable, whether growth is fast or slow, and thus provide indications on the longevity of the species. They are also particularly useful when age data are difficult to gather. The Iceland scallop (*Chlamys islandica* (O.F. Müller)) fishery in Iceland has so far been monitored with the use of the conventional dredge survey technique. Numerous studies have shown this method to be size selective and lacking in capture efficiency [1,2], which can lead to conservative population estimates and misrepresentations in the underlying population structure [3]. The negative impact it has on the habitat structure of the macrobenthic organisms also makes this technique unpopular [4,5]. A shift away from this traditional approach is to estimate abundance and size distributions from photographs and video surveys of the seafloor [6–8]. A comparative survey between dredge and video surveys of sea scallops conducted by [9] concluded that video surveys gave better estimates of shell height frequency distribution and densities.

Towed and tethered camera systems are dependent on a research vessel for operation, which increases their operational costs. In comparison, the capability of an autonomous underwater vehicle (AUV) to operate independently of a research vessel makes this survey technique potentially more cost effective [10]. AUVs can be deployed from any type of research vessel irrespective of their size, or simply deployed from the shore with minimal manpower. Shallow water habitats can also be surveyed where research vessels cannot operate. It is also more feasible to take sufficient replicates in a given area with this method. Repeating entire surveys to assess the confidence level of key population parameters such as density estimates is also possible, as shown by [11]. Some of the drawbacks of the AUV survey technique include limited battery power, little communication with the vehicle operator while in mission, and the need to surface up to obtain a correct geographic position fix via GPS. An Acoustic Doppler Current Profiler (ADCP), compass, and Inertial Navigation System (INS) are used to track position while submerged.

AUV photos have been used in the past to estimate lengths of groundfish by [12]. Fish size estimates were then related to the different substrate types and a logistic regression curve was used to determine the maturity of observed individuals. In a recent study by [10] an AUV based stereo-camera was used to assess the stocks of Ocean Perch *Helicolenus percoides* and estimate the lengths of individual fish to assess the length frequency distributions. Shell height estimation of scallop shells from video surveys has been done previously by [6,7,13,14] among others. Studies using towed camera systems include [15]. High resolution imagery has also been used to detect juvenile scallops of size 10 mm [16].

The scallop fishery in Iceland commenced in 1969. A combined effect of protozoan infestation and scallop dredging caused a steep descent in the stock index from 2000 to 2003, which led to a closure of the fishery in 2004 [17,18]. To develop the use of an AUV as a stock assessment tool is worthwhile for fragile habitats such as the current sampling grounds. In this study, a small AUV survey was carried out to study scallop populations in a defined site off West Iceland. Methods were developed for size estimation of individual scallop shells from the AUV photos. Repeated measurements were used to generate a confidence interval around the observed size distribution. A small-scale comparison was then carried out between the shell height frequency distributions obtained from the AUV survey and available data from nearby dredge sampling sites.

Materials and Methods

AUV Survey

A Gavia AUV (Figure 1) was used to sample a known scallop ground approximately 3 km north of Stykkisholmur, West Iceland (65°05.598′ N 22°45.500′ W) in November 2011. No specific permits were required for field work because this was an academic study and a collaborative work between University of Iceland and the Vör Marine Research Center at Breidafjordur, West Iceland. The survey region is not privately owned or classified as a protected area, and sampling did not involve endangered or protected species. The sampling location is close to an area monitored by the Marine Research Institute of Iceland on a regular basis through dredge surveys. A small-scale experimental survey was conducted that composed of 4 parallel lines of 300 m in length spaced 50 m apart. This was repeated 5 times within the survey location for variance estimation. This survey was first described in [11] who analyzed the survey for repeatability and for scallop abundance measurements. The survey description is repeated here for completeness.

The AUV was navigated at 2.0 m from the bottom at 500 rpm that roughly corresponds to 1.6 m/sec. The starting point of each survey was chosen at random. The planned mission time for each survey was approximately 20 min. Both digital photographs and side-scan sonar data were collected during the survey. The water depth at the sampling location ranged between 27 and 37 m. Four transects of approximately 300 m in length were extracted from each survey for analysis. Given the imprecise positioning information of this particular Gavia, the data were simply considered to be samples from possible transects within the survey area. Depth variations in the vehicle path caused the distance from the bottom at which the photographs were taken (altitude) to range from $1.5-2.5$ m. Given that the sea floor in the survey area is rather flat, these depth variations could be related to the strong current flow against which the vehicle had to maneuver, making it difficult to maintain a constant altitude from the bottom. Strong tidal currents can also lead to murky water conditions, in turn, affecting the visibility in the images. Images taken at greater distances from the bottom >2.5 m) cover more area on the seabed but tend to lack details. Identification of smaller scallops can also be more challenging with a wider field of view. Thus, for clear visibility, a subset of images taken at $1.8-2.2$ m from the bottom was extracted from the data set. The percentage of photographs that comprised this subset was dependent on the steadiness of the vehicle along the different transects. For instance, in survey 4 transect 2, 97% of the photos were taken within the defined range whereas in transect 3 only 40% were within $1.8-2.2$ m. A random sample of 25 images per transect were selected from this subset, which gave 100 samples in total per survey. The random samples of photographs were thus considered to be representative of the population in the area. A random selection was applied to avoid any bias in image selection.

The camera on the AUV is a Scorpion model SCOR-20CSO color digital camera with a resolution 800×600. The camera is located on the underside of the nose module and the lighting comes from a strobe comprised of 20 LEDs located on the underside of the control module. The camera was configured to take 3 frames/sec that ensures 50% overlap between the images [19]. The photos were stored in an 8-bit jpeg format. Each photo contains embedded metadata on time, GPS location, altitude (distance from the bottom), depth (distance from the surface), pitch (movement of the AUV along the y-axis causing a upward or downward tilt), roll (movement along the x-axis causing a sideways right or left tilt), and image properties such as brightness, exposure, gamma, white-balance, and shutter speed.

Shell identification & size estimation

The obtained images were enhanced for color balance and reduction of vignetting effect as described in [11]. Gaussian blur and linear filtering techniques were used to remove noise and sharpen the image. Color contrast was obtained with gamma correction. The image enhancement routine was implemented using EBImage routine in R statistical software (EBImage software version 3.10.0 developed by [20]). The associated metadata on time, latitude, longitude, altitude, depth, pitch, and roll of the vehicle were extracted for each image. Scallops were identified visually from all selected images. The main distinguishing features used for identification were the shape and color of the scallop shell, and the presence of a shadow around the valve (round edge) of the shell. Dead shells tend to break off at the hinges and decolorize. Therefore, scallop shells that had a brighter (whitish) appearance on the images were considered dead (Figure 2). Scallop shell size (height) was measured from the tip of the valve to the umbo (hinge) of the shell where the valves connect. Pixel points (x_i,y_i) at the tip and the base of each identified shell were recorded for height estimation. Prior to height estimation of objects from underwater photographs, the geometric distortions introduced by the unstable movement of the camera (due to the pitch and roll of the vehicle), and the camera lens need to be corrected [21].

The AUV has a downward facing camera. With the aid of a proper navigation system, which regulates the orientation (pitch, roll, yaw) of the vehicle, the camera takes photographs from the correct angle. This increases reliability in the estimation of the area covered off the seafloor by the survey. At a distance of 2 m from the bottom an image covers 1.9×1.6 m on the ground [11]. On a 800×600 resolution image each pixel represents 0.24×0.27 cm on the ground. The effect of pitch and roll on the area covered on the ground was calculated with the use of basic trigonometry, applied to each image. The angle, in deg, between the outside field of view and a straight line (θ) is given by:

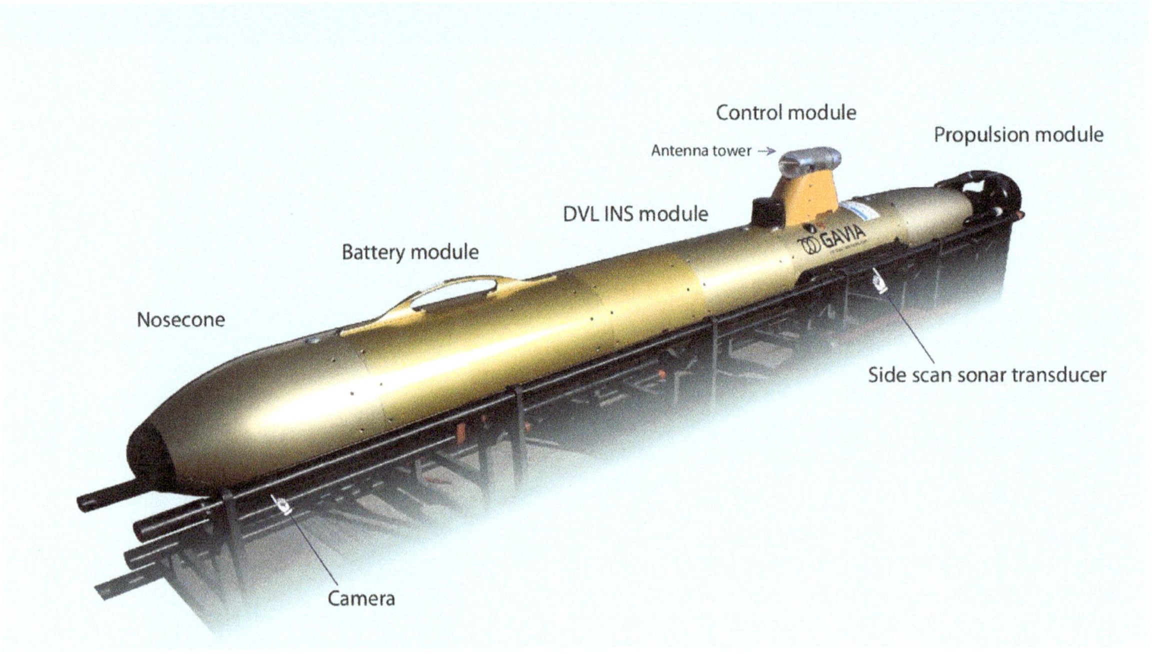

Figure 1. The Gavia autonomous underwater vehicle used for this study. Modules include nosecone with camera, battery module, doppler velocity log, control unit with tower and side-scan sonar, and propulsion module. The strobe is located under the control module.

$$\theta = \arctan\left(\frac{D/2}{H}\right) \times \frac{180}{\pi} \quad (1)$$

where D represents the horizontal or vertical dimension on the ground and H represents the height from the bottom at which the photo was taken.

Any change in the pitch and roll (ϕ) can now be accounted for and a new dimension (D_n) can be calculated by:

$$D_n = H \times \tan\left((\theta \pm \phi) \times \frac{\pi}{180}\right) \quad (2)$$

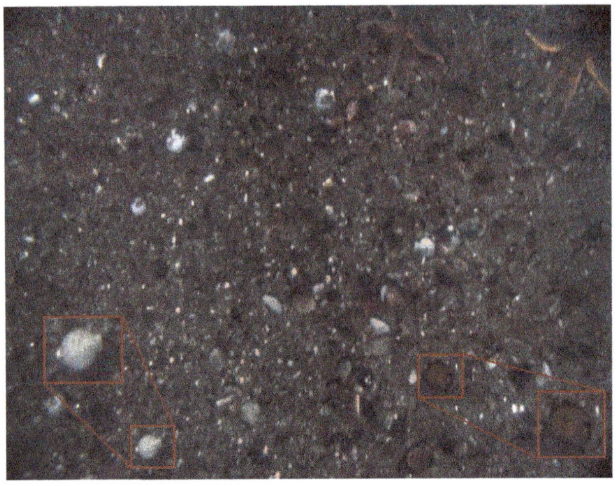

Figure 2. An enhanced color digital image of a scallop bed in Breidafjordur, Iceland, collected from the Gavia autonomous underwater vehicle. The image was taken at 1.96 m above the bottom and 28.9 m depth with a pitch and roll of 0.6 and 2 deg. The bottom right corner shows an enlargement of a live scallop. The bottom left corner shows an enlarged dead scallop shell that has decolorized. doi:10.1371/journal.pone.0109369.g002

The average pitch and roll in the dataset were approximately 1.3 and 2.3 deg, which resulted in an error of approximately 1% and 2% in the estimated horizontal and vertical dimensions of the image. These effects were thus considered negligible. None of the identified scallops were partially in the image therefore accounting for edge effects was not necessary in this analysis.

A frame with checkered lines was photographed underwater to study the inherent distortions introduced by the camera lens. The frame was photographed in a freshwater pool at a depth of approximately 1 m. A pincushion distortion effect was seen with the lines bending inwards. The bending of the straight lines along the corners indicate that the distortions caused by the camera lens are greater at the edges than the center of the image (Figure 3).

A second-order polynomial regression model was fitted to obtain a lens distortion correction. The identified model successfully predicted a undistorted reference grid (dependent variable) from the distorted matrix (independent variable) extracted from the underwater photo of the frame. The X and Y pixel points of the reference grid were modeled separately as follows:

$$Z_i = \beta_0 + \beta_1 X_i + \beta_2 X_i^2 + \beta_3 Y_i + \beta_4 Y_i^2 + \beta_5 X_i Y_i \quad (3)$$

where X_i and Y_i represent the coordinates of the distorted matrix.

The model fit is illustrated in Figure 4 where the fitted values (green) are superimposed on the reference grid. The models were then used to predict the distance d between the pixel points marked at the tip and base of each identified shell. This estimate of d was then scaled for the height from the bottom at which the photo was taken to obtain a size estimate for each identified shell:

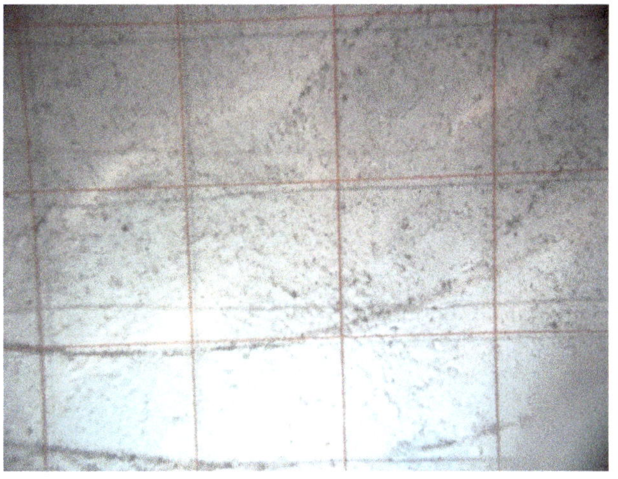

Figure 3. A checkered frame photographed underwater with the AUV camera for lens distortion correction where each square represents 15×15 cm.

$$L = d * (H/h) \tag{4}$$

where L represents the size of the scallop shell, h represents the height from the bottom at which the checkered frame was photographed and H represents the height from the bottom at which the scallop was photographed.

To validate this approach of shell height estimation, a ground-truthing experiment was conducted. Some live scallops were gathered from Breidafjordur, measured and aggregated into the following 4 height (cm) categories; 5.0 − 5.5, 5.5 − 6.0, 6.0 − 6.5 and 6.5 − 7.0. The categories were formed to test whether a difference of 0.5 cm in shell height can be detected. The scallops were then placed in separate sections in a net, planted along the intertidal zone, and photographed with the AUV camera at a depth of approximately 2 m during high tide. The heights of shells that were lying flat on the seafloor were predicted with the use of the second-order polynomial regression models defined above. These included 4 − 5 shells per size category.

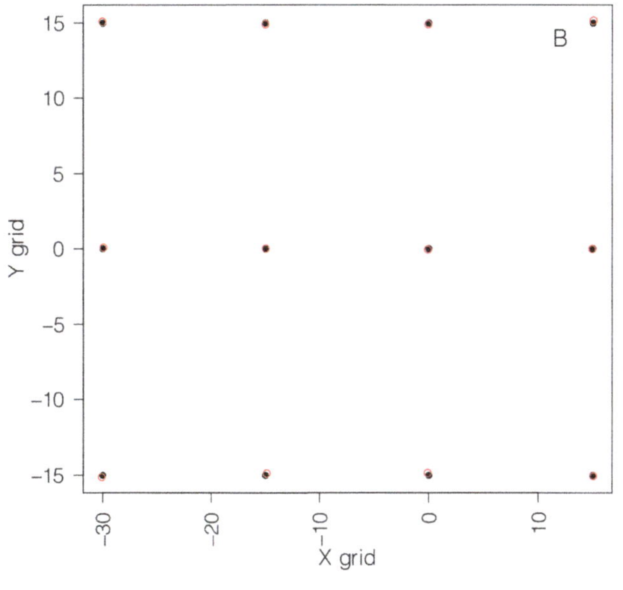

Figure 4. A undistorted reference grid (black) and prediction (red) from the distorted matrix extracted from the image of the checkered frame in Figure 3.

A nested design random effects model was used to compare the estimated scallop sizes across the 5 repeated surveys:

$$Y_{stfi} = \mu_{..} + \alpha_s + \beta_{t(s)} + \gamma_{f(st)} + \epsilon_{stfi} \tag{5}$$

where $\mu_{..}$ is a constant, α_s represents the random survey effect, $\beta_{t(s)}$ represents the random transect effect nested within survey, $\gamma_{f(st)}$ represents the random frame effect nested within the survey and transect, ϵ_{stfi} is the random error variable, all with expectations 0 and variances σ^2_α, σ^2_β, σ^2_γ and σ^2_ϵ respectively [22].

To determine whether the repeated surveys were significantly different, a full model was compared with a reduced model without the survey effect. The lme4 package [23] in the R statistical software [24] was used for random effects analyses.

Dredge Survey

The Marine Research Institute of Iceland conducts an annual dredge survey in the inner part of Breidafjordur to monitor the status of the scallop stock. A total of 120 sites are sampled with the use of a 470 kg sledge dredge of 1.5 m width. The length of each tow is approximately 0.4 nautical miles. A subsample of scallops are taken from each tow after weighing the total catch. Approximately 100 scallops are measured and the remainder are counted [17]. Survey data from September 2011 was used to select a tow that was close to the AUV sampling site. The start and end positions of the tow were 65°05.900′ N 22°43.630′ W and 65°05.660′ N 22°43.010′ W.

A smaller-scale monthly survey is also conducted at 10 sites where a smaller dredge of size 1 × 0.2 m is used. Dredging is done at a vessel speed of approximately 2 mph with a towing time of approximately 5 min. Tows are taken at random in the defined site and only the starting point of the tow is recorded. A sample of scallops from each tow are measured and counted. An estimate of the total number of scallops caught in the tow is not calculated. Data from the tow that was in close proximity to the AUV sampling site (~200 m) was used. The chosen tow was sampled in October 2011. The starting position of the tow was 65°05.557′ N 22°46.643′ W.

Results

The number of live scallops observed per image ranged from 0 − 12. Abundance estimation of scallops from these count data is presented in [11].

The images of scallops from the ground-truthing experiment were taken in ambient lighting. These images were thus of a better quality (Figure 5). A comparison of the known and estimated mean heights of the planted scallops showed that the heights were underestimated by approximately 0.5 cm in all 4 size categories (Figure 6). This could be attributed to pixel error. With each pixel representing 0.24 × 0.27 cm, clicking a pixel away on each side could introduce an error of approximately 0.5 cm. To account for this disparity, a simple linear regression model was used to predict the shell heights based on the relationship between the known and estimated mean shell heights of the planted scallops. The data thus transformed were used in the analysis below.

An analysis of variance between the full model and a reduced model without the survey effect yielded no significant difference

Figure 5. Iceland scallops photographed with a AUV camera at 1.94 m above the bottom, from the water surface. Approximate size scales, 6.0 and 6.5 cm, for known scallop sizes are also displayed.

($\chi^2 = 0.1521$) showing that the estimated lengths across the 5 repeated surveys were not significantly different i.e. there is no obvious inconsistency between the AUV surveys.

Models:

lme1: le.dat ~ 1+ (1 | survey:transect)+(1 | survey:transect:frame)

lme2: le.dat ~ 1+ (1 | survey)+(1 | survey:transect)+ (1 | survey:transect: frame)

	Df	AIC	BIC	logLik	Chisq	Chi Df	Pr(>Chisq)
lme1	4	3758.4	3779.9	−1875.2			
lme2	5	3758.4	3785.2	−1874.2	2.0513	1	0.1521

The variability in the data was dominated by the within-frame effect ($\hat{\sigma}^2_{\epsilon} = 0.535$) and the frame effect ($\hat{\sigma}^2_{\gamma} = 0.144$). The survey effect and transects within survey effect were not significant ($\hat{\sigma}^2_{\alpha} = 0.00675$, $\hat{\sigma}^2_{\beta} = 0.00196$).

The sample size of the 5 AUV surveys ranged from 289–341. The annual dredge tow and the small-scale dredge tow had sample sizes of 106 and 112 respectively. A comparison of the proportional shell height frequency distributions between the AUV and dredge surveys showed non-overlapping peaks (Figure 7). A broader size range of scallops was observed in the AUV survey. The estimated scallop shell heights from the AUV survey ranged from 3.8 − 9.3 cm with a mean of 6.6 cm. The repeated measurements were used to generate a confidence bound around the shell height distribution obtained from the AUV survey. The mean sizes of scallops observed in the small and annual dredge tows were 7.0 cm and 7.2 cm.

Discussion

An efficient survey scheme forms a fundamental component of a good fisheries management system. Improvements in data quality can be accomplished with enhanced data collection techniques and modern technological approaches. In this age of underwater robotics, autonomous underwater vehicles equipped with photographic and acoustic equipment are seen as novel tools that can be used to advance fisheries sampling techniques [10]. The time and cost effectiveness of the AUV survey technique makes it a novel approach, which can be utilized to gather enough data replicates in a given area. This survey was conducted on a small-scale with the aim to establish methodologies and develop the use of the Gavia AUV for management of macrobenthic organisms. In

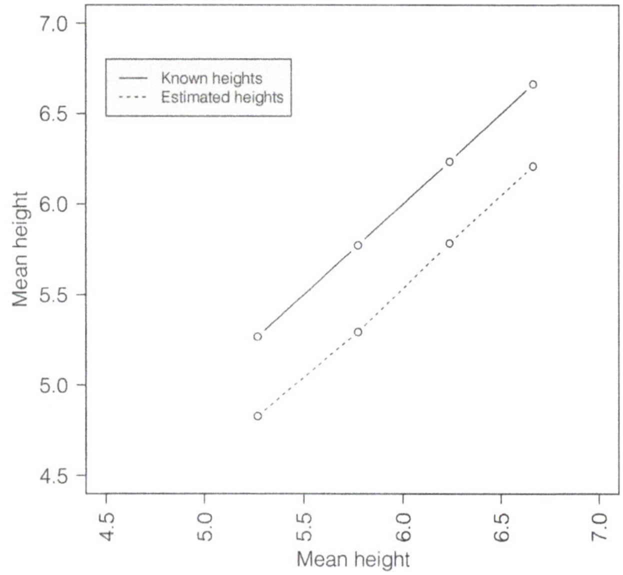

Figure 6. Known and estimated mean heights from 4 size categories of scallops planted along a inter-tidal zone for the ground-truthing experiment.

particular, a technique for scallop shell height estimation from AUV photos was presented. An analysis of variance showed that the scallop size estimates across the 5 repeated AUV surveys carried out in the study area were not significantly different. Precision is a desirable factor in marine survey data, and repeated measurements give confidence bounds that can be used to detect modes in size distributions with higher certainty.

Approximately 1 scallop was observed per square meter at the sampled location. The habitat type, as observed from the images, mainly appeared to be gravelly with scattered fragments of dead

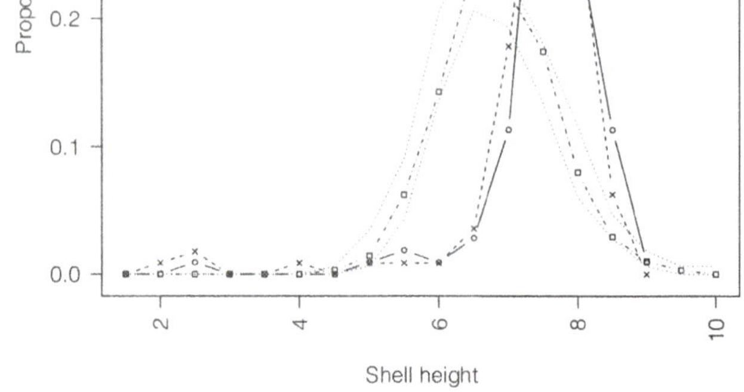

Figure 7. The proportional scallop shell height frequency distribution from the AUV survey with a 90% confidence bound (312 data points), the annual dredge tow (106 data points) and the small-scale dredge tow (112 data points).

shells. The observed dominance of larger individuals in the obtained shell height frequency distribution is characteristic of an unfished long-lived species, where larger individuals accumulate in a population over time. The Iceland scallop is a long-lived species that most commonly dwells in waters between $10-250$ m with a preferred physical environment characterized by strong currents and temperatures between -1.5 to 8°C [25]. Sexual maturity is related to food availability and surrounding temperature, and the size at maturity is estimated to be around $4-5$ cm shell height at the age of $5-7$ years [18].

Non-overlapping shell height frequency distributions were seen between the AUV and dredge surveys. A narrower size range of scallops was observed in the dredge sample than the AUV sample. This could reflect the size selectivity of a dredge and its bias towards larger individuals in a population, which is a known concern [3,7]. The dredge used for scallop surveys contains rings of approximately 6 cm in diameter. Hence, larger individuals dominate the height frequency distribution. Further, the tow from the annual dredge survey used for comparison was approximately 1 km away from the AUV survey location and falls in a slightly deeper area ($\sim$ 50 m) than the AUV survey. Scallop distribution is known to be correlated with water depth with a shift toward the deeper waters with increasing size [26]. It could be possible that bigger scallops were present in this deeper area. It was hard to determine the direction of the small-scale dredge tow as only the starting point of the survey was known. However, it was nearer to the AUV sampling site than the annual survey tow. It is well-known that standard tests comparing such frequency distributions can not be used due to the correlations in the data [27], but the difference here is compelling since the AUV finds many more scallops of smaller sizes. The outcome that the AUV finds scallops of a much wider size range highlights the bias of the dredge survey.

On the other hand, shell heights from the AUV survey could potentially be underestimated due to the low resolution of the photographs, e.g. due to pixel error. The edges of the shells appeared fuzzy and hence were difficult to detect with certainty. This pixel error could also be defined as operator error if the operator had the tendency to click on the inner edges of the shells. Only one operator identified and marked the scallops. Although, the model used for shell height estimation was validated with ground-truthing data on planted scallops of known sizes, the quality of the photos used for validation was much higher than the images from the actual survey. These calibration photos were taken in ambient lighting whereas the survey was conducted in low light and murky conditions. Additionally, identification of the shells in the survey was sometimes challenging due to the low resolution of the photos. The orientation of the shells could also be disputed at times. This would lead to the diameter of the shell being estimated instead of the height, a challenge also faced by [6].

The low resolution images made identification of smaller scallops difficult. Some small scallops were observed in the dredge survey. These could be juvenile scallops that tend to attach themselves to the inner part of the older shells and get collected in the dredge [28]. Overall, the observations of very few small scallops in both surveys could be attributed to low recruitment levels in the recent years due to the known scallop parasite outbreak in the region. Starfish *Asterias rubens*, which are the chief predators of scallops were also observed on a number of images where scallops were found. These predators are known to feed on juvenile scallops [28].

Basic image enhancement techniques were used here to improve the image quality; however, investigations into a better camera system are underway. Some of the problems faced here can be alleviated with higher resolution digital photos, but there will always be a need for image enhancement because a better camera will merely move the limits of detection towards higher altitudes (distance from the bottom at which photos are taken) and smaller organisms. The ground-truthing experiment showed that it was possible to detect a difference of 0.5 cm in shell heights. For the same altitude, this can be improved with better image quality. Given that it was seen that the vehicle position has some uncertainty, a better navigation system would also be desirable.

A small-scale comparison was made between the AUV and dredge survey data; however further investigations such as that carried out by [29] are necessary to evaluate any underlying bias and to validate how representative these surveys are of the true population. The preliminary results of this study, however, indicate that the AUV might provide a better estimate of population structure because a broader size range of scallops were seen. Nevertheless, a more detailed comparison with an improved camera system is needed prior to any firm conclusions.

Acknowledgments

We thank Einar Oli Gudmundsson and Simon Mar Sturluson for their valuable support in data collection. The data on dredge survey was obtained from the Marine Research Institute of Iceland, with valuable discussions on the data set held with Jonas Pall Jonasson and Hlynur Petursson. This research work was funded by the Icelandic Research Fund Rannis and the United Nations University - Fisheries Training Programme. The Gavia AUV used for this study is jointly owned by the University of Iceland and Vör Marine Research Center at Breidafjordur.

Author Contributions

Conceived and designed the experiments: WS EBÖ GS. Performed the experiments: WS EBÖ. Analyzed the data: WS GS. Contributed reagents/materials/analysis tools: WS EBÖ GS. Wrote the paper: WS EBÖ GS.

References

1. Jenkins S, Beukers-Stewart B, Brand A (2001) Impact of scallop dredging on benthic megafauna: a comparison of damage levels in captured and non-captured organisms. Mar Ecol Prog Ser 215: 297–301.
2. McLoughlin R, Young P, Martin R, Parslow J (1991) The australian scallop dredge: estimates of catching efficiency and associated indirect fishing mortality. Fish Res 11: 1–24.
3. Fifas S, Berthou P (1999) An efficiency model of a scallop (*Pecten maximus*, l.) experimental dredge: sensitivity study. ICES J Mar Sci 56: 489–499.
4. Collie J, Escanero G, Valentine P (2000) Photographic evaluation of the impacts of bottom fishing on benthic epifauna. ICES J Mar Sci 57: 987–1001.
5. Carbines G, Cole R (2009) Using a remote drift underwater video (duv) to examine dredge impacts on demersal fishes and benthic habitat complexity in foveaux strait, southern new zealand. Fish Res 96: 230–237.
6. Rosenkranz G, Byersdorfer S (2004) Video scallop survey in the eastern Gulf of Alaska, USA. Fish Res 69: 131–140.
7. Stokesbury K (2002) Estimation of sea scallop abundance in closed areas of Georges Bank, USA. T Am Fish Soc 131: 1081–1092.
8. Taylor R, Vine N, York A, Lerner S, Hart D, et al. (2008) Evolution of a benthic imaging system from a towed camera to an automated habitat characterization system. In: Oceans 2008. IEEE, pp. 1–7.
9. Giguère M, Brulotte S (1994) Comparison of sampling techniques, video and dredge, in estimating sea scallop (*Placopecten magellanicus*, gmelin) populations. J Shellfish Res 13: 25–30.
10. Seiler J, Williams A, Barrett N (2012) Assessing size, abundance and habitat preferences of the ocean perch *Helicolenus percoides* using a auv-borne stereo camera system. Fish Res 129–130: 64–72.
11. Singh W, Örnólfsdóttir E, Stefansson G (2013) A camera based autonomous underwater vehicle sampling approach to quantify scallop abundance. J Shellfish Res 32: 725–732.
12. Tolimieri N, Clarke M, Singh H, Goldfinger C (2008) Evaluating the seabed AUV for monitoring groundfish in untrawlable habitat. In: Marine habitat

mapping technology for Alaska. Alaska Sea Grant College Program, University of Alaska Fairbanks, Fairbanks, AK. pp. 129–141.
13. Stokesbury K, Carey J, Harris B, O'Keefe C (2010) High densities of juvenile sea scallop (*Placopecten magellanicus*) on banks and ledges in the central gulf of maine. J Shellfish Res 29: 369–372.
14. Harris BP, Stokesbury KD (2006) Shell growth of sea scallops (*Placopecten magellanicus*) in the southern and northern great south channel, usa. ICES J Mar Sci 63: 811–821.
15. Rosenkranz G, Gallager S, Shepard R, Blakeslee M (2008) Development of a high-speed, megapixel benthic imaging system for coastal fisheries research in Alaska. Fish Res 92: 340–344.
16. Carey J, Stokesbury K (2011) An assessment of juvenile and adult sea scallop, *Placopecten magellanicus*, distribution in the northwest atlantic using high-resolution still imagery. J Shellfish Res 30: 569–582.
17. Jonasson J, Thorarinsdottir G, Eiriksson H, Solmundsson J, Marteinsdottir G (2007) Collapse of the fishery for iceland scallop (*Chlamys islandica*) in Breidafjordur, west Iceland. ICES J Mar Sci 64: 298–308.
18. Garcia E (2006) The fishery for iceland scallop (*Chlamys islandica*) in the northeast atlantic. Adv Mar Biol 51: 1–55.
19. Gavia T (2010) Gavia autonomous underwater vehicle user manual. Technical report, Teledyne Gavia 110 pp.
20. Sklyar O, Pau G, Smith M, Huber W (2011) EBImage: Image processing toolbox for R. R package version 3.10.0.
21. Heikkila J, Silven O (1997) A four-step camera calibration procedure with implicit image correction. In: Computer Vision and Pattern Recognition, 1997. Proceedings IEEE Computer Society Conference. IEEE, pp. 1106–1112.
22. Kutner M, Nachtsheim C, Neter J, Li W (2005) Applied linear statistical models. McGraw-Hill, Irwin, Boston. 1091 pp.
23. Bates D, Maechler M, Bolker B, Walker S (2014) lme4: Linear mixed-effects models using Eigen and S4. URL http://CRAN.R-project.org/package=lme4. R package version 1.1-6.
24. R Development Core Team (2011) R: A Language and Environment for Statistical Computing. R Foundation for Statistical Computing, Vienna, Austria. URL http://www.R-project.org/. ISBN 3-900051-07-0.
25. Galand P, Fevolden S (2000) Population structure of *Chlamys islandica* in the northeast atlantic northern stocks compared with a southern relict population. Sarsia 85: 183–188.
26. Arsenault D, Himmelman J (1996) Ontogenic habitat shifts of the iceland scallop, *Chlamys islandica* (müller, 1776), in the northern gulf of st. lawrence. Can J Fish Aqua Sci 53: 884–895.
27. Hrafnkelsson B, Stefánsson G (2004) A model for categorical length data from groundfish surveys. Canadian Journal of Fisheries and Aquatic Sciences 61: 1135–1142.
28. Arsenault D, Himmelman J (1996) Size-related changes in vulnerability to predators and spatial refuge use by juvenile iceland scallops *Chlamys islandica*. Mar Ecol Prog Ser 140: 115–122.
29. Jacobson LD, Stokesbury KD, Allard MA, Chute A, Harris BP, et al. (2010) Measurement errors in body size of sea scallops (placopecten magellanicus) and their effect on stock assessment models. Fish Bull 108: 233–247.

3

High Mortality of Red Sea Zooplankton under Ambient Solar Radiation

Ali M. Al-Aidaroos[1]*, Mohsen M. O. El-Sherbiny[1,2], Sathianeson Satheesh[1], Gopikrishna Mantha[1], Susana Agustí[3,4], Beatriz Carreja[4], Carlos M. Duarte[1,3,4]

1 Department of Marine Biology, Faculty of Marine Sciences, King Abdulaziz University, Jeddah, Saudi Arabia, **2** Marine Science Department, Faculty of Science, Suez Canal University, Ismailia, Egypt, **3** The UWA Oceans Institute and School of Plant Biology, University of Western Australia, Crawley, WA, Australia, **4** Department of Global Change Research and LINC Global, IMEDEA (CSIC-UIB) Instituto Mediterráneo de Estudios Avanzados, Miquel Marqués, Esporles, Spain

Abstract

High solar radiation along with extreme transparency leads to high penetration of solar radiation in the Red Sea, potentially harmful to biota inhabiting the upper water column, including zooplankton. Here we show, based on experimental assessments of solar radiation dose-mortality curves on eight common taxa, the mortality of zooplankton in the oligotrophic waters of the Red Sea to increase steeply with ambient levels of solar radiation in the Red Sea. Responses curves linking solar radiation doses with zooplankton mortality were evaluated by exposing organisms, enclosed in quartz bottles, allowing all the wavelengths of solar radiation to penetrate, to five different levels of ambient solar radiation (100%, 21.6%, 7.2%, 3.2% and 0% of solar radiation). The maximum mortality rates under ambient solar radiation levels averaged (±standard error of the mean, SEM) 18.4±5.8% h^{-1}, five-fold greater than the average mortality in the dark for the eight taxa tested. The UV-B radiation required for mortality rates to reach ½of maximum values averaged (±SEM) 12±5.6 h^{-1}% of incident UVB radiation, equivalent to the UV-B dose at 19.2±2.7 m depth in open coastal Red Sea waters. These results confirm that Red Sea zooplankton are highly vulnerable to ambient solar radiation, as a consequence of the combination of high incident radiation and high water transparency allowing deep penetration of damaging UV-B radiation. These results provide evidence of the significance of ambient solar radiation levels as a stressor of marine zooplankton communities in tropical, oligotrophic waters. Because the oligotrophic ocean extends across 70% of the ocean surface, solar radiation can be a globally-significant stressor for the ocean ecosystem, by constraining zooplankton use of the upper levels of the water column and, therefore, the efficiency of food transfer up the food web in the oligotrophic ocean.

Editor: Inés Álvarez, University of Vigo, Spain

Funding: This project was funded by the Deanship of Scientific Research (DSR), King Abdulaziz University (grant no. 1-150/1433 HiCi). The funders had no role in study design, data collection and analysis, decision to publish, or preparation of the manuscript.

Competing Interests: The authors have declared that no competing interests exist.

* Email: aaidaroos@kau.edu.sa

Introduction

Zooplankton are essential components of the marine food web, relaying primary production from microbial algae to fish and seabirds [1]. Pressures disrupting zooplankton communities can, therefore, generate considerable impacts on the marine food web and, particularly, reduce the flow of primary production upwards in the food web. Hence, there have been significant efforts at examining the impacts of global stressors, such as ocean warming [2], ocean acidification [3] and pollutants [4] on marine zooplankton.

UV-B radiation is emerging as a largely overlooked but prevalent stressor in the marine environment [5,6], with increasing UV-B causing an increase in mortality across the taxa inhabiting the mixed layer of the ocean [6]. Concerns on increasing exposure of marine organisms to UV-B radiation derive from consideration of the combined effects of a global increase of the UV-B radiation incident in the ocean and increased penetration into ocean waters. The UV-B incident on the Earth surface has increased with the depletion of the stratospheric ozone layer four decades ago, with this increase being particularly strong over Antarctica but of global reach [7,8]. Ozone values decreased, on average, 10.97±2% in the Southern Hemisphere compared to 2.72±0.45% in the Northern Hemisphere between 1970 and 2012 [8]. Evidence for increased UV-B penetration derives from indications, still awaiting confirmation, that the oligotrophic gyres, with the most transparent waters to UV-B, may be expanding in size [9] and that chlorophyll *a* concentration has declined globally at rates of about 1% per year, related to increasing sea surface temperature and vertical stratification [10]. Indeed, impacts of UV-B radiation are greatest in oligotrophic, transparent seas. For instance UV-B levels sufficient to cause significant mortality of photosynthetic plankton have been reported to penetrate down to 60 m in the oligotrophic waters of the subtropical Atlantic Ocean [11] and to 26 m in the Mediterranean Sea [12]. Yet, there is still limited understanding of the vulnerability of zooplankton to solar and UV-B radiation, as most studies have been conducted in alpine lakes and north-temperate coastal waters.

A recent meta-analysis demonstrated that crustaceans, a dominant component of zooplankton communities, are highly vulnerable to UV-B radiation [6]. Previous studies on planktonic (often neustonic) crustacean zooplankton showed that exposure to UV-B lead to behavioral [13] and physiological responses [14–16] ultimately affecting a range of fitness components [17] conducive to increased mortality [18–20]. Impacts of UV-B radiation is expected to be highest in tropical oligotrophic marine regions [11], where high incident UV-B radiation combines with deep penetration to yield high UV-B doses to organisms inhabiting the mixed layer. Yet, the functional responses of zooplankton mortality to ambient solar radiation doses have been assessed only in productive, turbid coastal waters [14–20]. Here we show, based on experimental assessments of solar radiation dose-mortality curves on eight common taxa, the mortality of zooplankton in the oligotrophic waters of the Red Sea to increase steeply with ambient levels of solar radiation.

The Red Sea is particularly prone to high solar and UV-B radiation in the mixed layer because (1) this region receives very high solar and UV radiation, with UV indices in excess of 12, in spring [21,22]; (2) it is one of the most oligotrophic seas in the world, leading to high UV-B penetration reported to damage phytoplankton communities in the Gulf of Aqaba, Northern Red Sea [23]; and (3) the mixed layer is much shallower, at around 30 to 45 m depth [24,25], than typically observed in the open oligotrophic ocean, implying exposure to very high solar and UV-B doses for organisms inhabiting the mixed layer.

Materials and Methods

The sensitivity of the main zooplankton taxa in the Red Sea coastal ecosystem to solar radiation was assessed through experiments examining the dose-response curves relating mortality rates of zooplankton taxa to solar radiation. In short, the experiments involved exposing quartz bottles containing a number of individuals to five levels of solar radiation allowing100%, 21.6%, 7.2%, 3.2% and 0%, and assessing the mortality in the quartz bottles at regular time intervals until 90% of mortality was obtained. The mortality rates and cumulative doses received at the various treatments allowed calculation of dose-response curves, characterized by the asymptotic maximum mortality with increasing UV-B radiation and the UV-B radiation sufficient to achieve half of maximum mortality.

The experiments were conducted at the Obhur marine research station of King Abdulaziz University (Jeddah, Saudi Arabia, 21.71°N, 39.09°E) between June 1 and 7, 2013, the time of maximum solar and UV-B radiation in the region [21,22]. Underwater UV-B penetration down to 3 m depth was measured using an underwater Solar Light UV-B sensor (model PMA 2104) at 0.5 m depth intervals at the center of the Obuhr Creek, the site of zooplankton sampling, at the mouth of the Creek and 1 Km offshore into the Red Sea. Incident underwater UV-B radiation, the most damaging component of the solar radiation spectra [6], just below the water surface was measured during the experiments at 5 min intervals using a logging underwater Solar Light UV-B sensor (model PMA 2104) placed at the aquaculture pond where the experiments were conducted.

Organisms were collected at dusk or dawn from coastal waters of the Red Sea off the KAU marine station at the Obhur Creek, about 12 Km long and 300 m wide, inlet, using shallow (top 10 m) vertical plankton tows with a WP2 net fitted with a 150 or 500 μm mesh, depending on taxa. The net was fitted with a large, 20 L, plastic bag to collect the animals, thereby avoiding pressure stress to the organisms. Immediately upon collection, bags with the organisms were placed in a cooling box with seawater, to maintain in situ temperature and transported in the dark to environmental chambers. In these experimental chambers, set at the in situ seawater temperature, which varied between 31°C and 34°C during the experimental period, the organisms were transferred to buckets and maintained in aerated filtered seawater in dim light. Between 50 and 100 individuals, depending on the taxa, were isolated using sterilized Pasteur pipettes with extreme care to avoid damages so that live intact animals (male and female) were sampled. A total of 100 healthy individuals, as assessed by their motion patterns, were isolated using sterilized Pasteur pipettes with extreme care to avoid damages so that live intact animals (male and female) were sampled. The number of *Labidocera* sp. individuals was only 50 because this taxa was never present in sufficient abundance to isolated a larger number of individuals (Table 1).

The experiments were conducted with eight common zooplankton taxa, including six copepod taxa (*Acartia* sp., *Copilia* sp., *Centropages* sp., *Oncaea* sp., *Macrosetella* sp. and *Labidocera*sp.) and two decapods, *Lucifer* sp. and the nauplii of the Indian Ocean white prawn, *Fenneropenaeusindicus*, which was hatched at the aquaculture facility at the Obhur research station. For the first taxa tested, *Labidocera* sp., we used two duplicated quartz bottles containing 10 individuals each per treatment. However, we assessed that the experimental results were more robust if all 20 individuals were incubated in the same flask, so that all other taxa were tested by incubating 20 individuals contained in single quartz bottles for each treatment. Quartz flasks allowed the full spectrum of ambient solar radiation to penetrate into the bottles, across a range of solar radiation ranging from the full solar radiation incident below the water surface to darkness. The experiment involved, therefore, five light intensities: 100%, 21.6%, 7.2%, 3.2% and 0% of solar radiation. To produce these light levels quartz bottles were covered with multiple layers (0 through 3) of neutral screen to reduce the solar radiation incident on the bottles. The dark treatment, 0% of solar radiation, was achieved by enclosing the glass bottles in thick, dark plastic bags. All bottles were incubated, just under the water surface, submersed about 10 cm in an aquaculture pond at the research station, thereby maintaining the seawater temperature, which was recorded continuously using a calibrated HoBo light/temperature data logger, and fully exposed to the incoming solar radiation at 10 cm below the surface, thereby achieving the natural solar spectrum and photoperiod.

The experiments were initiated in the morning, typically around 8 am, and lasted, in general, less than 48 h, which sufficed to cause widespread mortality in the organisms receiving solar radiation. The number of dead organisms in each replicated bottle was counted at 4 to 6 h intervals during the daytime until mortality reached 90%. To do so, the experimental bottles were transported, inside a cooling box, into temperature-controlled chambers and the dead organisms counted, under dim light, as those that sank to the bottom and showed no motion. Upon counting, the bottles were returned to the experimental pond. Mortality rates (μ, units h^{-1}), were then calculated as the slope of regression equation describing the decline in the natural log of the number of surviving organisms over time. Mortality rates (μ, units h^{-1}) represent per capita rates and, therefore, represent the probability that anyone organism will die within one hour. If multiplied by 100, these represent mortality as % of individuals per hour. Accumulated UV-B doses were calculated by integrating the instantaneous values measured at 15 min intervals along the duration of the experiments. Where depletion of surviving organisms was steep, the mortality rates derived from the depletion curves contained uncertainty, as the logarithmic rates cannot be considered the sampling event when all organisms were dead and the 4 to 6 h

Table 1. Size and coloration of the zooplankton taxa tested and the mean (±SEM) experimentally determined maximum mortality (μ_{max}) under UV-B radiation and excess mortality relative to that in the dark (μ_{dark}) as well as the UV-B radiation, as percent of that just below the surface of the accumulated doses along the experiment, sufficient to raise mortality to 1/2 of μ_{max}.

Species	Length (μm)	Colour	μ_{max} (h^{-1})	μ_{dark} (h^{-1})	μ_{max} (h^{-1})	$UV\text{-}B_{1/2\ \mu max}$	
						%	KJ m^{-2}
Acartia sp.	1286±35	Brown	0.052±0.015	0.017	0.035	1.5±1.8	0.65±0.078
Lucifer sp.	8008±212	Yellow-transparent	0.187±0.077	0.053	0.134	8.1±11.3	0.83±1.16
Centropages sp.	1692±42	Red	>0.031	0	>0.31	>50	>15.7
Fenneropenaeus indicus	222±3	White body	0.53±0.07	0.024	0.506	5.3±2.6	0.19±0.09
Oncaea sp.	599±49	Yellow-orange	0.083±0.02	0.024	0.059	10.7±8.2	4.65±3.58
Macrostella sp.	1224±41	Dark-orange	0.04±0.02	0.018	0.022	2.2±2.6	0.95±1.1
Labidocera sp.	2273±34	Black	>0.016±0.01	0.096	0.064	>12.0	>0.53
Copilia sp.	3901±70	Red-brown (female) Transparent (male)	0.11±0.02	0.049	0.061	5.1±4.3	0.27±0.23

interval introduced a limitation in the precision of the mortality rate estimated. However, this estimate could not be shortened without introducing artifacts derived from excess handling and disturbance of the organisms. Catastrophic mortality at 100% UV-B radiation incident below the surface for *Labidocera* precluded resolving μ_{max}.

Dose-response curves were then constructed for each taxa tested as the relationship between the mortality rate (h^{-1}) and the UV-B radiation, as percent of the incident radiation or the accumulated dose (KJ m^{-2}), along the experiments was described using Michaelis-Menten equations of the form,

$$\mu = \frac{\mu_{max}\, UV-B}{UV-B_{\frac{1}{2}\mu\, max} + UV-B} \quad (1)$$

Where μ_{max} represents the asymptotic maximum mortality with increasing UV-B radiation and $UV\text{-}B_{1/2\mu max}$ represents the UV-B radiation sufficient to achieve half of maximum mortality (μ_{max}).

Results

The UV-B index at the surface during the experimental period averaged 11.9 (range 11.54 to 12.34) and the average (±SE) maximum daily values of the incident UV-B radiation just below the surface was 0.084±0.014 mW cm^{-2}. The PAR levels incident on the experimental bottles was also high, reaching, on average, maximum values of about 1,340 μmol quanta m^{-2} s^{-1} and integrated daily values of about 8.8 quanta m^{-2}. The water transparency increased rapidly offshore of the Creek, with the extinction coefficient for UV-B radiation at Obhur Creek, the site of zooplankton sampling, was 0.31±0.01 m^{-1}, 0.28±0.01 m^{-1} at the mouth of the Creek, and 0.14±0.04 m^{-1} at 1 Km offshore, corresponding to depths of 10% of incident UV-B radiation of 7.4 m, 8.22 m, and 16.4 m, respectively. Accordingly, the radiation incident on the experimental quartz bottles receiving 3.2% of the incident radiation treatment tested (see Methods) was equivalent to that received in situ at 18.5 m, 20.5 m and 41 m depth at each of the three coastal locations, respectively.

The species tested ranged 40-fold in body length and differed in pigmentation, from rather transparent organisms, such as *Lucifer* sp. and *Copilia* sp., to dark, red-pigmented organisms, such as *Macrosetella* sp. (Table 1). The mortality under ambient temperature in the dark also ranged broadly, from negligible in *Centropages* sp. to as high as 0.11 h^{-1} in *Copilia*sp., with an average of 3.5±1.1% h^{-1} across taxa (median 2.4% h^{-1}, Table 1). The number of surviving organisms declined over time, with the slowest decline in the dark and the steepest decline with increasing UV-B radiation received (Fig. 1). As a consequence, there were strong relationships between the mortality rate of the various species considered and the percent solar radiation, or the accumulated UV-B radiation, received over the experiment (Fig. 2). The shapes of the relationship between mortality rate and the percent UV-B irradiance or the accumulated UV-B radiation received different, however, among species (Fig. 2), with most species showing an asymptotic relationship between mortality rates and the percent solar radiation or the accumulated UV-B radiation along the experiment, which was best described by the Michaelis-Menten equation (Table 1), and the mortality rate of *Centropages* sp. showing a linear increase with UV-B over the range of UV-B radiation tested (Fig. 2).

The fitted Michaelis-Mensen equations allowed calculation of two key descriptors, the maximum mortality, μ_{max} (h^{-1}), under the

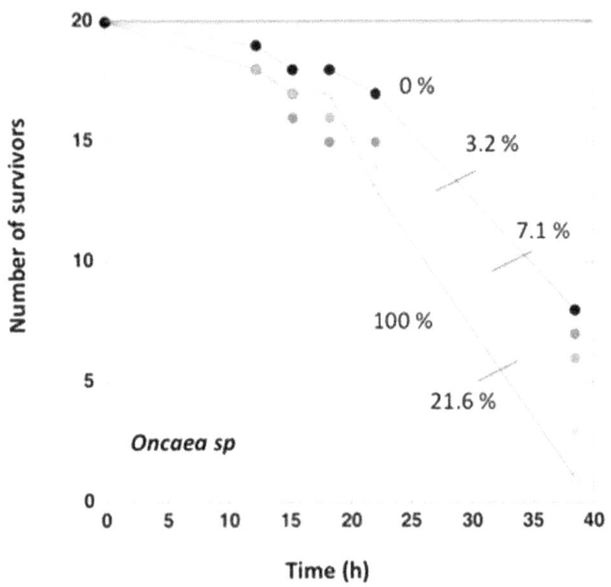

Figure 1. Sample depletion curves, for *Oncaea* sp., showing the decline in surviving individuals over time when experimentally exposed to different levels of UV-B radiation.

solar radiation incident underwater during the experiment, and the UV-B radiation at which mortality rates reached half of μ_{max}, UV-$B_{1/2\mu max}$ (Table 1, Figs. 2 and 3). The μ_{max} averaged 18.4±5.8% h^{-1} across taxa (median 13.5% h^{-1}), five-fold greater than the average mortality in the dark for the species assessed, and ranged almost 20-fold among taxa (Table 1), independently of their mortality rate in the dark ($p>0.05$). The nauplii of the prawn, *Fenneropenaeusindicus* exhibited the highest μ_{max} when exposed to ambient levels of solar radiation, while *Acartia* sp. and *Macrosetella* sp. exhibited the lowest μ_{max} (Table 1). The effect of solar radiation on mortality was represented by the excess μ_{max} relative to the mortality in the dark. Excess μ_{max} averaged 14.9±6.0% h^{-1} (median 6.25% h^{-1}), only slightly below the mean μ_{max}, indicating that most of the mortality experienced was attributable to solar radiation (Table 1).

UV-$B_{1/2\mu max}$ ranged from 1.5% to>50% of the UV-B radiation incident below the surface, averaging 12±5.6% of incident solar radiation across taxa (median 6.7%), corresponding to 12±5.6 KJ m^{-2} (median 0.74 KJ m^{-2}, Table 1). This indicates, when combined with the measured light extinction coefficients for UV-B, that the solar radiation sufficient to raise mortality to half of μ_{max} radiation reaches down to depths ranging from 4.9 m to 29.9 m across taxa (mean 19.2±2.7 m) in clear coastal waters and 2.2 to 13.5 m (mean 8.6±1.2 m) in the more turbid waters of Obhur Creek (Fig. 3).

Discussion

These results confirm that Red Sea zooplankton are highly vulnerable to ambient solar radiation, with preliminary experiments, removing UV-B radiation, demonstrating that mortality is induced by the UV-B component. Mortality rates increased steeply with moderate increases in UV-B radiation to reach maximum mortality levels at about 20% of the incident UV-B radiation below the surface. As a consequence, removing solar radiation reduced mortality rates down to, on average, 20.8% of mortality rates under maximum ambient solar radiation, much higher than the reduction in mortality rates to 60% under experimentally reduced UV-B radiation previously reported for crustaceans [6]. The much stronger mortality increase of the Red Sea zooplankton in responses to exposure to ambient solar radiation can be explained by the high UV-B doses underwater resulting from the combined high incident radiation and high water transparency allowing deep penetration of damaging UV-B radiation.

The experiments involved a limited number of individuals (N = 20) per treatment and, except for *Labidocera* sp. where duplicates were used (with N = 10), no replication within treatments, which represent a source of uncertainty and likely account for much of the variability in the derived dose-response parameters (Table 1). Whereas a larger number of organisms and replication would have possibly led to more robust results, this would have been possible only for the most abundant taxa. Use of different designs for different taxa would have added uncertainty in the comparisons among taxa, as the results for different species would have differed in power. On the other hand, use of a greater number of individuals per experiment, if possible across taxa, would have led to a longer time elapsed between collection and the onset of the experiment, as more time would have been required to sort the individuals, introducing additional artifacts, such as interference with the photoperiod and/or starvation and a general decline in condition of the organisms at the onset of the experiment.

Vulnerability to solar radiation varied widely across species, with *Centropages* sp. being the most resistant species to solar radiation and *Acartia* and *Macrosetella* being the most vulnerable ones, independently of size or pigmentation. Whereas relatively resistant species, such as *Centropages* sp., *Labidocera* sp., and *Oncaea*sp. tend to be dark colored, and transparent or white colored species, such as *F. indicus*, *Lucifer* sp. and *Copilia* sp., tend to be vulnerable, some colored species, such as *Acartia* sp. and *Macrosetella* sp. remained vulnerable. Strategies to increase zooplankton resistance to solar radiation include the accumulation of photoprotective pigments from ingested food [26] and the photoenzymatic repair of UV damages [27,28]. However, accumulation of photoprotective pigments, often carotenoids, red colored, or mycosporin-like aminoacids, increase the detectability of the zooplankton to visual predators and may, therefore, carry negative impacts [29]. This may be particularly significant for relatively large taxa, such as *Lucifer* sp., which body remains, therefore, near transparent. Yet, the results presented here indicate that pigmentation and existing capabilities for photoenzymatic repair are insufficient to protect Red Sea zooplankton from damages due to ambient solar radiation levels, as mortality rates reached half of maximum levels at a median UV-B level of only 6.7% of that incident below the surface.

That even highly pigmented zooplankton taxa in the Red Sea community studied remain vulnerable to ambient levels of solar radiation suggests that zooplankton must remain at depth to avoid damaging UV-B doses, likely constraining their day-night migration behavior, as observed in alpine lakes [13,30]. Specifically, our results suggest that zooplankton must remain below a median depth of 19 m in open coastal Red Sea waters to avoid UV-B exposure sufficient to raise mortality to 1/2 of μ_{max} (Fig. 3). This implies that only the most resistant among them can use the top half of the mixed layer during daylight, where most primary production occurs. The restriction on the food niche available to Red Sea zooplankton during daytime imposed by UV-B radiation may have important consequences in limiting the transference of production up the food web. In addition, changes in UV radiation, such as that found between more turbid coastal waters and more

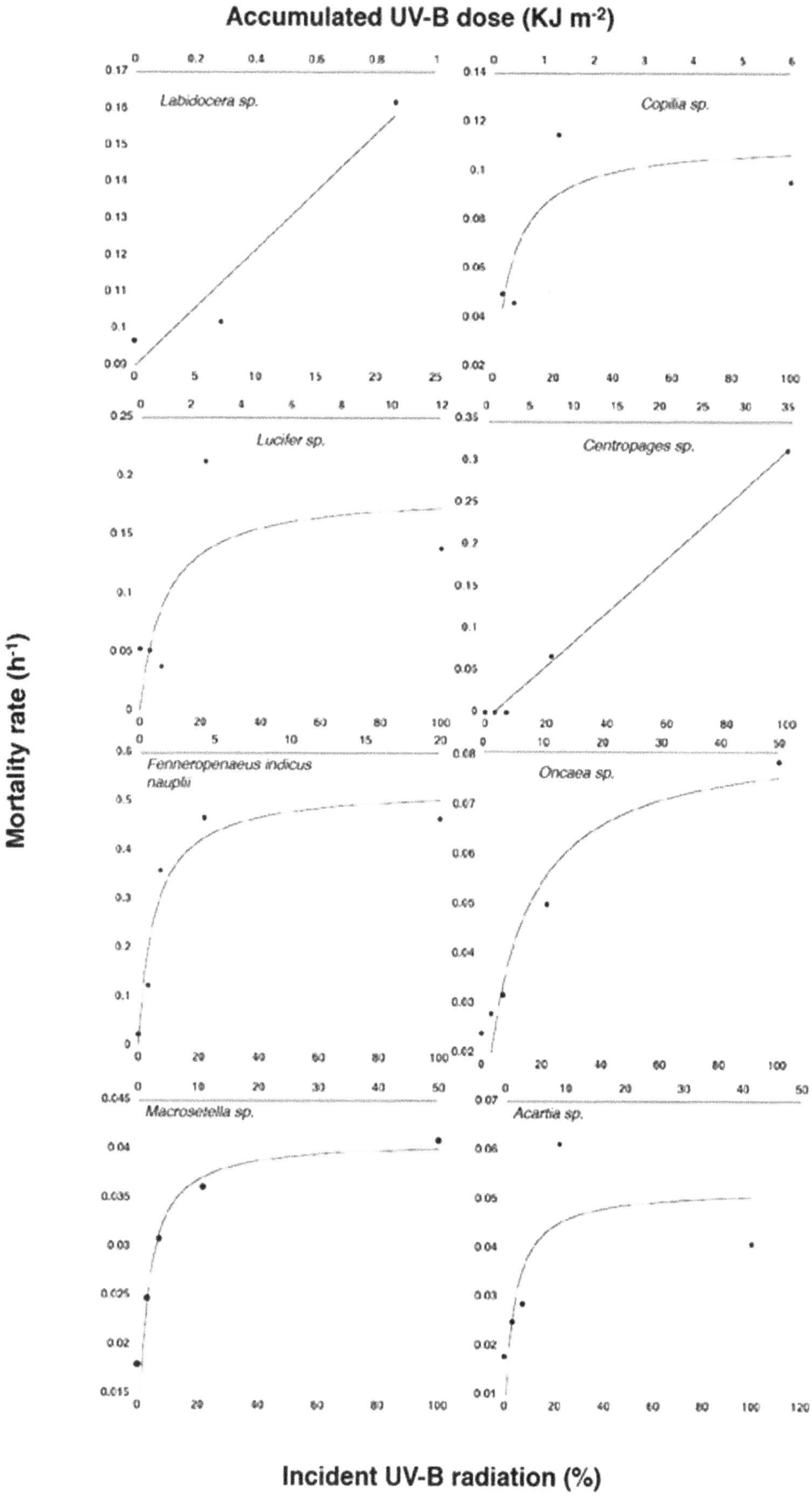
Accumulated UV-B dose (KJ m-2)
Labidocera sp.
Copilia sp.
Lucifer sp.
Centropages sp.
Fenneropenaeus indicus nauplii
Oncaea sp.
Macrosetella sp.
Acartia sp.
Mortality rate (h-1)
Incident UV-B radiation (%)

Figure 2. Dose-response curves describing the relationship between the mortality of zooplankton taxa and ambient UV-B levels (as % of UV-B incident below the surface or accumulated UV-B radiation along the experiments). The solid lines show the fitted linear regression of Michaelis-Menten equation (see Table 1 for parameters of the later). Catastrophic mortality at 100% UV-B radiation incident below the surface for *Labidocera* precluded resolving μ_{max}.

transparent open Red Sea waters, may lead to shifts in zooplankton communities by favoring more resistant taxa offshore.

These results are significant because they extend evidence of the significance of ambient UV-B levels as a stressor of zooplankton communities from high-mountain or high-latitude lakes, where most of the evidence of significant impacts of UV-B radiation originated [26,29,31,32], to the oligotrophic ocean. Evidence of UV-B impacts on marine zooplankton are much fewer, limited to laboratory studies or studies in relatively turbid waters [15,18]. Studies on the *Acartia* genus provided evidence of UV-B avoidance behavior in *Acartiahudsonica* [14] and *Acartiapacifica* [33] and demonstrated fitness costs of *Acartiatonsa* feed diets low in photoprotective pigments [17], and evidence that UV radiation reduces hatching success of the Arctic copepod *Calanusfinmarchicus* [20]. The analysis of dose-response levels of UV-B levels on a range of copepods in Puget Sound suggested they were resistant to ambient levels of UV-B radiation in those relatively turbid waters [18], similar to results from San Francisco Bay, where only 1% of UV-B radiation reaches to 0.5 m depth [15].

Hence, previous analyses of UV-B impacts on marine zooplankton communities suggested that these are relatively small or even negligible. However, these studies were conducted in relatively turbid waters, whereas the most vulnerable communities are expected to be those in tropical, oligotrophic waters where high incident solar radiation levels combine with extreme underwater penetration of UV-B radiation [11,12,34] to yield high UV-B doses. The steep increase in zooplankton mortality with ambient levels of solar radiation demonstrated here confirms the expectation that solar radiation is a significant stressor to zooplankton communities in oligotrophic waters. Because the oligotrophic ocean extends across 70% of the ocean surface, solar radiation can be a globally-significant stressor for the ocean ecosystem, by constraining zooplankton use of the upper levels of the water column and, therefore, the efficiency of food transfer up the food web in the oligotrophic ocean.

The results obtained here suggests that future efforts should focus on an improved understanding of the nature of the relationship between mortality rate and exposure to solar radiation, including examining the action spectra, i.e. the response to specific wavelengths of solar radiation [35], of mortality rates, and testing for the role of exposure time on mortality rates. The simple exponential model used here, the most parsimonious model provided the limited replication and the fact that survival was assessed at 4 h to 6 h intervals, assumes mortality rates to be constant, whereas other survival distributions, such as Weibull or Log linear, involve shifting mortality rates over time [36].

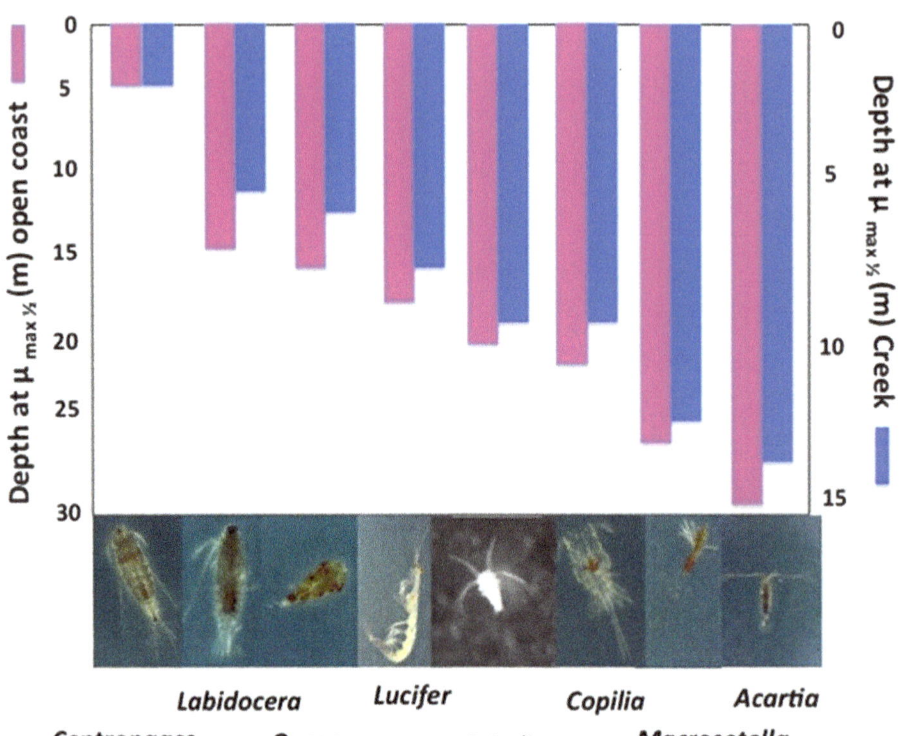

Figure 3. The depth at which sufficient UV-B radiation penetrates in the open coastal Red Sea waters (1 Km offshore from the Obhur Creek, magenta columns) and the Obhur Creek (blue columns) to raise mortality rates of different zooplankton taxa to 1/2 of μ_{max} (cf. Table 1).

Conclusions

The results presented here demonstrate that ambient solar radiation levels suffice to cause significant mortality on Red Sea zooplankton down to significant depths, in excess of 19 m, in the water column. Vulnerability to solar radiation varied widely across taxa, probably due to different repair mechanisms together with different pigmentation. The vulnerability of the Red Sea zooplankton community examined to solar radiation should lead to a severe restriction on the food niche available during daytime imposed by UV-B radiation, with important consequences in limiting the transference of production up the food web. As incident UV-B radiation has increased [7,8] and the oligotrophic regions of the ocean maybe expanding [9–10], the results presented here suggest that UV-B radiation must be considered as a significant stressor on a key node of the ocean food web, likely affecting the transference of production up the food web in the oligotrophic ocean.

Acknowledgments

The authors acknowledge with thanks the Deanship of Scientific Research (DSR), King Abdulaziz University for technical and financial support. We thank Mr. Muaadh Al-Nuwarah, Mr. Ahmed Al-Hajand Mr. Reny Devassy for their help in the field and the laboratory. We also thank Dr. Sambhu Chithambaran for providing facilities in the fish farm, Mr. Salman A. Al-Ahmadi and Mr. Bandar Al-Shareef for logistic assistance. B. Carreja was supported by a JAE fellowship of CSIC.

Author Contributions

Conceived and designed the experiments: AMA CMD. Performed the experiments: AMA MMOE SS GM CMD. Analyzed the data: CMD SS. Contributed reagents/materials/analysis tools: SA CMD. Wrote the paper: SA BC CMD.

References

1. Valiela I (1995) Marine ecologicalprocesses. Springer p. 686.
2. Mackas DL, Batten S, Trudel M (2007) Effects on zooplankton of a warmer ocean: recent evidence from the Northeast Pacific.ProgOceanogr 75: 223–252.
3. Riebesell U, Tortell PD (2011) Effects of ocean acidification on pelagic organisms and ecosystems. In: Gattuso JP, Hanson L, editors. Oxford University Press, Oxford. pp 99–121.
4. Dachs J, Méjanelle L (2010) Organic pollutants in coastal waters, sediments, and biota: a relevant driver for ecosystems during the anthropocene? Estuar coast 33: 1–14.
5. Häder DP, Kumar HD, Smith RC, Worrest RC (2007) Effects of solar UV radiation on aquatic ecosystems and interactions with climate change. Photochem Photobiol Sci 6: 267–285.
6. Llabrés M, AgustíS, Fernández M, Canepa A, Maurin F, et al. (2013) Impact of elevated UV-B radiation on marine biota: a meta-analysis. Global EcolBiogeogr 22: 131–144.
7. Weatherhead EC, Andersen SB (2006) The search for signs of recovery of the ozone layer. Nature 441: 39–45.
8. AgustíS, Llabrés M, Carreja B, Fernández M, Duarte CM (2014) Contrasting sensitivity of marine biota to UV-B radiation between Southern and Northern hemispheres. Estuaries and Coasts: DOI 10.1007/s12237-014-9790-9.
9. Polovina JJ, Howell EA, Abecassis M (2008) Ocean's least productive waters are expanding. Geophys Res Lett 35: L03618.
10. Boyce DG, Lewis MR, Worm B (2010) Global phytoplankton decline over the past century. Nature 466: 591–596.
11. Llabrés M, Agusti S (2006) Picophytoplankton cell death induced by UV radiation: evidence for oceanic Atlantic communities. Limnol Oceanogr 51: 21–29
12. Llabrés M, AgustíS, Alonso-Laita P, Herndl GJ (2010) *Synechococcus* and *Prochlorococcus* cell death induced by UV radiation and the penetration of lethal UVR in the Mediterranean Sea. Mar Ecol Prog Ser 399: 27–37.
13. Williamson CE, Fischer JM, Bollens SM, Overholt EP, Breckenridge JK (2011) Toward a more comprehensive theory of zooplankton diel vertical migration: Integrating ultraviolet radiation and water transparency into the biotic paradigm. LimnolOceanogr 56: 1603–1623.
14. Bollens SM, Frost BW (1990) UV light and vertical distribution of the marine planktonic copepod *Acartiahudsonica*Pinhey. J Exp Mar Biol Ecol 137: 89–93.
15. Speekmann CL, Bollens SM, Avent SR (2000) The effect of ultraviolet radiation on the vertical distribution and mortality of estuarine zooplankton. J Plankton Res 22: 2325–2350.
16. Ma Z, Li W, Gao K (2013) Impacts of UV radiation on respiration, ammonia excretion, and survival of copepods with different feeding habits. Hydrobiologia 701: 209–218.
17. Hylander S, Grenvald JC, Kiørboe T (2013) Fitness costs and benefits of ultraviolet radiation exposure in marine pelagic copepods. Funct Ecol doi: 10.1111/1365-2435.12159.
18. Damkaer DM, Dey DB, Heron GA, Prentice EF (1980) Effects of UV-B Radiation on near-Surface Zooplankton of Puget Sound. Oecologia 44: 149–158.
19. Chalker-Scott L (1995) Survival and Sex-Ratios of the Intertidal Copepod, *Tigriopus-Californicus*, Following Ultraviolet-B (290–320 nm) Radiation Exposure. Mar Biol 123: 799–804.
20. Browman HI, St-Pierre JF, Kuhn P (2003) Dose and dose-rate dependency in the mortality response of *Calanusfinmarchicus* embryos exposed to ultraviolet radiation. Mar Ecol Prog Ser 247: 297–302.
21. Elhadidy MA, Abdel-Nabi DY, Kruss P D (1990) Ultraviolet solar radiation at Dhahran, Saudi Arabia. Sol Energy 44: 315–319.
22. Khogali A, Al-Bar OF (1992) A study of solar ultraviolet radiation at Makkah solar station. Sol Energy 48: 79–87.
23. Boelen P, Post AF, Veldhuis MJW, Buma AGJ (2002) Diel patterns of UV-BR-induced DNA damage in picoplankton size fractions from the Gulf of Aqaba, Red Sea. Microb Ecol 44: 164–174.
24. Raitsos DE, Pradhan Y, Brewin RJW, Stenchikov G, Hoteit I (2013) Remote Sensing the Phytoplankton Seasonal Succession of the Red Sea. PLOS ONE 8: e64909.
25. Maillard C, Soliman G (1986) Hydrography of the Red-Sea and exchanges with the Indian-Ocean in summer. OceanolActa 9: 249–269.
26. Hansson L-A, Hylander S, Sommaruga R (2007) Escape from UV threats in zooplankton: a cocktail of behavior and protective pigmentation. Ecology 88: 1932–1939.
27. Hansson L-A, Hylander S (2009) Effects of ultraviolet radiation on pigmentation, photoenzymatic repair, behavior, and community ecology of zooplankton. Photochem Photobiol Sci 8: 1266–1275.
28. Souza MS, Hansson L-A, Hylander S, Modenutti B, Balseiro E (2012) Rapid Enzymatic Response to Compensate UV Radiation in Copepods. PLOS ONE 7: e32046.
29. Hylander S, Boeing WJ, Granéli W, Karlsson J, Einem JV, et al. (2009) Complementary UV protective compounds in zooplankton. Limnol Oceanogr 54: 1883–1893.
30. Hylander S, Hansson LA (2010) Vertical migration mitigates UV effects on zooplankton community composition. J Plankton Res3 2: 971–980.
31. Rautio M, Bonilla S, Vincent WF(2009) UV photoprotectants in arcticzooplankton. Aquat Biol 7: 93–105.
32. Sommaruga R (2010) Preferential accumulation of carotenoids rather than of mycosporine-like amino acids in copepods from high altitude Himalayan lakes. Hydrobiologia 648: 143–156.
33. Zengling MA, Wei L, Kunshan G (2010) Horizontal migration of *Acartia pacifica*Steuer (copepoda) in response to UV-radiation. J Photochem Photobiol B Biol 101: 233–237.
34. Tedetti M, SempéréR, Vasilkov A, Charrière B, Nérini D, et al. (2007) High penetration of ultravioletradiation in thesoutheastPacificwaters. Geophy Res Lett 34: L12610.
35. Cullen JJ, Neale PJ, Lesser MP (1992) Biologicalweightingfunctionfortheinhibition of phytoplanktonphotosynthesisbyultravioletradiation. Science 258: 646–650.
36. Wilson DL. (1994) The analysis of survival (mortality) data: fitting Gompertz, Weibull, and logistic functions. Mech. Ageing Dev. 1994 74: 15–33.

4

Stable Isotopes in Fish Eye Lenses as Potential Recorders of Trophic and Geographic History

Amy A. Wallace*, David J. Hollander, Ernst B. Peebles

College of Marine Science, University of South Florida, St. Petersburg, Florida, United States of America

Abstract

We evaluated eye lenses as potential recorders of stable isotope histories in fish because they consist of metabolically inert optical proteins that are deposited in successive, concentric circles (laminae) much like otolith circuli and tree rings. We conducted four different tests on lenses from red snapper, red grouper, gag, and white grunt. The first test was a low-resolution screening of multiple individuals (4–5 radial groups of laminae per lens, all species except white grunt). Along the radial axis, all individuals exhibited substantial isotopic variability. Red snapper individuals separated into two groups based on $\delta^{15}N$ and gag separated into two groups based on $\delta^{13}C$. Two gag with the greatest variation were chosen for high-resolution temporal analysis using individual laminae from their second eye lenses. The first-order patterns from the high-resolution analysis generally mimicked patterns from the low-resolution screening of grouped laminae, yet the high-resolution plots revealed early-life details that were not apparent in the low-resolution screenings. For the third test, left- versus right-eye variation was compared using high-resolution methods. White grunt left- and right-eye radial isotopic patterns were almost identical for both $\delta^{13}C$ and $\delta^{15}N$, suggesting the variations observed among individual fish were not artifacts. The final test evaluated intra-laminar variation; multiple samples were analyzed from different parts of the same lamina. Seven laminae from three individuals of two species were analyzed in this manner; variations among laminae were found to be much higher than variations within laminae. However, nominal intra-laminar variations were comparable to nominal differences between left and right lenses, suggesting intra-laminar variation established measurement precision. Eye lens isotopes appear to be useful for reconstructing the isotopic histories of individual fish; these histories can be compared with spatially-derived isoscapes to reconstruct individual histories for site fidelity, movement and trophic position.

Editor: André Chiaradia, Phillip Island Nature Parks, Australia

Funding: AAW was supported by a fellowship from National Oceanic and Atmospheric Administration (NOAA) Fisheries (grant NA10NMF4550468). The research was also supported by the Gulf of Mexico Research Initiative through an award to C-IMAGE, the Center for Integrated Modeling and Analysis of the Gulf Ecosystem. The funders had no role in study design, data collection and analyses, decision to publish, or preparation of the manuscript.

Competing Interests: The authors have declared that no competing interests exist.

* Email: awallac4@mail.usf.edu

Introduction

Internally recorded stable-isotope histories can be used to recreate lifelong trends in animal diets, trophic dependences, and movement within isotopically variable landscapes [1–5]. Such lifelong records are difficult to obtain because most tissues undergo metabolic turnover, and this turnover places limits on the retrospective time period that can be investigated [2,3,5,6,7].

Otoliths, fish scales, and vertebrae are tissues currently used to provide life-history records for fish. Otoliths are used to determine fish age and to reconstruct aquatic environments [8,9]. Otoliths grow during the entire life of the fish and consist of calcium carbonate deposited within a proteinaceous matrix. Otolith protein is generally not abundant enough (3–4% by weight) to allow retrospective $\delta^{15}N$ comparisons [10]. Fish scales grow during most of the life of the fish and provide a nonlethal way to measure isotopic ratios. Fish scales overlap old organic material with new material during growth in a phenomenon called "overplating" [11,12,13]. Overplating affects isotopic ratios by combining isotopic values from new and old material; scales may not provide a true representation of isotopic information on a temporal scale [11]. Some researchers have explored the use of vertebrae in sharks for stable-isotope analyses with encouraging results [14–16]. The collagen extracted from cartilaginous vertebrae is preserved because cartilage has low metabolic activity (little or no turnover) and it deposits in sequential layers [17]. However, this approach cannot be applied to the vertebrae of bony fishes because bone collagen is reworked over time [2,3]. Although this paper focuses on fish, it should be noted that there are a number of isotopic records for terrestrial animals as well. These include claws, hooves, hair, feathers, and teeth [18,19]. Although these tissues are isotopically conservative, the period of time represented by the record varies with tissue wear (e.g. claws, hooves) and replacement (e.g. feathers, hair) [18,19]. Some tissues can be sampled non-invasively on a regular basis to create a long-term isotopic record, but require recapture of the same animal on a regular basis [18,19].

In this paper, we evaluate fish eye lenses as potential recorders of the isotopic histories of individual fish; the eye lenses of squid (invertebrates) also appear to have this potential [20,21]. Vertebrate eye lenses are largely composed of different forms of a structural protein, crystallin, that develop in layers (laminae) at the outer lens as the animal grows [22,23]. During the mid-to-late gastrula stage, lens cells derive from surface ectoderm and differentiate into two cell types, lens epithelial cells and fiber cells [24,25], although Bloemendal considered all lens cells to be epithelial [26]. At the surface of the lens, epithelial cells form a one-cell-thick outer layer (lens epithelium) adjacent to the fiber cells. Fiber cells remove their cytoplasmic organelles, including the cellular nucleus and its DNA, to establish and maintain optical transparency within the lens [22–25,27]. This process is a specialized form of apoptosis known as "attenuated" apoptosis, where the organelles are removed from the cell over a period of days, leaving behind the crystallin and other proteins [25,28]. Attenuated apoptosis is different from "classical" apoptosis in which everything, including the cell membrane, is degraded [25].

Fish eye lenses are spherical, and lens fiber cells are laid down in concentric layers much like the layers of an onion [22–24]. Fiber cells increase in length as the animal grows, but the increase in length is a step function of lens radius rather than a continuous increase, and this step function creates the physical discontinuities that are visible between adjacent laminae [22,27]. The oldest part of the lens is at the center (lens nucleus); the nuclear fiber cells are among the oldest cells in the body [25]. The outermost margin of the lens is the youngest material in the lens. *De novo* protein synthesis ceases following attenuated apoptosis [22]. Crystallins and other lens proteins are rich in both carbon and nitrogen, and thus are suitable for $\delta^{13}C$ and $\delta^{15}N$ analysis; this contrasts with otoliths, which are nitrogen-poor. Eye lenses thus have the potential to allow reconstruction of lifetime $\delta^{13}C$ and $\delta^{15}N$ histories and their associated ecological interpretations.

To investigate variation in fish eye-lens stable isotopes, we selected fish species that occur in geographic locations that had already been the subject of isotopic mapping (isotope maps are known as "isoscapes"). This allowed comparison of any observed isotopic trends within the fish eye lenses with a known range and pattern of isotopic variation documented by Radabaugh et al. [29]. Four species (red snapper, red grouper, gag, white grunt) were selected that have life histories that are geographically bracketed between the polyhaline coastal zone and the outer continental shelf, a region that coincides with the Radabaugh et al. isoscapes (we avoided estuarine, strongly estuary-dependent, and highly migratory species). We were otherwise limited to specimens that became available near the time of analysis, as no specimens were collected for the sole purpose of conducting the present study. We expected trophic positions to increase with age (as mouth gape increases), which generally causes $\delta^{13}C$ and $\delta^{15}N$ to increase with age [30,31].

We conducted four tests to investigate variation in fish eye-lens isotopes ($\delta^{13}C$, $\delta^{15}N$). First, we conducted a low-resolution comparison of a small number (4–5) of eye-lens layers (groups of laminae) to determine if there was enough isotopic variation during life to warrant further study, with the specific objective being to compare lifetime isotopic variation with isoscapes recently produced by Radabaugh et al. [29], therein possibly providing insight into changes in fish movement and trophic position during life. After discovering that substantial, measurable isotopic variation did exist within the lenses, we used the remaining, second lens to conduct high-resolution comparisons, dissecting (delaminating) the second lens into as many thin layers (hereafter referred to as "laminae") as was practical, and comparing the high-resolution results with the coarse patterns from the first test. We then compared high-resolution patterns in the left and right lenses of an additional specimen to investigate repeatability. Finally, we examined intra-laminar variation to identify the level of inherent variability (precision) in the method.

Materials and Methods

Ethics Statement

No fish were collected or killed for the purpose of this study. All tissues were collected post-mortem according to a protocol (IS00000504) approved by the University of South Florida Institutional Animal Care and Use Committee. None of the sampled species are protected except by recreational and commercial harvest regulations. Fish were obtained from surveys conducted by the Florida Fish and Wildlife Commission (FWC) and from a licensed charter-fishing vessel operating within state and federal regulatory guidelines in waters no more than 80 km distant from John's Pass, Madeira Beach, Florida.

Lens Collection and Dissection

Four red grouper (*Epinephelus morio*) and three gag (*Mycteroperca microlepsis*) were obtained from an FWC survey conducted 9–10 July, 2013. Eight red snapper (*Lutjanus campechanus*) were obtained from John's Pass charter vessels on 10 September, 2013 and one white grunt (*Haemulon plumierii*) was obtained from a charter vessel on 11 June, 2013. Whole eyes were removed by severing the sclera at its junction with the optic nerve and by severing the rectus (orbital) muscles near their junction with the sclera. Eyes were wrapped in aluminum foil, placed in plastic bags on ice, and frozen at −40°C upon return to the laboratory.

Eyes were thawed individually before dissection. After thawing, an incision was made with a scalpel to create a flap in the cornea, which was folded back to allow removal of the lens using forceps. Exterior tissue and vitreous material were manually removed using a deionized water rinse. The rinsed lens, which contained the lens nucleus, cortex, and lens epithelium together as one cohesive unit (whole lens), was then placed on a glass petri dish where successive layers of cortical laminae were separated using two pairs of fine-tip forceps under a dissecting stereomicroscope, leaving the lens nucleus as the final tissue in the analyzed series. The lens nucleus did not visibly delaminate further when crushed. Because the total number of cortical laminae varied with lens diameter (age), the convention for numbering laminae started with the lens nucleus as 1, with assigned numbers increasing toward the outer lens margin. Lamina removal (delamination) started near one lens pole and proceeded to the other, always starting the removal process at the same pole (note that the "anterior-posterior" polar axis referred to in studies of the human eye applies to forward-oriented eyes; most fish have laterally oriented eyes, and thus the homolog of the human anterior-posterior polar axis is usually mediolateral in fish, with the anterior pole and associated primary sutures being oriented laterally). Some forms of crystallin are water-soluble [26,32]; deionized water was used sparingly to assist delamination.

After each delamination, an ocular micrometer was used to measure the diameter at the equator (to nearest 0.1 mm); the equator was identified as the largest diametric measurement that could be made perpendicular to the (mediolateral) polar axis. Lamina position is the radial midpoint of the lamina (in mm) on the equator, where the midpoint is the lens radius after lamina removal plus half the thickness of the removed lamina.

Lens Sample Processing

Four types of isotopic variation were plotted: 1) low-resolution screening plots, 2) high-resolution plots for selected, screened individuals, 3) plots of left vs. right eyes, and 4) plots of variation within individual laminae (intra-laminar variation). For low-resolution screening, eye lenses were separated into 4–5 sections using fine-tip forceps, with multiple laminae intentionally included within each section; the number of true laminae constituting each

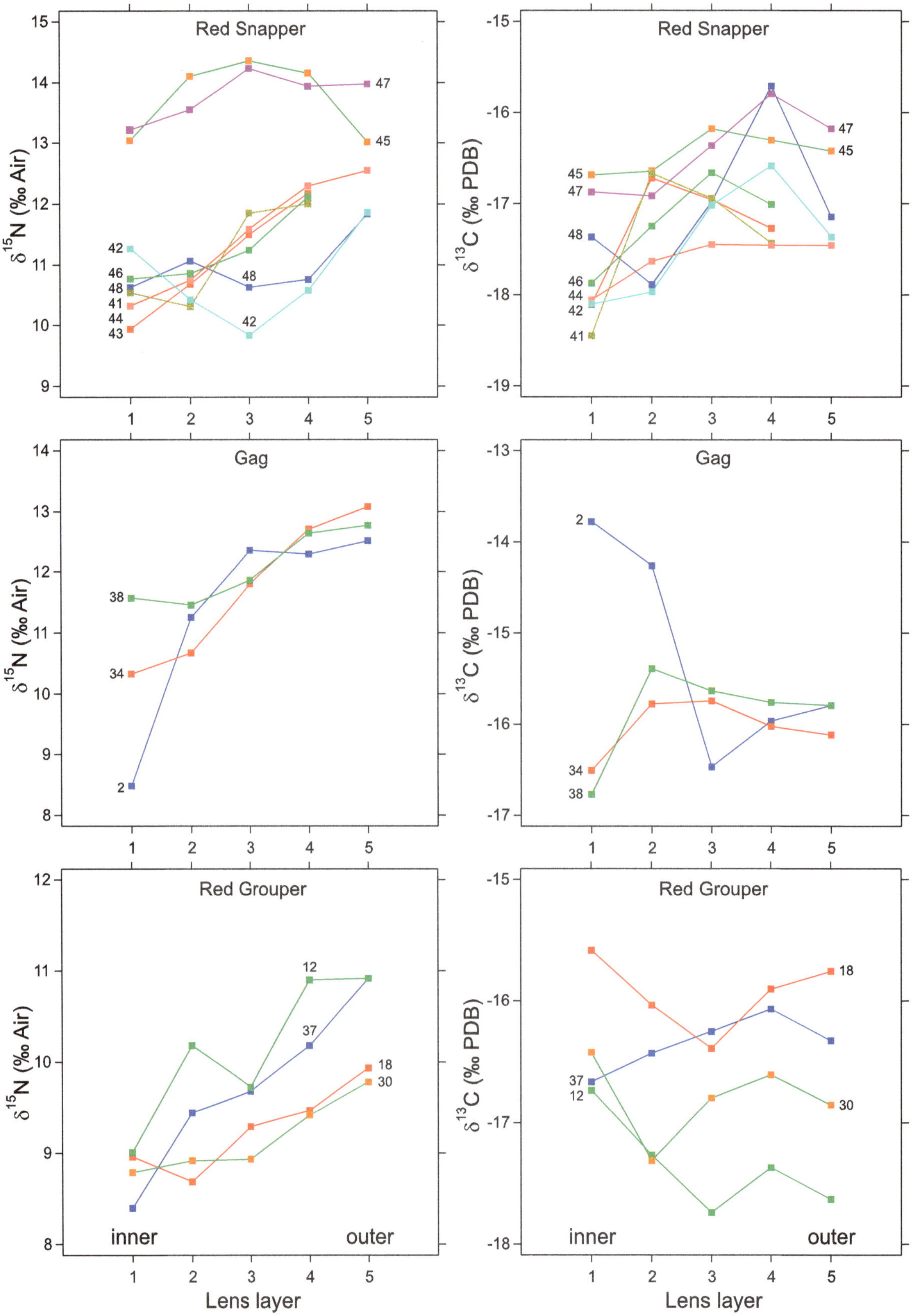

Figure 1. Low-resolution screening. Rapid screening of specimens to detect large-scale differences in isotopic history. Each plotted line is a different specimen; lines are labeled with specimen numbers.

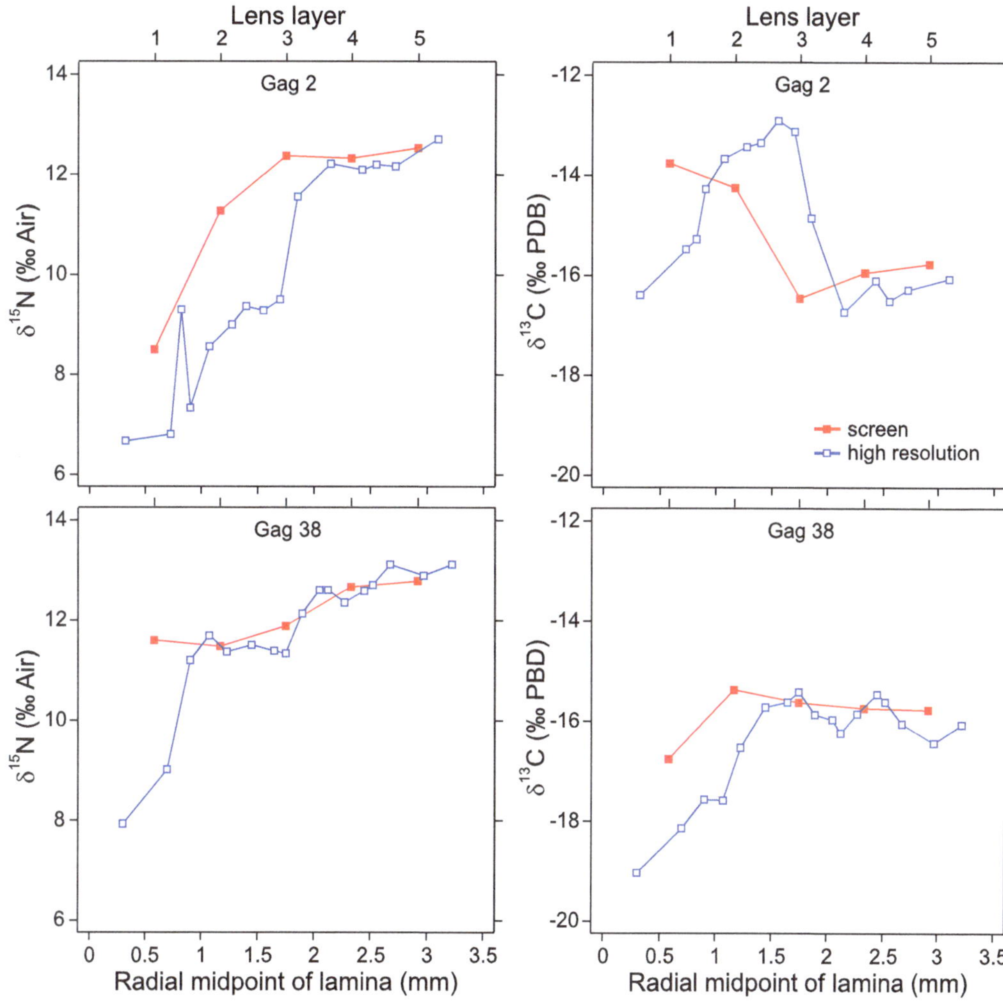

Figure 2. Low- and high-resolution comparison. The low-resolution screening method (analysis of coarse lens layers) is compared with the high-resolution method (analysis of individual laminae) for gag 2 (upper panels) and gag 38 (lower panels). The leftmost observation in the high-resolution method is for the lens nucleus; the remaining observations are for individual laminae. Gag 2 was 704 mm total length (TL); gag 38 was 832 mm TL. Both were female.

section was not known. For all other tests, an effort was made to subdivide the lenses into the smallest thicknesses that were manually practical ("laminae"). Individual sections and laminae were stored separately in 1-dram glass vials. Laminae became desiccated in <1 h at 25°C, after which most samples were homogenized using a mortar and pestle; exceptions were laminae that were haphazardly selected for intra-laminar variation analysis, which were subdivided first and then homogenized.

Dried laminae were weighed to the nearest µg on an analytical balance. A dry weight of 300–600 µg of material was placed in tin capsules for combustion and isotopic analysis. $^{13}C/^{12}C$ and $^{15}N/^{14}N$ and C:N were measured in replicate using a Carlo-Ebra NA2500 Series II elemental analyzer coupled to a continuous-flow ThermoFinnigan Delta+XL isotope ratio mass spectrometer at the University of South Florida College of Marine Science in St. Petersburg, Florida. The lower limit of quantification was 12 µg C or N. Calibration standards were NIST 8573 and NIST 8574 L-glutamic acid standard reference materials. Analytical precision, obtained by replicate measurements of NIST 1577b bovine liver, was ±0.11 ‰ for $\delta^{13}C$ and ±0.19 ‰ for $\delta^{15}N$ (average standard deviations of n = 30 replicates). Results are presented in standard notation (δ, in ‰) relative to international standards Pee Dee Belemnite (PDB) and air:

$$\delta^{j/i}X = \frac{\left({}^{j}X/{}^{i}X\right)_{sample}}{\left({}^{j}X/{}^{i}X\right)_{standard}} - 1$$

where X is the element and j and i are each an isotope of X.

Data Location and Analysis

All data are published at the Gulf of Mexico Research Initiative Information and Data Cooperative (https://data.gulfresearchinitiative.org/data/R1.x135.120:0006). In all comparisons except for the intra-laminar comparison, replicate isotope measurements were averaged and plotted for descriptive purposes. In the intra-laminar comparison, the objective was to determine if inter-laminar variation was greater than intra-laminar variation; individual replicate isotope measurements were compared among laminae using one-way ANOVA. Individual laminae were compared using 95% Tukey HSD intervals as a multiple range

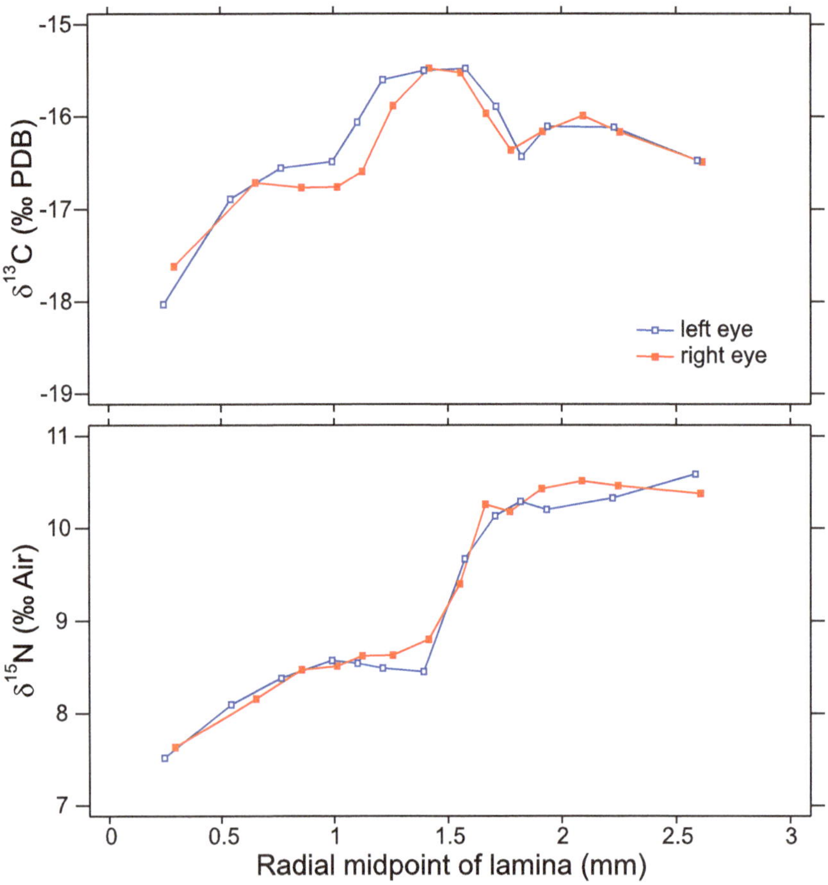

Figure 3. High-resolution comparisons of left and right eyes of white grunt specimen 2.

test (Statgraphic Centurion, v. 16.2.04). In all comparisons, means and replicates were plotted without any exclusion.

Results

Low-resolution screening analyses were conducted for eight red snapper, four red grouper, and three gag. All individuals had variable isotopic values among lens layers (Figure 1). The majority of values were centered within ranges reported by Radabaugh et al. [29] for trawl-caught demersal fishes on the West Florida Shelf (-19– -14 ‰ δ^{13}C; 7–13.5‰ δ^{15}N). Red snapper appeared to form two groups based on their δ^{15}N values; fish 45 and 47 (Figure 1) had consistently high δ^{15}N values. There were no clear groupings in red grouper. Gag 2 had higher δ^{13}C values for its innermost layer than the other two gag, and all three gag had isotopic values that converged in the outermost layers.

Gag 2 and 38 were analyzed using high-resolution analysis to further evaluate the difference recorded in the innermost layers. The second lens from gag 2 (3.1 mm radius) was dissected into 14 laminae plus a lens nucleus of 0.65 mm radius, indicating an average laminar thickness of 175 µm. The second lens from gag 38 (3.2 mm radius) was dissected into 16 laminae plus a lens nucleus of 0.60 mm radius, indicating an average laminar thickness of 163 µm. When data from the two lenses were combined, dry weight (DW, in mg) increased as $DW = 3.24r^3$ ($n = 32$, $R^2 = 0.99$, $p < 0.0001$), where r is lens radius in mm. The mean dry density of these two gag lenses was 0.89 mg mm^{-3}. Laminar thickness varied at a sub-annual scale in these two specimens, but did not have a strong positive or negative overall relationship with lens radius (slope $p > 0.05$).

The exponential (cubic) increase in average laminar weight that accompanies increasing lens size translates into a mass-balanced bias toward outer laminae when groups of laminae are processed together as a collective layer (e.g., during low-resolution screening analysis). The innermost layer from the screening analysis was most likely to be misrepresentative of patterns observed in the high-resolution analysis (Figure 2). Nevertheless, the screening method indicated different incipient δ^{15}N values for gag 2 and 38, and this difference was borne out by the high-resolution analysis, albeit at a nominally lower difference (2.5 vs. 3.0 ‰).

The left lens of white grunt 2 was dissected into 12 laminae plus the lens nucleus. The radius of the left lens nucleus was 0.50 mm and the radius of the whole left lens was 2.85 mm, indicating a mean laminar thickness of 196 µm. The right lens was dissected into 13 laminae plus the lens nucleus. The radius of the right lens nucleus was 0.60 mm and the radius of the whole right lens was 2.88 mm, indicating a mean laminar thickness of 175 µm. Isotopes in both eyes were centered within ranges reported by Radabaugh et al. [29], and δ^{15}N increased during life; this latter trend was observed in 15 of the 16 fish examined (94%, with red snapper 45 being the sole exception). Isotopic patterns were very similar between left and right eyes (Figure 3), yet plotted pairings of radial midpoints in Figure 3, average laminar thicknesses (196 vs. 175 µm), and the difference in total lamina count (12 vs. 13) were not in perfect agreement. One-way ANOVA indicated inter-laminar variation (Figure 4) was much larger than intra-laminar

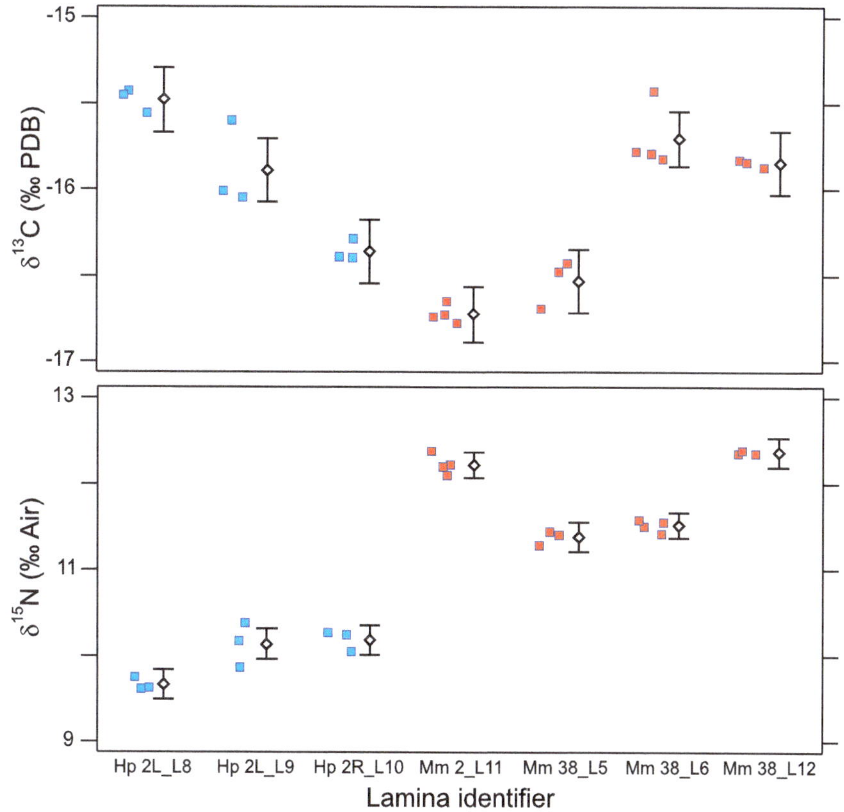

Figure 4. Intra- and inter-laminar comparison. Comparison of intra- and inter-laminar isotopic variation in white grunt specimen 2, left eye (Hp 2L) and right eye (Hp 2R) and gag specimens 2 and 38 (Mm 2 and Mm 38). Lamina number (e.g., L8 = lamina 8) is indicated within lamina identifiers. Data points have been horizontally jittered for clarity, and are summarized by means (diamonds) with 95% Tukey HSD intervals.

variation for both δ^{13}C [$F(6,16) = 40.6$, $p<0.0001$] and δ^{15}N [$F(6,16) = 232.4$, $p<0.0001$].

A total of 197 samples were analyzed (72 sections, 125 laminae). The range of mass C:N was 2.72–3.37. The mass C:N mean value was 2.97±0.12 (molar C:N range = 3.17–3.39; mean = 3.46±0.14). The C:N ratio was consistently <3.5; therefore, no lipid correction or normalization was needed because the lipid concentrations were uniformly low [33].

Discussion

The results indicate low-resolution isotopic screening has strong potential as an index of broad-scale changes in the isotopic history of individual fish. Broad-scale changes were evident in all species, yet red snapper and gag appeared to originate from distinct groups that converged later in life (all fish were collected from the same geographic region). The magnitude of isotopic variation observed in both the low- and high-resolution tests (Figures 1–3) matches the scale of variation in isoscapes produced by Radabaugh et al. [29] for the West Florida Shelf, the region from which the fish in our tests were collected. The low-resolution screening tests were rapid and cost-effective, but sometimes missed important life events. In the case of the gag (Figure 2), the missed detail involved early life history; for gag 2, the low-resolution analysis had a starting value of −13.8 ‰ for δ^{13}C while the high-resolution analysis was variable, starting at −16.4 and increasing to −12.9 ‰. The high-resolution approach contains the most information that can be obtained practically using the manual delamination method.

The white grunt's left and right eyes identified very similar lifelong trends (Figure 3), suggesting the temporal trends observed in the other comparisons (e.g., Figure 1) were repeatable and were not artifacts. The scale of variability between left and right eyes was similar to that of intra-laminar variation. Most intra-laminar variation was less than the instrument error calculated using bovine liver (±0.11 ‰ for δ^{13}C; ±0.19 ‰ for δ^{15}N,). Small, intra-laminar variation likely imparts a limit to the precision of the overall method.

All of the values and trends in the eye-lens isotopes are consistent with expected life-history patterns and trophic positions, as interpreted using published δ^{13}C and δ^{15}N isoscapes for demersal fish muscle [29]. Isoscapes visually portray spatial isotopic variation within natural systems [34]. While many terrestrial isoscapes exist, a need exists for developing more marine isoscapes [35]. When eye-lens isotopes are compared to observed and modeled isoscapes for demersal fish muscle on the West Florida Shelf [29,36], it appears that the two red snapper with higher δ^{15}N (Figure 1) traveled from the northern end of a δ^{15}N latitudinal gradient, and those with lower δ^{15}N traveled from or remained within more southerly waters. Isoscape trends in δ^{13}C of demersal fishes are more depth-related (higher values in shallow water) than latitude-related [29]. Red snapper eye-lens δ^{13}C remained below −15.5‰ throughout life, which is consistent with the red snapper's known preference for deeper water on the middle and outer portions of West Florida Shelf. The three gag may have started life at different locations, but spent a lengthy period of time together before collection. Unlike the red snapper, the gag uses shallow inshore waters as nursery habitat [37]. The elevated $\delta^{13}C$ in gag 2 (Fig. 2) is indicative of residence in shallow water. However, because $\delta^{13}C$ for demersal fish in inshore waters varies along the coastline as a statistical function of water clarity/depth (photosynthetically active radiation at bottom), particulate organic carbon concentration, and sea-surface temperature [36], not all gag can be expected to reside in high-$\delta^{13}C$ locations while in shallow nursery habitats.

All of the preceding geographic interpretations are subject to future corrections that accommodate shifts in trophic level and associated changes in total trophic fractionation. The increase in $\delta^{15}N$ that was observed in most (94%) specimens is consistent with increasing trophic level. The trend in the sole exception, red snapper 45, would be explained if the fish moved south later in life [29,36]. Among the four species, the gag (a large, piscivorous grouper) had relatively high $\delta^{15}N$ values at capture; this is consistent with expected differences in trophic position among species.

Directions for future work include, but are not limited to, the following areas of study: 1) eye lens aging by synchronizing lamina radial position with otolith-based or known captive age; relationships between fish eye-lens diameter and age, lens wet-weight and age, and lens dry-weight and age have already been established [38–41], 2) validating diet relationships using compound specific stable isotope analysis of the both the trophic and source amino acids within different laminae [42–44], 3) determination of long-term geographic histories for individual fish by combining isotope histories, knowledge of how trophic level changes with age (trophic growth rate), and output from isoscape models [36]; modeled isoscape outputs must be synchronized with time periods within eye lenses for accuracy because spatial stationarity cannot be assumed [36,45], 4) comparing isotopic fractionation of eye lens tissues with fractionation of other tissues [46,47]. Studying captive fish may be particularly definitive both for aging eye lenses and for validating diet relationships with eye-lens isotopes. However, the production and accurate interpretation of long-term geographic histories for individual fish require that eye-lens aging and trophic growth rate studies be completed first.

While our ability to rigorously interpret geographic histories from the eye-lens isotope data is limited by the present lack of the aforementioned studies, the patterns of variation within the eye lenses (Figs. 1–4) indicate that the overall approach is promising. There is no other known source of simultaneous $\delta^{13}C$ and $\delta^{15}N$ temporal records within bony fish. The approach presented here is not necessarily limited to bony fish, and may also be applied to other vertebrates as long as the spatiotemporal distribution of apoptosis within the lens is taken into consideration. In fish, TUNEL assays (terminal deoxynucleotidyl transferase dUTP nick-end labeling) indicate that even recently formed laminae can be apoptotic and thus isotopically conservative [24], whereas humans and other higher vertebrates may contain large proportions of pre-apoptotic (potentially non-conservative) cells in the outer cortical region of the lens.

Acknowledgments

Dr. Theodore Switzer (FWC) assisted us in efforts to obtain specimens from FWC fish surveys. Jenny Fenton (USF) assisted during sample collection and preparation, and Ethan Goddard (USF) assisted with stable isotope analysis. Hubbard's Marina kindly granted access to their docks and patrons during sample collection. The manuscript benefited substantially from thoughtful, anonymous contributions made during the review process. The research presented here partially fulfills requirements for the first author's doctoral degree.

Author Contributions

Conceived and designed the experiments: EBP. Performed the experiments: AAW. Analyzed the data: AAW EBP. Contributed reagents/materials/analysis tools: DJH. Wrote the paper: AAW EBP. Provided advice on study design and data interpretation: DJH. Reviewed first draft of the manuscript: DJH.

References

1. Tieszen LL, Boutton TW, Tesdahl KG, Slade NA (1983) Fractionation and turnover of stable carbon isotopes in animal tissues: implications for $\delta^{13}C$ analysis of diet. Oecologia 57(1–2): 32–37.
2. Hobson KA, Clark RG (1992) Assessing avian diets using stable isotopes I: Turnover of ^{13}C in tissues. Condor 94(1): 181–188.
3. Hobson KA, Clark RG (1992) Assessing avian diets using stable isotopes II: Factors influencing diet-tissue fractionation. Condor 94(1): 189–197.
4. Post DM (2002) Using stable isotopes to estimate trophic position: Models, methods, and assumptions. Ecology 83(3): 703–718.
5. Phillips DL, Eldridge PM (2006) Estimating the timing of diet shifts using stable isotopes. Oecologia 147(2): 195–203.
6. Thompson RC, Ballou JE (1956) Studies of metabolic turnover with tritium as a tracer: V. The predominantly non-dynamic state of body constituents in the rat. J Biol Chem 223(2): 795–809.
7. Fry B, Arnold C (1982) Rapid C-13/C-12 turnover during growth of brown shrimp (*Penaeus aztecus*). Oecologia 54(2): 200–204.
8. Jones CM (1986) Determining age of larval fish with the otolith increment technique. Fish Bull 84:91–103.
9. Thorrold SR, Campana SE, Jones CM, Swart PK (1997) Factors determining $\delta^{13}C$ and $\delta^{18}O$ fractionation in aragonitic otoliths of marine fish. Geochim Cosmochim Acta 61(14):2909–2919.
10. Campana SE (1999) Chemistry and composition of fish otoliths: pathways, mechanisms and applications. Mar Ecol Prog Ser 188:263–297.
11. Hutchinson JJ, Trueman CN (2006) Stable isotope analyses of collagen in fish scales: limitations set by scale architecture. J Fish Biol 69:1874–1880.
12. Trueman CN, Moore A (2007) Use of the stable isotope composition of fish scales for monitoring aquatic ecosystems. In: Dawson TE, Siegwolf RTW, (eds.)Terrestrial EcologyVolume1.
13. Woodcock SH, Walther BD (2014) Trace elements and stable isotopes in Atlantic tarpon scales reveal movements across estuarine gradients. Fish Res 153:9–17.
14. Estrada JA, Rice AN, Natanson LJ, Skomal GB (2006) Use of isotopic analysis of vertebrae in reconstructing ontogenetic feeding ecology in white sharks. Ecology 87(4): 829–834.
15. Kerr LA, Andrews AH, Cailliet GM, Brown TA, Coale KH (2006) Investigations of $\Delta^{14}C$, $\delta^{13}C$, and $\delta^{15}N$ in vertebrae of white shark (*Carcharodon carcharias*) from the eastern North Pacific Ocean. Environ Biol Fishes 77(3–4): 337–353.
16. Werry JM, Lee SY, Otway NM, Hu Y, Sumpton W (2011) A multi-faceted approach for quantifying the estuarine-nearshore transition in the life cycle of the bull shark, *Carcharhinus leucas*. Mar Freshw Res 62(12): 1421–1431.
17. Walker JL, Macko SA (1999) Dietary studies of marine mammals using stable carbon and nitrogen isotopic ratios of teeth. Mar Mamm Sci 15(2): 314–334.
18. Dalerum F, Angerbjörn A (2005) Resolving temporal variation in vertebrate diets using naturally occurring stable isotopes. Oecologia 144:647–658.
19. Ethier DM, Kyle CJ, Kyser TK, Nocera JJ (2010) Variability in the growth patterns of the cornified claw sheath among vertebrates: implications for using biogeochemistry to study animal movement. Can J Zool 88:1043–1051.
20. Parry MP (2003) The trophic ecology of two ommastrephid squid species, *Ommastrephes bartramii* and *Sthenoteuthis oualaniensis*, in the North Pacific sub-tropical gyre. Doctoral dissertation, University of Hawaii at Manoa.
21. Hunsicker ME, Essington TE, Aydin KY, Ishida B (2010) Predatory role of the commander squid *Berryteuthis magister* in the eastern Bering Sea: Insights from stable isotopes and food habits. Mar Ecol Prog Ser 415: 91–108.
22. Nicol JAC (1989) The eyes of fishes. Oxford: Oxford University Press. 308p.
23. Horwitz J (2003) Alpha-crystallin. Exp Eye Res 76(2): 145–153.
24. Dahm R, Schonthaler HB, Soehn AS, van Marle J, Vrensen GFJM (2007) Development and adult morphology of the eye lens in the zebrafish. Exp Eye Res 85(1): 74–89.
25. Wride MA (2011) Lens fibre cell differentiation and organelle loss: many paths lead to clarity. Philos Trans R Soc Lond B Biol Sci 366(1568): 1219–1233.
26. Bloemendal H (1982) Lens proteins. CRC Crit Rev Biochem 12(1): 1–38.
27. Vihtelic TS (2008) Teleost lens development and degeneration. Int Rev Cell Mol Biol 269: 341–373.
28. Dahm R (1999) Lens fibre cell differentiation—a link with apoptosis? Ophthalmic Res 31(3): 163–183.
29. Radabaugh KR, Hollander DJ, Peebles EB (2013) Seasonal $\delta^{13}C$ and $\delta^{15}N$ isoscapes of fish populations along a continental shelf trophic gradient. Cont Shelf Res 68: 112–121.

30. Karpouzi VS, Stergiou KI (2003) The relationships between mouth size and shape and body length for 18 species of marine fishes and their trophic implications. J Fish Biol 62:1353–1365.
31. McMahon KW, Hamady LL, Thorrold SR (2013) Ocean ecogeochemistry: a review. Oceanography and Marine Biology: An Annual Review 51:327–373.
32. Sharma KK, Santhoshkumar P (2009) Lens aging: Effects of crystallins. Biochim Biophys Acta 1790(10): 1095–1108.
33. Post DM, Layman CA, Arrington DA, Takimoto G, Quattrochi J, et al. (2007) Getting to the fat of the matter: models, methods and assumptions for dealing with lipids in stable isotope analyses. Oecologia 152:179–189.
34. West JB, Bowen GJ, Dawson TE (2010) Preface: Context and background for the topic and book. In: West JB, Bowen GJ, Dawson TE, Tu KP, editors. Isoscapes: Understanding movement, pattern, and process on Earth through isotope mapping. New York: Springer. pp. v–xi.
35. Hobson KA, Barnett-Johnson R, Cerling T (2010) Using isoscapes to track animal migration. In: West JB, Bowen GJ, Dawson TE, Tu KP, editors. Isoscapes: Understanding movement, pattern, and process on Earth through isotope mapping.New York:Springer pp. 273–298.
36. Radabaugh KR, Peebles EB (2014) Multiple regression models of δ^{13}C and δ^{15}N for fish populations in the eastern Gulf of Mexico. Cont Shelf Res 84: 158–168.
37. Fitzhugh GR, Koenig CC, Coleman FC, Grimes CB, Sturges W (2005) Spatial and temporal patterns in fertilization and settlement of young gag (*Mycteroperca microlepis*) along the West Florida Shelf. Bulletin of Marine Science 77(3): 377–396.
38. Burkett RD, Jackson WB (1971) The eye lens as an age indicator in freshwater drum. Am Midl Nat 85(1): 222–225.
39. Siezen RJ (1989) Eye lens aging in the spiny dogfish (*Squalus acanthias*): 1. Age-determination from lens weight. Curr Eye Res 8(7): 707–712.
40. Al-Hassan LAJ, Al-Sayab AA (1994) Eye lens diameter as an age indicator in the catfish, *Silurus triostegus*. Pak J Zool 26(1): 81–82.
41. Al-Mamry J, Jawad L, Al-Busaidi J, Al-Mamari A, Al-Mamry S, et al. (2012) The use of eye lens diameter and weight in age determination in *Siganus canaliculatus* (Park, 1797) (Perciformes, Siganidae) collected from the Arabian Sea coasts of Oman. Natura Montenegrina 11(1): 73–78.
42. Popp BN, Graham BS, Olson RJ, Hannides C, Lott MJ, et al. (2007) Insight into the trophic ecology of yellowfin tuna, *Thunnus albacares*, from compound-specific nitrogen isotope analysis of proteinaceous amino acids. In: Dawson TE, Siegwolf RTW, editors. Terrestrial Ecology, Volume I. Elsevier pp. 173–190.
43. Chikaraishi Y, Ogawa NO, Kashiyama Y, Takano Y, Suga H, et al. (2009) Determination of aquatic food-web structure based on compound-specific nitrogen isotopic composition of amino acids. Limnol Oceanogr Methods 7:740–750.
44. Ellis G (2012) Compound-specific stable isotopic analysis of protein amino acids: Ecological applications in modern and ancient systems. Doctoral dissertation, University of South Florida, Tampa, FL.
45. Haché S, Hobson KA, Bayne EM, Van Wilgenburg SL, Villard MA (2014) Tracking natal dispersal in a coastal population of a migratory songbird using feather stable isotope (δ^{2}H, δ^{34}S) tracers. PLOS ONE DOI: 10.1371/journal.pone.0094437.
46. Caut S, Angulo E, Courchamp F (2009) Variation in discrimination factors (Δ^{15}N and Δ^{13}C): the effect of diet isotopic values and applications for diet reconstruction. J Appl Ecol 46:443–453.
47. Martinez del Rio C, Carleton SA (2012) How fast and how failthful: the dynamics of isotopic incorporation into animal tissues. J Mammal 93(2):353–359.

5

In Situ Mortality Experiments with Juvenile Sea Bass (*Dicentrarchus labrax*) in Relation to Impulsive Sound Levels Caused by Pile Driving of Windmill Foundations

Elisabeth Debusschere[1,2]*, Bert De Coensel[3], Aline Bajek[4], Dick Botteldooren[3], Kris Hostens[1], Jan Vanaverbeke[2], Sofie Vandendriessche[1], Karl Van Ginderdeuren[1], Magda Vincx[2], Steven Degraer[2,5]

1 Bio-environmental research group, Institute for Agricultural and Fisheries Research, Oostende, Belgium, **2** Biology Department, Ghent University, Ghent, Belgium, **3** Department of Information Technology, Ghent University, Ghent, Belgium, **4** Ecloserie Marine de Gravelines, Gravelines, France, **5** Operational Directorate Natural Environment (OD Nature), Royal Belgian Institute of Natural Sciences (RBINS), Brussels, Belgium

Abstract

Impact assessments of offshore wind farm installations and operations on the marine fauna are performed in many countries. Yet, only limited quantitative data on the physiological impact of impulsive sounds on (juvenile) fishes during pile driving of offshore wind farm foundations are available. Our current knowledge on fish injury and mortality due to pile driving is mainly based on laboratory experiments, in which high-intensity pile driving sounds are generated inside acoustic chambers. To validate these lab results, an *in situ* field experiment was carried out on board of a pile driving vessel. Juvenile European sea bass (*Dicentrarchus labrax*) of 68 and 115 days post hatching were exposed to pile-driving sounds as close as 45 m from the actual pile driving activity. Fish were exposed to strikes with a sound exposure level between 181 and 188 dB re 1 μPa^2.s. The number of strikes ranged from 1739 to 3067, resulting in a cumulative sound exposure level between 215 and 222 dB re 1 μPa^2.s. Control treatments consisted of fish not exposed to pile driving sounds. No differences in immediate mortality were found between exposed and control fish groups. Also no differences were noted in the delayed mortality up to 14 days after exposure between both groups. Our *in situ* experiments largely confirm the mortality results of the lab experiments found in other studies.

Editor: Z. Daniel Deng, Pacific Northwest National Laboratory, United States of America

Funding: Elisabeth Debusschere acknowledges an IWT predoctoral grant (Agency for Innovation by Science and Technology, 111217). http://www.iwt.be/subsidies/sb/steun. Bert De Coensel is a postdoctoral fellow of the Research Foundation-Flanders (FWO-Vlaanderen) http://www.fwo.be/. Co-author Aline Bajek is employed by Ecloserie Marine de Gravelines. Ecloserie Marine de Gravelines provided support in the form of salary for author Aline Bajek, but did not have any additional role in the study design, data collection and analysis, decision to publish, or preparation of the manuscript. The specific role of this author is articulated in the 'author contributions' section. The funders had no role in study design, data collection and analysis, decision to publish, or preparation of the manuscript.

Competing Interests: Co-author Aline Bajek is employed by Ecloserie Marine de Gravelines. There are no patents, products in development or marketed products to declare. This does not alter the authors' adherence to all the PLOS ONE policies on sharing data and materials.

* Email: Elisabeth.Debusschere@ilvo.vlaanderen.be

Introduction

The increasing demand for renewable energy has led to innovative techniques and numerous ambitious projects. Offshore wind energy is particularly popular across the North Sea bordering countries. However, the construction of offshore wind farms - and the pile driving activities in particular - involve strong impulsive sounds, which are potentially harmful to marine fishes and more specifically to the early life stages of fishes, i.e. eggs, larvae and young juveniles [1,2,3,4]. These early life stages are important as prey for pelagic fishes and for the recruitment to the adult fish populations, which stresses the need to understand how underwater sounds affect their fitness. Remarkably, many offshore wind farms are being installed or planned without extensive quantitative data on the physiological impact of strong impulsive pile driving sounds on fishes [2,4].

In light of environmental impact control, it is necessary to establish sound level thresholds for pile driving at which fishes don't get injured. A first step in this process is to assess the sound level range causing immediate or delayed mortality [2]. Such assessments have been performed for Chinook salmon (*Oncorhynchus tschawytscha*), hybrid striped bass (white bass *Morone chrysops* x striped bass *Morone saxatilis*) and common sole (*Solea solea*), through lab experiments, using different methods to generate high-intensity 'pile driving' sounds in acoustically controlled chambers [5,6,7]. Juvenile Chinook salmon of 103 mm±8.75 (SD) exhibited mortal injuries, which include any mortality or injuries that can lead to mortality, at relatively high single-strike sound exposure levels (SEL_{ss}) of 187 dB re 1 μPa²s, leading to a cumulative sound exposure level (SEL_{cum}) of 220 dB re 1 μPa²s for 1920 strikes [6]. The hybrid striped bass of two different sizes; average 1.3 g and 17.2 g, exhibited mortal injuries at a SEL_{ss} of 180 dB re 1 μPa².s for 960 strikes, resulting in a SEL_{cum} of 210 dB re 1 μPa².s [7]. For common sole larvae that were exposed to a SEL_{ss} of 186 dB re 1 μPa².s for 100 strikes, leading to a SEL_{cum} of 206 dB re 1 μPa².s, no difference in mortality was found between control and exposed groups up to 7 days after exposure [5]. Since these laboratory experiments have not yet been verified in the field, there is a need to expose juvenile fishes to the sound exposure levels present in the immediate vicinity of the pile driving activity in order to examine direct or delayed mortality.

In the current study, a 'worst-case scenario' *in situ* field experiment was carried out to fill the gaps in the establishment of sound level thresholds for young fishes. Sound pressure was measured alongside a piling platform (45 m from the pile) and immediate and delayed mortality in young sea bass (*Dicentrarchus labrax*) were assessed.

Material and Methods

1 Piling vessel and study location

To examine the impact of pile driving, this field study was performed on board of the jack-up piling vessel Neptune DP2 (GeoSea). Location was the Lodewijckbank, Belgian Part of the North Sea (Figure 1). This field experiment is part of the scientific research within the zone for renewable energy issued by the Belgian Ministry for the North Sea as described in the environmental permit NV Eldepasco [8].

Access to the deck side opposite to the pile driving activity was granted to perform the experiment, as close as 45 m from the monopile. Each monopile was designed for its specific position in the wind farm and varied in length, diameter and weight. Two monopiles were installed per trip using a hydraulic piling hammer (IHC Hydrohammer B.V.). Monopiles C8 (Lat N 51.637648, Long E 2.9003345, WGS84) and B3 (Lat N 51.629995, Long E 2.926765) were driven into the seabed during the first trip, monopiles G7 (Lat N 51.60667, Long E 2.881473) and G8 (Lat N 51.602782, Long E 2.877282) during the second trip (Table 1, Figure 1).

2 Fish characteristics and preparation

European sea bass (*Dicentrarchus labrax*) is commercially exploited in the Southern North Sea and Mediterranean Sea, both through fisheries and aquaculture. Sea bass is a well-studied species, especially concerning larval growth, development and skeletal formation [9,10,11,12,13]. Swim bladder formation in sea bass larvae starts 6 **d**ays **p**ost **h**atching (dph) and is complete at 16 days. Sea bass is a physoclist round fish with an opening between the mouth and swim bladder only during the first days of the swim bladder formation [14]. Thereupon, buoyancy is controlled by gases retrieved from the blood, hence volume changes cannot be performed abruptly [15].

As sea bass eggs and larvae are available year round in the Ecloserie Marine de Gravelines (France), this largely facilitates the use of this species in experiments. This study was carried out in accordance with the Belgian Council for Laboratory Animal Science (BCLAS) guidelines. The experimental protocol was approved by the ethical committee of the Institute for Agricultural and Fisheries Research (ILVO) (Permit Number: 2012/178). The field study did not involve endangered or protected species.

Sea bass of 45 dph (hatched at the Ecloserie Marine de Gravelines) were incubated in cylindro-conic linear low-density polyethylene (LLDPE) containers of 9.5 L at ILVO. They were provided with aeration and a flow through of UV-sterilized sea water on a half-closed recirculation system. The fingerlings were kept in this aquaculture unit until the experiment could take place on board of the piling vessel. They were fed twice a day with Aglonorse (300-500 μm) or MariCo Start 1.5 mm (Coppens). The water temperature in the cultivation aquaria was 20.1°C±0.5, with a salinity of 30.2 ppt±0.1 and a pH of 8.1±0.1.

Fingerlings of 68 dph were used for the first *in situ* experiments on board of the piling vessel, while young sea bass of 115 dph from another batch were used during the second trip. Sea bass fingerlings were transported to the piling vessel during the first trip in 2 L containers and the second trip in 10 L buckets, each provided with oxygen tablets (JBL). After 3.5 hours, the seawater in the containers was renewed on board and a continuous air supply was provided. The sea bass were fed with Aglonorse (300–500 μm) or MariCo Start 1.5 mm (Coppens). Debris and dead fish were removed from the container and seawater was renewed on a daily basis to ensure good water conditions. Fish were checked twice a day and if they were in poor condition (showing illness, stress or decreased activity) they were not used in the experiments. Humane endpoints were incorporated if the fish would display behavioural abnormalities or haemorrhage. The fish would be humanely sacrificed by transferring them into an overdose anaesthetic ((2 ml of 5 g benzocaine dissolved in 25 ml aceton)/ 1 L seawater).

3 Acoustic equipment

Sound pressure was measured using a Brüel & Kjaer hydrophone (type 8104, voltage sensitivity 47.7 $\mu V.Pa^{-1}$, charge sensitivity 0.391 $pC.Pa^{-1}$, 10 m cable). The hydrophone was connected to the charge channel of a Brüel & Kjaer portable amplifier (Nexus type 2690-0S). The measurement chain was completed with a multi-channel portable recorder (Tascam DR-680). The signal was recorded in 1-channel WAVE format (.wav) on Compact Flash cards of 16 GB (SanDisk Ultra) with a sampling rate of 44 100 Hz at 24 bit.

4 Experimental setup and treatment

The experimental unit existed of two parts: a stainless steel frame holding a case with the field recorder and amplifier; and a similar frame holding the six 500 ml vials with the fish (Figure 2). The vials were made of poly 4-methyl, 1- pentene (PMP) with an acoustic impedance of 1.84 Rayl, which is as close as possible to the impedance of seawater (1.56 Rayl) [16,17]. The hydrophone was attached to the stainless steel cables holding the frames in such a way that it was hanging unobstructed 0.3 m above the vials. The whole experimental unit was lowered with a crane submerging the lowest frame to a depth of 2.5 m below the water surface.

Dead sea bass release toxins which negatively affect the survival of the others. To counteract this phenomenon, each treatment group (exposed and control) was subdivided into six vials per experiment. The number of fingerlings in the vials was based on their oxygen demand and on oxygen availability in the seawater in the vials, furthermore they were randomly assigned to a vial and group. During the first trip, each vial had a density of 20 individuals (68 dph; wet weight 42.8±15.8 mg; standard length 16.8±1.5 mm), whilst the density during the second trip was reduced to 2 individuals per vial (115 dph; 1613.3±472.5 mg; 47.9±4.5 mm). After transferring the fish into the vials, each vial was filled with care until a bubble of seawater would form on top of the vial. Then, the screw cap was screwed on the vial with a

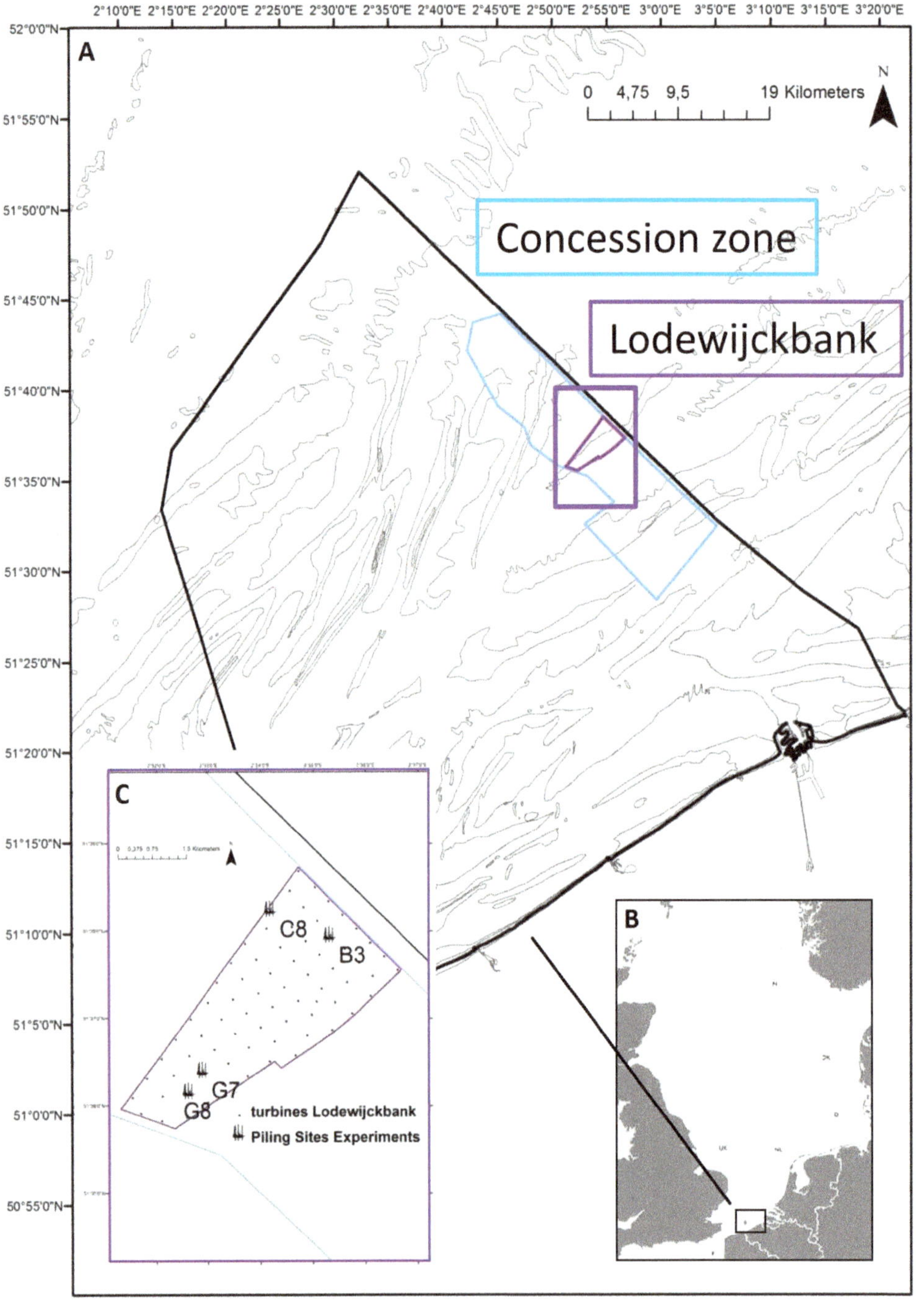

Figure 1. The offshore renewable energy zone in the Belgian part of the North Sea (a). North Sea exclusive economic zones (b). The Northwind concession on the Lodewijckbank includes 72 monopiles of ~5 m diameter (c). The experiment was repeated at four monopiles (C8, B3, G7, G8).

small overspill of seawater as a consequence. Each vial was checked for air bubbles before it went into the sea. After ~1.5 hours (approximate duration of 1 monopile hammering event) the vials were inspected for immediate mortality, after which the fish were put together per two vials into a 2 L container with air supply. The same procedure was followed with the control group in the absence of pile driving and repeated for the following exposed and control groups, resulting in 4 experiments each. The total sample size was 528 individuals with 120 fish in each treatment group in the first two experiments and 12 fish per group in the third and fourth experiment.

After three days on board of the piling vessel, the fingerlings were transported back to the laboratory (see 2) and placed in cylindro-conical aquaria (9.5 L, first trip) or rectangular aquaria (30 L, second trip) per treatment and experiment. Delayed mortality was monitored twice a day during the following 14 days after exposure. 72 fish and four fish were followed for delayed mortality for each group respectively from the first and second

Table 1. Characteristics of the four monopiles and number of hammering strikes, energy and time of pile driving for each monopile.

Monopile number	C8	B3	G7	G8
Trip	1	1	2	2
Date	19/06/2013	19/06/2013	27/08/2013	28/08/2013
Start time (h)	1:47	18:38	22:39	11:15
Weight (tons)	471	357	346	346
Diameter (m)	5	5	5.2	5.2
Steel thickness (mm)	50	50	50	50
Length (m)	62	57	56	56
Depth in seafloor (m)	33	32	30	33
Total strikes	2282	2331	3249	2964
Maximum energy/strike (kJ)	1173	867	1069	1162
Total energy (kJ)	2333436	2276948	2333240	2526331
Total pile-driving time (h)	1:29	1:14	1:45	1:23
Net hammering time (h)	0:57	0:59	1:17	1:10
Time of day	1:47	18:38	11:15	22:39

trip, while the other 48 and eight of each group were stored in liquid nitrogen to analyse sub-lethal effects, and more precisely effects on stress hormone levels (Debusschere et al., in prep). At the end of the monitoring period, all fish were humanely euthanized.

5 Sound analyses

Metrics for the sound pressure p were calculated using Matlab R2012b (version 8.0).

The combination of single strike sound exposure level, cumulative sound exposure level and the total number of impulse events can be successfully correlated to the severity of fish injuries [6,18]. In addition to these sound pressure metrics, the peak sound pressure level and the 1/3-octave band containing most energy were also calculated.

- The sound pressure level (SPL) is defined as the logarithmic measure of the instantaneous sound pressure within a given time interval. The unit is dB re 1 μPa.
- The peak sound pressure level (SPL_{peak}) is defined as the level associated to the maximum absolute value of the instantaneous sound pressure within a given time interval. The unit is dB re 1 μPa.
- The single strike sound exposure level (SEL_{ss}) is defined as the level associated to the integral of the squared sound pressure over the duration of a single impulse event. The unit is dB re 1 μPa^2.s. Impulse events were detected on the basis of the SPL time series. The detection threshold was set to 170 dB, and single events were selected using a temporal window around the times that the threshold is exceeded, 0.20 s before the threshold and 0.50 s after. The minimum time between two events was set to 0.50 s. These optimal values were set after visual inspection of the recordings.
- The cumulative sound exposure level (SEL_{cum}) is defined as the decibel sum of the sound exposure level over a number of individual impulse events. The time between the individual impulse events is excluded using this procedure.

Given the average SEL_{ss} and the number of impulse events, the cumulative levels can be more practically (p) calculated [5,6] as:

$$SEL_{cum,p} = SEL_{ss} + 10\log_{10}(\text{number of impulse events}) \quad (1)$$

6 Statistical analyses

The Plymouth Routines In Multivariate Ecological Research (PRIMER) programme, version 6.1.12 with PERMANOVA add-on software, was applied for statistical analyses ([19]). A significance level of $p < 0.05$ was used in all tests. The univariate permutational ANOVA's (PERMANOVA) were carried out with a 2 - factor design including treatment (T) and age (A) to analyse immediate mortality, while the delayed mortality over 14 days was analysed with a 3 - factor design, including treatment (T), age (A), and days after exposure (D) analysed in this order. The data were not transformed and Euclidean distance similarity matrices were applied. Due to the low replicate number the Monte Carlo P-value was preferred over the permutation P-value [20]. A simple linear regression analysed the potential linear relationship between total pile driving energy (Table 1) and SEL_{cum} (Table 2) using R version 2.15.1. Non-parametric Spearman rank tests were carried out to examine a correlation between the monopile characteristics (weight, length, penetration depth) and the pile driving characteristics (total energy, total strikes and maximum energy per strike). Furthermore, a non-parametric Spearman rank test was performed with penetration depth at each strike and energy per strike.

Results

1 Sound parameters

No correlation was found between the monopile characteristics (weight, length, penetration depth, Table 1) and the pile driving characteristics (total energy, total strikes, and maximum energy per strike). Also, no linear relationship was found between the

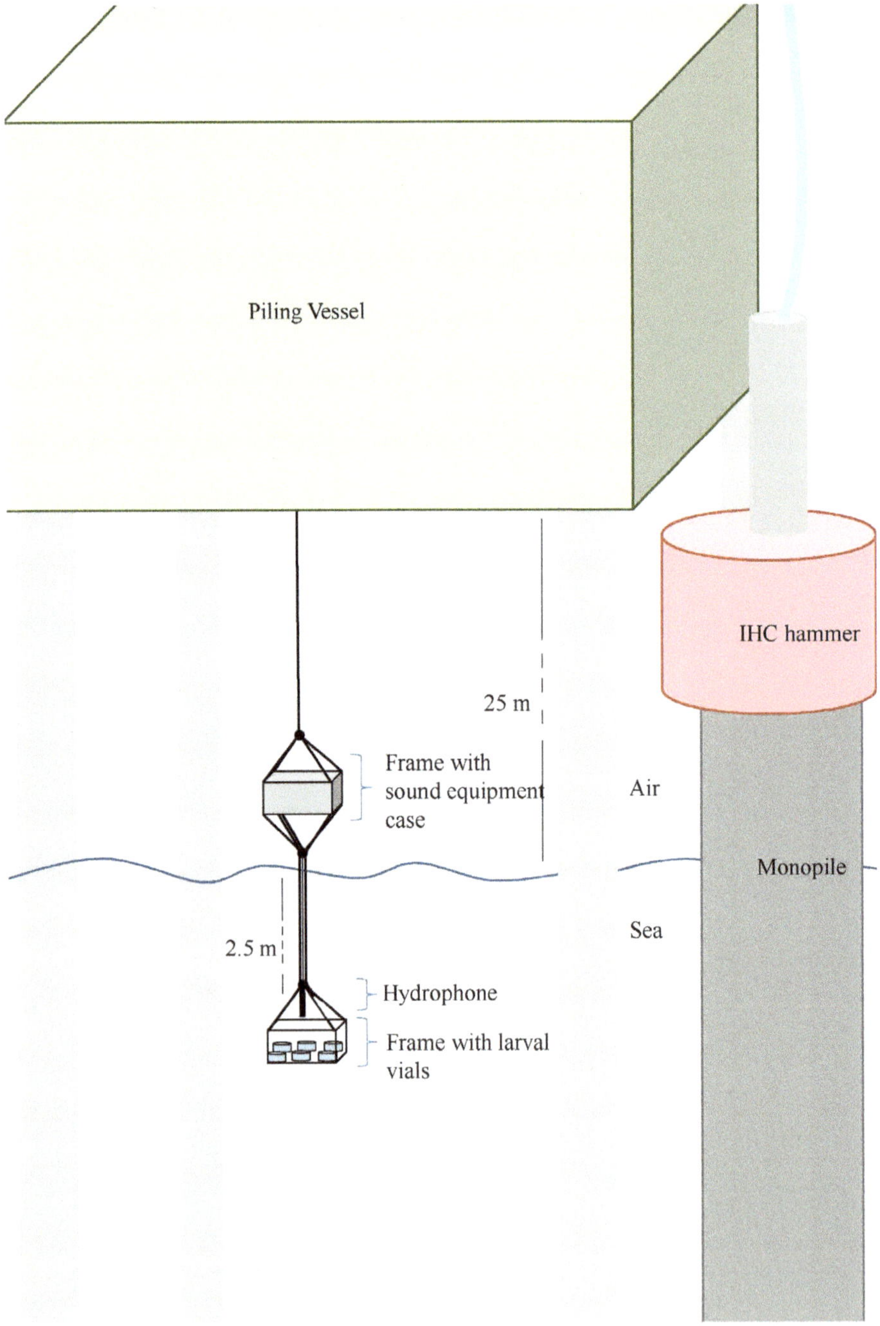

Figure 2. Experimental unit at the side of the piling vessel opposite to the monopile and hammering unit. The piling vessel is jacked-up 25 m above the sea surface on four steel piles. The upper frame above the sea surface holds the sound recording equipment. The lower frame holds the fish in vials. The hydrophone is attached between the steel cables of the lower frame.

cumulative sound exposure level calculated for the total number of strikes ($SEL_{cum, p}$) (Table 2) and the total energy necessary for one monopile ($R^2 = 0.14$). On the other hand, the energy needed per strike was positively correlated with the penetration depth, which means that the type of sediment layers that have to be penetrated forms a key factor in terms of energy requirements (Spearman's rank correlation coefficient 0.94 (C8); 0.90 (B3); 0.94 (G7); 0.92 (G8)).

The pile driving sound levels that were measured during the four experiments at 2.5 m below the water surface reached on average $SEL_{ss} = 181–188$ dB re 1 $\mu Pa^2 s$, rose to $SPL_{peak} = 210–211$ dB re 1 μPa^2, and led to $SEL_{cum} = 215–222$ dB re 1 $\mu Pa^2.s$, with 1739 up to 3067 strikes per monopile (Table 2). The SPL_{peak} was constant across the four monopiles (see example in Figure 3), while B3 (experiment 2) had higher SEL_{ss} and SEL_{cum} values compared to the other three monopiles. The dominant energy during exposure (SEL_{ss}) was present at 125–200 Hz, although no steep decline was recorded towards the higher frequencies (Figure 4a). For the experiments with control groups that were carried out in the absence of pile-driving activities on the piling

Table 2. Sound pressure metrics measured at 45 m during pile-driving of the four monopiles and for the control groups of each experiment.

Monopiles	**C8**	**B3**	**G7**	**G8**
Trip number	1	1	2	2
Sound metrics to which the fish were exposed to				
total strikes exposed to*	1739	2312	3067	2959
time between peaks (s)	1.49	1.49	1.39	1.41
SEL_{ss} mean (dB re 1 μPa^2.s)	183	188	181	183
SEL_{ss} max (dB re 1 μPa^2.s)	185	191	185	187
SEL_{ss} min (dB re 1 μPa^2.s)	160	159	173	157
SPL_{peak} (dB re 1 μPa)	210	210	211	211
SEL_{cum} (dB re 1 μPa^2.s)	215	222	217	218
$SEL_{cum,p}$ (dB re 1 μPa^2s)	215	222	216	218
1/3 octave band with most energy (Hz)	200	160	125	200
Control replicate	**1**	**2**	**3**	**4**
SPL (dB re 1 μPa)	138	128	145	136
1/3 octave band with most energy (Hz)	31.5	25	40	100

* Different from total number of strikes in table 1 as net time of exposure is less than total hammering time.

vessel, only sound pressure level could be measured. The SPL varied between 127 and 145 dB re 1 μPa (Figure 4b).

2 Sea bass survival

During the first trip with the 68 dph fish (120 fish per treatment group), no immediate mortality was observed for the control groups, whereas a mortality of 1.7 and 5% was found for the exposure experiments 1 and 2, respectively (Figure 5). The dead fish were only seen in one of the six vials in experiment 1 and in two vials in experiment 2. During the second trip (experiments 3 and 4) with the 115 dph fish (12 fish per treatment group), no mortality was observed immediately after the experiments. The difference in immediate mortality (Monte Carlo p-value = 0.11) between control and exposed groups was not statistically significant. Transportation back to the lab did not cause direct mortality. At the end of the 14-day monitoring period of trip one, 9% of the exposed fish and 10.1% of the control fish died. As only four fish per group of the second trip were followed for delayed mortality, a dead fish caused a mortality increase of 25%. One fish of the exposed group died on day four and one fish of the control group on day 13, the latter jumped out of the aquarium and was not included in the statistical analysis. Delayed mortality did not differ significantly between the control and exposed groups of both trips (Monte Carlo p-value = 0.39).

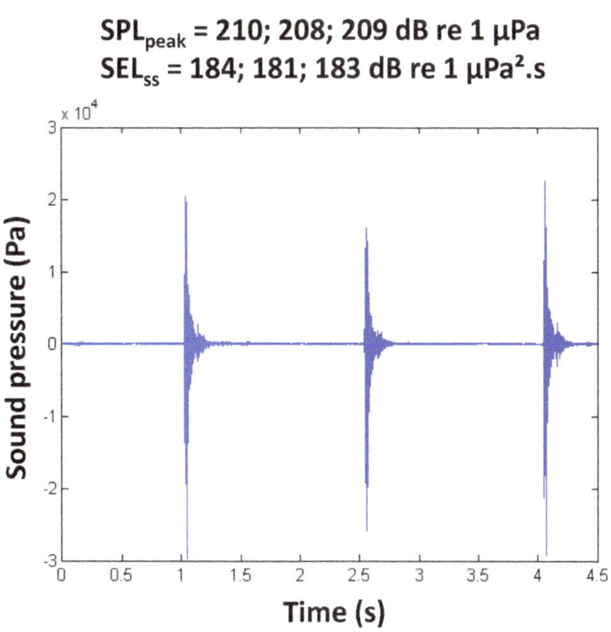

Figure 3. Detail of three consecutive piling strike signals. Detail of three consecutive piling strike signals recorded in the field for underwater sound pressure during experiment 3 (G7) (strike number 3005–3007 at 4892–4896.5 s).

Discussion

1 Field exposure

As far as we know, this work on the effects of sound on fish presents the first *in situ* field study carried out as close as 45 m from offshore pile driving of monopiles. As such, our study is very valuable to validate the mortality results of laboratory experiments presented in other studies [5,6,15,18]. Although, the underwater sound levels during the experiments with the control groups were above the background levels in the North Sea [21], due to sound radiation through the four jack-up piles of the piling vessel and the presence of the working vessels, these sound levels were unlikely to cause immediate or delayed mortality [4,22,23].

Both ages of sea bass that were used in our experiments are considered new fingerlings, but juveniles of 68 dph seem to be more sensitive in general (higher mortality) compared to juveniles of 115 dph. Still, both age groups largely survived the exposure to the high sound pressure levels that were exhibited over a complete piling session.

Exposure in the field to such sound levels did not cause significantly increased mortality during the first 14 days after exposure of sea bass *Dicentrarchus labrax* of 68 and 115 dph. This

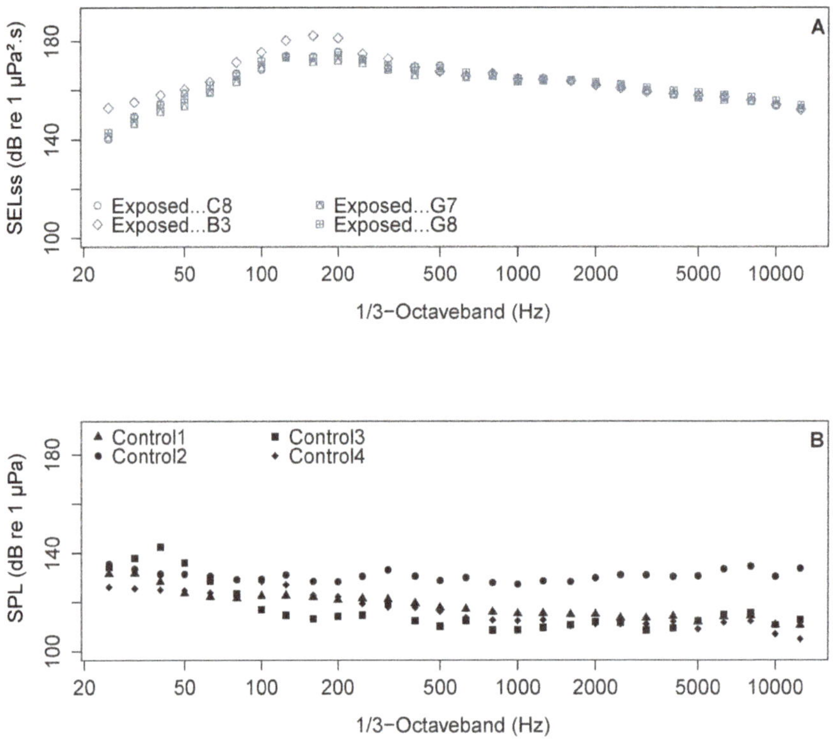

Figure 4. Measured frequency spectra in the presence and absence of pile driving. Mean SEL_{ss} (A) of the total recorded piling strikes versus 1/3 octave bands and SPL (B) of the control groups versus 1/3 octave bands. C8, B3, G7, G8 are the groups exposed to pile-driving; Control 1–4 are the control groups.

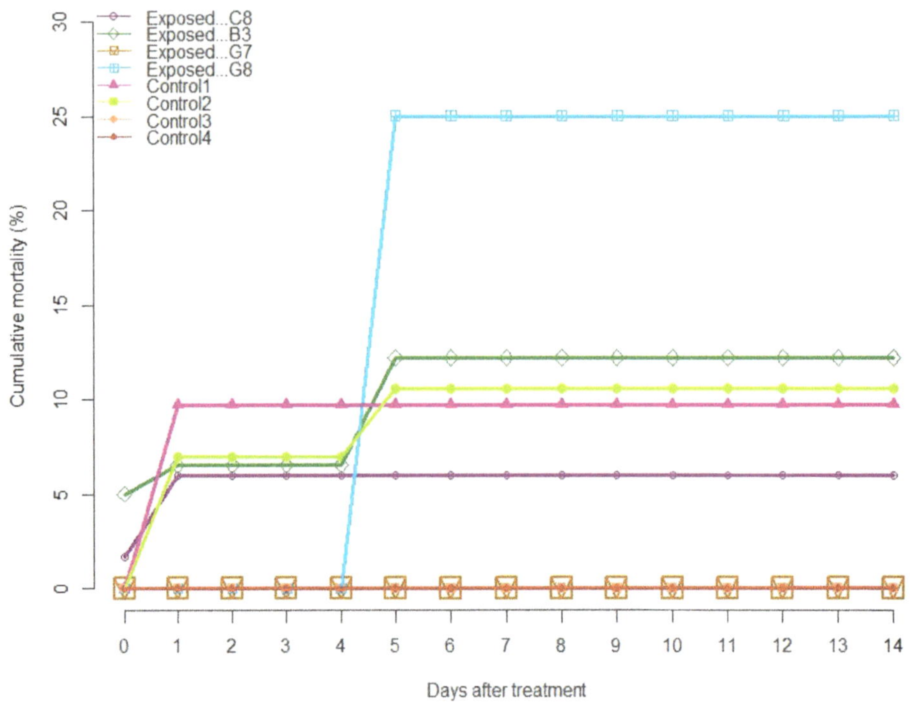

Figure 5. Cumulative mortality during 14 days after the experiment. Trip 1 (C8, B3, Control 1–2) and trip 2 (G7, G8, Control 3–4). All fish were transported back to the laboratory on day three.

result strengthens the trend seen in the recent lab studies with the survival of common sole larvae (*Solea solea*) [5], juvenile Chinook Salmon (*Oncorhynchus tshawytscha*) [6], lake sturgeon (*Acipensen fulvescens*), Nile tilapia (*Oreochromis niloticus*), hogchoker (*trinectes maculatus*) [15], and hybrid striped bass (white bass *Morone chrysops* x striped bass *Morone saxatilis*) [7]. The recorded sound pressure levels (SEL_{ss}, SEL_{cum}, SPL_{peak}) were comparable to the values measured in laboratory experiments using a high intensity controlled impedance fluid filled wave tube (HICI-FT) [6,7,15,18] or a larvaebrator [5]. Accordingly, our sound measurements (SEL_{ss}, SEL_{cum}, SPL_{peak}) confirm the sound pressure levels that have been used in the lab to mimic real time pile driving sound levels. Bearing in mind the expensive and challenging logistics to execute field experiments in addition to the difficulties to control all environmental parameters, laboratorial studies are a good approach.

2 Sound pressure thresholds

Exposure to sound pressures can lead to internal injuries. Sound pressure thresholds for the onset of injury in juvenile Chinook salmon (*Oncorhynchus tshawytsch*) were recommended as: SEL_{cum} = 210 dB re 1 μPa^2.s derived from 960 strikes, each strike having a SEL_{ss} of 180 dB re 1 μPa^2.s [6]. These values have been exceeded during pile driving of each monopile in our experiments. In an experiment using a HICI-FT, it was shown that after exposure to pile driving sounds, more individuals of a physoclist fish species were injured and the injuries per fish were more severe compared to a physostomous fish and a flatfish species, the latter lacking a swim bladder [6,15]. These observations were strengthened by another experiment in the HICI-FT with the physoclist fish, the hybrid striped bass (white bass *Morone chrysops* x striped bass *Morone saxatilis*) [7]. Physoclist species are slower to change the volume of their swim bladder, which increases their vulnerability to high sound pressure levels [24]. Furthermore, more severe injuries and a higher number of injuries were reported in the larger size group of hybrid striped bass (white bass *Morone chrysops* x striped bass *Morone saxatilis*) physoclist fish (mean size 17.2 g) compared to smaller size group (mean size 1.3 g), [7]. This is in contrast to the current hypothesis that fishes less than 2 g are more susceptible to injury than larger fish when exposed to impulsive pile driving [7]. Internal sublethal injuries could also be present in the physoclist sea bass we used in our experiments, but this was beyond the scope of the present study.

Another experiment with the HICI-FT showed that fishes can heal from injuries as post-exposure time increased, after they were exposed to 960 strikes of 187 dB re 1 μPa^2.s (SEL_{cum} = 217 dB re 1 μPa^2.s) [7,18]. However, internal injuries can also lead to mortality [6]. Mortal injuries in physoclistous hybrid striped bass of 1.3 g appeared at energy exposures with SEL_{ss} of 180 dB re 1 μPa^2.s and SEL_{cum} of 210 dB re 1 μPa^2.s for 960 strikes [7]. Similar levels did not result in increased mortality in our experiments. The mortality we noted, in any of the experiments, is more likely to be attributed to handling stress than to internal injuries. The sound pressure threshold causing mortality in juvenile sea bass less than 2 g lies above SPL_{peak} of 211 dB re 1 μPa, SEL_{cum} of 222 dB re 1 μPa^2.s derived from 2312 strikes, and SEL_{ss} of 188 dB re 1 μPa^2.s.

3 Acoustic particle motion

As shown above, the presence of sublethal or lethal injuries is correlated with the presence and type of a swim bladder, and sound 'pressure' seems to be the main sound component that induces the injuries. Nevertheless, the other underwater sound component 'acoustic particle motion' may also have an impact on the fishes, e.g. hearing damage, behavioural and masking studies [2,25]. Although it is unlikely that particle motion has a direct effect on mortality, which is the topic of this paper, it is also influenced by pile-driving and literature is very scarce on particle displacement, velocity and acceleration in the ocean. Particle motion is a highly directional quantity and fishes may be able to determine the sound source direction [26]. While only those fishes that have evolved towards an acoustic coupling between their swim bladder (or other gas-filled structures) and their ear, can sense sound pressure, all fishes are able to sense particle motion [27,28].

Shallow water acoustics are characterized by propagation complexities, like direct transmission, reflection and re-radiation [29,30]. Therefore, potential effects of changes in the acoustic particle motion will only be seen in the near field of a sound source. Hence, it is recommended to measure both sound components in assessing all the effects of underwater sound.

4 Underwater sound in a wider perspective

The Marine Strategy Framework Directive (MSFD, Directive 2008/56/EC) aims at a good environmental status (GES) by 2020 for all European waters. One of the eleven descriptors concerns the impact of anthropogenic underwater noise. At present, there is insufficient knowledge on the impact of impulsive sounds to establish proper sound level thresholds for the marine environment. The US Fisheries Hydro-acoustic Working Group (FHWG) formulated interim criteria for maximum sound levels fishes could be exposed to without causing non-auditory tissue damage. The maximum SEL_{cum} for fishes weighing <2 gram was set at 183 dB re 1 μPa^2.s and for fishes>2 gram at 187 dB re 1 μPa^2.s [30]. Also, SEl_{ss} and the number of strikes should be taken into account, and the importance of particle motion should be explored further to develop sound criteria.

The sound pressure levels that were measured in this field study only represent a snapshot of the acoustic near field, with acoustic characteristics measured at a depth of 2.5 m below the sea surface and 45 m away from the sound source. Moreover, these *in situ* experiments can be seen as a worst case scenario in terms of exposure time (number of strikes) and sound levels. In the real world, young sea bass will be drifting with the currents through the wind farm construction zones, which influences the residence time in these zones. Most likely, the encounter time to very high sound levels (pressure and particle motion) will be too short to induce any physiological effects, except during slack tide when the water currents decrease to a minimum.

This *in situ* field study confirms and validates the mortality results found by laboratory experiments and contributes to our general understanding of the effects of pile driving sounds, a first step in the assessment process to establish sound criteria.

Supporting Information

Data Figure S1 SEL_{ss} in 1/3 octave band of three consecutive piling strike signals. This file serves as an example of the SEL_{ss} spectrum in 1/3 octave bands (averaged over 0.7 s intervals) for three consecutive piling strike signals.

Data Figure S2 Mean SEL_{ss} and SPL spectrum in 1/3 octave bands in presence and absence of pile driving. This file contains the averaged SEL_{ss} values per 1/3 octave bands of the four sampled monopiles and the SPL values per 1/3 octave bands in the absence of pile driving.

Data Figure S3 Monitoring of the immediate and delayed mortality of the exposed and control groups.

This file contains the daily mortality observed in the exposed and control groups up to 14 days after the experiment. In addition, mortality is expressed in mortality per day and cumulative mortality.

Acknowledgments

The authors would like to thank Northwind NV and its contractor GeoSea NV for their collaboration and support during the field experiments. Pieter Tomas, Luc Dekoninck, David Vuylsteke, Fernand Delanghe, and Jan Ranson are acknowledged for their technical support.

Author Contributions

Conceived and designed the experiments: ED DB KH SV SD. Performed the experiments: ED KH KVG. Analyzed the data: ED JV BDC. Contributed reagents/materials/analysis tools: BDC AB JV. Wrote the paper: ED BDC DB KH JV SV KVG MV SD. Taught the authors how to rear sea bass and provided the sea bass: AB.

References

1. Southall BL, Bowles AE, Ellison WT, Finneran JJ, Gentry RL, et al. (2007) Marine Mammal Noise Exposure Criteria: Initial Scientific Recommendations. Aquatic Mammals 33: 121.
2. Popper AN, Hastings MC (2009) The effects of anthropogenic sources of sound on fishes. Journal of Fish Biology 75: 455–489.
3. Van der Graaf A, Ainslie MA, Brensing AM, Dalen J, Dekeling RPA, et al. (2012) European Marine Strategy Framework Directive - Good Environmental Status (MSFD GES): Report of the Technical Subgroup on Underwater noise and other forms of energy.
4. Popper AN, Hastings MC (2009) The effects of human-generated sound on fish. Integrative Zoology 4: 43–52.
5. Bolle LJ, de Jong CAF, Bierman SM, van Beek PJG, van Keeken OA, et al. (2012) Common Sole Larvae Survive High Levels of Pile-Driving Sound in Controlled Exposure Experiments. Plos One 7.
6. Halvorsen MB, Casper BM, Woodley CM, Carlson TJ, Popper AN (2012) Threshold for Onset of Injury in Chinook Salmon from Exposure to Impulsive Pile Driving Sounds. Plos One 7.
7. Casper BM, Halvorsen MB, Matthews F, Carlson TJ, Popper AN (2013) Recovery of Barotrauma Injuries Resulting from Exposure to Pile Driving Sound in Two Sizes of Hybrid Striped Bass. Plos One 8.
8. R DS, A V (2008) MER - Offshore Windturbinepark Band zonder Naam NV ELDEPASCO. Zeebrugge: Arcadis Belgium. 407 p.
9. Roncarati A, Meluzzi A, Melotti P, Mordenti O (2001) Influence of the larval rearing technique on morphological and productive traits of European sea bass (Dicentrarchus labrax L.). Journal of Applied Ichthyology 17: 244–246.
10. Koumoundouros G, Maingot E, Divanach P, Kentouri M (2002) Kyphosis in reared sea bass (Dicentrarchus labrax L.): ontogeny and effects on mortality. Aquaculture 209: 49–58.
11. Sfakianakis DG, Georgakopoulou E, Kentouri M, Koulmoundouros G (2006) Geometric quantification of lordosis effects on body shape in European sea bass, Dicentrarchus labrax (Linnaeus, 1758). Aquaculture 256: 27–33.
12. Zouiten D, Ben Khemis I, Masmoudi AS, Huelvan C, Cahu C (2011) Comparison of growth, digestive system maturation and skeletal development in sea bass larvae reared in an intensive or a mesocosm system. Aquaculture Research 42: 1723–1736.
13. Sfakianakis DG, Katharios P, Tsirigotakis N, Doxa CK, Kentouri M (2013) Lateral line deformities in wild and farmed sea bass (Dicentrarchus labrax, L.) and sea bream (Sparus aurata, L.). Journal of Applied Ichthyology 29: 1015–1021.
14. van der Kooij J, Righton D, Strand E, Michalsen K, Thorsteinsson V, et al. (2007) Life under pressure: insights from electronic data-storage tags into cod swimbladder function. Ices Journal of Marine Science 64: 1293–1301.
15. Halvorsen MB, Casper BM, Matthews F, Carlson TJ, Popper AN (2012) Effects of exposure to pile-driving sounds on the lake sturgeon, Nile tilapia and hogchoker. Proceedings of the Royal Society B-Biological Sciences 279: 4705–4714.
16. Bradley DL, Wilson WD (1966) The acoustic impedance (pc) of sea water is presented as a function of temperature, pressure and salinity. Maryland: United States Naval Ordnance Laboratory. 66–146 66–146.
17. Thompson CM (1990) Development of a Structurally Rigid, Acoustically Transparent Plastic. Journal of the Acoustical Society of America 87: 1138–1143.
18. Casper BM, Popper AN, Matthews F, Carlson TJ, Halvorsen MB (2012) Recovery of Barotrauma Injuries in Chinook Salmon, Oncorhynchus tshawytscha from Exposure to Pile Driving Sound. Plos One 7.
19. Clarke K, Gorley R (2006) PRIMER V6: User Manual/Tutorial. Primer-E. Plymouth.
20. Anderson MJ (2005) PERMANOVA: a FORTRAN computer program for permutational multivariate analysis of variance. Department of Statistics, University of Auckland, New Zealand.
21. Haelters J, Norro A, Jacques T (2009) Underwater noise emission during the phase I construction of the C-Power wind farm and baseline for the Belwind wind farm. In Degraer S & Brabant R (Eds) (2009) Offshore wind farms in the Belgian part of the North Sea State of the art after two years of environmental monitoring. pp. 17–37.
22. Neo YY, Seitz J, Kastelein RA, Winter HV, Ten Cate C, et al. (2014) Temporal structure of sound affects behavioural recovery from noise impact in European seabass. Biological Conservation 178: 65–73.
23. Wysocki LE, Davidson JW, Smith ME, Frankel AS, Ellison WT, et al. (2007) Effects of aquaculture production noise on hearing, growth, and disease resistance of rainbow trout Oncorhynchus mykiss. Aquaculture 272: 687–697.
24. Mann DA, Hawkins AD, Jech JM (2007) Active and passive acoustics to locate and study fish. In: Webb JF, Popper AN, Fay RR, editors. Fish bioacoustics: Springer. pp. 279–310.
25. Tasker ML, Amundin M, Andre M, Hawkins A, Lang B, et al. (2010) Marine Strategy Framework Directive Task Group 11 Report Underwater noise and other forms of energy. 64 p.
26. Popper AN, Fay RR (2011) Rethinking sound detection by fishes. Hearing Research 273: 25–36.
27. Hastings MC, Popper AN (2005) Effects of sound on fish. Sacramento, CA: California Department of Transportation 82 p.
28. Wysocki LE, Codarin A, Ladich F, Picciulin M (2009) Sound pressure and particle acceleration audiograms in three marine fish species from the Adriatic Sea. Journal of the Acoustical Society of America 126: 2100–2107.
29. Ainslie MA, De Jong CAF, Dol HS, Blacquière G, Marasini C (2009) Assessment of natural and anthropogenic sound sources and acoustic propagation in the North Sea. The Hague: TNO. 110 p.
30. Oestman R, Buehler D, Rodkin R (2011) Technical Guidance for Assessment and Mitigation of the Hydroacoustic Effects of Pile Driving on Fish. San Francisco: Caltrans.

6

Marine Communities on Oil Platforms in Gabon, West Africa: High Biodiversity Oases in a Low Biodiversity Environment

Alan M. Friedlander[1,2]*, Enric Ballesteros[3], Michael Fay[2,4,5], Enric Sala[2,3]

1 Fisheries Ecology Research Laboratory, Department of Biology, University of Hawaii, Honolulu, Hawaii, United States of America, **2** Pristine Seas, National Geographic Society, Washington, DC, United States of America, **3** Centre d'Estudis Avançats-CSIC, Blanes, Spain, **4** Wildlife Conservation Society, Bronx, New York, United States of America, **5** Special Advisor, Presidence de la République, Libreville, République Gabonaise

Abstract

The marine biodiversity of Gabon, West Africa has not been well studied and is largely unknown. Our examination of marine communities associated with oil platforms in Gabon is the first scientific investigation of these structures and highlights the unique ecosystems associated with them. A number of species previously unknown to Gabonese waters were recorded during our surveys on these platforms. Clear distinctions in benthic communities were observed between older, larger platforms in the north and newer platforms to the south or closer to shore. The former were dominated by a solitary cup coral, *Tubastraea* sp., whereas the latter were dominated by the barnacle *Megabalanus tintinnabulum*, but with more diverse benthic assemblages compared to the northerly platforms. Previous work documented the presence of limited zooxanthellated scleractinian corals on natural rocky substrate in Gabon but none were recorded on platforms. Total estimated fish biomass on these platforms exceeded one ton at some locations and was dominated by barracuda (*Sphyraena* spp.), jacks (Carangids), and rainbow runner (*Elagatis bipinnulata*). Thirty-four percent of fish species observed on these platforms are new records for Gabon and 6% are new to tropical West Africa. Fish assemblages closely associated with platforms had distinct amphi-Atlantic affinities and platforms likely extend the distribution of these species into coastal West Africa. At least one potential invasive species, the snowflake coral (*Carijoa riisei*), was observed on the platforms. Oil platforms may act as stepping stones, increasing regional biodiversity and production but they may also be vectors for invasive species. Gabon is a world leader in terrestrial conservation with a network of protected areas covering >10% of the country. Oil exploration and biodiversity conservation currently co-exist in terrestrial and freshwater ecosystems in Gabon. Efforts to increase marine protection in Gabon may benefit by including oil platforms in the marine protected area design process.

Editor: Christina A. Kellogg, U.S. Geological Survey, United States of America

Funding: This work was funded by the Waitt Foundation, Blancpain, Davidoff Cool Water, National Geographic Society, and the Wildlife Conservation Society. The funders had no role in study design, data collection and analysis, decision to publish, or preparation of the manuscript.

Competing Interests: The authors received funding from commercial sources (Blancpain and Davidoff Cool Water). This does not alter the authors' adherence to PLOS ONE policies on sharing data and materials.

* Email: alan.friedlander@hawaii.edu

Introduction

Gabon, West Africa, is situated in the Guinea Current Large Marine Ecosystem (GCLME), one of the world's most productive ocean regions, which is rich in fisheries resources, petroleum production, and is crucial to the lives of the region's 300 million coastal residents [1]. Despite this importance, the marine biodiversity of Gabon is poorly known owing to limited financial resources, lack of regional expertise, and a greater emphasis on extractive resources. Previous research has focused on the deep-sea biodiversity of the continental margin associated with oil and gas exploration [2], or related to continental shelf and slope fisheries resources [3]. The continental margins possess small methane-rich cold-see sites and associated non-biogenic carbonate reefs [1,4,5]. However, most of Gabon's shelf area is sandy or muddy, with limited hard bottom habitat. The dozens of oil platforms along the shelf may provide the only hard substrate for thousands of km^2 [3].The marine life on these structures has never been studied and here we report on the first scientific surveys of Gabon's oil platforms.

The sea bottom of the Gulf of Guinea is characterized by terrigenous and sandy muds with low habitat and seascape diversity [6]. The Congo River, the second most voluminous river in the world after the Amazon, supplies a large amount of particulate and dissolved organic matter to the ocean, both at the surface and at the seafloor through the Congo Canyon and into the offshore waters of Gabon [7]. In addition, the Ogooué River in central Gabon has an average discharge rate of 4,758 m^3 s^{-1}, making it the third most important river in west equatorial Africa [8,9]. Although the water discharge is one order of magnitude lower compared to the Congo River, the particulate load of the two rivers is similar [10]. Collectively these river systems greatly influence the marine ecosystem of Gabon.

The shelf off Gabon is dominated by sandy, sand-shell and gravel bottoms, becoming muddy toward the shelf [3]. Factors contributing to the enrichment in nutrients are related to seasonal up-welling, the discharge from the Congo and Ogooué rivers, and shelf-break upwelling [11]. A strong primary productivity gradient exists from north to south with the south showing higher productivity [12]. Nearshore waters off Gabon are generally warmer, with lower salinity and higher oxygen concentrations compared with more offshore environments [3]. These gradients are likely important to the distribution and abundance of marine organisms along the continental shelf of Gabon.

Ocean productivity in Gabon is dominated by seasonal equatorial upwelling and is strongly influenced by freshwater and sediments from the Congo and Ogooué rivers, particularly in nearshore waters [13]. The Benguela Current is a strong biogeographic barrier that prevents species from the southwestern region of the African continent and the Indian Ocean from colonizing the Gulf of Guinea [14,15]. As a result of these oceanographic processes, the marine communities of the Gulf of Guinea are considered unique and a hotspot of marine biodiversity due primarily to the high proportion of endemic species [16]. Despite the high productivity of the region and its high degree of endemism, we know very little about Gabon's subtidal communities, with the exception of demersal fishes targeted by the large, industrial trawl fishery [3]. There are no true coral reefs in Gabon, and most of the largely unmapped reefs are rocky [17]. Except for a checklist of scleractinian corals for the region [17], there have been no studies of Gabon's subtidal reefs.

Extensive oil exploration and development began onshore in Gabon in the 1950s but offshore exploration did not begin until the 1960s [18]. There are currently >40 offshore oil platforms in Gabon yet virtually nothing is known about the marine biological communities associated with these platforms. Oil and gas platforms act as artificial reefs on continental shelves and provide hard substrate in open water that might otherwise be unavailable to marine organisms requiring such habitat [19,20]. As a result of the limited reefs and the high biodiversity potential of the region, these platforms may provide a unique and important habitat for marine communities in Gabon.

Offshore oil and gas platforms are among the largest artificial structures in the ocean [21]. Worldwide there are >7,500 oil platforms [22,23], and these structures are colonized by diverse ecological communities that have, in some instances, been shown to enhance biodiversity and fisheries production [24–26]. Approximately 40,000 oil and gas wells have been drilled in the northern Gulf of Mexico (GOM) since the 1940s [27]. There are currently about 3,600 oil and/or gas production platforms in the GOM and these platforms act as hard substratum upon which many reef organisms can settle and grow in a region where very little hard substratum now exists, nor has existed in recent geological time [27,28]. Oil and gas platforms in southern California harbor diverse fish and invertebrate populations that are considered *de facto* artificial reefs and serve as important habitat for a number of species [29–31]. Off the north-western Australia continental shelf, a diverse range of taxa were observed on oil industry structures, including reef-dependent species and transient pelagics [23]. These platforms provide habitat that potentially increases the growth and survival of individuals, affording shelter for protection from predation and spawning substrate, and acting as a visual attractant for organisms not otherwise dependent on hard bottom, such as pelagic species [32]. However, hazards to shipping, exclusion of certain fisheries (e.g., trawling), spread of alien species, and potential spills and leaks have all been cited as some of the negative consequences associated with oil platforms, particularly older ones that are no longer in production [33–35].

The objectives of this study were to describe the marine communities on oil platforms in Gabon, determine the biogeographic affinities of the associated species, and discuss the implications of these platforms for marine conservation.

Methods

The Government of Gabon granted all necessary permission to conduct this research. No vertebrate sampling was conducted and therefore no approval was required by the Institutional Animal Care and Use Committee.

Oil platforms

The oil platforms off Gabon investigated in this study sit on sandy and muddy bottoms at depths ranging from 26 to 102 m (Table 1, Fig. 1). The platforms had three to four vertical pillars linked by crossbeams at different depths. The shallowest crossbeams were at depths between 6–10 m, and the second shallowest ('deep' crossbeam hereafter) between 13–24 m (Table 1). There were a number of deeper crossbeams depending on the size and depth of the platform but these were beyond scuba depths and were not surveyed. We conducted quantitative surveys of benthic communities and fish assemblages along the two shallowest crossbeams at each of the 10 oil platforms surveyed. Since these crossbeams were not independent samples, values from benthic and fish surveys were averaged to derive density estimates for each platform.

Benthic cover

Benthic cover was quantified *in situ* on the crossbeams using 0.25 m×0.25 m quadrats divided into 25 sub-quadrats of 5×5 cm [36,37]. Eight quadrats were positioned haphazardly along every crossbeam, resulting in 24–32 quadrats for every depth and platform. Algae and other benthic cover (sponges, hydrozoans, soft and hard corals, barnacles, mollusks, echinoderms, and ascidians) within each quadrat were identified to the lowest possible taxonomic level. The percentage of sub-quadrats in which a species appeared was recorded and used as a measure of occurrence. A highly abundant taxon that appeared in all 25 subquadrats would produce a presence of 100%, whereas the total lack of a taxon would produce a presence of 0% [36]. The final abundance of a taxon within each depth and oil platform was then calculated as the mean of the percentage presence values of the quadrats sampled. The sum abundance of all quadrats exceeded 100% due to the combined presence of several taxa on the same sub-quadrat. Field visual survey methods did not allow identification of the smallest species (turf and filamentous algae, hydrozoans), which were sometimes pooled into groups during visual estimations. Vagile crustaceans (e.g. decapods) were not quantified in the quadrats, although the presence of lobsters during the survey was noted since they were conspicuous and commercially important.

Fish assemblages

We conducted underwater visual fish censuses on oil platforms using two methods. First, we estimated the numerical density and size of individuals of each species along 4-m wide × 4-m tall belt transects along the 'shallow' crossbeams and the 'deep' crossbeams on each oil platform. Depth and length of transects varied corresponding to the length of the crossbeam where the transects were located (Table 1). Crossbeam lengths were measured upon completion of each quantitative survey. Because many of the schooling pelagic species (e.g., jacks, barracudas) were loosely associated with the platform structure, we estimated the numbers

Table 1. Descriptions of oil platforms sampled along the coast of Gabon, West Africa.

Date	Platform no.	Company	Install year	Lat.	Long.	Bottom depth (m)	Deep* crossbeam			Shallow beam			Flare
							Depth (m)	Length (m)	Area (m^{-2})	Depth (m)	Length (m)	Area (m^{-2})	
9-Oct-12	1	Total Gabon	1973	−1.509	8.827	43	19.0	16.0	64.0	6.0	13.0	52.0	Yes
9-Oct-12	2	Total Gabon	1980	−1.738	8.738	45	17.0	19.6	78.4	6.0	16.6	66.4	No
10-Oct-12	3	Total Gabon	1975	−1.103	8.594	62	19.0	18.3	73.0	6.0	15.0	60.0	No
10-Oct-12	4	Total Gabon	1974	−1.155	8.475	102	20.2	18.1	72.5	7.1	12.4	49.5	No
13-Oct-12	5	Vaalco	2001	−3.717	10.436	45	13.3	13.0	52.0	-	-	-	No
13-Oct-12	6	Vaalco	2001	−3.877	10.590	45	13.2	13.0	52.0	-	-	-	No
15-Oct-12	7	Perenco	1975	−2.039	9.105	45	23.8	6.0	24.0	6.7	9.0	36.0	No
18-Oct-12	8	Perenco	1975	−1.357	8.892	26	15.0	10.0	40.0	-	-	-	No
18-Oct-12	9	Perenco	1975	−1.177	8.703	31	20.0	5.0	78.5	10.0	5.0	78.5	No
18-Oct-12	10	Perenco	1975	−1.033	8.592	50	19.0	20.0	80.0	8.0	18.0	72.0	Yes

* 'Deep' is the second shallowest crossbeam of each platform.

and sizes of these species that were visible during the dive to calculate their standing stock around each platform since they were underrepresented on quantitative transects. Although estimates associated with this method are constrained by schooling behavior, water clarity, and species identification, it complements the finer resolution belt transects by providing valuable information on highly vagile, and commercially important species that would have otherwise not been quantified.

Length estimates of fishes from both survey methods were converted to mass (M) using the following length–mass relationship: $M = aTL^b$, where the parameters a and b are constants for the allometric growth equation and TL is total length in cm [38]. From belt transects, all biomass estimates were converted to metric tons per hectare (t ha^{-1}) and numerical abundance estimates were converted to number of individuals m^{-2}. All fishes observed were categorized into four trophic groups (piscivores, carnivores, planktivores, and herbivores; [39,40]), and classified into province and regional affinities based on [41].

Statistical analysis

Non-metric multi-dimensional scaling (nMDS) analysis was used to examine differences in benthic communities and fish assemblages among oil platforms (PRIMER v.5 – [42]). Separate Bray–Curtis similarity matrices were created using abundance of benthic components at the lowest possible taxonomic level and fish numerical density (no. m^{-2}). Prior to conducting the nMDS, benthic data were arcsin square root transformed, while fish density was square root transformed. The primary taxa vectors driving the ordination (Pearson correlation Product movement correlations ≥0.9 for benthos, ≥for 0.7 for fishes) were overlaid on the nMDS plot to visualize the major taxa that explained the spatial distribution patterns observed

Benthic taxon and fish species diversity were calculated from the Shannon-Weaver diversity index [43]: $H' = -\sum (p_i \ln (p_i))$, where p_i is the proportion of all individuals counted that were of taxa i. Fish trophic biomass data did not conform to parametric statistical assumptions despite various transformations, therefore comparisons among trophic groups were conducted using a Kruskal-Wallis rank-sum test (H) with Dunn's test for unplanned multiple comparisons [44].

To describe the pattern of variation in community structure (patterns of distribution of biomass/abundance of high-level functional groups within the community) and their relationship to environmental-human gradients we used linear ordination methods. Linear models are appropriate for these data because a preliminary detrended correspondence analysis showed short gradient lengths (<2 SD) [45]. To explore the spatial distribution of community structure across oil platforms and its relationship with environmental variables (see below), we performed a direct gradient analysis (redundancy analysis: RDA) on the log-transformed community and environmental data matrices using the ordination program CANOCO for Windows version 4.0 [46]. The RDA introduces a series of explanatory (environmental) variables and resembles the model of multivariate multiple regression, allowing us to determine what linear combinations of these environmental variables determine the gradients. We pooled data from all taxa into the following community groups to facilitate the large-scale analysis: biomass of fish trophic groups (e.g., piscivores, carnivores, planktivores, and herbivores) and higher order benthic groups (e.g., Rhodophyta, Porifera, Cnidaria-Anthozoa, Cnidaria-Hydrozoa, Crustacea, Mollusca-Bivalvia).

The environmental data matrix included the following variables: latitude, longitude, depth, age of oil platform, platform area, distance offshore, and distance from the Ogooué and Congo river

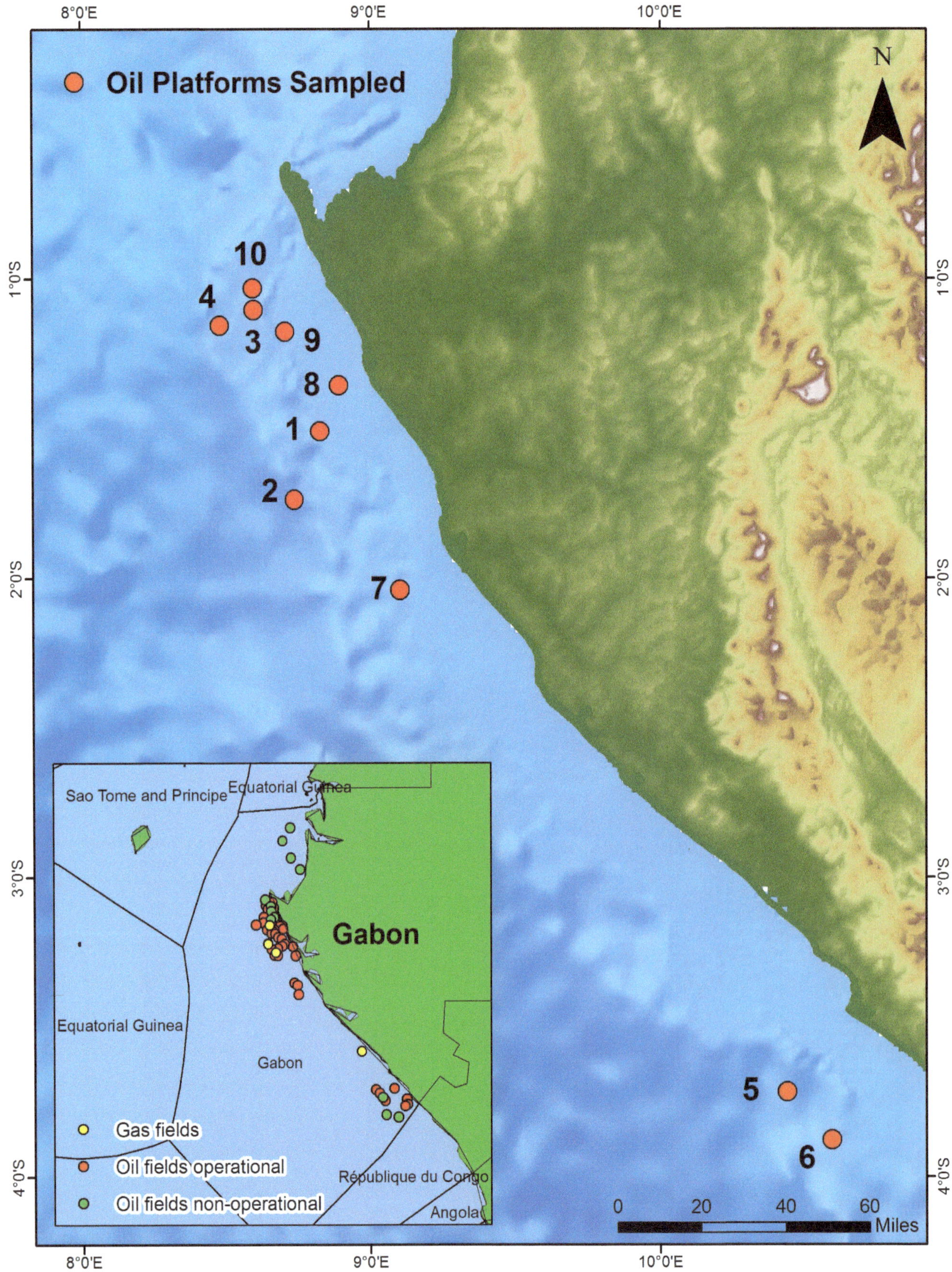

Figure 1. Oil platforms sampled along the coast of Gabon, West Africa. Platform numbers and descriptions are found in Table 1.

mouths. Correlations among environmental variables were examined using Spearman's rank-order correlation coefficients. Distance to the Congo and Ogooué rivers were significantly correlated with latitude and longitude (all $p<0.05$) and were excluded from the analysis. Longitude and depth, which were significantly correlated with latitude and platform area respectively ($p<0.05$), also were excluded from the analysis. To rank environmental variables in their importance for being associated with the structure of communities, we used a forward selection where the statistical significance of each variable was judged by a Monte-Carlo permutation test [47].

Results

Benthic composition

A total of 45 benthic taxa were recorded within quadrats on oil platforms sampled during our expedition (Table S1). Anthozoans made up 32.2% of the mean percent presence, followed by Cirriped crustaceans (19.9%), sponges (10.9%), bivalve mollusks (10.2%), and hydrozoans (8.3%). A solitary, undetermined cup coral (*Tubastraea* sp.) was the dominant species on many of the platforms, occurring on all platforms and accounting for 30.7% of the mean percent presence of all taxa (Table 2, Fig. 2). A barnacle, *Megabalanus tintinnabulum*, comprised 18.6% of the mean percent presence of all taxa and was present on 80% of the platforms. A sponge, *Haliclona* sp., accounted for 9.2% of the mean percent presence of all taxa, followed by an unidentified hydrozoan (9.2%), and two bivalve mollusks – *Crassostrea gassar* (7.6%) and *Dendostrea frons* (6.4%).

Based on benthic community composition, the platforms formed three distinct clusters in ordination space (Fig. 3). The clusters dominated by *Tubastraea* sp. were highly concordant and well separated from the platforms to the south (5 and 6) and the two inshore platforms (8 and 9). These southern and inshore platforms were dominated by the barnacle *Megabalanus tintinnabulum*, but had much more diverse assemblages compared to the ones dominated by *Tubastraea* sp. The average number of taxa was highest on the inshore platforms ($\overline{X} = 18.0 \pm 5.7$) and lowest on the *Tubastraea* dominated ones ($\overline{X} = 14.8 \pm 2.4$). Diversity was highest on the southern platforms ($\overline{X} = 2.0 \pm 0.1$) and lowest on the *Tubastraea* dominated platforms ($\overline{X} = 1.5 \pm 0.4$).

Although not present on quadrats, we commonly observed two commercial species of lobsters on the platforms: spiny lobsters (*Panulirus regius*) and slipper lobsters (*Scyllarides herklotsii*). Spiny lobsters were found in groups of 2–4 individuals and were of small size (10–15 cm total body length), while slipper lobsters were often found alone or in pairs and attained larger adult sizes (20–25 cm total length). No estimates of lobsters by platform are provided as we did not use a specific protocol for quantifying these species.

Fish assemblage characteristics

Based on all quantitative and qualitative surveys we observed a total of 65 fish species associated with oil platforms during this study (Table S2). Of these, 22 (34%) are new records for Gabon and 4 (6%) are new to the tropical West African coast. Assemblage richness consisted of 32% tropical eastern Atlantic (TEA) endemics and 31% amphi-Atlantic species (those found on both sides of the Atlantic), followed by circumtropical species (17%). Based on quantitative surveys, amphi-Atlantic species accounted for 73% of the numerical abundance, and 76% of the biomass observed on oil platforms along the coast of Gabon. Species endemic to the TEA only accounted for 21% of the numerical abundance and 16% of the biomass observed on quantitative surveys, while circumtropical species accounted for an additional 4% of the abundance and 2% of the biomass. Of the 65 species observed, only 38 (42%) were recorded on our quantitative surveys. Most of the species not observed on quantitative surveys were either pelagic or encountered at the bases of the platforms, associated with the seabed, and therefore not encountered on transects on the crossbeams.

The number of fish species on transects ranged from a low of 7 on platform 6 to a high of 22 on platform 1 ($\overline{X} = 14.6 \pm 4.1$). Higher numbers of species were observed on the platforms to the north, which were also the oldest platforms with the highest abundance of *Tubastraea* sp. The number of individuals observed on transects ranged from 0.9 m^{-2} on platform 2 to 8.5 m^{-2} on platform 10 ($\overline{X} = 3.9 \pm 2.3$), with no apparent pattern observed. Fish biomass among platforms also showed no clear pattern, with the highest biomass observed on platform 6 (10.8 t ha^{-1}) and the lowest on platform 8 (0.9 t ha^{-1}) with a mean of 4.3 (± 3.3). Fish diversity varied without pattern and averaged 1.1 (± 0.3) with the lowest diversity on platform 6 ($H' = 0.5$) and highest on platform 9 ($H' = 1.6$). Unlike the benthic communities, the fish assemblages associated with oil platforms did show spatial separation, with a small number of species accounting for much of the spatial variability (Fig. 3).

Table 2. Percent occurrence of major taxa found on oil platforms in Gabon.

Group	Taxa	Mean % (sd)	% freq.
Cnidaria-Anthozoa	*Tubastraea* sp.	47.34 (39.50)	100
Crustacea-Cirripedia	*Megabalanus tintinnabulum*	28.63 (33.33)	80
Sponge	*Haliclona* sp.	14.23 (15.61)	80
Cnidaria-Hydrozoa	*Hidrarians* unidentified	12.73 (16.29)	90
Mollusca-Bivalvia	*Crassostrea gasar*	7.61 (9.52)	100
Mollusca-Bivalvia	*Dendostrea frons*	6.36 (9.30)	60
Red alga	*Antithamnionella elegans*	6.15 (7.36)	60
Green alga	*Bryopsis plumosa*	5.25 (6.92)	50
Bryozoa	*Pentapora*-like	3.03 (5.10)	30
Red alga	Red algal turf	2.36 (5.15)	40

The mean is the percentage of the sub-quadrats were the taxa was observed. % freq. is the percentage of platforms upon which this taxon was observed (n = 10).

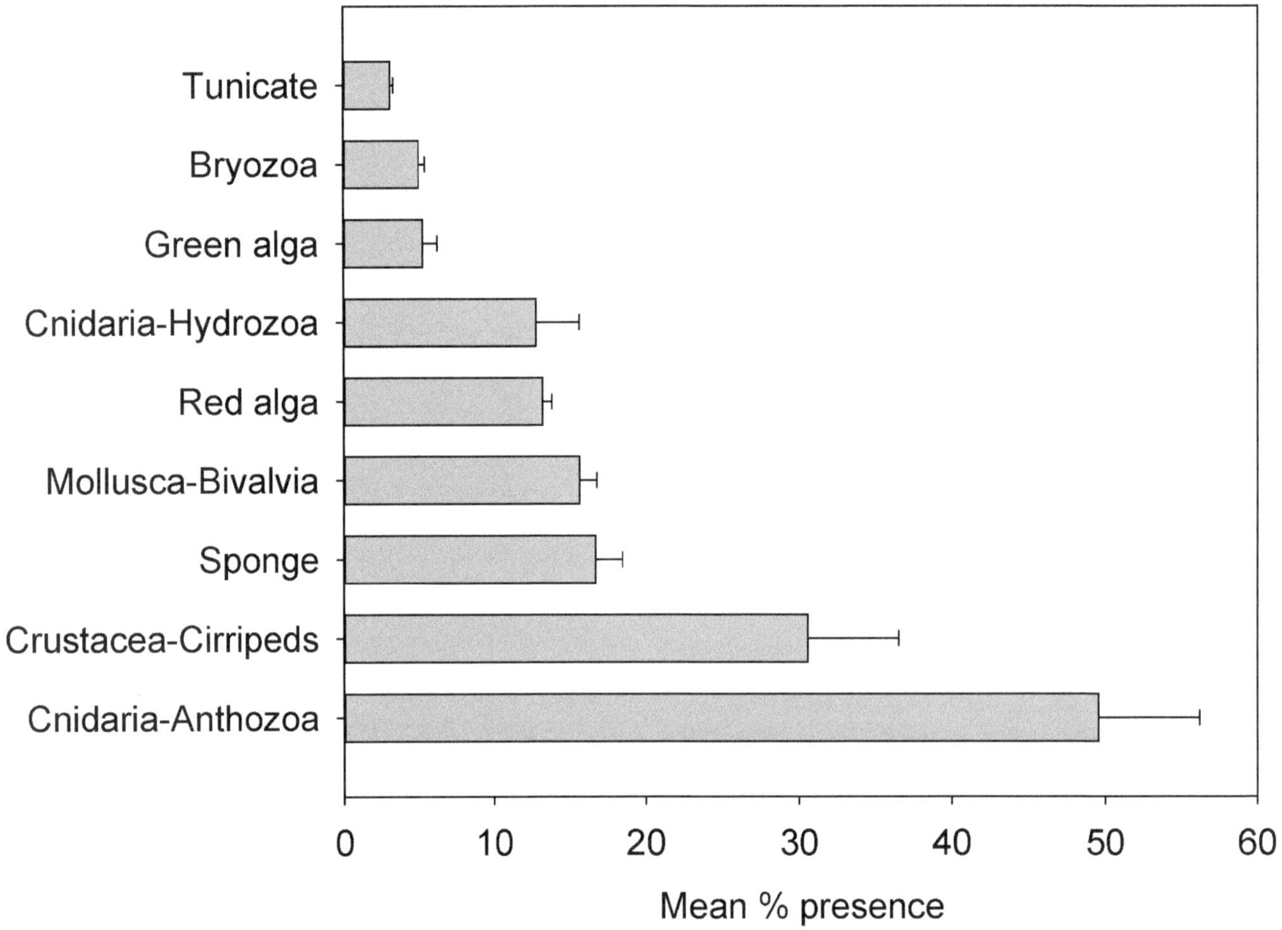

Figure 2. Benthic taxa on transects associated with oil platforms in Gabon, West Africa. Values are mean percent presence and standard error of the mean.

Fish trophic structure

Based on transect data, there were significant differences in the biomass and abundances among trophic groups observed on oil platforms (Fig. 4). Planktivores accounted for 74.0% of the numerical abundance and 36.8% of the overall biomass on platform transects. Planktivore abundance was highest at the two most southern platforms, which was also where the highest primary productivity occurs. Piscivores were the next most important trophic group by weight, accounting for 35.8% of the total biomass but 15.7% of the total abundance, with the highest biomass observed at the three north platforms (4, 9, and 10). Herbivores accounted for an additional 20.6% of the total biomass but only 4.3% of the numerical abundance. More than 80% of the total herbivore biomass was observed on the two most southern platforms. Invertivores accounted for 6.9% of the biomass and 5.9% of the abundance with no apparent pattern among platforms.

Fish species composition

The creolefish, *Paranthias furcifer*, was the dominant fish species both numerically (32.2%) and by weight (26.3%) on oil platform transects (Table 3 and 4). The brown chromis, *Chromis multilineata*, was the second most important species numerically (27.1%) but only accounted for 3.4% of the total biomass. This species was followed in abundance by the African sergeant, *Abudefduf hoefleri*, comprising 13.4% of the total assemblage by numbers and 7.2% by weight. The second most important species by weight was the Bermuda chub, *Kyphosus sectator*, accounting for 23.5% of the total biomass, followed by blue runner, *Caranx crysos* (14.9%).

Associated fish assemblage

Since many of the vagile pelagic schooling species were not recorded on transects, we estimated total abundance and sizes of these species over the course of the dive that were associated with each oil platform. Nine species, primarily jacks and barracuda, made up the vast majority of schooling pelagic species associated with these platforms (Table 5). Estimates of standing stock were based on the estimated numbers in each school and the median size of individuals in the school. The standing stock of these pelagic species averaged 0.83 t (±0.24) per platform with a minimum of 0.57 tonnes on platform 8 and a maximum of 1.18 tonnes on platform 1. Great barracuda (*Sphyraena barracuda*) accounted for 33.2% of the average standing stock, followed by longfin jack crevalle (*Caranx fischeri* - 33.0%), and rainbow runner (*Elagatis bipinnulata* - 12.7%)

Fish spawning and courtship

On October 12, 2012, at 3 pm local time, we observed pair spawning of red snapper, *Lutjanus dentatus*, on platform no. 4 off

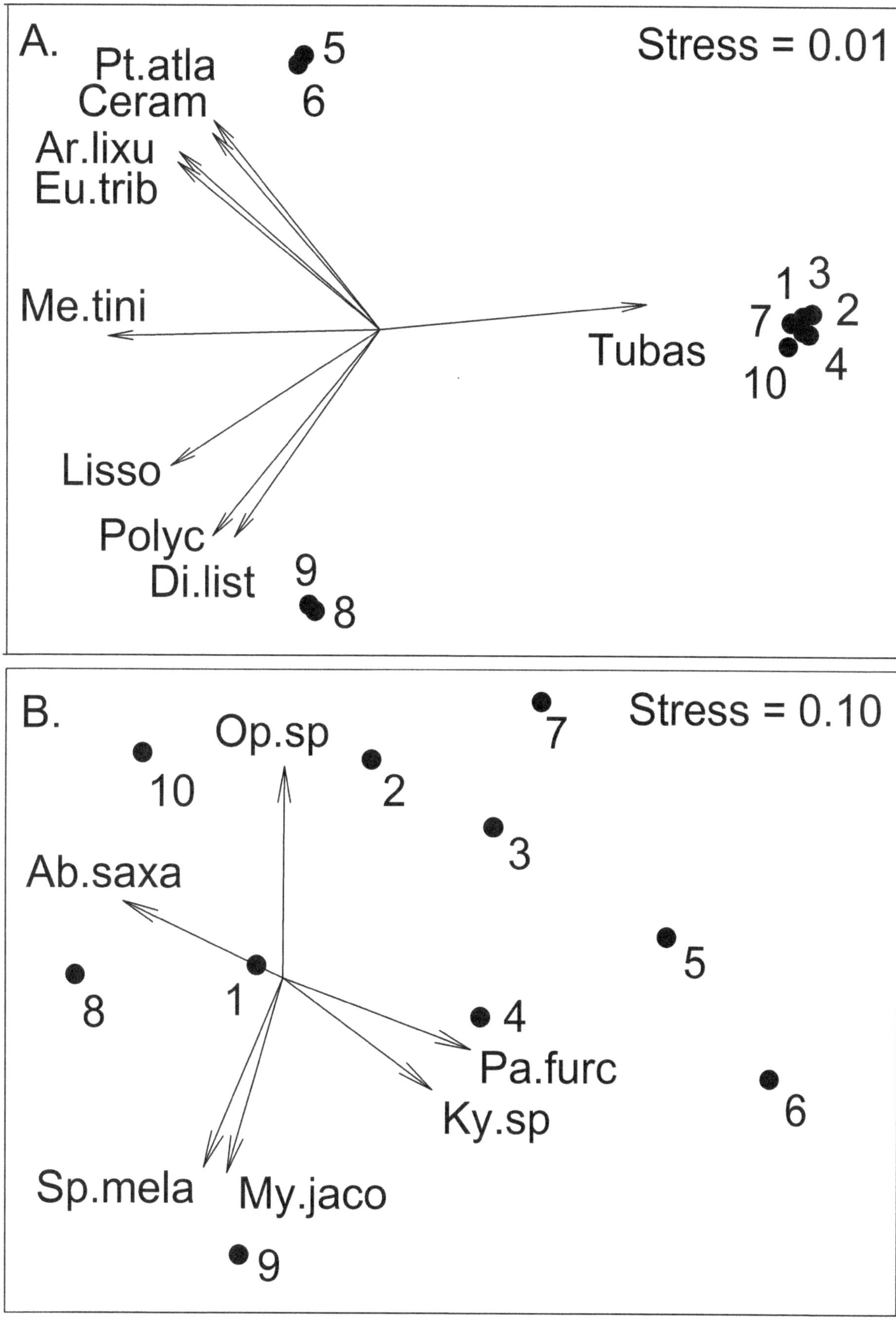
A.
Stress = 0.01
Pt.atla
Ceram
Ar.lixu
Eu.trib
Me.tini
Tubas
Lisso
Polyc
Di.list
1
2
3
4
5
6
7
8
9
10
B.
Stress = 0.10
Op.sp
Ab.saxa
Pa.furc
Ky.sp
Sp.mela
My.jaco
1
2
3
4
5
6
7
8
9
10

Figure 3. Non-metric multidimensional scaling plot of benthic communities and fish assemblages associated with oil platforms in Gabon. Vectors are the primary taxa driving the ordination (Pearson Product movement correlations ≥0.9). A. benthic communities species codes: Ar.lixu = *Arbacia lixula v. africana*, Ceram = *Ceramiaceae* unidentified, Di.list = *Diplosoma listerianum*, Eu.trib = *Eucidaris tribuloides v. africana*, Liss = *Lissoclinum* sp., Me.tini = *Megabalanus tintinnabulum*, Poly = Polyclinidae unidentified, Pt.atla = *Pteria atlantica*, Tubas = *Tubastraea* sp. B. Fish assemblage species codes: Ab.saxa = *Abudefduf saxatilis*, Ky.sp = *Kyphosus* sp., My.jaco = *Myripristis jacobus*, Op.sp = *Ophioblennius* sp., Pa.furc = *Paranthias furcifer* Sp.mela = *Spicara melanurus*.

Cape Lopez. A male (~70 cm) and a female (~50 cm) swam together from deeper water towards the surface, forming a spiral, and swimming apart while releasing eggs and sperm at 10 m depth. We observed two consecutive spawning events; the first outside of the platform structure, and the second within the structure. Immediately after spawning, dozens of creolefish swam towards the spawn cloud, to eat the eggs. We also observed the courtship behavior of the yellow jack, *Caranx bartholomaei*, at several platforms. Yellow jacks formed schools of up to several hundred individuals, most with typical silver body coloration. In the afternoon, a small number of individuals developed a dark coloration, and silver-dark pairs formed and swam together away from the school. Although we did not observe spawning, we believe this is courtship behavior that has been identified for other species of jacks (e.g., [48]).

Community structure

Ordination of biomass/abundance of functional groups among oil platform communities revealed three groups of sites mainly consisting of: (1) platforms dominated by anthozoans and red algae (2) those dominated by piscivorous feeding fishes and sponges, and (3) platforms dominated by planktivorous and herbivorous fishes along with hydroids and bivalves (Fig. 5). The first two axes of the RDA biplot explained 78% of the functional group variance and 97% of the functional group-environment relationship (Table 6). The main factors involved in this ordination are platform age and area, which are orthogonal to one another in ordination space (Fig. 5, Table 7). Platform age and latitude showed similar ordination trajectories, reflecting the history of expansion of oil exploration in Gabon starting in Port Gentil and moving south to the border with Congo. Platform area is correlated with water depth and distance offshore, with assemblages becoming more diverse with increasing platform area, water depth, and distance from shore.

Discussion

This work represents the first scientific surveys of the marine ecosystems on oil platforms off West Africa. We found an extraordinary diversity and biomass of fishes and rocky invertebrates on these platforms that are in sharp contrast to much of Gabon's marine environment, which is dominated by soft sediment communities. Quantitative transect surveys were most useful in describing the biodiversity and abundance of "resident" fishes and benthic communities on oil platforms but they underestimate the standing stock of pelagic species that are not as closely associated with the platforms as other species. The estimated fish biomass on oil platforms exceeded 1 ton and was dominated by pelagic species (barracuda, rainbow runner, jacks). Large snappers (primarily *L. dentatus*) were often deeper and not easily observed, so the actual standing stock of fish biomass on these platforms is likely much higher than our estimates suggest. Elsewhere, the potential production around platforms is high as indicated by estimates of biomass of cod (*Gadus morhua*) and saithe (*Pollachius virens*) around oil platforms in the North Sea, which range from 7–12 tonnes [49].

Owing to the known prevalence of resource species on these platforms, spear fishermen frequent them, particularly out-of-service ones near the population centers of Libreville and Port Gentil. Videos and anecdotal information suggest substantial hunting of goliath grouper (*Epinephelus itajara*), barracuda, cobia (*Rachycentron canadum*), Guinean pompano (*Trachinotus maxillosus*), cubera snapper (*Lutjanus agennes*), jacks (Carangidae) and several species of carcharhinid sharks. Despite this fishing pressure, fish biomass at some platforms is larger than is reported in the literature for most tropical reef fishes, and even higher than many pristine reefs surveyed in the Pacific [50–52]. This high standing stock of fishes is possible because of the high productivity of the region, and because many of the platforms are largely unfished since they are de facto MPAs owing to security restrictions. As on pristine reefs, top predators account for a large part of the fish biomass. Before the systematic spearfishing of the large groupers, snapper, jacks, and sharks, biomass and dominance of predators must have been even higher.

Most of the fish biomass on the platforms was composed of pelagic species with broad biogeographic distributions. However, much of the observed species richness consisted of demersal species, many of which had distinct and unique assemblages. Studies at nearby São Tomé and Príncipe found the archipelago fish fauna to consist of both western and eastern Atlantic species and noted that the easterly flowing Equatorial currents (the seasonal Equatorial Counter Currents and the subsurface Equatorial Undercurrent) link the western Atlantic and the eastern Atlantic at this latitude [53,54]. Likewise, the marine invertebrate fauna of São Tomé and Príncipe is known to consist of a mix of the two faunal regions [17,55–59]. Oil platforms off Gabon may therefore help extend this unique assemblage onto the West African shelf.

Despite its proximity to the African continent, these platforms harbor a unique fish fauna, with a large proportion of the assemblage composed of amphi-Atlantic species. Several amphi-atlantic species, which, in the Eastern Atlantic occur only around oceanic islands (e.g., *Epinephelus adscensionis, Paranthias furcifer, Mulloidychtis martinicus, Bodianus pulchellus, Chromis multilineata, Gnatholepis thomsoni, Melychthis niger*, [53]) were observed on the oil platforms off Gabon. The wrasse *Thalassoma newtoni* was considered endemic to the islands of São Tomé but has recently been recorded from Senegal [60]. We observed this species on all of the platforms surveyed, where it was very common. These facts highlight how these platforms act like small islands in an otherwise featureless seascape across the continental shelf of Gabon. We observed spawning and spawning behavior for two commercially important snappers on the platforms, and they are likely spawning sites for numerous other species as well. Therefore, these platforms are likely sources for the replenishment of other platforms and the scarce reefs in the region, acting like distinct populations in a metapopulation [61]. They could be of great value to fisheries conservation for commercially harvested fish species that associated with these platforms.

Some species of invertebrates from our surveys have not been successfully identified to the species level and taxonomists are

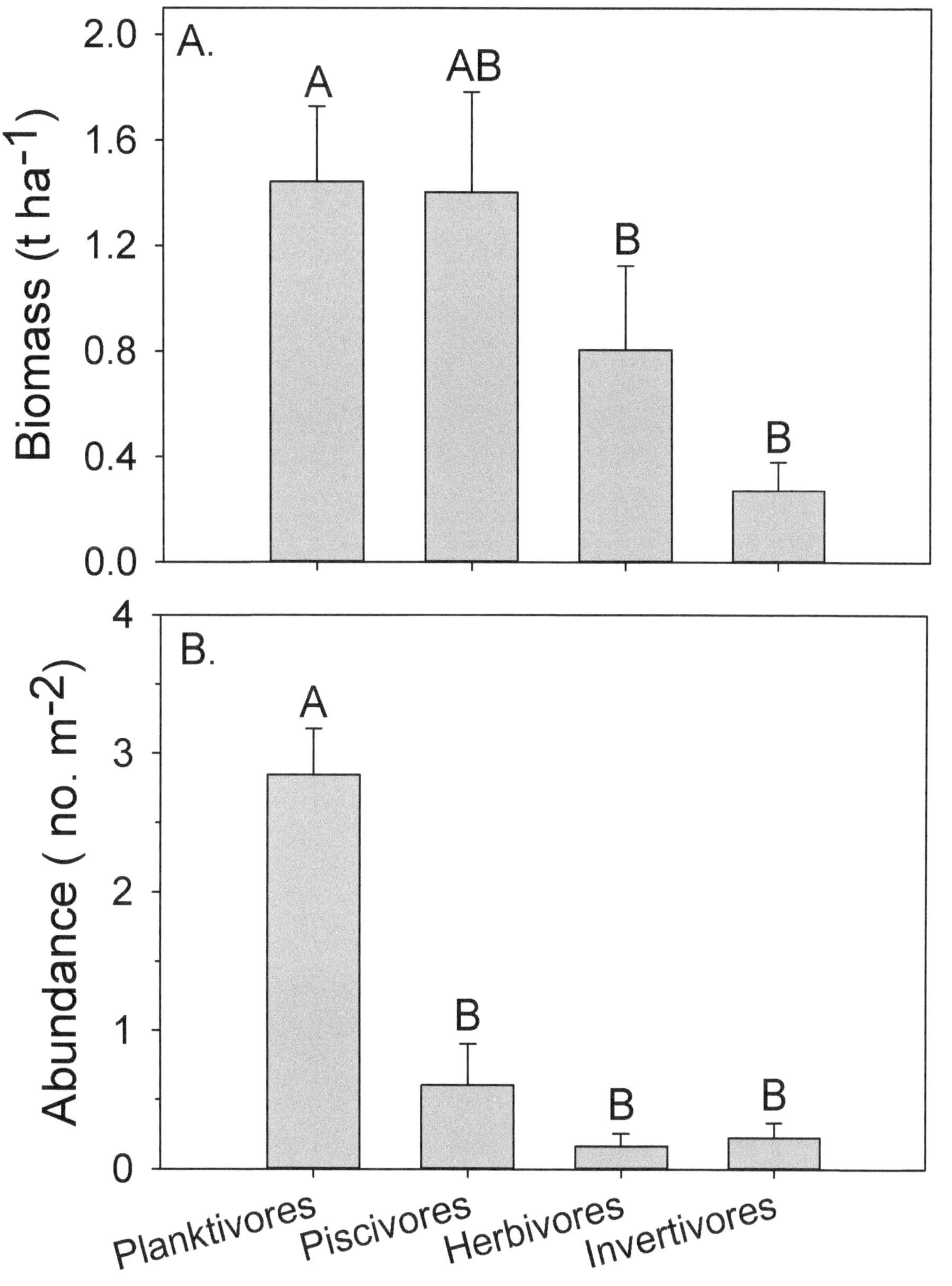

Figure 4. Comparisons of trophic (a) biomass (t ha^{-1}) and (b) abundance (no. m^{-2}) at oil platforms in Gabon. Values are means and standard error. Kruskal-Wallis Rank Sum comparisons among trophic groups were statistically different for biomass ($H = 12.3$, $p = 0.007$) and abundance ($H = 35.3$, $p < 0.001$). Trophic groups with the same letter are not significantly different (Dunn's unplanned multiple comparisons procedures, $\alpha = 0.05$).

Table 3. Biomass (g m^{-2}) of the top ten species observed on quantitative transects on oil platforms in Gabon, West Africa.

Family	Species	Common name	Distribution	Trophic	Mean (sd)	%
Groupers	*Paranthias furcifer*	Creolefish	Atlantic	Plank	112.7 (149.3)	26.3
Chubs	*Kyphosus sp*	Bermuda Chub	Atlantic	Herb	101.0 (213.5)	23.5
Jacks	*Caranx crysos*	Blue Runner	Atlantic	Pisc	63.8 (127.7)	14.9
Damselfishes	*Abudefduf hoefleri*	African Sergeant	West Africa	Plank	30.8 (32.5)	7.2
Snappers	*Lutjanus fulgens*	Golden African Snapper	West Africa	Pisc	16.2 (48.6)	3.8
Damselfishes	*Chromis multilineata*	Brown Chromis	Atlantic	Plank	14.4 (14.8)	3.4
Surgeonfishes	*Acanthurus monroviae*	Monrovia Doctorfish	Atlantic	Herb	13.9 (14.9)	3.2
Barracuda	*Sphyraena barracuda*	Great Barracuda	Circumtropical	Pisc	11.1 (15.3)	2.6
Jacks	*Caranx fischeri*	Longfin Crevalle Jack	West Africa	Pisc	10.8 (12.4)	2.5
Hawkfishes	*Cirrhitus atlanticus*	West African Hawkfish	West Africa	Invert	7.4 (7.6)	1.7

Values are means and one standard deviation. Trophic groups - Plank = planktivores, Herb = herbivores, Pisc = piscivores, Invert = invertivores. Distributions derived from [82,83]. % is the percent contribution of each species to total fish biomass.

currently working on them. Although several species of scleractinian corals have been reported from Gabon [10], no true coral reefs are found. Despite the fact that oil platforms provide suitable substrate for the settlement of corals in other locations, no hermatypic corals were recorded on the platforms during our surveys. The ahermatypic dendrophyllid *Tubastraea* is circumtropical in distribution but likely a relatively recent invader to the Gulf of Guinea, possibly coming from the Indo-Pacific via the Panamenian region (Óscar Ocaña, pers. comm.). The taxonomy of the genus *Tubastraea* is currently under debate.

Conservation implications

Artificial reefs are increasingly being applied in mitigation efforts for natural systems – including their use as physical barriers to discourage illegal trawling in seagrass beds in Western Europe [62]. Oil platforms have been described as "de facto marine protected areas" [63], because they exclude trawl fishing and their large internal spaces offer shelter to fishes and other organisms. Platforms are complex structures, involving numerous crossbeams and large interstitial spaces, which increase overall habitat complexity and likely support high reef fish diversity and abundance.

The presence of oil platforms in the marine environment can create a number of environmental and social problems. The sinking of the Deepwater Horizon drilling rig created the largest marine oil spill in history (nearly 5 million barrels) and resulted in significant ecological and economic damage to the GOM region [64,65]. Smaller spills and leaks from offshore platforms commonly occur around the world, and as oil exploration and extraction moves into ever deeper water and into stormier and icier seas, potential risks will increase [66,67]. Invasive and exotic species have been reported on oil platforms around the world, where they compete with native species for space [20,28,68]. This is also an issue with decommissioned platforms that may be moved to a distant locations and enhance the spread of these exotic and invasive species [69]. Offshore platforms can also be a source of chronic stress that can lead to sub-lethal impacts on resident benthic organisms, resulting in loss of biodiversity [70].

Although often controversial, rigs-to-reef programs that allow decommissioned oil platforms to stay in place have gained some support by governmental and non-governmental organizations

Table 4. Numerical abundance (no. m-2) of the top ten species observed on quantitative transects on oil platforms in Gabon, West Africa.

Family	Species	Common Name	Distribution	Trophic	Mean (sd)	%
Groupers	*Paranthias furcifer*	Creolefish	Atlantic	Plank	1.25 (1.68)	32.2
Damselfishes	*Chromis multilineata*	Brown Chromis	Atlantic	Plank	1.05 (1.13)	27.1
Damselfishes	*Abudefduf hoefleri*	African Sergeant	West Africa	Plank	0.52 (0.55)	13.4
Jacks	*Caranx* juveniles	Jacks	Atlantic	Pisc	0.31 (0.99)	8.0
Damselfishes	*Abudefduf saxatilis*	Sergeant-Major	Atlantic	Plank	0.14 (0.18)	3.6
Snappers	*Lutjanus fulgens*	Golden African Snapper	West Africa	Pisc	0.11 (0.34)	2.8
Chubs	*Kyphosus sp*	Bermuda Chub	Atlantic	Herb	0.10 (0.18)	2.6
Jacks	*Caranx crysos*	Blue Runner	Atlantic	Pisc	0.07 (0.14)	1.8
Damselfishes	*Stegastes imbricatus*	Cape Verde Gregory	West Africa	Herb	0.06 (0.08)	1.5
Wrasses	*Thalassoma newtoni*	Newton Wrasse	West Africa	Invert	0.06 (0.09)	1.5

Values are means and one standard deviation. Trophic groups - Plank = planktivores, Herb = herbivores, Pisc = piscivores, Invert = invertivores. Distributions derived from [82,83]. % is the percent contribution of each species to total fish abundance.

Table 5. Estimates of abundance and sizes of large schooling resource species associated with oil rigs in Gabon.

Species	Platform 1	Platform 2	Platform 3	Platform 4	Platform 5	Platform 6	Platform 7	Platform 8	Platform 9	Platform 10
Caranx bartholomaei										100
Yellow Jack										(40–45)
Caranx crysos					100	100				300
Blue Runner					(40)	(40)				(40–45)
Caranx fischeri	200	200	100	100	75	75	100	75	75	
Longfin Crevalle Jack	(50–60)	(50–60)	(50–60)	(50–60)	(50–75)	(50–75)	(60–75)	(65–70)	(65–75)	
Caranx hippos			50	50			75			
Crevalle Jack			(50–65)	(50–65)			(60–70)			
Caranx latus			50							
Horse-eye Jack			(60–65)							
Elagatis bipinnulata	100	100	50	75	50	50	75			150
Rainbow Runner	(65–70)	(65–70)	(65–75)	(65–75)	(75)	(75)	(65–80)			(70–80)
Rachycentron canadum								2		
Cobia								(120–130)		
Sphyraena barracuda	200	200	100	100	50	50		100	100	100
Great Barracuda	(80–100)	(80–100)	(75–100)	(75–100)	(75–90)	(75–90)		(80–100)	(80–100)	(75–100)
Trachinotus ovatus	50		25		75	75				
Pompano	(45–50)		(45–50)		(50–65)	(50–65)				

Values are approximate number of individuals associated with the oil rig with minimum and maximum size (TL [total length] cm) in parentheses.

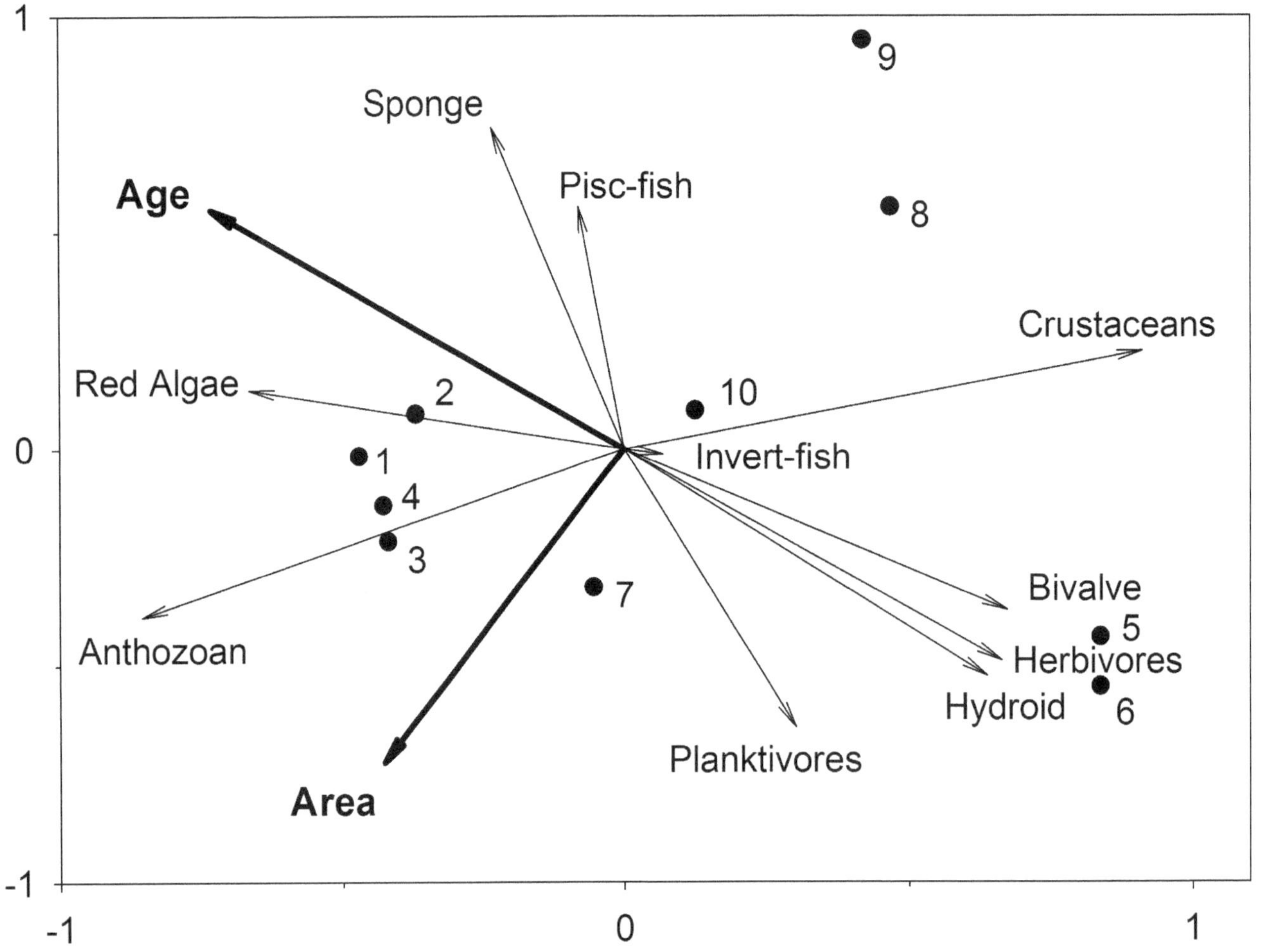

Figure 5. Biplot of results of redundancy analysis on biological (biomass of fish trophic groups and higher order benthic groups [e.g, Red alga, Sponge, Cnidaria-Anthozoa, Cnidaria-Hydrozoa, Crustacea-Cirripedia, Mollusca-Bivalve) and environmental data (e.g., age of platform, platform area).

[71]. They are particularly effective when they serve as no-take reserves and included in a marine protected area network [72,73]. Preliminary evidence indicates that decommissioned rigs in shallower waters can help rebuild declining fish stocks and therefore have high conservation value. Love and colleagues provided data that support the beneficial role of rigs for southern California's bocaccio rockfish (*Sebastes paucispinis*) populations [74]. They estimated that, of the average number of juveniles that survive annually across the geographic range of the species, approximately 20% (equivalent to ~430,000 individuals) were supported by eight southern Californian rigs.

Table 6. Results of redundancy analysis (RDA) on log-transformed data on fish biomass of trophic groups and abundance of higher-level benthic groups with environmental variables (e.g., age of platform, distance offshore, latitude, platform area).

Axes	**1**	**2**	**3**	**Total variance**
Eigenvalues	0.59	0.19	0.01	1
FG-environment correlations	0.96	0.97	0.56	
Cumulative percentage variance				
of FG data	58.92	77.93	79.15	
of FG-environment relation	73.74	97.32	98.87	
Sum of all eigenvalues				1
Sum of all canonical eigenvalues				0.80

FG = functional groups.

Table 7. Conditional effects of Monte-Carlo permutation results on the redundancy analysis (RDA).

Variable	Lambda A	P	F
Age	0.40	**0.01**	5.4
Area	0.25	**0.01**	5.0
Offshore	0.09	0.13	1.9
Lattitude	0.06	0.17	1.6

Variables in bold are significantly correlated with the RDA axes.

Population persistence in the deep sea relies heavily on connectivity between deep-sea communities on isolated reefs [75]. Over large distances, small, isolated reefs may act as stepping stones within an inhospitable matrix of soft sediment [19]. Propagules that settle on to artificial reefs in locations relatively isolated from natural reefs would likely not have found suitable natural habitat before perishing. In this case, artificial reefs could potentially increase biomass production by increasing settlement [19].

The addition of artificial reefs in the deep sea is likely to increase ecological connectivity, which will have important biogeographical consequences [19]. In the Gulf of Mexico, where the amount of natural hard substratum is limited, oil platforms contribute substantially to local and regional abundance of reef habitat and the abundance of reef-associated fishes [76–80]. Platforms have been shown to facilitate the expansion of coral populations in the Gulf of Mexico and therefore they possess an intrinsic environmental value through the presence of coral populations [20]. However it has also been shown that these platforms can act as vectors for invasive species.

Rocky bottoms are extremely scarce on the continental shelf of Gabon [3]. The addition of artificial reefs may contribute substantially to local and regional abundance of hard substrate habitat and the abundance of reef-associated fishes [76–80] and is likely to increase ecological connectivity [75]. Moreover, if fish populations are limited by the amount of available habitat, then the addition of suitable artificial habitat increases the environmental carrying capacity, resulting in a sustained increase in population biomass (the "production" hypothesis; [17]). Polovina and Sakai [81] showed that regional catch per unit effort of the Pacific giant octopus (*Octopus dofleini*) increased after the addition of artificial reefs in northern Japanese waters, and therefore supporting fisheries production. However, fish observed on artificial reefs may simply have been attracted to those locations from surrounding habitat (the "attraction" hypothesis; [17]). Our results suggest that these platforms increase local production through enhanced settlement, increased reproductive output, and likely through reduced natural and fishing mortality.

The Government of Gabon made history in 2002 when they placed nearly 11% of their land area under permanent protection in 13 National Parks. To date three marine protected areas have been designated within this region (Akanda National Park, Mayumba National Park, and Pongara National Park) and are located in inshore territorial and coastal waters, covering ~1,745 km^2; of these only Mayumba National Park is entirely a no-take MPA, whereas Akanda and Pongara National Parks are terrestrial in nature and have marine buffer components. In February 2014, President Ali Bongo Ondimba announced the creation of a system of Marine National Parks covering 20% of Gabon's exclusive economic zone. Between oil exclusion zones and Marine National Parks all of the oil infrastructure as of 2014 will be in no-take zones. These platforms should become core areas for replenishment of the marine parks and some of Gabon's fisheries.

Supporting Information

Table S1 List of macroalgae and invertebrates observed on oil platforms in Gabon.

Table S2 List of fish species observed on oil platforms in Gabon. Species ordered phylogenetically based on Eschmeyer 2013. Provinces based on [51,66–67], NWA = Northwestern Atlantic, SWA = Southwestern Atlantic, MAR = Mid-Atlantic Ridge, NEA = Northeast Atlantic, TEA = Tropical Eastern Atlantic. New = new record for Gabon.

Acknowledgments

This expedition was made possible by the Waitt Foundation, National Geographic Society, and the Wildlife Conservation Society. Thanks to the crew of the Waitt Institute research vessel for making our work safe and easy. Thanks to President Ali Bongo Ondimba, White Lee, Koumba Koumbila, Parcs Gabon, Total Gabon, Perenco and Vaalco for making it possible for us to survey Gabon's waters. The following people provided assistance with species taxonomy: Serge Gofas (mollusks), Óscar Ocaña (corals), María Jesús Uriz (sponges), and Xavier Turon (ascidians).

Author Contributions

Conceived and designed the experiments: AMF EB MF ES. Performed the experiments: AMF EB. Analyzed the data: AMF EB MF ES. Contributed reagents/materials/analysis tools: AMF EB MF ES. Wrote the paper: AMF EB MF ES.

References

1. McGlade JM, Cury P, Koranteng KA, Hardman-Mountford NJ (2002) The Gulf of Guinea Large Marine Ecosystem: environmental forcing and sustainable development of marine resources. Amsterdam: Elsevier Science. 428 p.
2. Sibuet M, Vangriesheim A (2009) Deep-sea environment and biodiversity of the West African Equatorial margin. Deep Sea Res. Part II Top Stud Oceanogr 56: 2156–2168.
3. Bianchi G (1992) Study of demersal assemblages of the continental shelf and upper slope of Congo and Gabon, based on the trawl surveys of the RV "Dr. Fridtjoff Nansen". Mar Ecol Prog Ser 85: 9–23.
4. Ondréas H, Olu K, Fouquet Y, Charlou JL, Gay A, et al. (2005) ROV study of a giant pockmark on the Gabon continental margin. Geo-Marine Letters 25: 281–292.

5. Roy OL, Caprais JC, Fifis A, Fabri MC, Galeron J, et al. (2007) Cold-seep assemblages on a giant pockmark off West Africa: spatial patterns and environmental control. Mar Ecol 28: 115–130.
6. Bentahila Y, Hebrard O, Ben Othman D, Luck JM, Seranne M, et al. (2006) Gulf of Guinea continental slope and Congo (Zaire) deep-sea fan: Sr–Pb isotopic constraints on sediments provenance from ZaiAngo cores. Mar Geol 226: 323–332.
7. Baudin F, Disnar JR, Martinez P, Dennielou B (2010) Distribution of the organic matter in the channel-levees systems of the Congo mud-rich deep-sea fan (West Africa). Implication for deep offshore petroleum source rocks and global carbon cycle. Mar Petrol Geol 27: 995–1010.
8. Vanden Bossche JP, Bernacsek GM (1990) Source book for the inland fishery resources of Africa: Vol. 2. CIFA Technical Paper, no. 18/1. Rome: FAO. 411 p.
9. Reid GM (1996) Ichthyogeography of the Guinea–Congo rain forest, West Africa. Proc Roy Soc Edinb B 104: 285–312.
10. Seranne M, Bruguier O, Moussavou M (2008) U–Pb single zircon grain dating of Present fluvial and Cenozoic aeolian sediments from Gabon: consequences on sediment provenance, reworking, and erosion processes on the equatorial West African margin. Bull Societé Géologique de France 179: 29–40.
11. Longhurst AR, Pauly D (1987) Ecology of tropical oceans. San Diego: Academic Press. 407 p.
12. Voituriez B, Herbland A (1982) Comparaison des systèmes productifs de l'Atlantique tropical Est: dômes thermiques, upwelling côtier et upwelling équatorial. Rapp P-V, Reun Cons Int Explor Mer 180: 114–130.
13. Lœuff PL, von Cosel R (1998) Biodiversity patterns of the marine benthic fauna on the Atlantic coast of tropical Africa in relation to hydroclimatic conditions and paleogeographic events. Acta Oecologica 19: 309–321.
14. Wauthy B (1983) Introduction à la climatologie du Golf de Guinée. Oceanogr Trop 18: 103–138.
15. Arnaud E, Arnaud PM, Intès A, Le Loeuff P (1976) Transport d'invertébrés benthiques entre l'Afrique du Sud et Sainte-Hélène par les laminaires (Phaeophyceae), Bull Mus nat Hist nat Paris sér 3e Ecol génér 30: 49–55.
16. Roberts CM, McClean CJ, Veron JEN, Hawkins JP, Allen GR, et al. (2002) Marine biodiversity hotspots and conservation priorities for tropical reefs. Science 295: 1280–1284.
17. Laborel J (1974) West African reef corals, and hypothesis on their origin. Proceedings of the Second International Symposium on Coral Reefs' Brisbane Vol. 1: 425–443.
18. Lee ME, Alonso A, Dallmeier F, Campbell P, Pauwels OSG (2006) The Gamba complex of protected areas: an illustration of Gabon's biodiversity. Bull Biol Soc Wash 12: 229–241.
19. Scarborough-Bull A (1989) Some comparisons between communities beneath the petroleum platforms off California and in the Gulf of Mexico. In: Reggio, V.C. (Ed.), Petroleum structures as artificial reefs: A compendium. Fourth Int. Conf. on Artificial Habitats for Fisheries, Rigs-to-Reefs Special Session, Miami, FL: US Dept. Interior, Minerals Mgt. Service, New Orleans, OCS Study MMS 89-0021. pp. 47–50.
20. Atchison AD, Sammarco PW, Brazeau DA (2008) Genetic connectivity in corals on the Flower Garden Banks and surrounding oil/gas platforms, Gulf of Mexico. J Exp Mar Biol Ecol 365: 1–12.
21. Hamzah BA (2003) International rules on decommissioning of offshore installations: some observations. Mar Policy 27: 339–348.
22. Parente V, Ferreira D, dos Santos EM, Luczynskic E (2006) Offshore decommissioning issues: deductibility and transferability. Energ Policy 34: 1992–2001.
23. Pradella N, Fowler AM, Booth DJ, Macreadie PI (2013) Fish assemblages associated with oil industry structures on the continental shelf of north-western Australia. J Fish Biol 84: 247–255.
24. Bohnsack JA (1989) Are high densities of fishes at artificial reefs the result of habitat limitation or behavioral preference? Bull Mar Sci 44: 631–645.
25. Jørgensen T, Løkkeborg S, Soldal AV (2002) Residence of fish in the vicinity of a decommissioned oil platform in the North Sea. ICES J Mar Sci 59: S288–S293.
26. Macreadie PI, Fowler AM, Booth DJ (2011) Rigs-to-reefs: will the deep sea benefit from artificial habitat? Front Ecol Environ 9: 455–461.
27. Sammarco PW, Atchison AD, Boland GS (2004) Expansion of coral communities within the Northern Gulf of Mexico via offshore oil and gas platforms. Mar Ecol Prog Ser 280: 129–143.
28. Sammarco PW, Porter SA, Cairns SD (2010) A new coral species introduced into the Atlantic Ocean - *Tubastraea micranthus* (Ehrenberg 1834) (Cnidaria, Anthozoa, Scleractinia): An invasive threat? Aquat Invasions 5: 131–140.
29. Wolfson A, van Blaricom G, Davis N, Lewbel GS (1979) The marine life of an offshore oil platform. Mar Ecol Prog Ser 1: 81–89.
30. Love MS, Caselle JE, Snook L (2000) Fish assemblages around seven oil platforms in the Santa Barbara Channel area. Fish Bull 98: 96–117.
31. Helvey M (2002) Are southern California oil and gas platforms essential fish habitat? ICES J Mar Sci: Journal du Conseil 59 (suppl): S266–S271.
32. Bohnsack JA, Johnson DL, Ambrose RF (1991) Ecology of artificial reef habitats. In: Seaman W Jr, Sprague LM, editors.Artificial habitats for marine and freshwater fisheries.New York: Academic Press. pp. 61–108.
33. Olsgard F, Gray JS (1995) A comprehensive analysis of the effects of offshore oil and gas exploration and production on the benthic communities of the Norwegian continental shelf. Mar Ecol Prog Ser 122: 277–306.
34. Schroeder DM, Love MS (2004) Ecological and political issues surrounding decommissioning of offshore oil facilities in the Southern California Bight. Ocean Coast Manage 47: 21–48.
35. Page HM, Dugan JE, Culver CS, Hoesterey JC (2006) Exotic invertebrate species on offshore oil platforms. Mar Ecol Prog Ser 325: 101–107.
36. Sala E, Ballesteros E (1997) Partitioning of space and food resources by three fishes of the genus *Diplodus* (Sparidae) in a Mediterranean rocky infralittoral ecosystem. Mar Ecol Prog Ser 152: 273–283.
37. Cebrian E, Ballesteros E, Canals M (2000) Shallow rocky bottom benthic assemblages as calcium carbonate producers in the Alboran Sea (Southwestern Mediterranean). Oceanol Acta 23: 311–322.
38. Froese R, Pauly D (2011) FishBase. World Wide Web electronic publication. Available: www.fishbase.org. Accessed 2014 May 5.
39. Floeter SR, Ferreira CEL, Dominici-Arosemena A, Zalmon IR (2004) Latitudinal gradients in Atlantic reef fish communities: trophic structure and spatial use patterns. J Fish Biol 64: 1680–1699.
40. Ferreira CEL, Floeter SR, Gasparini JL, Ferreira BP, Joyeux JC (2004) Trophic structure patterns of Brazilian reef fishes: a latitudinal comparison. J Biogeogr 31: 1093–1106.
41. Almada VC, Toledo JF, Brito A, Levy A, Floeter SR, et al. (2013) Complex origins of the Lusitania biogeographic province and northeastern Atlantic fishes. Front Biogeogr 5: 20–28.
42. Clarke KR, Gorley RN (2001) PRIMER version 5: user manual/tutorial. Plymouth, UK: Primer-E Ltd. 91 p.
43. Ludwig JA, Reynolds JF (1988) Statistical Ecology. New York, New York: John Wiley and Sons. 337 p.
44. Zar JH (1999) Biostatistical analysis, 4th edition. India: Pearson Education. 662 p.
45. ter Braak C, Šmilauer P (2002) CANOCO Reference Manual and CanoDraw for Windows User's Guide: Software for Canonical Community Ordination (version 4.5). Section on permutation methods. Microcomputer Power (Ithaca NY, USA). 500 p.
46. ter Braak C (1994) Canonical community ordination. Part I: Basic theory and linear methods. Ecoscience 1: 127–140.
47. ter Braak C, Verdonschot P (1995) Canonical correspondence analysis and related multivariate methods in aquatic ecology. Aquat Sci 55: 1–35.
48. Sala E, Aburto O, Paredes G, Thompson G (2003) Spawning aggregations and reproductive behavior of reef fishes in the Gulf of California. Bull Mar Sci 72: 103–121.
49. Soldal AV, Svellingen I, Jørgensen T, Løkkeborg S (2002) Rigs-to-reefs in the North Sea: hydroacoustic quantification of fish in the vicinity of a "semi-cold" platform. ICES J Mar Sci: Journal du Conseil 59 (suppl): S281–S287.
50. Sandin SA, Smith JE, DeMartini EE, Dinsdale EA, Donner SD, et al. (2008) Degradation of coral reef communities across a gradient of human disturbance. PLoS ONE 2008; 3(2):e1548.
51. Friedlander AM, Ballesteros E, Beets J, Berkenpas E, Gaymer CF, et al. (2013) Effects of isolation and fishing on the marine ecosystems of Easter Island and Salas y Gómez, Chile. Aquat Conserv 23: 515–531.
52. Friedlander AM, Zgliczynski BJ, Ballesteros E, Aburto-Oropeza O, Bolaños A, et al. (2012) The shallow-water fish assemblage of Parque Nacional Isla del Coco, Costa Rica: structure and patterns in an isolated, predator-dominated ecosystem. Rev Biol Trop (Suppl. 3): 321–338.
53. Wirtz P, Ferreira CEL, Floeter SR, Fricke R, Gasparini JL, et al. (2007) Coastal fishes of São Tomé and Príncipe islands, Gulf of Guinea (Eastern Atlantic Ocean) – an update. Zootaxa 1523: 1–48.
54. Afonso P, Porteiro FM, Santos RS, Barreiros JP, Worms J, et al. (1999) Coastal marine fishes of São Tomé Island (Gulf of Guinea). Arquipélago Life Mar Sci 17: 65–92.
55. Scheltema RS (1971) The dispersal of the larvae of shallow-water benthic invertebrate species over long distances by ocean currents. In: Crisp PJ, editor.Fourth European Marine Biology Symposium.Cambridge: Cambridge University Press. pp. 7–28.
56. Scheltema RS (1995) The relevance of passive dispersal for the biogeography of Caribbean molluscs. Am Malac Bull 11: 99–115.
57. Wirtz P (2001) New records of marine invertebrates from the Cape Verde Islands. Arquipélago Life Mar Sci 18A: 81–84.
58. Wirtz P (2003) New records of marine invertebrates from São Tomé Island. J Mar Biol Assoc UK 83: 735–736.
59. Wirtz P (2004) Four amphi-Atlantic shrimps new for São Tomé and Príncipe (eastern central Atlantic). Arquipélago Life Mar Sci 21: 83–85.
60. Wirtz P (2012) Seven new records of fish from Ngor Island, Senegal. Arquipélago, Life Mar Sci 29: 77–81.
61. Kritzer JP, Sale PF (2010) Marine metapopulations. San Diego, California: Elsevier Academic Press. 576 p.
62. González-Correa JM, Bayle JT, Sánchez-Lizaso JL, Valle C, Sánchez-Jerez P, et al. (2005) Recovery of deep *Posidonia oceanica* meadows degraded by trawling. J Exp Mar Biol Ecol 320: 65–76.
63. Schroeder DM, Love MS (2004) Ecological and political issues surrounding decommissioning of offshore oil facilities in the Southern California Bight. Ocean Coast Manage 47: 21–48.
64. White HK, Hsing PY, Cho W, Shank TM, Cordes EE, et al. (2012) Impact of the Deepwater Horizon oil spill on a deep-water coral community in the Gulf of Mexico. Proc Nat Acad Sci 109: 20303–20308.

65. Sumaila UR, Cisneros-Montemayor AM, Dyck A, Huang L, Cheung W, et al. (2012) Impact of the Deepwater Horizon well blowout on the economics of US Gulf fisheries. Can J Fish Aquat Sci 69: 499–510.
66. Jernelöv A (2010) The threats from oil spills: Now, then, and in the future. Ambio 39: 353–366.
67. Trevors JT, Saier MH (2010) The legacy of oil spills. Water Air Soil Poll 211: 1–3.
68. Rooker JR, Holt GJ, Pattengill CV, Dokken Q (1997) Fish assemblages on artificial and natural reefs in the Flower Garden Banks National Marine Sanctuary, USA. Coral Reefs 16: 83–92.
69. Yeo DC, Ahyong ST, Lodge DM, Ng PK, Naruse T, et al. (2010) Semisubmersible oil platforms: understudied and potentially major vectors of biofouling-mediated invasions. Biofouling 26: 179–186.
70. Street GT, Montagna PA (1996) Loss of genetic diversity in Harpacticoida near offshore platforms. Mar Biol 126: 271–282.
71. Macreadie PI, Fowler AM, Booth DJ (2012) Rigs-to-reefs policy: can science trump public sentiment? Front Ecol Environ 10: 179–180.
72. Schroeder DM, Love MS (2002) Recreational fishing and marine fish populations. CalCOFI Rep 43: 182–190.
73. Caselle JE, Love MS, Fusaro C, Schroeder D (2002) Trash or habitat? Fish assemblages on offshore oilfield seafloor debris in the Santa Barbara Channel, California. ICES J Mar Sci 59: S258–S265.
74. Love MS, Schroeder DM, Lenarz W, MacCall A, Bull AS, et al. (2006) Potential use of offshore marine structures in rebuilding an overfished rockfish species, bocaccio (*Sebastes paucispinis*). Fish Bull 104: 383–390.
75. Cowen RK, Sponaugle S (2009) Larval dispersal and marine population connectivity. Ann Rev Mar Sci 1: 443–66.
76. Gallaway BJ, Lewbel GS (1981) The ecology of petroleum platforms in the northwestern Gulf of Mexico: a community profile. Washington, DC: FWS 10BS-82/27, Open File Report 82–03. USFWS Office of Biology Services.
77. Stanley DR, Wilson CA (1989) Utilization of offshore platforms by recreational fishermen and SCUBA divers off the Louisiana coast. Bull Mar Sci 44: 767–775.
78. Stanley DR, Wilson CA (1990) A fishery-dependent based study of fish species composition and associated catch rates around oil and gas structures off Louisiana. Fish Bull 88: 719–730.
79. Scarborough-Bull A, Kendall JJ Jr (1994) An indication of the process: offshore platforms as artificial reefs in the Gulf of Mexico. Bull Mar Sci 55: 1086–1098.
80. Shipp RL, Bortone SA (2009) A perspective of the importance of artificial habitat on the management of red snapper in the Gulf of Mexico. Rev Fish Sci 17: 41–47.
81. Polovina JJ, Sakai I (1989) Impacts of artificial reefs on fishery production in Shimamaki, Japan. Bull Mar Sci 44: 997–1003.
82. Floeter SR, Rocha LA, Robertson DR, Joyeux JC, Smith-Vaniz WF, et al. (2008) Atlantic reef fish biogeography and evolution. J Biogeogr 35: 22–47.
83. Wirtz P, Brito A, Falcón JM, Freitas R, Fricke R, et al. (2013) The coastal fishes of the Cape Verde Islands–new records and an annotated check-list. Spixiana 36: 113–142.

7

Transcriptome Analysis of the *Trachinotus ovatus*: Identification of Reproduction, Growth and Immune-Related Genes and Microsatellite Markers

Xie Zhenzhen[1☯], Xiao Ling[1☯], Wang Dengdong[1☯], Fang Chao[1], Liu Qiongyu[1], Li Zihao[1], Liu Xiaochun[1], Zhang Yong[1]*, Li Shuisheng[1]*, Lin Haoran[1,2]

1 State Key Laboratory of Biocontrol, Institute of Aquatic Economic Animals, and the Guangdong Province Key Laboratory for Aquatic Economic Animals, School of Life Sciences, Sun Yat-Sen University, Guangzhou, China, **2** College of Ocean, Hainan University, Haikou, Hainan, China

Abstract

Background: The *Trachinotus ovatus* (Teleostei, Carangidae) is an economically important marine fish species in the world. However, the lack of genomic information regarding this species limits our understanding of the genetics and biological mechanisms in *Trachinotus ovatus*. In this study, high throughput transcriptome sequencing was used to obtain comprehensive genomic information in *Trachinotus ovatus*.

Principal Findings: Transcriptome sequencing was performed by using Illumina paired-end sequencing technology. The 98,534,862 high quality reads were yielded, and were *de novo* assembled into 156,094 unigenes with an average sequence length of 1179 bp. Transcriptome annotation revealed that 75,586 and 67,923 unigenes were functionally annotated in the NCBI non-redundant database and Swiss-Prot protein database, respectively. Functional analysis demonstrated that 67,923 unigenes were grouped into 25 Cluster of Orthologous Groups (COG) functional categories, 37,976 unigenes were clustered into 61 Gene Ontology (GO) terms, and 38,172 unigenes were assigned to 275 different Kyoto Encyclopedia of Genes and Genomes (KEGG) pathways. Based on the transcriptome dataset, a large number of unigenes associated with reproduction, growth and immunity were identified. Furthermore, a total number of 38,794 simple sequence repeats (SSRs) were discovered and 16 polymorphic loci were characterized in *Trachinotus ovatus*.

Conclusion/Significance: The present study is the first transcriptome analysis of a fish species belonging to the genus *Trachinotus* and provides a valuable genomic resource for novel gene discovery, gene expression and regulation studies, and the identification of genetic markers in *Trachinotus ovatus* and the other fish of the genus *Trachinotus*.

Editor: Jingfa Xiao, Beijing Institute of Genomics, China

Funding: This work was supported by the National 863 Program of China (No. 2012AA10A407) and Special Fund for Agro-scientific Research in the Public Interest (201403008). The funders had no role in study design, data collection and analysis, decision to publish, or preparation of the manuscript.

Competing Interests: The authors have declared that no competing interests exist.

* Email: lshuish@mail.sysu.edu.cn (LSS); lsszy@mail.sysu.edu.cn (LHR)

☯ These authors contributed equally to this work.

Introduction

There are 20 species of carangid fish that belong to the genus *Trachinotus*. They are distributing all over the world and possess very desirable biological characteristics such as fast adaptation to controlled conditions, rapid growth, and so on. Therefore, fish species belonging to the genus *Trachinotus* have long been considered as promising candidates for mariculture [1]. According to the Food and Agriculture Organization of the United Nations (FAO) fishery statistics, the global aquaculture production of the carangid fish had exceeded 113,000 tons in 2012, with a value of approximately 458 million US dollars [2].

Trachinotus ovatus (Linnaeus 1758), the most popular cultured species of the genus *Trachinotus*, mainly inhabits in tropical and subtropical waters of the eastern Atlantic Ocean, and is also found in the Mediterranean and along the African coast, including offshore islands [3]. Due to its fast growth rate, pleasing flavor and increased market demand, *Trachinotus ovatus* is recognized as one of the most economically important marine fish species. With the wide use of large-scale marine cage culture in the Asia-Pacific region, including China and Southeast Asian countries, production of this species has increased in momentum in recent years.

However, with the expansion of large-scale culture and the neglect of fishery management and conservation, the genetic diversity of *Trachinotus ovatus* appears to be declining. Several biological changes including early age of sexual maturation, low growth rate, and increasing disease susceptibility have been found in cultured *Trachinotus ovatus* [4]. Since 2001, frequent disease occurrence has been reported in cultured *Trachinotus ovatus* each year in the southern coast of China, resulting in great economic losses [5,6].

In spite of its significant economic importance, there is very little existing basic research on *Trachinotus ovatus*. Although some DNA markers (amplified fragment length polymorphism and microsatellite markers) have been developed [7,8], the genetic diversity and population structure of *Trachinotus ovatus* cannot be effectively assessed with the existing genetic information. With the exception of a few studies on growth performance and feeding in confined environments [9–11], there is currently only limited information on other important traits such as reproduction, growth and immunity in *Trachinotus ovatus*. Moreover, owing to the lack of genomic data, molecular-genetic association studies remain unexplored in this species. Therefore, all areas of research, particularly the basic research on the genetics and on the gene networks involved in the regulation of reproduction, growth and immune response should be strengthened to support the sustained development of the *Trachinotus ovatus* industry.

Despite the rapid development of high-throughput sequencing technologies, genome sequencing is still expensive and takes long time. Transcriptome sequencing is a good choice for quickly and cost-effectively obtaining large-scale genetic information and functional genes in non-model species that do not have reference genomic information. Transcriptome analysis has now been widely applied in the fields of ecology, genetics, and molecular biology, playing valuable roles in gene discovery, gene expression and regulation, developing molecular markers and so on [12–16].

In the present study, using the Illumina paired-end sequencing technology, transcriptome analysis on various tissues of *Trachinotus ovatus* was performed. The genes involved in biological processes associated with reproduction, growth, and immunity were analyzed, and a set of microsatellite markers was developed. Our study provides abundant genomic information for future research into the genetics and molecular biology of *Trachinotus ovatus*.

Materials and Methods

Sample Preparation and RNA Extraction

Ten two-year-old *Trachinotus ovatus* with body weight of approximately 1000 ± 100 g were obtained from the Daya Bay Aquaculture Center, Guangdong, China. Fish were anesthetized and sacrificed by decapitation. Tissue samples were collected immediately and snap frozen in liquid nitrogen. All animal experiments were conducted in accordance with the guidelines and approval of the respective Animal Research and Ethics Committees of Sun Yat-Sen University.

Total RNA was isolated from different tissues of female and male *Trachinotus ovatus* (brain, pituitary, liver, gill, heart kidney, kidney, spleen, testes, and ovaries) using Trizol Reagent (Invitrogen, USA) according to the manufacturer's instructions. The concentration of total RNA was estimated by measuring the absorbance at 260 nm using a 2100 Bioanalyzer (Agilent Technologies, USA), and the RNA integrity was checked by ethidium bromide staining of 28S and 18S ribosomal bands on a 1% agarose gel. Equal volumes of RNA from each tissue were pooled and used for cDNA synthesis.

cDNA Library Construction and Sequencing

The cDNA library construction and sequencing was carried out as described by Li [17]. Briefly, Poly (A) mRNA was isolated from 20 ug of total RNA using oligo-dT beads (Qiagen, Germany), and was broken into short fragments (200 nt) in fragmentation buffer. The short fragments were used to synthesize first-strand cDNA by using random hexamer-primed reverse transcription, and then the second-strand cDNA was synthesized using RNase H and DNA polymerase I. The double-stranded cDNAs were purified using the QIAquick PCR extraction kit (Qiagen, Germany). After washing with EB buffer, the purified products were used for end reparation poly (A) addition and ligated to sequencing adapters. Following agarose gel electrophoresis, the adaptor-ligated fragments (200 bp ± 25 bp) were further extracted on the agarose gel and enriched by PCR to construct the final cDNA library, which was sequenced on the Illumina HiSeq 2000 sequencing platform using single-end paired-end technology in Beijing Genomics Institute-Shenzhen, Shenzhen, China. The original data process to sequences, base-calling and quality value calculation were achieved by the Illumina Genome Analyzer Pipeline (version 1.6), and 100 bp paired-end reads were obtained.

Illumina Reads Processing and Assembly

A Perl program was written to select clean reads. Low-quality reads that were more than 50% bases with quality lower than 20 in one sequence, ambiguous reads containing more than 5% unknown bases, and reads containing adaptor sequences were removed. Then the clean reads were assembled using Trinity to construct unique consensus sequences [18].

Functional Annotation and Classification

Using the BLASTx tool which was developed to evaluate the similarities between two sequences [19], the filtered transcripts were search against the National Center for Biotech-nology Information (NCBI) non-redundant protein (Nr) database (http://www.ncbi.nlm.nih.gov/), the Swiss-Prot protein database (http://www.expasy.ch/sprot) [20], the Cluster of Orthologous Groups (COG) (http://www.ncbi.nlm.nih.gov/COG/) [21], and the Kyoto Encyclopedia of Genes and Genomes (KEGG) pathway database [22] with an E-value cut off of 1e-5. Protein sequences from the databases which had the highest similarity scores were used as the functional annotation for the related unigene. Blast2GO was applied to get the annotation results of unigenes in Gene Ontology (GO) database (http://www.geneontology.org/) [23]. The WEGO software (http://wego.genomics.org.cn/cgibin/wego/index.pl) [24], a statistics tool, was then used to classify the annotation results of unigenes in GO database. Meanwhile, the unigenes were also aligned to the KEGG database to annotate the signal pathways.

Simple Sequence Repeat (SSR) Detection and Primer Design

Potential SSR markers were detected among the 156,094 unigenes using the MIcroSAtellite identification tool (MISA) (http://pgrc.ipk-gatersleben.de/misa/) [25]. We searched for SSRs with motifs ranging from di- to hexa-nucleotides in size. Given difficulties in distinguishing genuine mono-nucleotide repeats from polyadenylation products and some mono-nucleotide repeats that were generated by base mismatches or sequencing errors, mono-nucleotide repeats were discarded in this study. According to the MISA results, primer pairs flanking each SSR locus were designed using the Primer3 program (http://www.broadinstitute.org/genome_software/other/primer3.html) [26].

Survey of SSR Polymorphism

Fin samples were randomly collected from 30 six-month old *Trachinotus ovatus* individuals, which were the progeny of one broodstock population in the Daya Bay Aquaculture Center, Guangdong, China. All the samples were placed in absolute ethanol and kept frozen at −20°C until DNA extraction. Polymorphism was tested in the 30 *Trachinotus ovatus* individuals. Genomic DNA was extracted from the fin samples using the salting-out procedure [27]. PCR was performed on a Veriti Thermal Cycler in a total volume of 20 uL containing 0.4 uM of each primer, 10× PCR buffer (Genstar, China), and 100 ng DNA. Cycling conditions consisted of initial denaturation at 94°C for 5 min, 35cycles of 45 s at 94°C,40 s at the annealing temperature, 40 s at 72°C, and a final cycle of 5 min at 72°C. Allele sizes were estimated according to the pBR322 DNA/MspI marker (TianGen, China) after PCR products were separated on 8% denaturing polyacrylamide gel. The expected and observed heterozygosities together with an analysis of Hardy-Weinberg equilibrium (HWE) and linkage disequilibrium were calculated using GENEPOP 4.0 [28]. Null allele frequencies were calculated with MICRO-CHECKER2.2.3 [29]. The significant values for all multiple tests were corrected by the sequential Bonferroni's procedure [30]. Polymorphism information content (PIC) was calculated using the PIC-CALC 0.6 software.

Results and Discussion

Sequencing, Assembly and Evaluation

Using the Illumina sequencing platform, a total of 99,714,868 raw reads were produced. After strict quality control and data filtration, 98,534,862 cleaned reads were harvested, with an average length of 100 bp (Table 1). All cleaned reads generated in this study have been deposited in the NCBI Sequence Read Archive (SRA) database (accession number: SRA161953).

Using the Trinity assembly program, 156,094 unigenes were generated by *de novo* assembly, with an average length of 1179 bp and an N50 of 2404 bp (Table 1). The mean length of the unigenes in this study was longer than those assembled in other teleost fish such as *Misgurnus anguillicaudatus* (387 bp) [31], *Oncorhynchus mykiss* (662 bp) [32], *Scophthalmus maximus* (671 bp) [33] using *de novo* transcriptome assembly or 454 sequencing. Furthermore, the number of assembled unigenes and their length distribution were also analyzed. About 53,930 unigenes were >1,000 bp in length (Figure 1).

In order to assess the extent of coverage of the assembled unigenes and of evaluating the effects of coverage depth on unigene assembly, the ratio of the assembled unigene length to *Danio rerio* ortholog length against coverage depth was calculated. As shown in Figure 2A, the coding regions of most of *Danio rerio* orthologs could be effectively covered by the assembled *Trachinotus ovatus* unigenes. There were 10,219 unigenes with the ratio within 0.8 to 1.2, providing abundant long transcripts for gene function studies in *Trachinotus ovatus*. Additionally, it was observed that coverage depth was not directly proportional to coverage of the coding regions, although increased coverage depth could improve the extent of transcript coverage to a certain degree. The total percentage of *Danio rerio* ortholog coding sequences that were covered by all *Trachinotus ovatus* unigenes was also plotted. As shown in Figure 2B, 1,679 *Danio rerio* orthologs could be covered by *Trachinotus ovatus* unigenes with a percentage more than 80%. In addition, there were about 2,337 *Danio rerio* orthologs with coverage varying from 50% to 80%, and 2,725 orthologs with coverage below 20%. These data indicate that the assembled unigenes in this current study are of high quality of the present assembly.

Annotation of Unigenes

Using the BLASTx algorithm (E-value $<10^{-5}$), all assembled unigenes was searched against the databases of NCBI Nr, Swiss-Prot, COG, and KEGG. The annotation results were demonstrated through the Venn diagram (Figure 3). Of the 156,094 unigenes, 75,586 and 67,923 had homologous sequences in the Nr and Swiss-Prot protein databases, while 27,798 and 38,127 unigenes could be classified by COG and KEGG databases, respectively. In total, 15,145 unigenes could be simultaneously annotated by all four databases.

A BLASTx top-hit species distribution showed that 50,324 unigenes exhibited similarity to the sequences of *Oreochromis niloticus*, 8,173 to the sequences of *Danio rerio*, 3,322 to the sequence of *Dicentrarchus labrax*, and 1,342 to the sequence of *Salmo salar* (Figure 4). These results reveal that the gene content is phylogenetic conserved between *Trachinotus ovatus* and *Oreochromis niloticus*, and indicate that *Trachinotus ovatus* is more closely related to *Oreochromis niloticus*, both of which belong to the order Perciformes.

COG, GO and KEGG Classification

For functional prediction and classifications, all unigenes were aligned to the COG database. Together, 67,923 unigenes were grouped into 25 COG classifications (Figure 5). The largest cluster was the general function prediction only (18.31%), followed by replication, recombination and repair (9.41%), transcription (8.28%), cell cycle control, cell division and chromosome partitioning (7.68%), signal transduction mechanisms (6.68%), translation, ribosomal structure and biogenesis (6.24%), function unknown (5.39%), posttranslational modification, protein turnover, and chaperones (5.14%) and carbohydrate transport and metabolism (5.13%) (Table S1).

Gene Ontology is a standardized system for gene functional classification, containing three domains which are classified by biological process, cellular components and molecular functions of gene products [34]. Gene ontology analysis of our dataset showed that a total of 22,481 unigenes were annotated. Of these, 17,838 were grouped into cellular component, 18,064 were grouped into molecular function, and 18,823 were grouped into biological process (Figure 6). In the category of biological process, a large proportion of unigenes were related to cellular progress and metabolic progress. Within the category of cellular component, unigenes related to cell and cell part represented the largest clusters. Under the category of molecular function, a high

Table 1. Statistics of the assembled transcripts and unigenes.

	Total number	Total Nucleotides base pair (bp)	Average length(bp)	N50
Raw Reads	99,714,868	9,971,486,800	100	-
Clean Reads	98,534,862	9,853,486,200	100	-
Unigenes	156,094	184,107,954	1179	2404

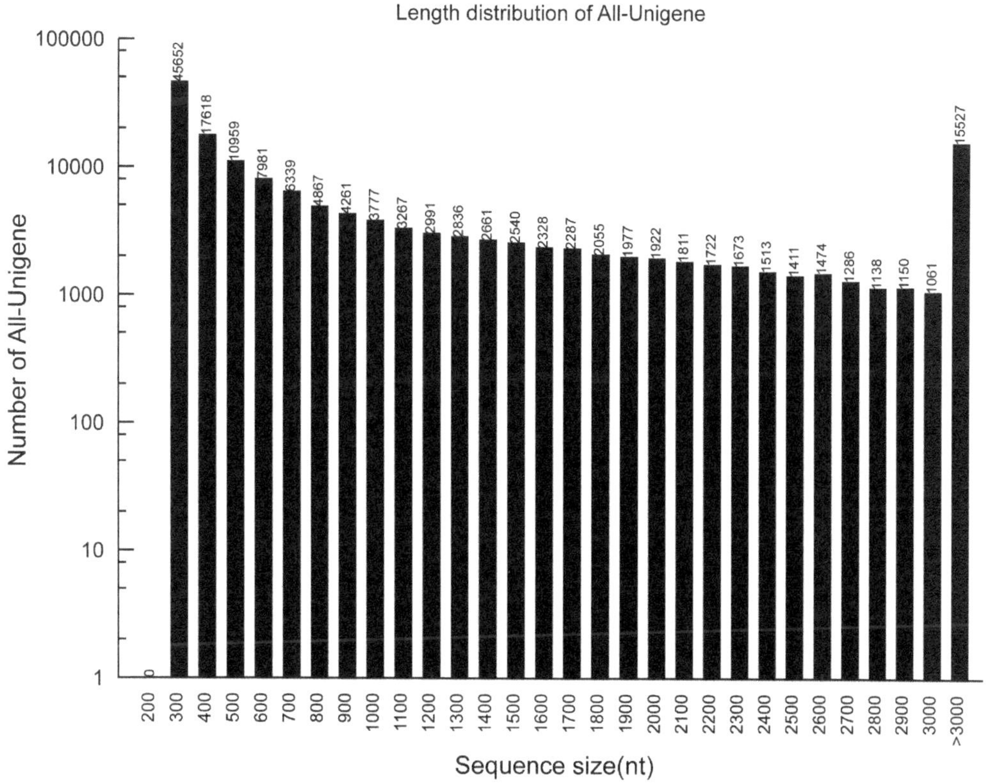

Figure 1. Length distribution of all unigenes of *Trachinotus ovatus*.

percentage of unigenes was involved in binding and catalytic activity.

For further identification of the biological pathways in *Trachinotus ovatus*, we mapped the assembled sequences to the reference of typical pathways in the KEGG database. The 38,172 unigenes were matched to 275 different KEGG pathways (Table S2). Among these sequences, 6,639 were classified into metabolism groups, mostly involving in purine metabolism (776), glycerophospholipid metabolism (409), pyrimidine metabolism (358), and glycerolipid metabolism (341). The second largest cluster is the genetic information processing (5,364), involving degradation (1,146), repair (247), transcription and replication (172) and so on. There were 14,981, 9,849, and 1,339 sequences that were classified into cellular processing, human disease and environment information, respectively.

Identification of Sequences Related to Reproduction

Fish reproduction is the precondition for large-scale commercial culture and breeding. The artificial reproductive manipulation in *Trachinotus ovatus* has been established. However, the regulatory mechanisms behind reproductive development in this species are poorly understood and there is a lack of effective techniques for reproductive manipulation, leading to the abuse of hormone in current culture practices. Reproductive activities in fish are mainly controlled by gonadotropin-releasing hormone (GnRH)-gonadotropic hormone (GtH) axis [35]. GnRH agonists and exogenous GtH preparations have been widely used to induce ovulation and spermiation in cultured fish [36]. Therefore, understanding the regulation of GnRH and GtH is highly important for understanding reproductive activities in *Trachinotus ovatus*. In the transcriptome data, the transcripts for three subtypes of GnRH and different GtH subunits, including the complete coding DNA sequence of follicle-stimulating hormone subunit beta, luteinizing hormone subunit beta were indentified (Text S1). Moreover, the transcripts for kisspeptin, tachykinin3, aromatase, estrogen receptors, and androgen receptors were also found (Text S1). Kisspeptin and tachykinin3 are considered to be the key regulators of GnRH secretion in vertebrates [37]. Aromatase, estrogen receptors, and androgen receptors are critical factors in gonad responding to the actions of GnRH and GtH [38]. These sequences provide valuable information for elucidating the regulatory mechanisms of reproductive axis in *Trachinotus ovatus*. Further studies will likely focus on the role of the GnRH-GtH axis in the regulation of reproductive activities and on establishing highly effective techniques for the induction of gamete maturation, ovulation, and spermiation in *Trachinotus ovatus*.

In addition, *Trachinotus ovatus* is a unique species without any obvious phenotypic characteristics that can be used to distinguish the males and females. Difficulty in identifying the sexes has greatly hampered the artificial breeding program. In order to isolate sex-specific molecular markers, identification of genes related to sex determination and differentiation is necessary. Through the GO analysis, there were about 714 unigenes related to the reproductive regulation (Table S3), including forkhead box L2 that and doublesex and mab-3 related transcription factor 1 (DMRT1), which are key genes in the control of sex determination and differentiation in vertebrates [39]. The Realtime PCR analysis revealed that the expression of DMRT1 was much higher in the testes than in the ovaries (Figure S1), which is consistent with the findings in other fish species [40–42]. Studies have found that overexpression of DMRT1 could induce a female-to-male sex-reversal in Nile tilapia [43]. A linkage map in zebrafish has revealed the sex determination locus harboring DMRT1 gene

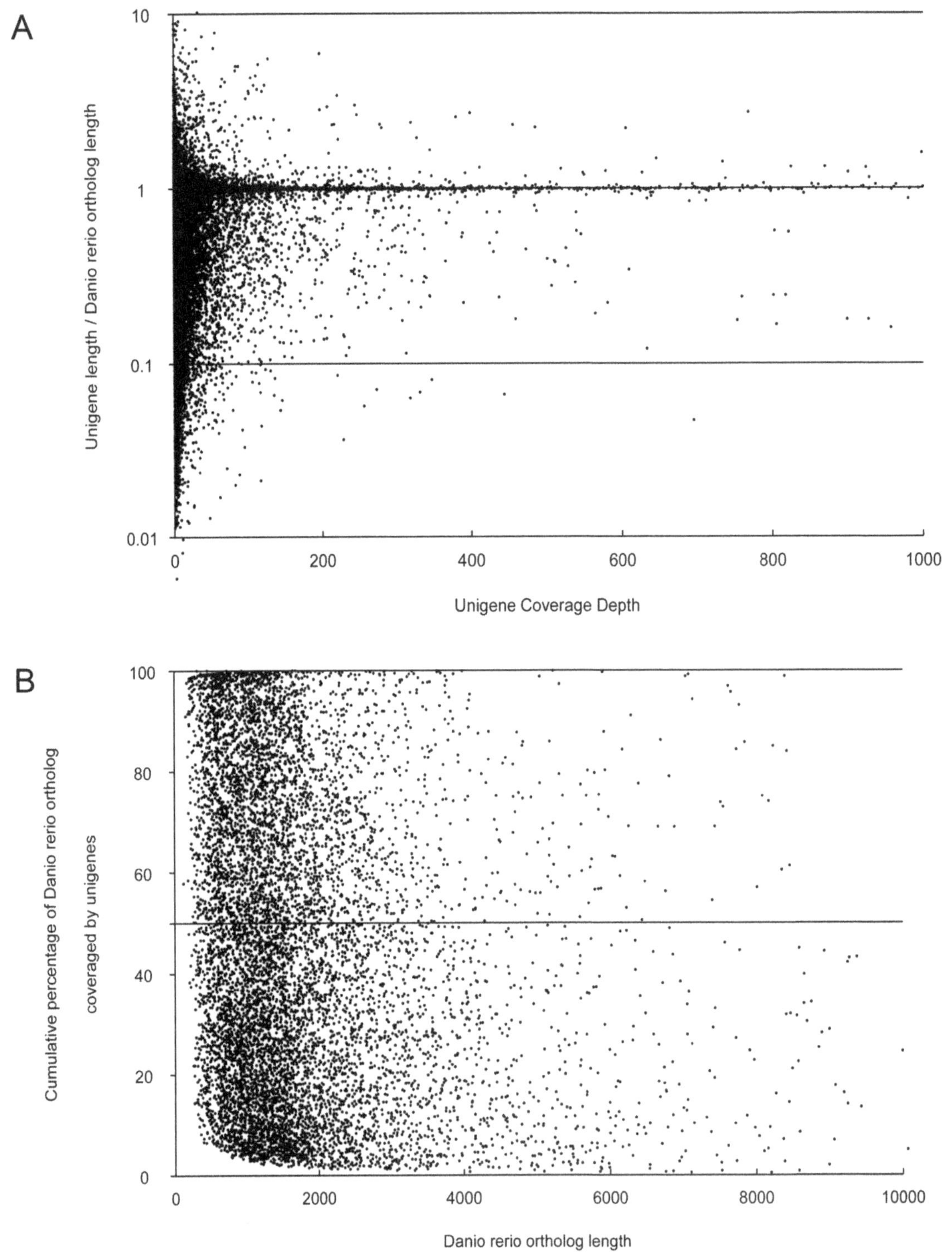

Figure 2. Comparison of *Trachinotus ovatus* unigenes to *Danio rerio* orthologous coding sequences. (A) The ratio of *Trachinotus ovatus* unigene length to *Danio rerio* ortholog length was plotted against *Trachinotus ovatus* unigene coverage depth. (B) The total percentage of *Danio rerio* ortholog coding sequence covered by all *Trachinotus ovatus* unigenes.

[44]. Therefore, studying the roles of DMRT1 and other sex related genes in sex determination and differentiation will be useful in finding sex-specific markers that can help solve the problem of sex identification in *Trachinotus ovatus*.

Identification of Sequences Related to Growth and Metabolism

Improving the growth rate of cultivated fish and selecting fish with desired traits including rapid growth are major objectives in

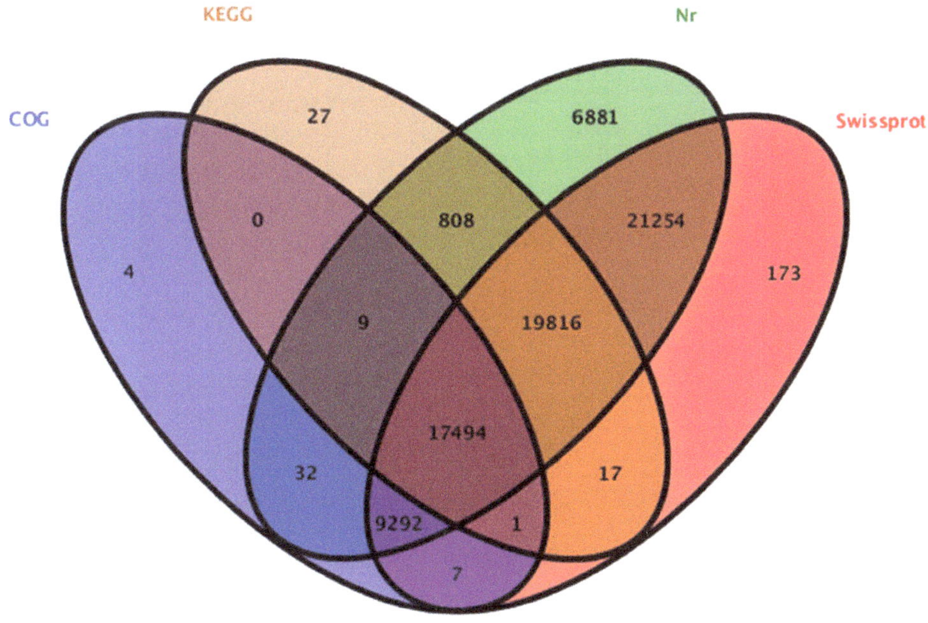

Figure 3. Venn diagram of annotation results against Nr, SwissProt, COG and KEGG databases. The number in each color block indicates the number of unigenes that is annotated by single or multiple databases.

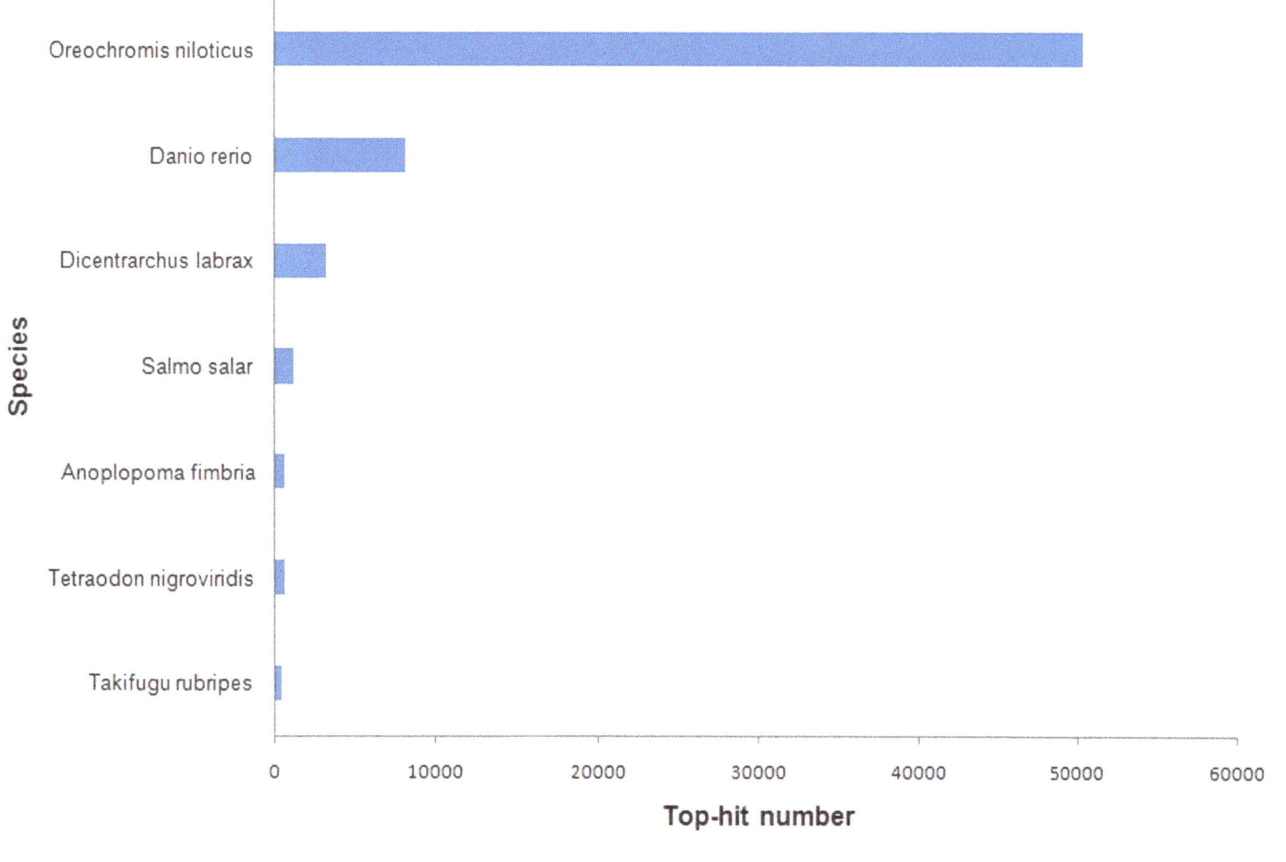

Figure 4. Top-hit species distribution for sequences from *Trachinotus ovatus* submitted BLASTX against the NCBI-Nr database. Only fish species were present in this study.

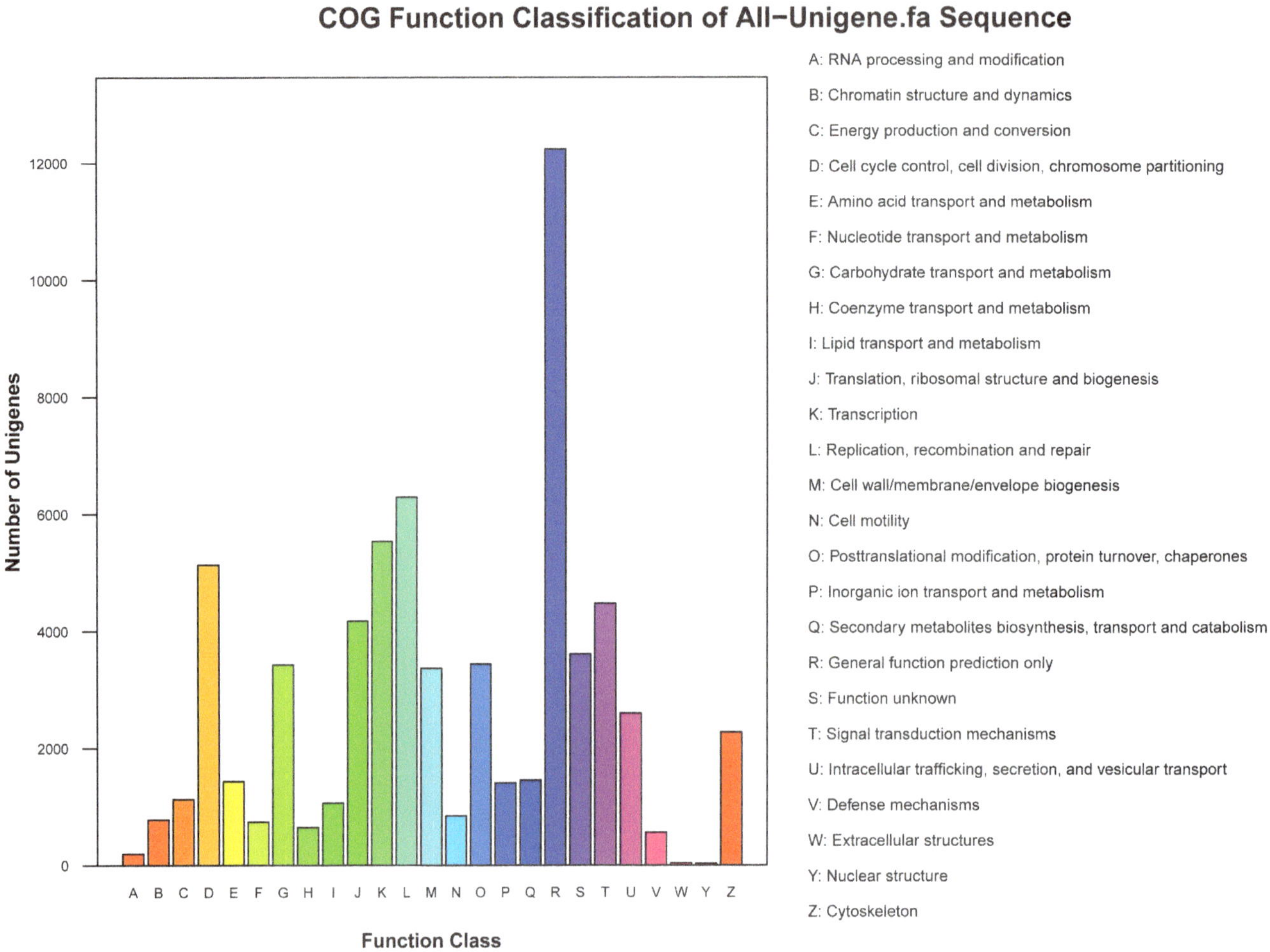

Figure 5. Clusters of Orthologous Groups (COG) functional classification of the *Trachinotus ovatus* transcriptome.

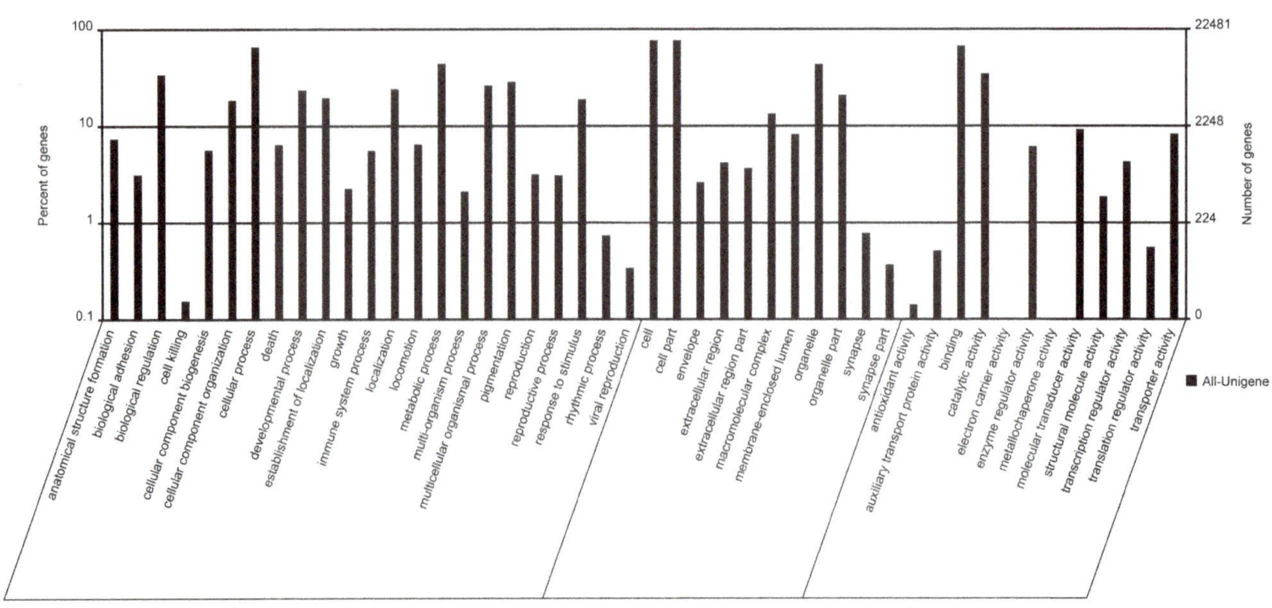

Figure 6. Gene Ontology (GO) analysis and functional classification of the *Trachinotus ovatus* transcriptome.

aquaculture. Elucidation of the regulatory mechanisms behind growth control will help achieve these goals. Similar to other vertebrates, fish growth is also controlled by growth hormone/insulin-like growth factor I (GH/IGFI) axis [45]. In this study, the genes coding for GH, IGFs and their receptors were identified in the transcriptome of *Trachinotus ovatus* (Text S2). Meanwhile, the signal transducers associated with GH/GHR signaling pathway and IGF/IGFR signaling pathway were also found (Text S2, Table S4). Furthermore, we also discovered genes that regulate GH and IGFs, such as the growth hormone releasing hormone (GHRH), pituitary adenylate cyclase-activating polypeptide (PACAP), somatostatins, insulin-like growth factor-binding proteins (IGFBP) (Text S2). GHRH and PACAP are the main regulators that promote GH release, while somatostatins play an inhibitory role [46]. IGFBP act as the carrier proteins for IGF [47]. With these sequences, a full view of the endocrine regulatory network of growth axis in *Trachinotus ovatus* can be constructed. Nevertheless, further molecular biology and biochemical studies are required to confirm the roles that these gene products play in *Trachinotus ovatus* growth.

Growth is a highly complex process that involves the regulation of appetite, muscle growth, weight gain and, protein and lipid metabolism. We found a few genes related to appetite and muscle development, including neuropeptide Y (NPY), pro-opiomelanocortin (POMC), leptin and myostatin (Text S2). NPY and POMC exert opposite roles in stimulating or inhibiting feeding, respectively [48]. Leptin has been recognized to regulate energy intake and energy expenditure [49]. Myostatins act as negative regulators of muscle growth [50]. Polymorphisms in myostatin genes have been identified in several fish species and are associated with growth traits [51–52]. These may be developed to growth-related markers for the molecular marker-assisted breeding.

In addition, signaling pathways involving carbohydrate, protein, and lipid metabolism were found in the KEGG pathway analysis (Table S2). Studies of these metabolic signaling pathways in combination with previous studies on growth performance and digestion [9–11] will expand our understanding of growth control in *Trachinotus ovatus* and will help optimize the artificial feeding for its aquaculture.

Identification of Sequences Related to Immunity

Aquaculture practices have been found to reduce genetic variability in hatchery-reared stocks, which may result in loss of disease resistance [53–54]. In recent years, frequent outbreaks of infectious diseases in fish farming have become a great threat to the *Trachinotus ovatus* industry [5–6]. However, we have little knowledge on the immune system of this species. In order to establish efficient defenses against pathogenic infections, the first step is to understand the immune system of *Trachinotus ovatus* and to identify genes and pathways involved in its immune response. GO classification revealed that 4,249 and 1,250 unigenes belonged to two subcategories: response to stimulus and immune system process respectively (Table S5, Table S6). KEGG pathways analysis revealed 29 immune-related pathways, including the toll-like receptor (TLR) signaling pathway, antigen processing and presentation, intestinal immune network for Immunoglobulin A (IgA) production, natural killer cell mediated cytotoxicity and so on (Table S7). These signaling pathways provide comprehensive information for the understanding of the *Trachinotus ovatus* immune system.

The toll-like receptor family is an important group of pattern-recognition receptors which are expressed on antigen-presenting cells, participating in innate immune responses and the subsequent promotion of adaptive immune responses [55]. We identified 10

Table 2. Distribution of identified SSRs using the MISA software.

Motif	Repeat numbers												Total	%
	4	5	6	7	8	9	10	11	12	13	14	≥15		
Di-	0	0	6560	4079	3480	4610	3121	655	23	0	0	1	22529	58.07
Tri-	0	6299	3434	2235	116	0	5	1	1	1	2	4	12098	31.19
Tetra-	1701	1095	144	3	3	0	0	0	1	0	0	1	2948	7.60
Penta-	656	56	7	2	2	3	1	0	0	0	0	1	728	1.88
Hexa-	391	27	22	5	4	5	5	3	8	0	5	16	491	1.27
Total	2748	7477	10167	6324	3605	4618	3132	659	33	1	7	23	38794	100.00
%	7.08	19.27	26.21	16.30	9.29	11.90	8.07	1.70	0.09	0	0.02	0.06	100.00	-

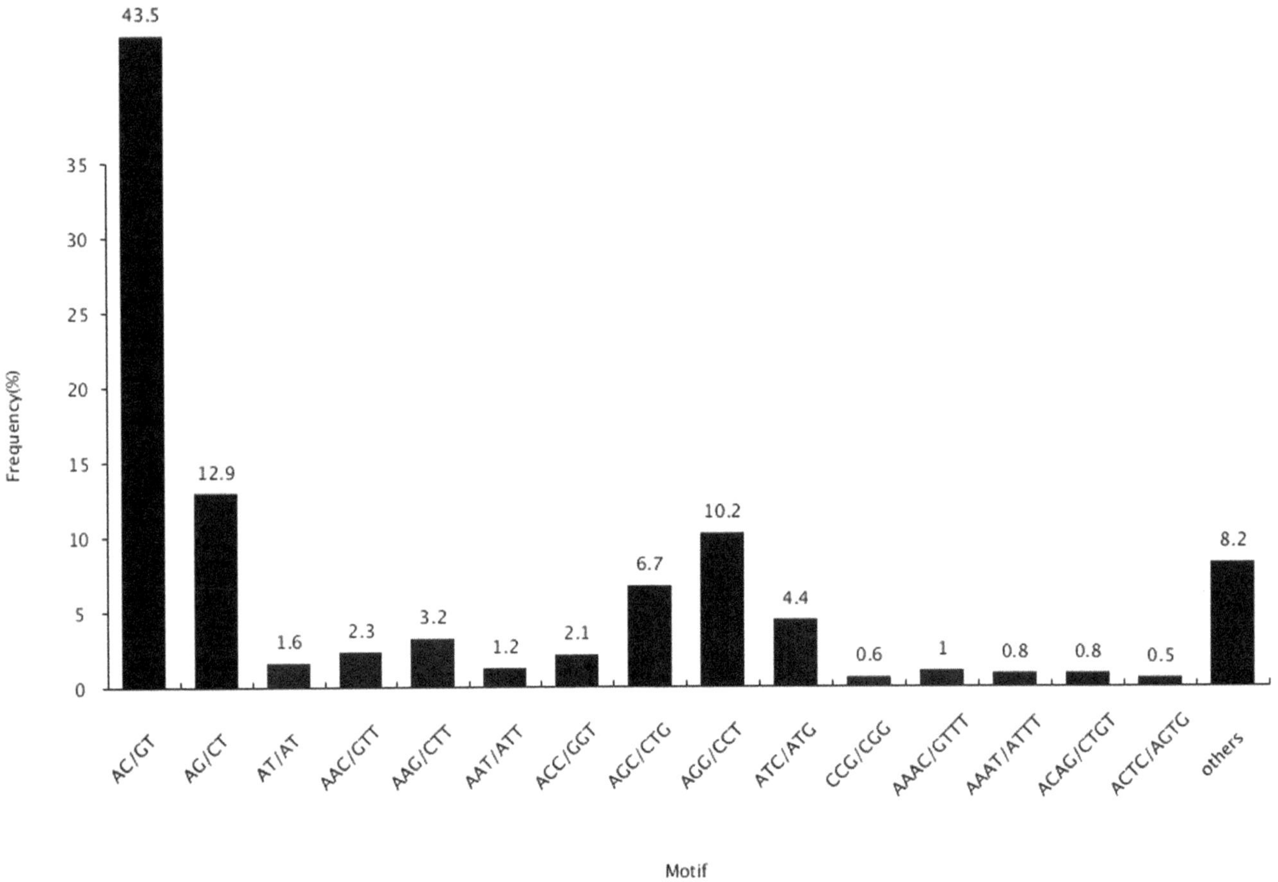

Figure 7. Frequency of classified repeat types of SSRs.

different TLR transcripts in our transcriptome dataset. Moreover, we also found the genes belong to the TLR signaling pathway, such as myeloid differentiation factor 88 (Text S3). The major histocompatibility complex (MHC) are important molecules that help the immune system recognize foreign substances by binding peptide fragments derived from pathogens and presenting them to T cells [56]. Transcripts encoding MHC class I and II were also identified in the transcriptome dataset (Text S3). MHC genes have been considered as candidate markers associated with disease resistance as well, as it were found to be highly polymorphic in teleosts [57–59]. Further analysis of TLR and MHC genes will provide insights into *Trachinotus ovatus* immune defense. Additionally, some transcripts coding for cytokines were discovered in this study, including 18 different interleukins (Text S3). Cytokines play important roles in mediating immune responses in vertebrates. Many recombinant cytokines have been approved for clinical use [60], and these small soluble proteins also have the potential to be used as immunostimulants in aquaculture. However, their primary structures are not conserved among species, and their cDNA sequences are not easily cloned in non-model fish species by traditional homology cloning strategies. These sequence information will facilitate functional studies of cytokines in *Trachinotus ovatus*.

SSR Discovery: Distribution and Frequencies

Simple sequence repeats, also known as microsatellites, are tandem repeats of 1–6 nucleotides found in all prokaryotic and eukaryotic genomes [61]. They have been widely used in genetic linkage map construction, population genetic studies, molecular marker-assisted breeding and so on [62]. As there is no genome sequence information, transcriptomes are a useful alternative for SSR markers discovery of in *Trachinotus ovatus*. In our study, all of assembled sequences were applied to exploit potential SSRs using MISA software [25]. In total, 38,794 SSRs were recognized in 30,295 sequences (19.41%), and 6,616 sequences (4.24%) contained more than one SSR. There were 2,118 SSRs present in compound formation. Di-nucleotide repeats were the most abundant SSR motif (58.07%), followed by tri- (31.19%), tetra- (7.60%), penta- (1.88%) and hexa- (1.27%) nucleotide repeats (Table 2). A summary on the frequency of SSRs with different numbers of tandem repeats were presented in Figure 7. SSRs with six tandem repeats occurred at the highest frequency (26.21%), followed by those with five tandem repeats (19.27%), seven tandem repeats (16.30%), and nine tandem repeats (11.90%). Among the di-nucleotide repeats, the AT/GT motif accounted for the majority of SSRs (43.5%), followed by the AG/CT di-nucleotide repeat motif (12.9%), and the AT/TA di-nucleotide repeat motif (1.6%). The most frequent motifs in the tri-nucleotide repeats were ACG/CTG (10.2%).

Polymorphism Test of SSR Markers

To detect the genetic polymorphism of these microsatellite markers in the transcriptome, 331 pairs of SSR primers were designed for validation. In our study, 290 primer pairs were successfully amplified using *Trachinotus ovatus* genomic DNA (File S8). A large number of these microsatellite-containing

Table 3. Characteristics of 16 polymorphic microsatellite loci.

Unigene ID	SSRs	No. of alleles	Ho	He	PIC	BLASTX, sequence description
0102325	(TG)9	8	0.165	0.835	0.780	similar to KIAA1486 protein
0102780	(CA)9	3	0.663	0.337	0.292	No
0041000	(CTG)20	7	0.352	0.648	0.576	CCAAT/enhancer-binding protein beta 2
0096447	(GT)9	5	0.292	0.708	0.650	transforming protein RhoA-like
0141384	(AC)9	14	0.152	0.848	0.807	No
0095395	(TG)10	4	0.261	0.740	0.674	No
0120853	(CA)9	5	0.349	0.651	0.581	No
0118815	(CAGCTC)12	8	0.210	0.790	0.744	similar to expressed protein
0128444	(TG)10	4	0.281	0.719	0.655	No
0019285	(CA)9	5	0.257	0.743	0.682	E3 ubiquitin-protein ligase RNF220-like
0090156	(CAGCTC)12	7	0.226	0.774	0.728	serine/threonine protein kinase
0110112	(TG)9	2	0.492	0.508	0.375	No
0141617	(CA)10	4	0.352	0.648	0.570	retinoic acid receptor responder protein 3-like
0106655	(TG)9	4	0.457	0.543	0.460	cell adhesion molecule 4-like
0091717	(GT)9	4	0.335	0.666	0.589	protein FAM40B-like
0108204	(AC)10	10	0.246	0.754	0.716	golgin subfamily B member 1-like

Ho, observed heterozygosity; HE, expected heterozygosity; PHWE, Hardy-Weinberg probability; PIC, polymorphic information content.

transcripts could be annotated, most of which were associated with the metabolic pathways and immune system. Using the 290 primer pairs, 30 *Trachinotus ovatus* fish were analyzed, and 16 highly polymorphic microsatellite loci were found (Table 3). The rate of polymorphic microsatellites isolated in this study was low, perhaps because the tested individuals came from the same fish farm. More polymorphic microsatellites may be developed if more geographically distant populations were examined. Of the 16 polymorphic markers, the number of alleles per locus ranged from 2 to 14, with an average of 5.8. The observed and expected heterozygosities ranged from 0.152 to 0.663 and from 0.337 to 0.848, with an average of 0.318 and 0.682, respectively. The PIC values of 13 microsatellite loci were higher than 0.5, while that of the other 3 loci were between 0.29 and 0.46. Amongst the 16 SSR markers developed, 10 were located in unigenes annotated with known functions. The mammalian counterparts of these unigenes are involved in the regulation of various important cellular processes. For example, CCAAT/enhancer-binding protein beta 2 is an important transcription factor that controls the expression of genes involved in immune responses [63]. Retinoic acid receptor responder protein 3 is thought to act as a tumor suppressor or growth regulator [64]. Fam40b is essential for the differentiation of murine embryonic stem cells [65]. The potential links between these expressed sequence tag-simple sequence repeats (EST-SSRs) and some interesting phenotypes should be explored in further study.

However, although the remaining 6 polymorphic microsatellite loci were came from transcribed regions, their sequences showed no similarity to any known protein when BLAST searches were performed on nucleotide and protein database. A possible explanation is that these microsatellite-containing sequences belong to microsatellite-containing transcripts. Another possible explanation may be that some sequences may be derived from poorly conserved untranslated regions. EST-SSRs have been recognized to be conserved and have cross-species transferability [66]. Therefore, these transcriptome-derived microsatellite markers are of high value, and can be employed for *Trachinotus ovatus* population genetic studies and linkage map construction in the future.

Conclusion

For the first time, the whole transcriptome was studied in a fish species belonging to the genus *Trachinotus*. A large number of genes related to reproduction, growth and metabolism, and immune response were identified in the transcriptome dataset, providing abundant genomic data for future studies on the molecular mechanisms behind important physiological processes in *Trachinotus ovatus*. At the same time, many microsatellites were discovered and 16 polymorphic microsatellite loci were developed in this study, which are useful markers for genetic studies and marker-assisted selection in *Trachinotus ovatus*.

Supporting Information

Figure S1 Relative mRNA amounts of DMRT1 in adult gonads of *Trachinotus ovatus*.

Table S1 The functional classification of the COG classes for the transcriptome of *Trachinotus ovatus*.

Table S2 KEEG pathway analysis for the transcriptome of *Trachinotus ovatus*.

Table S3 GO terms associated with reproduction.

Table S4 IGF/IGFR signaling pathways.

Table S5 GO terms associated with response to stimulus.

Table S6 GO terms associated with immune system process.

Table S7 KEEG pathway analysis for immune response.

Table S8 Information of 290 SSRs primer pairs derived from unigenes.

Text S1 Putative sequences related to reproduction in the transcriptome of *Trachinotus ovatus*.

Text S2 Putative sequences related to growth and metabolism in the transcriptome of *Trachinotus ovatus*.

Text S3 Putative sequences related to immunity in the transcriptome of *Trachinotus ovatus*.

Author Contributions

Conceived and designed the experiments: XZZ LSS ZY. Performed the experiments: XZZ XL LQY LZH. Analyzed the data: XZZ WDD FC LSS. Contributed reagents/materials/analysis tools: LXC LHR. Wrote the paper: XZZ LSS ZY.

References

1. Jory DE, Iversen ES, Lewis RH (1985) Culture of fishes of the genus *Trachinotus* (Carangidae) in the western Atlantic: prospects and problems. J World Aquacult Soc 16(1–4): 87–94.
2. FAO website, Global Aquaculture Production (online query). Available: http://www.fao.org/fishery/statistics/global-aquaculture-production/query/en. Accessed 2014 May 9.
3. Froese R, Pauly D (2014) Fishbase, a Global Information System on Fishes: http://www.fishbase.org. World Wide Web electronic publication. Accessed 2014 April.
4. Ou YJ, Ji L, Li JE, Fan CY, Wang G (2013) Correlation analysis of major morphometric traits and body weight of selective group at different month ages of *Trachinotus ovatus*. Journal of Fisheries of China 37(7): 961–969 (in Chinese).
5. Zhou YC, Zhu CH, Zhang B, Su YQ (2001) Isolation and prevention of the pathogen causing large scale death on *Trachinotus ovatus*. Mar Sci 25 (4): 40–44 (in Chinese).
6. Wang R, Feng J, Su Y, Ye L, Wang J (2013) Studies on the isolation of *Photobacterium damselae* subsp. *piscicida* from diseased golden pompano (*Trachinotus ovatus* Linnaeus) and antibacterial agents sensitivity. Vet Microbiol 162(2–4): 957–963.
7. Peng M, Chen XH, Jiang WM, Chen XL, Yang CL, et al. (2011) Optimization of AFLP Fingerprinting and Screening of Primer Pairs in Genetic Diversity Analysis of *Trachinotus ovatus*. Southwest China Journal of Agricultural Sciences 24 (3): 1152–1155 (in Chinese).
8. Sun LY, Zhang DC, Jiang SG, Guo HY, Zhu CY (2011) Isolation and characterization of 21 polymorphic microstatellites in golden pompano *Trachinotus ovatus*. Conser Genet Res 5(4): 1107–1109.
9. Zhang Q, Yu HR, Tong T, Tong WP, Dong LF, et al. (2014) Dietary supplementation of Bacillus subtilis and fructooligosaccharide enhance the growth, non-specific immunity of juvenile ovate pompano, *Trachinotus ovatus* and its disease resistance against *Vibrio vulnificus*. Fish Shellfish Immun 38(1): 7–14.
10. Ma ZH, Guo HY, Zheng PL, Wang L, Jiang SG, et al. (2014) Ontogenetic development of digestive functionality in golden pompano *Trachinotus ovatus* (Linnaeus 1758). Fish Physiol Biochem (in press).
11. Lin SM, Mao SH, Guan Y, Lin X, Luo L (2012) Dietary administration of chitooligosaccharides to enhance growth, innate immune response and disease resistance of *Trachinotus ovatus*. Fish Shellfish Immun 32(5): 909–13.
12. Garg R, Patel RK, Tyagi AK, Jain M (2011) *De novo* assembly of chickpea transcriptome using short reads for gene discovery and marker identification. DNA research 18(1): 53–63.
13. Blanca J, Cañizares J, Roig C, Ziarsolo P, Nuez F, et al. (2011) Transcriptome characterization and high throughput SSRs and SNPs discovery in *Cucurbita pepo* (Cucurbitaceae). BMC genomics 12(1): 104.
14. Ji PF, Liu GM, Xu J, Wang XM, Li JT, et al. (2012) Characterization of common carp transcriptome: sequencing, *de novo* assembly, annotation and comparative genomics. PloS one 7(4): e35152.
15. Ma KY, Qiu GF, Feng JB, Li JL (2012) Transcriptome analysis of the oriental river prawn, *Macrobrachium nipponense* using 454 pyrosequencing for discovery of genes and markers. PloS one 7(6): e39727.
16. Gao ZX, Luo W, Liu H, Zeng C, Liu XL, et al. (2012) Transcriptome analysis and SSR/SNP markers information of the blunt snout bream (*Megalobrama amblycephala*). PloS one 7(8): e42637.
17. Li P, Deng WQ, Li TH, Song B, Shen YH (2013) Illumina-based *de novo* transcriptome sequencing and analysis of Amanita *exitialis basidiocarps*. Gene 532(1):63–71.
18. Grabherr MG, Haas BJ, Yassour M, Levin JZ, Thompson DA, et al. (2011) Full-length transcriptome assembly from RNA-Seq data without a reference genome. Nat Biotechnol 29(7): 644–652.
19. Altschul SF, Gish W, Miller W, Myers EW, Lipman DJ (1990) Basic local alignment search tool. J Mol Biol 215(3): 403–410.
20. Boeckmann B, Bairoch A, Apweiler R, Blatter MC, Estreicher A, et al. (2003) The SWISS-PROT protein knowledgebase and its supplement TrEMBL in 2003. Nucleic Acids Res 31(1):365–370.
21. Tatusov RL, Fedorova ND, Jackson JD, Jacobs AR, Kiryutin B, et al. (2003) The COG database: an updated version includes eukaryotes. BMC Bioinformatics. 4:41.
22. Kanehisa M, Araki M, Goto S, Hattori M, Hirakawa M, et al. (2008) KEGG for linking genomes to life and the environment. Nucleic Acids Res 36: D480–D484.
23. Conesa A, Götz S, García-Gómez JM, Terol J, Talón M, et al. (2005) Blast2GO: a universal tool for annotation, visualization and analysis in functional genomics research. Bioinformatics 21(18): 3674–3676.
24. Ye J, Fang L, Zheng Hk, Zhang Y, Chen J, et al. (2006) WEGO: a web tool for plotting GO annotations. Nucleic Acids Res 34 (suppl 2): W293–W297.
25. Thiel T, Michalek W, Varshney RK, Graner A (2003) Exploiting EST databases for the development of cDNA derived microsatellite markers in barley (*Hordeum vulgare*L.) Theor Appl Genet 106(3): 411–422.
26. Rozen S, Skaletsky H (2000) Primer3 on the WWW for general users and for biologist programmers. Methods Mol Biol 132: 365–386.
27. Howe JR, Klimstra DS, Cordon-Cardo C (1997) DNA extraction from paraffin-embedded tissues using a salting-out procedure: A reliable method for PCR amplification of archival material. Histol Histopathol 12(3): 595–601.
28. Raymond M, Rousset F (1995) GENEPOP (Version 1.2): population genetics software for exact tests and ecumenicism. J Hered 86 (3): 248–249.
29. Van Oosterhout C, Hutchinson WF, Wills DPM, Shipley P (2004) MICRO-CHECKER: software for identifying and correcting genotyping errors in microsatellite data. Mol Ecol Notes 4(3): 535–538.
30. Rice WR (1989) Analyzing tables of statistical tests. Evolution 43(1): 223–225.
31. Long Y, Li Q, Zhou BL, Song GL, Li T, et al. (2013) *De novo* assembly of mud loach (*Misgurnus anguillicaudatus*) skin transcriptome to identify putative genes involved in immunity and epidermal mucus secretion. PLoS One 8(2): e56998.
32. Salem M, Rexroad CE, Wang J, Thorgaard GH, Yao J (2010) Characterization of the rainbow trout transcriptome using Sanger and 454-pyrosequencing approaches. BMC Genomics 11: 564.
33. Pereiro P, Balseiro P, Romero A, Dios S, Forn-Cuni G, et al. (2012) High-throughput sequence analysis of turbot (*Scophthalmus maximus*) transcriptome using 454-pyrosequencing for the discovery of antiviral immune genes. PloS one 7(5): e35369.
34. Ashburner M, Ball CA, Blake JA, Botstein D, Butler H, et al. (2000) Gene Ontology: tool for the unification of biology. Nat Genet 25(1): 25–29.
35. Levavi-Sivan B, Bogerd J, Mañanós EL, Gómez A, Lareyre JJ (2010) Perspectives on fish gonadotropins and their receptors. Gen Comp Endocrinol 165(3): 412–437.
36. Mylonas CC, Fostier A, Zanuy S (2010) Broodstock management and hormonal manipulations of fish reproduction. Gen Comp Endocrinol 165(3): 516–534.
37. Rance NE, Dacks PA, Mittelman-Smith MA, Romanovsky AA, Krajewski-Hall SJ (2010) Modulation of body temperature and LH secretion by hypothalamic KNDy (kisspeptin, neurokinin B and dynorphin) neurons: a novel hypothesis on the mechanism of hot flushes. Front Neuroendocrinol 34(3): 211–227.
38. Yaron Z, Levavi-Sivan B (2011) Endocrine Regulation of Fish Reproduction. In: Farrell A.P., (ed.), Encyclopedia of Fish Physiology: From Genome to Environment, volume 2, pp. 1500–1508. San Diego: Academic Press.
39. Baroiller JF, D'Cotta H, Saillant E (2009) Environmental effects on fish sex determination and differentiation. Sex Dev 3(2–3): 118–135.
40. Guo YQ, Cheng HH, Huang X, Gao S, Yu HS, et al. (2006) Gene structure, multiple alternative splicing, and expression in gonads of zebrafish Dmrt1. Biochem Biophys Res Commun 330(3): 950–957.
41. Veith AM, Schäfer M, Klüver N, Schmidt C, Schultheis C, et al. (2006) Tissue-specific expression of dmrt genes in embryos and adults of the platyfish *Xiphophorus maculatus*. Zebrafish 3(3): 325–337.
42. Ijiri S, Kaneko H, Kobayashi T, Wang DS, Sakai F, et al. (2008) Sexual dimorphic expression of genes in gonads during early differentiation of a teleost fish, the Nile tilapia *Oreochromis niloticus*. Biol Reprod 78(2): 333–341.
43. Wang DS, Zhou LY, Kobayashi T, Matsuda M, Shibata Y, et al. (2008) Doublesex- and Mab-3-related transcription factor-1 repression of aromatase transcription, a possible mechanism favoring the male pathway in tilapia. Endocrinology 151(3): 1331–1340.

44. Bradley KM, Breyer JP, Melville DB, Broman KW, Knapik EW, et al. (2011) An SNP-Based Linkage Map for Zebrafish Reveals Sex Determination Loci. G3 (Bethesda) 1(1):3–9.
45. Rousseau K, Dufour S (2007) Comparative aspects of GH and metabolic regulation in lower vertebrates. Neuroendocrinology 86(3): 165–174.
46. Li WS, Lin HR (2010) The endocrine regulation network of growth hormone synthesis and secretion in fish: emphasis on the signal integration in somatotropes. Sci China Life Sci 53(4): 462–470.
47. Duan CM, Xu QJ (2005) Roles of insulin-like growth factor (IGF) binding proteins in regulating IGF actions. Gen Comp Endocrinol 142(1–2): 44–52.
48. Valassi E, Scacchi M, Cavagnini F (2008) Neuroendocrine control of food intake. Nutr Metab Cardiovasc Dis 18(2): 158–168.
49. Morton GJ, Schwartz MW (2011) Leptin and the central nervous system control of glucose metabolism. Physiol Rev 91(2): 389–411.
50. Gabillard JC, Biga PR, Rescan PY, Seiliez I (2013) Revisiting the paradigm of myostatin in vertebrates: insights from fishes. Gen Comp Endocrinol 194: 45–54.
51. Nadjar-Boger E, Funkenstein B (2011) Myostatin-2 gene structure and polymorphism of the promoter and first intron in the marine fish *Sparus aurata*: evidence for DNA duplications and/or translocations. BMC Genet 12: 22.
52. Sun YH, Yu XM, Tong JG (2012) Polymorphisms in Myostatin Gene and Associations with Growth Traits in the Common Carp (*Cyprinus carpio* L.). Int J Mol Sci 13(11): 14956–14961.
53. Lind CE, Evans BS, Knauer J, Taylor JU, Jerry DR (2009) Decreased genetic diversity and a reduced effective population size in cultured silver-lipped pearl oysters (*Pinctada maxima*). Aquaculture 286 (1–2): 12–19.
54. Meng ZN, Yang S, Fan B, Wang L, Lin HR (2012) Genetic variation and balancing selection at MHC class II exon 2 in cultured stocks and wild populations of orange-spotted grouper (*Epinephelus coioides*). Genet Mol Res 11(4): 3869–3881.
55. Akira S, Uematsu S, Takeuchi O (2006) Pathogen recognition and innate immunity. Cell 124(4): 783–801.
56. Neefjes J, Jongsma ML, Paul P, Bakke O (2011) Towards a systems understanding of MHC class I and MHC class II antigen presentation. Nat Rev Immunol 11(12): 823–836.
57. Langefors A, Lohm J, Grahn M, Andersen O, von ST (2001) Association between major histocompatibility complex class IIB alleles and resistance to *Aeromonas salmonicida* in Atlantic salmon. Proc Biol Sci 268(1466): 479–485.
58. Zhang YX, Chen SL, Liu YG, Sha ZX, Liu ZJ (2006) Major histocompatibility complex class IIB allele polymorphism and its association with resistance/susceptibility to *Vibrio anguillarum* in Japanese flounder (*Paralichthys olivaceus*). Mar Biotechnol (NY) 8(6): 600–610.
59. Xu TJ, Chen SL, Ji XS, Sha ZX (2009) Molecular cloning, genomic structure, polymorphism and expression analysis of major histocompatibility complex class IIA and IIB genes of half-smooth tongue sole (*Cynoglossus semilaevis*). Fish Shellfish Immun 27(2): 192–201.
60. Sirko A, Vanek T, Gora-Sochacka A, Redkiewicz P (2011) Recombinant cytokines from plants. Int J Mol Sci 12(6): 3536–3552.
61. Zane L, Bargelloni L, Patarnello T (2002) Strategies for microsatellite isolation: a review. Mol Ecol 11(1): 1–16.
62. Selkoe KA, Toonen R J (2006) Microsatellites for ecologists: a practical guide to using and evaluating microsatellite markers. Ecol Lett 9(5): 615–629.
63. Campion CG, Labrie M, Grosset AA, St-Pierre Y (2014) The CCAAT/Enhancer-Binding Protein Beta-2 Isoform (CEBPβ-2) Upregulates Galectin-7 Expression in Human Breast Cancer Cells. PloS one 9(5): e95087.
64. DiSepio D, Ghosn C, Eckert RL, Deucher A, Robinson N, et al. (1998) Identification and characterization of a retinoid-induced class II tumor suppressor/growth regulatory gene. Proc Natl Acad Sci U S A 95(25): 14811–14815.
65. Wagh V, Doss MX, Sabour D, Niemann R, Meganathan K, et al. (2014) Fam40b is required for lineage commitment of murine embryonic stem cells. Cell Death Dis 5: e1320.
66. Xie F, Sun G, Stiller JW, Zhang B (2011) Genome-wide functional analysis of the cotton transcriptome by creating an integrated EST database. PLoS One 6(11): e26980.

8

Rapid Fishery Assessment by Market Survey (RFAMS) – An Improved Rapid-Assessment Approach to Characterising Fish Landings in Developing Countries

William T. White[1]*, Peter R. Last[1], Dharmadi[2], Ria Faizah[2], Umi Chodrijah[2], Rik C. Buckworth[3], Catherine M. Dichmont[3]

1 CSIRO Oceans & Atmosphere Flagship, Hobart, Tasmania, Australia, 2 Research Centre for Fisheries Management and Conservation, Agency for Marine and Fisheries Research and Development, Ministry of Marine Affairs and Fisheries, Jakarta, Indonesia, 3 CSIRO Oceans & Atmosphere Flagship, Dutton Park, Queensland, Australia

Abstract

The complex multi-gear, multi-species tropical fisheries in developing countries are poorly understood and characterising the landings from these fisheries is often impossible using conventional approaches. A rapid assessment method for characterising landings at fish markets, using an index of abundance and estimated weight within taxonomic groups, is described. This approach was developed for contexts where there are no detailed data collection protocols, and where consistent data collection across a wide range of fisheries types and geographic areas is required, regardless of the size of the site and scale of the landings. This methodology, which was demonstrated at seven fish landing sites/fish markets in southern Indonesia between July 2008 and January 2011, provides a rapid assessment of the abundance and diversity in the wild catch over a wide variety of taxonomic groups. The approach has wider application for species-rich fisheries in developing countries where there is an urgent need for better data collection protocols, monitoring future changes in market demographics, and evaluating health of fisheries.

Editor: Konstantinos I Stergiou, Aristotle University of Thessaloniki, Greece

Funding: This project was funded by the Australian Centre for Agricultural Research (ACIAR, website: aciar.gov.au) and the Wealth from Oceans Flagship (website: www.csiro.au/Organisation-Structure/Flagships/Wealth-from-Oceans-Flagship.aspx), project number FIS2006\142. The funders had no role in study design, data collection and analysis, decision to publish, or preparation of the manuscript.

Competing Interests: The authors have declared that no competing interests exist.

* Email: william.white@csiro.au

Introduction

Indonesian fisheries are among the largest and most productive worldwide, and are critical to the nation's economic development and in providing food resources to millions of people. In 2004, production from marine capture fisheries in Indonesia was approximated at 4.5 million tonnes with an estimated gross value of US $3.13 million [1]. The total marine capture fisheries production in Indonesia increased to 5.4 million tonnes in 2010 [2]. The high level of employment, which in 2003 was estimated at 3.3 million people directly employed in marine capture fisheries [1], is an important indicator of the value of fisheries to Indonesia. This places Indonesian marine capture fisheries among the top five worldwide in terms of fisheries production. The level of fishing effort in Indonesia is also increasing rapidly, with the number of motorised marine vessels increasing from an estimated 348,425 in 2007 to 390,770 in 2009 [3].

The majority of Indonesia's capture fisheries are considered to be fully or overexploited despite the fact that marine capture production in Indonesia is increasing annually [4]. One of the greatest problems facing Indonesia's Ministry of Marine Affairs and Fisheries is the quality of fishery statistics available. Since Indonesia's fisheries are both multi-gear and multi-species fisheries operating over 80,000 km of coastline, it would not be feasible to collected catch statistics from all catches. As a result, catch statistics are obtained by averaging daily catch from a small subset of vessels and multiplying by the number of vessels and number of fishing days to obtain annual catches [5]. This method has many obvious weaknesses, with a major problem being that illegal, unregulated and unreported (IUU) catches are not accounted for in these statistics [3,6,7].

Catches of many small-scale fisheries are often overlooked, but the cumulative catches from this sector can be very high, particularly in biologically rich and densely populated countries such as Indonesia. In developing countries these small scale-fisheries are also extremely important to the local economy and account for a substantial proportion of the overall capture production [8,9]. Despite this, there are very few studies of multi-gear, multi-species fisheries in tropical regions of the world [10,11,12], hindering the ability to develop adequate management strategies for such areas. [12] undertook comprehensive surveys of the main fish market at Manado, North Sulawesi (Indonesia), which was dominated by small-scale coral reef fisheries. These market surveys recorded species-level composition data where possible and allocated fishes into 10 cm size-classes. This approach allows for an accurate assessment of the landings at this particular market, but obtaining this level of information at much larger and demographically more complex markets would not be possible. There is a need for a more rapid, standardised method of collecting landings information from fish markets or landing sites in tropical regions with diverse fisheries.

In this paper, we present a new Rapid Assessment Protocol: Rapid Fisheries Assessment by Market Survey (RFAMS) for assessing catch composition at fish landing sites and markets in the circumstances where no detailed data collection protocols exist. Present methods of recording catch composition data in developed and small-scale fisheries are not suitable for countries such as Indonesia which have very large, complex, multi-gear and multi-species fisheries. We emphasise the importance of collecting accurate taxonomic information, albeit not to species level, to enable biodiversity to be assessed. The major aim of this new data collection methodology is to provide data-poor regions with a tool for being able to efficiently characterise the landings at different fish landing sites and markets regardless of their physical size and the scale of the landings. This method is considered to be widely applicable to fisheries researchers and managers in data-poor regions where there is no existing, or minimal formalised, market data collection protocols. The data collected using this approach will also provide important benchmarks for monitoring future changes in these fisheries.

Methods

1. Ethics Statement

All marine life examined in this study were landed from small-scale fisheries in Indonesia and were already dead upon inspection. Permission to undertake surveys in Indonesia was granted by the Research Centre for Fisheries Management and Conservation in Jakarta as part of a collaborative project funded by the Australian Centre for International Agricultural Research (project code FIS2006\142).

2. Study Area

Seven fish landing sites were surveyed across five provinces in southern Indonesia, i.e. West Java (Pelabuhanratu), Central Java (Cilacap and Sadeng), East Java (Muncar and Pacitan), Bali (Kedonganan) and West Nusa Tenggara (Tanjung Luar in Lombok) (Fig. 1). Three of these sites, Muncar, Pacitan and Sadeng, were surveyed 3 or 4 times, while the remaining sites were surveyed between 6 and 9 times during the study. Across these 7 sites, a total of 60 daily surveys were completed between July 2008 and January 2011.

Pelabuhanratu is a medium-sized port on the south coast of West Java with a diverse range of fisheries operating, from *bagans* (light traps) to oceanic longlining. Cilacap is a large port on the south coast of Central Java which is dominated by a tuna longline fishery. Sadeng is a small port on the south coast of Central Java, east of Cilacap, with only small landings from several different fisheries. Pacitan is a small port on the southwest coast of East Java which is dominated by a trolling fishery targeting small tuna. Muncar is a large port on the east coast of East Java which is dominated by a purse-seine fishery for lemuru (sardines). Kedonganan is a medium to small landing site at Jimbaran Bay in southern Bali which has a diverse range of fisheries targeting a variety of fish and invertebrate groups. Tanjung Luar is a medium-sized landing site on the east coast of Lombok which is best known for its longline fishery targeting large sharks and large mobulid bycatch from the tuna gillnet fisheries.

3. The Rapid Fisheries Assessment by Market Survey (RFAMS) methodology

The data sheets used for RFAMS can be found in Datasheet S1. The initial information recorded is the fish market (or landing site/port), data recorder (person or group), date and time of survey (e.g. 7:00–9:00 am). The total estimated weight of the landings was classified into one of five categories, i.e. <100 kg, 100–500 kg, 500 kg to 1 tonne, 1–10 tonnes and>10 tonnes. Where possible, a total estimated weight was also provided calculated by summing the estimated weights in each of the taxonomic categories (see below). Where possible, the fleet size was recorded by determining the approximate number of boats in each of six categories based on boat size (note GT = gross tonnes): <5 GT, 5–10 GT, 10–30 GT,>30 GT, small with outboards and small without outboards.

To capture data on the broad habitats being fished, the estimated percentage contribution of the catches in each predetermined bathome (depth-related habitat zone) was also recorded. Generalised bathomes were used in this study: 1) freshwater; 2) estuarine; 3) coastal; 4) coral reef; 5) shelf demersal; 6) slope demersal; 7) inshore pelagic; 8) oceanic; 9) aquaculture; 10) intertidal. Likewise, to capture data on the various fisheries involved in the landings, the estimated percentage contribution of the catches in each predetermined fishing method was also recorded. The predetermined fishing methods used for Indonesia in this study were: 1) cast net; 2) diver/spear; 3) dredge; 4) gillnet; 5) handline; 6) longline; 7) traps; 8) purse seine; 9) tidal trap; 10) trawl; 11) beach seine; 12) *bagan* (light trap); 13) by hand; 14) trammel net; 15) hand net; 16) dynamite/poisoning.

Catch composition of the landings was divided into 6 major taxonomic groupings: Cephalopods (e.g. squid, octopus), Other Molluscs (e.g. bivalves, clams), Crustaceans (e.g. prawns, crayfish and crabs), Other (e.g. marine mammals, sea turtles, algae), Elasmobranchs (e.g. sharks, rays) and Teleosts (bony fishes). For each of these major groupings, their percentage contribution to the landings (calculated from estimated weights of each of the taxonomic categories for all groups) and an estimate of the number of species observed was recorded.

Within each of the major taxonomic groupings, catch information was recorded on a family-level basis (e.g. Octopodidae, Penaeidae), except in a few cases where broader categories were used due to difficulties in accurate family-level identification during rapid surveys (e.g. squids, algae) (see Appendix S1). In the case of elasmobranchs, catch information was recorded at a species level due to the detailed knowledge available on this group from the Indonesian fish markets surveyed in this study [13]. In the case of the most abundant or most diverse teleost families, catch data was also recorded for the major genera or species groups, i.e. Carangidae (*Caranx/Carangoides*, *Decapterus*, *Scomberoides*, other carangids), Scombridae (*Thunnus/Euthynnus*, *Katsuwonus/Sarda*, *Scomberomorus/Acanthocybium*, other scombrids). The predetermined taxonomic categories provided in the data sheets (Datasheet S1) were based on surveys undertaken in Indonesia prior to this study (e.g. see [13]) as well as general knowledge of the fish groups exploited in South-east Asia. Blank spaces are also provided for other families encountered which are not listed.

Within each of the taxonomic categories, the data recorded were: an index of abundance, bathome, fishing method and estimated weight. An index of abundance (0–11) was recorded based on a log scale: 0 = 0; 1 = 1; 2 = 2–3; 3 = 4–9; 4 = 10–27; 5 = 28–81; 6 = 82–251; 7 = 250–749; 8 = 750–1999; 9 = 2000–5999; 10 = 6000–19,999; 11 = >20,000 individuals. A log scale index was used as it provides a good indication of abundance and is simple and quick to record which is important in fish markets where large numbers of some fish groups are landed. The corresponding number for the relevant bathome and fishing method for each taxonomic category was recorded (see above). For example, tuna caught by pelagic longlines would have a bathome

Figure 1. Map of eastern Indonesia. Fish landing sites surveyed in this study.

of 8 (oceanic) and fishing method of 6 (longline). If there was multiple bathomes or fishing methods for a particular taxonomic category, only the major bathome and fishing gear was recorded. A total weight (kg) for each taxonomic category recorded was estimated. Estimation of weight was usually determined using known container weights that fish were stored in (e.g. Styrofoam boxes, cane baskets, plastic tubs and drums).

4. Multivariate analyses

The average estimated weight (kg) data for each family (or proxy taxonomic grouping, e.g. squids) at each of the 7 landing sites on each of the survey trips were used to construct a resemblance matrix, following fourth-root transformation, employing the Bray-Curtis similarity coefficient. The resulting matrix was then subjected to non-metric multidimensional scaling (nMDS) ordination and hierarchical cluster analysis using Primer v6 package [14]. A one-way Analysis of Similarities (ANOSIM) was employed to test whether the catch composition of the landings at each site were significantly different. Where significant differences exist, the magnitude of the R-statistic (0, no significant differences; 1, differences highly significant) was used to ascertain the extent to which the catch compositions of a priori groups differed. Multivariate Dispersion (MVDISP) was used to determine the degree of dispersion of each of the sites on ordination plots [15] and Similarity Percentages (SIMPER) was employed to determine the families that typified particular groups and/or contributed most to the dissimilarities between groups [16].

The same approach was also followed to analyse the catch composition by major taxonomic group (Cephalopods, Other molluscs, Crustaceans, Other, Elasmobranchs and Teleosts), by functional group (e.g. reef fish-herbivore, pelagic rays, inshore pelagics) and by a bathome/gear combination (e.g. coral reef/handline, shelf demersal/gillnet). The 'functional group' category was based roughly on the ecological niche that each family best fits, e.g. surgeonfishes (Acanthuridae) into reef fish-herbivore, devilrays and manta rays (Mobulidae) into pelagic rays (see Table S1 for full list). Any bathome/gear combination used which was recorded less than 10 times were put into a single category called 'Rare' (see Table S2 for full list).

Results and Discussion

1. Understanding dynamics of the survey sites

In order to undertake RFAMS successfully, it was important to understand the basic dynamics of the fish landing site or market being surveyed. The most critical information is the timing and duration of landings. Preliminary information on the dynamics of the sites surveyed in this study provided key information on when to commence the survey and approximate daily duration of landings at each site. Landing sites and/or markets varied greatly not only in their catch composition but also in their dynamics. For example, Tanjung Luar is a moderately large landing site on the east coast of Lombok where catches are brought into the site directly from fishing boats or from neighbouring islands via trucks. This market commences at about 6:00 am, usually begins to slow around 8:00 am, and by 10:00 or 11:00 am is completely empty with no more activity. In contrast, the larger fishing port of Cilacap has landings commence around 8:00 am but continues off and on for most of the day, with some catches of prawns not being landed until 7:00 or 8:00 pm. Without accurate information on timing of landings at the sites being surveyed, large components of the catches can be completely missed.

The spatial dynamics at each site also need to be taken into account on each new survey trip to ensure no changes in landing sites have occurred. For example, at Cilacap, the area where catches from the trammel net fishery are landed changed multiple times between 2008 and 2011. Local fisheries officers were able to provide information where landings had been relocated. At this same location, the tuna landings from the tuna longline fishery, which often dominate landings at this site, was relocated to a different location during one survey trip due to dredging in the harbour. Without the local knowledge of these events, it could easily have been assumed no tuna were landed during that particular survey.

2. Taxonomic accuracy

One of the most important facets of RFAMS was the ability to capture good biodiversity information from the catches. Such detailed information is currently not often captured in official catch statistics for tropical fisheries with the vast majority of minor catches lumped into "other fish" categories. However, to obtain this information, a sound knowledge of the landed biota is required. Obtaining accurate species level information is not always possible in such complicated multi-species and multi-gear fisheries without prior taxonomic effort. Nonetheless, family-level or group-level taxonomic data will still provide some useful biodiversity information. Basic taxonomic training is required for users of this method and identification tools such as line drawings and regional photographic guides are useful. In this study, material for a guide to Indonesian fishes was used [17]. Line drawings and keys for the family-level groupings which can be used in Indonesia and adjacent developing countries can be found in Appendix S1. Such guides should ideally be tailored to a specific region based on taxa likely to be encountered, but in their absence broader regional guides are still useful identification tools.

3. Acquisition of catch data

The use of the log-scale index of abundance for each taxonomic category was a rapid and simple method of recording abundance information from such large and complex catches. Capturing abundance indices is a minimum requirement when recording catch data in RFAMS. In the RFAMS conducted in Indonesia, it was also possible to record the bathome, fishing method and an estimated weight for each taxonomic category. This allows for a much more detailed assessment of catch data as it provides good information on the fisheries involved, as well as a volumetric estimate for each taxonomic category. However, it may not be possible to easily record all of these fields in much larger markets where volumes are much larger and/or harder to estimate. In these cases, an index of abundance may be the only field that can be recorded for each category.

In some instances, large quantities of fish were observed in the landing sites which also consisted of a diverse array of fish families and species. For example, the landings from trammel nets at Cilacap often contained several hundred kilos of small fishes (mostly <100 mm length). In such cases, sorting through the whole landing was not practical so the proportions of each family were obtained from a 10–20 kg subset of the catch. These proportions could then be used to estimate the total contribution of each of those families to the total landings present. Some rarer and less commercially important species may be missed in this exercise, but sub-sampling was needed to obtain useful information from these landings without sacrificing too much time examining the entire catch.

4. Important considerations of the RFAMS methodology

It is important to understand any limitations or other considerations of RFAMS so that the resulting analyses don't misinterpret the data collected or draw incorrect conclusions. The main considerations to be taken into account when using RFAMS are discussed below.

Number of data recorders required. While some landing sites are small and can easily be surveyed by one or two people, some landing sites are far larger and more complex and require a different approach. At the landing site of Muncar, the landings are dominated by very large quantities of sardines (*lemuru*) from the purse seine fishery and are spread out over a wide area. During peak fishing times, unloading occurs very quickly to enable the next boat access to the shore and trucks transporting these landings are constantly coming and going. To be able to adequately survey the daily landings at this site, at least four data recorders are required at the various landing areas.

Standing stock. Although some sites have a 100% turnover of landings from day to day, e.g. Tanjung Luar, Sadeng and Pacitan, other sites have a lower turnover of landings with a number of fish remaining in the market area for many days at a time, e.g. Kedonganan and Pelabuhanratu. This can present a problem if surveys are undertaken over consecutive days with some of the fish being recorded multiple times. In busy and complex fish markets, it can be difficult to discriminate freshly landed fish from 'standing stock' which may have been there for multiple days. It is not always possible to access this information from vendors.

Transhipment and translocation of catches. At some landing sites, produce is transported from other nearby locations either by sea or by land. For example, at Pelabuhanratu, large quantities of bigeyes (Priacanthidae) and moon fish (Menidae) are trucked to the market from East Java. Whenever possible, these components should be recorded separately. When dealing with multi-species and multi-gear fisheries, it is very difficult to obtain accurate capture locality information for all catch brought into the landing site or market. At some locations, catches from adjacent islands are landed separately from local catches, but once brought into the market area they are all combined. In order to accurately record capture fishing areas of all of the landings, a large number of data recorders positioned at all landing areas would be required which is not feasible in most cases.

Undertaking surveys for only short periods of time. Since surveys were conducted for short periods of time (usually 1–4 days) at each of the sites per survey trip, temporal variability in landings cannot be fully captured. Economic conditions at the time of a survey trip can influence landings. For example, at Cilacap, landings during some survey trips were much smaller than expected due to increased fuel prices which resulted in many fishers not going to sea but turning to earning income on land, e.g. agriculture. Periods of bad weather is another factor which can strongly influence the size of landings with rough sea conditions preventing smaller vessels from heading to sea to fish.

5. Multivariate analysis of the market survey data

When the average daily estimated weights of each family-level taxonomic category (e.g. Lutjanidae, Penaeidae) for each site on each trip were subjected to MDS ordination, the samples from each site formed relatively discrete groupings in most instances (Fig. 2). The samples from Kedonganan and Tanjung Luar formed the tightest groupings by far which is reflected by the low MVDISP values of 0.44 and 0.69, respectively. In comparison, the samples from Pacitan, Muncar and Sadeng were the most widely dispersed, reflected by high MVDISP values (1.68–1.75). A one-way ANOSIM showed that, overall, the samples for each site differed significantly from each other ($P<0.01$; R-statistic $= 0.823$), and pairwise comparisons between sites showed that all sites differed significantly from each other ($P<0.05$; R-statistic>0.751), except Pacitan and Sadeng which did not differ significantly ($P>0.05$).

Since ANOSIM showed that the sites were significantly different from one another, SIMPER analyses on the same data set were also performed to determine the families which typified each site, and which families were most responsible for causing the difference between two sites (Table 1). It should be noted that while the most abundant families are often the most typifying and distinguishing groups in this type of analysis, they also need to be consistently typifying or distinguishing across all samples (e.g. one large catch of a group on only one occasion will rarely be an important family, even if its total biomass is very large). Table 1

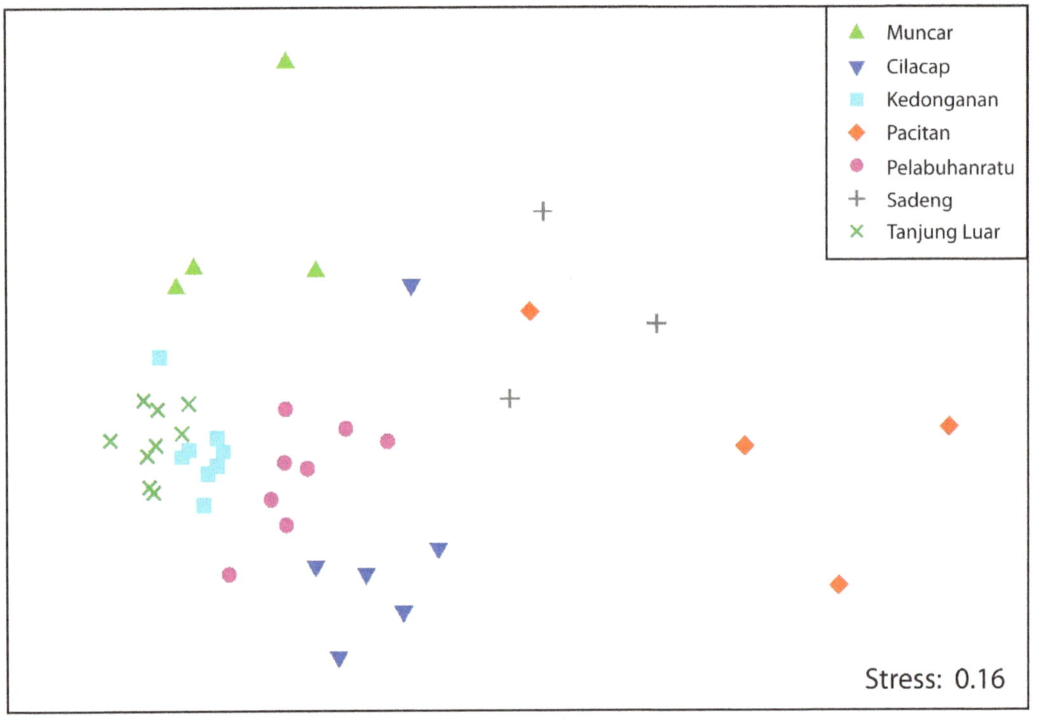

Figure 2. nMDS of family-level data at each site. Non-metric multidimensional scaling ordination of average daily estimated weights of each of the family-level taxonomic categories at each of the 7 sites on each of the survey trips.

summarises the families which typify each site, with Scombridae (tunas), Carangidae (trevallies) and Clupeidae (sardines) the most reoccurring families, and which distinguish each site.

Cluster analysis of the same data also shows that each of the sites differ from one another in terms of catch composition (Fig. 3), but also allows a slightly better way of determining which sites are most different from others. For example, all but two of the Pacitan and Sadeng samples separate very quickly from the samples for the other sites at nearly 30% similarity, with the remaining two falling out with the Cilacap samples (reflecting the only two sampling occasions where large landings of scombrids were landed at these two sites). Pelabuhanratu and Muncar samples separate out at about 45–50% similarity from the Kedonganan and Tanjung Luar samples, which in turn separate out clearly from each other at close to 60% similarity (Fig. 3).

When the average daily estimated weights of each major taxonomic groups (e.g. Teleosts, Elasmobranchs) and also each functional group (e.g. reef fish – herbivore, inshore pelagics) for each site on each trip were subjected to MDS ordination, samples for each site formed relatively discrete groupings, particularly for functional groups. However, they were not as clearly delineated as in Fig. 2 when family-level categories were used. Although ANOSIM showed that in both cases, overall the samples for each site were significantly different, the R-statistic was much lower than when using family-level data (0.67–0.68 vs. 0.82). This suggests that obtaining more taxonomically accurate catch composition data allows for better characterisation of the markets.

When the average daily estimated weights of the landings from each of the major bathome/gear combinations (see Table S2) at each site on each survey trip were subjected to MDS ordination, the samples for each site formed very discrete groupings (Fig. 4). The Kedonganan samples were the most tightly formed groupings (MVDISP = 0.48), indicating that across all of the survey trips, the landings within each of the bathome/gear combinations varied the least. Tanjung Luar and Pelabuhanratu samples also were relatively tight, with MVDISP values of 0.87 and 1.07, respectively. In comparison, the Pacitan and Muncar samples were the most dispersed (MVDISP = 1.65 and 1.66), highlighting the wide variations in landings from various fisheries recorded at these sites. ANOSIM showed that sites were very significantly different from each other ($P<0.001$; R-statistic = 0.87), which highlights that the gear used and bathomes fished are important determinants of the landings at each site. While this is to be expected, is does highlight that RFAMS is providing useful information on the multi-gear fisheries at these sites. SIMPER analysis of this data was also performed and those bathome/gear combinations that typified a site and those that distinguished between two sites are shown in Table 2.

6. Conclusions

Despite the fact that a large proportion of the capture production of the world's fisheries comes from small-scale activities in developing countries, fisheries management in these regions are usually based on approaches and techniques from developed countries with little or no adaptation [18]. These approaches and the science supporting them have generally been applied to large stocks of few species, and are largely unsuitable for rich multispecies fisheries. A major impediment to fisheries science and management in developing countries is that fishery statistics are very difficult to collect due to a lack of well-trained staff and suitable data collection methods; this is further exacerbated as they are also typically diverse multi-gear and multi-species. In the case of Indonesia, a study commissioned by the Ministry of Fisheries [19] stated that the historical data collection methodology was obsolete and the information collected was of questionable quality.

Table 1. Results of the Similarity of Percentages (SIMPER) analyses of the daily estimated weight of family-level taxonomic categories at each site on each of the survey trips.

Survey site	Important families	Muncar	Cilacap	Kedonganan	Pacitan	Pelabuhanratu	Sadeng	Tanjung Luar
Muncar	1	*Clupeidae*						
	2	*Scombridae*						
	3	*Leiognathidae*						
Cilacap	1	Clupeidae	*Scombridae*					
	2	Scombridae	*squid*					
	3		*Penaeidae*					
Kedonganan	1	Clupeidae	Clupeidae	*Scombridae*				
	2	Lethrinidae	Lethrinidae	*Clupeidae*				
	3	Penaeidae	Lutjanidae	*Lutjanidae*				
Pacitan	1	Clupeidae	Scombridae	Clupeidae	*Scombridae*			
	2	Scombridae	squid	Lutjanidae	*Coryphaenidae*			
	3	Leiognathidae	Istiophoridae	Lethrinidae	*Carangidae*			
Pelabuhanratu	1	Clupeidae	Scombridae	Menidae	Clupeidae	*Scombridae*		
	2	Scombridae	Clupeidae	Bivalves	Trichiuridae	*Clupeidae*		
	3	Cnidaria	Menidae	Lethrinidae	Menidae	*Carangidae*		
Sadeng	1	Clupeidae	Scombridae	Clupeidae	-	Clupeidae	*Scombridae*	
	2	Leiognathidae	squid	Lethrinidae		Menidae	*Carangidae*	
	3	Cnidaria	Mobulidae	squid		Trichiuridae		
Tanjung Luar	1	Clupeidae	Scombridae	Mobulidae	Carcharhinidae	Trichiuridae	squid	*Scombridae*
	2	Mobulidae	Caesionidae	Carcharhinidae	Clupeidae	Menidae	Mobulidae	*Carcharhinidae*
	3	Cnidaria	Hemiramphidae	Clupeidae	Caesionidae	Carcharhinidae	Caesionidae	*Carangidae*

Italics indicate the families that typified a site; regular text are those families which distinguished two sites.

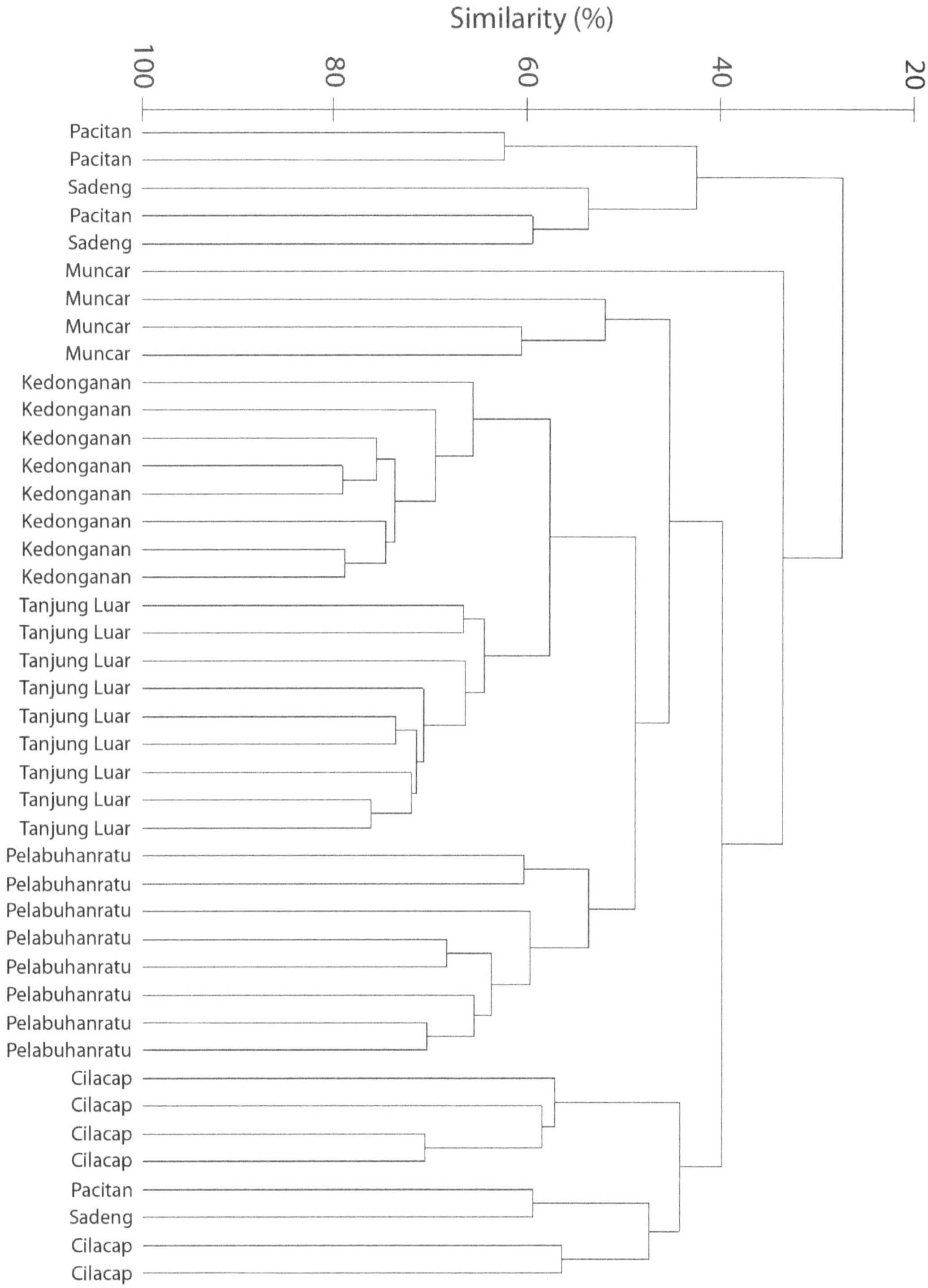

Figure 3. Cluster analysis of family-level data at each site. Average daily estimated weights of each of the family-level taxonomic categories at each of the 7 sites on each of the survey trips.

As a result, it was concluded that using this data to obtain Maximum Sustainable Yield (MSY) estimates is flawed.

The health of fisheries varies considerably throughout Southeast Asia. Hence, there is an urgent need to improve catch statistics across the region. The region is too large and faunally diverse to adopt fishery independent assessment methods used elsewhere. Fish landings and markets offer a cost effective and viable data surrogate for wild catch information in developing countries, particularly those characterised by multi-gear and multi-species fisheries. RFAMS provides an important, standardised approach to characterising fish landing sites and markets in developing countries. This approach was proposed in the late 1990s during surveys of fish markets in the Philippines led by one of us (PL). The initial aim of this survey method was to determine

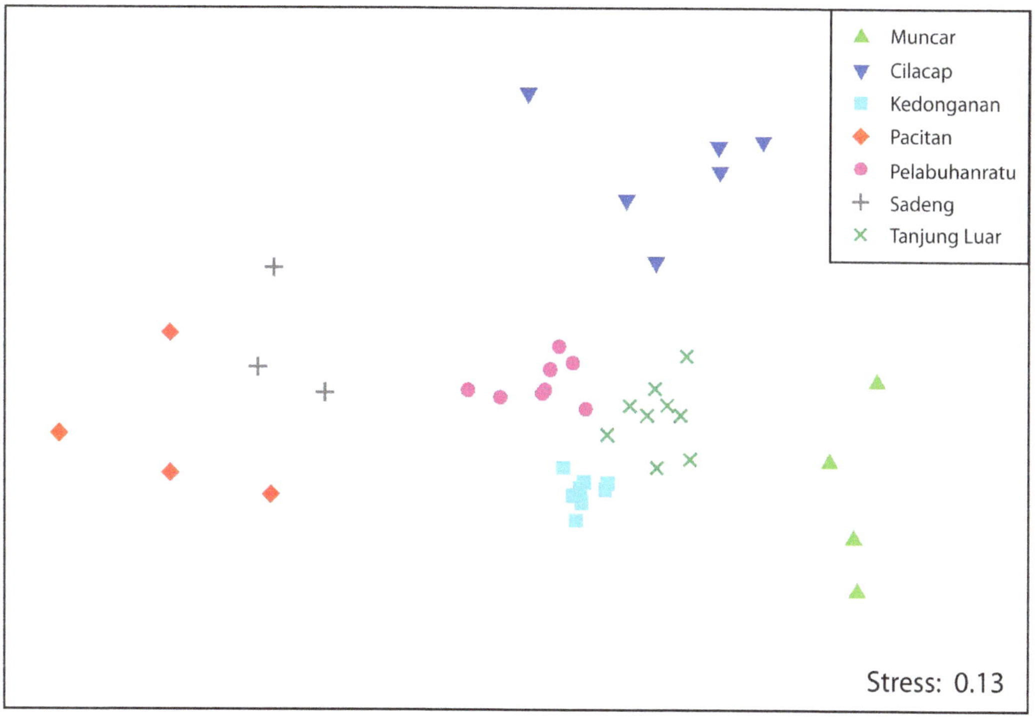

Figure 4. nMDS of bathome/gear type data at each site. Non-metric multidimensional scaling ordination of daily estimated weights of the landings of each of the bathome/gear types at each of the 7 sites on each survey trip.

whether it could be used to provide an indication of ecosystem and/or fishery health. For example, the species or taxonomic groups present or absent in a market are likely to partly reflect the level of fishing pressure in a particular area or fishery; for example, less commercially valuable species such as lizardfishes (Synodontidae) and ponyfishes (Leiognathidae) become increasingly more abundant in overfished demersal habitats at the expense of more valuable species such as snappers (Lutjanidae) and emperors (Lethrinidae) [20,21]. Similarly, heavily fished areas/regions are likely to experience a trophic shift with decreasing proportions of top order predators and increasing proportions of lower trophic levels in the catch. This method alone will not provide the necessary information to determine ecosystem health, but will provide crucial baseline information required to support more focused investigations. This pilot study was subsequently adapted and implemented herein on a much larger scale in southern Indonesia.

The RFAMS approach is important for those contexts where there are no detailed and accurate data collection protocols. During this study, RFAMS was shown to effectively characterise the landings at each of the different landing sites and/or fish markets surveyed. This method allowed for data to be collected in a consistent, standardised manner across a wide range of sites, regardless of the size of the site and the scale of the landings. The multivariate analyses highlighted the way the resulting data can be used to characterise various sites based on catch data for each taxonomic category, fishing method and bathomic combination.

One of the main challenges of this methodology is the basic taxonomic knowledge required to allocate the landings to family-level (or proxy, e.g. squids) categories. However, multivariate analysis highlighted that the sites are better characterised from one another when using landings data for family-level categories, rather than functional groups (e.g. reef fish-herbivore) or major taxonomic groups (e.g. crustaceans). Species-level data should provide a better discrimination of sites, but such detailed information would be impossible to collect due to the high species richness at most landing sites. For example, at the Kedonganan fish market, an estimated average of 88 teleost species was observed each day. The speed of turnover at sites surveyed would also hinder species-level information to be collected. There is usually a constant movement of fish reaching and leaving these sites so data needs to be collected rapidly. Identification guides are needed for users of this method to be able to assign landings into the family-level categories. A line drawing guide, which includes basic identifying features, for Indonesian fish families is now available in [17]. This guide can be used effectively in adjacent regions in South-east Asia; although the species may differ, the families present will be very similar.

This novel approach for assessing landings at fish markets provides an important capacity building tool for developing countries. If used in the proposed standardised manner, the data from RFAMS can be used to ground-truth catch statistics from particular landing sites or fish markets. For example, at Tanjung Luar, the results from RFAMS for sharks were compared with local fisheries data and provincial (*Dinas Perikanan*) data for the times they overlapped. This comparison highlighted that the local fisheries data was similar to that collected in RFAMS, but the provincial data suggest far greater volumes and appears to be erroneous. Data from these surveys also provide important baseline information which can be used as a benchmark against which future changes in catch and market demographics can be assessed.

The data collected by RFAMS is also extremely useful for future studies in the regions surveyed. The data can be explored to show where particular groups are consistently landed and in what quantities, and the fishing methods dominating particular sites,

Table 2. Results of the Similarity of Percentages (SIMPER) analyses of the average daily estimated weight of landings for each of the bathome/gear combinations (see Table S2) for each at each site on each of the survey trips.

Survey site	Important families	Muncar	Cilacap	Kedonganan	Pacitan	Pelabuhanratu	Sadeng	Tanjung Luar
Muncar	1	*InsPel:PurSei*						
Cilacap	1	InsPel:PurSei	*Oce:Lng*					
	2	Oce:Lng	*Oce:Gil*					
Kedonganan	1	InsPel:PurSei	Oce:Lng	*InsPel:Gil*				
	2	InsPel:Gil	CorRef:Hln	*CorRef:Hln*				
	3		InsPel:Gil	*Oce:Gil*				
Pacitan	1	InsPel:PurSei	Oce:Hln	InsPel:Gil	*Oce:Hln*			
	2	Oce:Hln	Oce:Lng	CorRef:Hln				
	3			Oce:Gil				
Pelabuhanratu	1	InsPel:PurSei	Oce:Hln	CorRef:Hln	InsPel:Gil	*Oce:Hln*		
	2	Oce:Hln	Oce:Lng	Oce:Hln	SloDem:Lng	*InsPel:Gil*		
Sadeng	1	InsPel:PurSei	Oce:Lng	InsPel:Gil	-	InsPel:Gil	*Oce:Hln*	
	2	Oce:Hln	Oce:Hln	CorRef:Hln		Oce:Gil		
	3			Oce:Gil		SloDem:Lng		
Tanjung Luar	1	InsPel:PurSei	Cor:Hln	Oce:Lng	InsPel:Gil	Oce:Hln	InsPel:Gil	*InsPel:Gil*
	2	InsPel:Gil	Oce:Lng	Est:Pot	Oce:Gil	Oce:Lng	Oce:Gil	*Oce:Gil*
	3		InsPel:Gil				Oce:Lng	*CorRef:Hln*

Italics indicate the bathome/gear combinations that typified a site; regular text are those that distinguished between two sites. Bathomes: CorRef = coral reef; Est = estuarine; InsPel = inshore pelagic; Oce = oceanic; SheDem = shelf demersal; SloDem = slope demersal. Gear: Gil = gillnet; Hln = handline; Lng = longline; Pot = trap or pot; PursSei = purse seine.

etc. This can provide important information to direct future research and to allow comparisons to be made with other regions.

7. Recommendations and future efforts

The RFAMS method is an extremely useful tool for recording catches at fish landing sites in tropical developing countries where the catches are highly diverse with many fishing types involved. However, a potentially major issue is whether a large portion of the catches is being missed for any particular reason, e.g. catches by-passing the surveyed part of the landing site, sale direct from boats prior to catches entering landing site, etc. This needs to be determined on a case by case basis depending on the landing site being surveyed. Thus, one of the critical first steps of using this method is to develop a good understanding of the dynamics of the landing site or sites being surveyed. Using this method without understanding how a landing site operates can lead to significant underestimation of landings and potentially erroneous data being collected.

Landing sites can, and typically do, vary substantially in how they operate. Two landing sites which have similar catch composition and quantities may vary in other ways which can strongly influence the data being collected. A good example is timing of the landings. One landing site may have all catches landed in a two hour period while a second could have a similar period of landing followed by a second landing/unloading of catches much later in the day. This latter landing could simply be missed if it is assumed they operate at similar times. Similarly, one landing site may have certain components of the catch (e.g. high quality tuna of certain sizes) by-passing the typical landing site area. This process needs to be understood before data collection using the RFAMS method should be commenced.

The aim of this paper was to trial a rapid data collection method for complicated fisheries in data poor regions in a standardised manner. This study has demonstrated that RFAMS can be used to collect catch composition data in a standardised manner from different landing sites in Indonesia, but how this will be adopted by fisheries managers is the next major step. For the Indonesian context, RFAMS should ideally first be used at various fish landing sites and ports within each of the provinces for short period and compared with the fisheries statistics currently captured for the same period. This would highlight discrepancies in catch information and indicate where problems may be occurring. In this context, the RFAMS could be used as a quality control mechanism which could be extremely beneficial for local fisheries authorities. The adoption of RFAMS as a useful tool for helping to quality control and potentially standardising fisheries catch data collection would be highly beneficial to Indonesia and adjacent developing nations with large complex fisheries.

Supporting Information

Table S1 Functional groups. List of the families (or species groups, e.g. squids) recorded during the RFAMS in Indonesia and the corresponding functional group each is allocated to.

Table S2 Major bathome and gear combinations recorded during this study. Those combinations recorded less than 10 times during the study were allocated to 'Rare'. Abbreviations used in MDS analyses are also included.

Datasheet S1 The Rapid Fisheries Assessment by Market Survey (RFAMS) data sheets. Used to capture the daily landings data and landing site characteristics.

Appendix S1 Line drawing guide. Guide and keys to the marine, family-level categories used in RFAMS. Suitable for Indonesia and adjacent developing countries.

Acknowledgments

We would like to thank Dr Chris Barlow (ACIAR) and Dr David Smith (CSIRO) who provided invaluable support for this project. We wish to thank Ms Dian Oktaviani, Mr Thomas Ralph Mahulette, Mr. Anthony Sisco Panggabean (Research Center for Fisheries Management and Conservation, Jakarta) and Dr David Milton, Mr Gary Fry (CSIRO) who assisted in the numerous survey trips undertaken. The following people provided invaluable assistance at the fish landing sites on most survey trips: Mr J. Rianto (Ocean Fishing Port Fisheries Office, Cilacap); Andi (Ocean Fishing Port, Pelabuhanratu Fish Landing Office); I Nyoman Sudarta (Kedonganan fish market); and Mr Lalu Eka Armusadi (Tanjung Luar Fisheries Office). In addition, we also thank the numerous people at each of the landing sites who endured us during all of our survey trips and assisted with our data collection. We would also like to thank Daniel Gledhill and John Pogonoski (CSIRO) who assisted with editing of the manuscript; Dr Ron West (University of Wollongong) for his support as project leader on this Indonesian survey; Dr Purwanto and Dr Subhat Nurhakim (Research Center for Fisheries Management and Conservation, Jakarta) and Dr Ir. Agus Apun Budhiman and Dr. Besweni (Directorate General of Capture Fisheries, Jakarta) for their project support.

Author Contributions

Conceived and designed the experiments: PL WW. Performed the experiments: D RF UC PL CD. Analyzed the data: WW CD RB. Contributed reagents/materials/analysis tools: CD RB. Wrote the paper: WW PL CD RB D.

References

1. FAO (2004–2013) Fishery and Aquaculture Country profiles. Indonesia. Fishery and Aquaculture Country Profiles. In: FAO Fisheries and Aquaculture Department [online]. Rome. Updated 5 August 2004.
2. FAO (2013) FIGIS. FishStat (Database). Available: http://data.fao.org/ref/babf3346-ff2d-4e6c-9a40-ef6a50fcd422.html?version=1.0. Accessed 2013 May.
3. FAO (2012) The State of World Fisheries and Aquaculture 2012. Rome: FAO.
4. Mous PJ, Pet JS, Arifin Z, Djohani R, Erdmann MV, et al. (2005) Policy needs to improve marine capture fisheries management and to define a role for marine protected areas in Indonesia. Fisheries Manag Ecol 12: 259–268.
5. Pet-Soede C, Machiels MAM, Stam MA, Van Densen WLT (1999) Trend in an Indonesian coastal fishery based on catch and effort statistics and implications for the perception of the state of the stocks by fisheries officials. Fish Res 42: 41–56.
6. Dudley RG, Harris KC (1987) The fisheries statistics system of Java, Indonesia: operational realities in a developing country. Aquacult Fish Manage 18: 365–374.
7. Venema SC (ed) (1996) Report on the Indonesia/FAO/DANIDA Workshop on the assessment of the potential of the marine fishery resources of Indonesia. GCP/INT/575/DEN. FAO Fisheries Technical Paper 338.
8. Jennings S, Kaiser MJ, Reynolds JD (2001) Marine Fisheries Ecology. Oxford: Blackwell Science.
9. Béné C (2005) Small-scale fisheries: assessing their contribution to rural livelihoods in developing countries. FAO Fisheries Circular No. 1008. Rome: FAO.
10. Zeller D, Booth S, Davis G, Pauly D (2007) Re-estimation of small-scale fishery catches for US flag-associated island areas in the western Pacific: the last 50 years. Fish Bull 105: 266–277.
11. Houk P, Rhodes K, Cuetos-Bueno J, Lindfield S, Fread V, et al. (2012) Commercial coral-reef fisheries across Micronesia: a need for improving management. Coral Reefs 31: 13–26.
12. Tokeshi M, Arakaki S, Daud JRP (2013) Consuming diversity: analysis of seasonal catch patterns in multi-species artisanal reef fisheries in North Sulawesi, eastern Indonesia. Pac Sci 67: 1–13.

13. White WT, Last PR, Stevens JD, Yearsley GK, Fahmi, et al. (2006) Economically Important Sharks and Rays of Indonesia. Canberra: ACIAR Publishing.
14. Clarke KR, Gorley RN (2006) PRIMER v6 User Manual/Tutorial. Plymouth, United Kingdom: Primer-E.
15. Somerfield PJ, Clarke KR (1997) A comparison of some methods commonly used for the collection of sublittoral sediments and their associated fauna. Mar Environ Res 43: 145–156.
16. Clarke KR (1993) Non-parametric multivariate analyses of changes in community structure. Aust J Ecol 18: 117–143.
17. White WT, Last PR, Dharmadi, Faizah R, Chodrijah U, et al. (2013) Market fishes of Indonesia. ACIAR Monograph Series, ACIAR Publishing, Canberra, 438 pp.
18. Mahon R (1997) Does fisheries science serve the needs of managers of small stocks in developing countries? Can J Fish Aquat Sci 54: 2207–2213.
19. PCI (2001) Study on fisheries development policy formulation, Vol II. Review and Analysis of Policies and Performances and Recommendations. Pacific Consultants International under Jakarta Fishing Port/Market Development Project (Phase IV: JBIC Loan No. IP-403). Jakarta: PCI.
20. Thresher RE, Sainsbury KJ, Gunn JS, Whitelaw AW (1986) Life history strategies and recent changes in population structure in the lizardfish genus, *Saurida*, on the Australian Northwest Shelf. Copeia 1986: 876–885.
21. Harris AN, Poiner IR (1991) Changes in species composition of demersal fish fauna of Southeast Gulf of Carpentaria, Australia, after 20 years of fishing. Mar Biol 111: 503–519.

Rockpool Gobies Change Colour for Camouflage

Martin Stevens*, Alice E. Lown, Alexander M. Denton

Centre for Ecology and Conservation, College of Life and Environmental Sciences, University of Exeter, Penryn Campus, Penryn, Cornwall, United Kingdom

Abstract

Camouflage is found in a wide range of species living in numerous habitat types, offering protection from visually guided predators. This includes many species from the intertidal zone, which must cope with background types diverse in appearance and with multiple predator groups foraging at high and low tide. Many animals are capable of either relatively slow (hours, days, weeks) or rapid (seconds and minutes) colour change in order to better resemble the background against which they are found, but most work has been restricted to a few species or taxa. It is often suggested that many small intertidal fish are capable of colour change for camouflage, yet little experimental work has addressed this. Here, we test rock gobies (*Gobius paganellus*) for colour change abilities, and whether they can tune their appearance to match the background. In two experiments, we place gobies on backgrounds of different brightness (black or white), and of different colours (red and blue) and use digital image analysis and modelling of predator (avian) vision to quantify colour and luminance (perceived lightness) changes and camouflage. We find that gobies are capable of rapid colour change (occurring within one minute), and that they can change their luminance on lighter or darker backgrounds. When presented on backgrounds of different colours, gobies also change their colour (hue and saturation) while keeping luminance the same. These changes lead to predicted improvements in camouflage match to the background. Our study shows that small rockpool fish are capable of rapid visual change for concealment, and that this may be an important mechanism in many species to avoid predation, especially in complex heterogeneous environments.

Editor: Peter A. Todd, National University of Singapore, United States of America

Funding: M.S. and A.E.L. were supported by a Biotechnology and Biological Sciences Research Council (BBSRC) David Phillips Research Fellowship (BB/G022887/1). The funders had no role in study design, data collection and analysis, decision to publish, or preparation of the manuscript.

Competing Interests: The authors have declared that no competing interests exist.

* Email: martin.stevens@exeter.ac.uk

Introduction

Predator-prey interactions have played a substantial role in shaping the diversity of life, leading to many adaptations and counter-adaptations for attack and defence [1–3]. Perhaps the most widespread defence is camouflage, preventing an object from being detected or recognised by an observer [4]. On a basic level, camouflage is intuitively simple, often involving matching the general appearance of the background environment. Yet this ignores a rich complexity to the subject because many different types of camouflage are thought to exist, from background matching and disruptive coloration to countershading and masquerade (e.g. [5–7]), and camouflage can also be optimised in a variety of ways.

Almost certainly, the most common form of camouflage in nature, and the basis for many other types of concealment, is background matching, where an animal resembles the general colour and pattern of the background [4,8]. In nature, animals could exhibit background matching through a variety of mechanisms, including genetic adaptation over long time periods, phenotypic plasticity during development, behavioural choice of substrates, using materials and decorations from the environment, and colour change over different time scales (e.g. [9–13]). The ability to change colour over a short time frame, physiological colour change, is based on the redistribution of pigment within chromatophore cells [14]. Investigating animals that can change colour is a particularly useful method to study how camouflage works and is tuned to different environments because researchers can manipulate the background on which individuals are found and investigate how the camouflage is changed in response (e.g. [9,15]). This approach can also yield insights into how the visual systems of animals work and interpret information in the environment, and how action is mediated via visual pathways [16].

While a great deal of recent research has investigated the different types of camouflage that may exist and how they work, mostly in artificial systems (e.g. [5,6,17–19]), comparatively little work has studied camouflage optimisation and tuning in real animals. The most extensive work to date has focussed on cephalopods, which show remarkable abilities for rapid colour and pattern change in response to specific features of the environment, such as pattern contrast and edge information [9,16]. Studies of colour change for camouflage have also been undertaken in, for example, chameleons, flatfish, and crabs (e.g. [15,20–22]). However, outside of studies on cephalopods, relatively little work has directly quantified how effective rapid (here defined as changes occurring in seconds or minutes) colour change for camouflage is, and how quickly this can occur. Little work has also directly investigated the exact form that colour change takes in terms of changes in colour and brightness. A major problem has been that conventional methods using spectrometry to quantify coloration require extensive handling of specimens, leading to stress induced colour change, and are also slow and hence unable to quantify rapid colour change.

Fish make an ideal group to study colour change for camouflage because it is widely reported that many species have this ability [23], and they occur in a wide range of habitats and on many backgrounds. The ability to change colour over short term, often for concealment, is thought to be widespread among teleost fish, both in the marine and freshwater environments [24,25], and enables them to occupy a greater range of backgrounds and to cope with heterogeneous habitats. A variety of goby species have been observed to change colour, both to match their backgrounds in order to maintain camouflage [26], and during breeding phases [27]. However, only one study by Fries in 1942 [26] has specifically conducted experiments investigating camouflage by these fish on different backgrounds, with gobies reported to become paler and less red when on blue colours, darker and more red on red backgrounds, and more yellow on yellow backgrounds, owing to changes in chromatophore cells. However, colour change was monitored by human eye, without quantifying it objectively, how it affected match to the background, or how fast it occurred.

In this paper we study the colour change abilities of the abundant and widely distributed rock goby (*Gobius paganellus*) when placed on backgrounds of different colours and brightness, to test whether they can change appearance for camouflage, and how quickly they do this. We use digital image analysis and predator vision modelling to quantify the speed and extent of colour change. Rockpool and intertidal fish such as gobies are excellent candidate species for this type of study for a variety of reasons. First, the environment in which they live is highly changeable, with a wide range of background types existing even over very small areas. In addition, physical disturbance of tides and waves will often push individuals over a range of backgrounds against which they are viewed by predators. Furthermore, owing to the tidal nature of the environment, the fish are under intense predation pressure, at low tide from birds and at high tide from other groups, including larger fish. Therefore, there may be a major advantage in being able to rapidly change colour for camouflage.

Materials and Methods

Gobies were collected by dip net in the intertidal zone from Gyllyngvase beach, Falmouth, Cornwall, UK (50° 8′33.4690″N, −005° 04′07.9716″W) between July 2013 and September 2013 for experiment 1 (40 individuals), and between October 2013 and July 2014 (40 individuals) for experiment 2. Once caught, fish were kept in fresh seawater in a grey bucket in order to minimise colour change prior to use. Both experiments were conducted *in situ* on Gyllyngvase beach under natural light conditions in shallow trays lined with waterproof paper (see below). All work was conducted under approval from the University of Exeter Biosciences ethics committee (application 2013/149). The field location (specified above) where the experiments were conducted and fish collected is public land and no further licences or permits were needed. Fish were kept no longer than 2 hours, and all individuals were returned unharmed to their original rockpool area after being tested. Rock gobies are not an endangered or protected species.

Experimental Background Creation

Our aim in experiment 1 was to test whether gobies are capable of changes in their luminance when placed on a black or white background. In experiment 2, we aimed to test for changes in colour. Our design here was intended to minimise perceived differences in brightness of the backgrounds by the fish, and to test whether gobies change their actual colours as opposed to just changes in luminance. We used red and blue as two colours at different ends of the visual spectrum that gobies are likely to be able to discriminate (see Discussion). In both experiments we also aimed to test how quickly any changes in appearance occurred. Background colours for both experiments were created by printing colours at 300 dpi on waterproof paper (HP LaserJet Tough paper; Hewlett Packard, Palo Alto, USA), with a Hewlett Packard Colour LaserJet 2605 dn printer. Before the experiments, gobies were placed on an intermediate grey background for at least 15 minutes as a standard background for all individuals. This was to provide the same starting point at the beginning of the experiment, and to remove some of the individual variation that would otherwise exist owing to individuals being found on different rockpool substrates on collection. To produce an intermediate grey midway between white and black we followed past approaches [15] and printed a range of grey squares of different intensity (pixel) values from black through to white made in Photoshop Elements 5.0 (Adobe Systems Inc., San Jose, USA). We measured the reflectance of each square using an Ocean Optics (Dunedin, FL, USA) USB2000+ spectrometer, held at 45° to normal, with illumination by a PX-2 pulsed xenon lamp, and calculated the average reflectance of each square across 400–750 nm (we excluded UV light because the paper and the print toner reflect little UV light), followed by plotting image pixel value against reflectance. We then calculated the mid grey value based on a ratio scale [28]. For experiment 1, we simply printed the darkest black and used the white paper for the two experimental backgrounds. For experiment 2, the red and blue colours were calculated to be the same average brightness across the visible spectrum [15]. This was achieved by photographing a range of different red and blue colours printed on the same paper (as above), followed by measuring their reflectance values in longwave (LW), mediumwave (MW), and shortwave (SW) images.

Experimental Procedure

Fish were placed in a 24 cm wide × 34 cm long × 5 cm deep (internal measurements) white tray that had been covered with a background of midpoint grey paper, calibrated as described above. The tray was filled 2 cm deep with fresh seawater and a spirit level was used to ensure trays were flat and the water level accurate to prevent variation in colour measurements due to water depth, and to ensure all areas of the tray had sufficient water to minimise stress to the fish by ensuring that even the largest individuals were fully submerged, while at the same time keeping water depth low so as to not affect the colour analyses. Fish were given 15 minutes to acclimatize on the grey background then photographed (see below) in the control tray before being transferred individually into a secondary experimental tray of 28.5 cm wide × 39 cm long × 7 cm deep (internal measurements) divided into eight compartments by thin plastic barriers attached with silicone sealant glue. These compartments were either four white and four black for experiment 1, or four red and four blue for experiment 2. Transfer of the fish between trays was done as quickly as possible and with a net in order to minimise any stress associated with capture and handling. Fish were unable to see each other, although barriers were not completely sealed and water was able to flow around the tray. Although this also meant that chemical cues could potentially transfer among individuals, any such effects should not produce directional colour changes in line with responses to background colours and brightness. The experimental trays also ensured that pairs of fish were tested under the same water conditions (e.g. temperature), and the relatively small size of the compartments prevented fish from swimming around too much, which would have made photography difficult.

Experimental trials were undertaken in blocks, with a single block consisting of a pair of fish, with one fish placed on each background colour, and with those individuals approximately matched by size to remove bias that may occur due to variation in colour change with individual size. Twenty fish were tested on each background colour for each experiment (40 fish in total per experiment). Fish were subsequently photographed again at 1–2, 10 and 60 minutes while remaining in the tray to establish the extent of colour change over time. Photos were taken using a Nikon D90 SLR camera, which had undergone a quartz conversion to enable ultraviolet sensitivity (Advanced Camera Services, Norfolk, UK) and fitted with a Nikon 105 mm Nikkor lens. In both experiments photographs were taken in human visible (400–700 nm) and ultraviolet (300–400 nm). For the human visible photos a UV/IR blocking filter was used (Baader UV/IR 2″ Cut Filter) and a UV pass filter was used during the ultraviolet photographs (Baader U 2″ Cut Filter). All photographs included a Spectralon 40% grey reflectance standard (Labsphere, Congleton, UK) next to the tray and a ruler. Due to changing light conditions and reflectance from the water surface, a black and silver photographic umbrella (Neewer, Guangdong, China) was used to shade the trays from direct sunlight.

Image Analysis

Images were taken in RAW format with manual white balance and fixed aperture settings. Images were then linearized with regards to light intensity based on camera responses to a set of eight Spectralon grey standards with reflectance values ranging from 2 to 99% (in custom programs written in Image J) in order to correct for the non-linear responses in image values many cameras produce in response to changes in light levels [29]. Image values were then equalised with regards to the 40% grey standard, and each image channel (LW, MW, SW and UV) scaled to reflectance, where 255 on an 8-bit scale is equal to 100% reflectance [29].

We wanted to analyse colour change with regards to one of the likely main predator groups of rockpool fish: shore birds. To obtain data corresponding to avian vision, we transformed the reflectance based image based on spectral sensitivity data from the peafowl (*Pavo cristatus*) [30] using a polynomial mapping technique to convert from camera to avian colour space [15,29,31,32]. The likely predators of rockpool fish include a range of shorebird species found at the intertidal zone. Previous work has shown that these are likely to have a 'violet' sensitive system [33], with the UV cone type shifted in sensitivity to slightly longer wavelengths than species that fall into the 'ultraviolet' group (although violet sensitive species can still detect UV light). Although gulls are likely to be predators of rockool fish too, and seem to have a UV visual system [33], the relatively low levels of UV involved in the backgrounds and fish should mean that differences in the perception between these systems is small. The peafowl is often used as a model species for modelling birds that fall into the violet group.

Once calibrated, the outline of each goby was drawn around by hand using Image J and the region of interest (ROI) saved. Each image layer was measured to acquire values for photon catch. We then calculated a series of metrics to analyse the appearance of each goby. Saturation (the amount of a given colour compared to white light) was defined as the distance an object is in a tetrahedral colour space from the achromatic grey point [34]. Larger distances equate to colours that appear more saturated. We next derived a measure of colour type, or hue. Here, we followed past approaches that have defined hue based on a ratio of the relative photoreceptor stimulation in different parts of the light spectrum [15,35,36]. Broadly, this approach, whereby colour types are defined in terms of a ratio of the different channels present, is based on the way that opponent colour channels are thought to work in animal vision, and in practical terms is a way of defining a colour type in an intuitive and readily interpretable manner [15]. In experiment 1, we had no *a priori* reason to expect particular changes in colour of fish because all the backgrounds used were achromatic shades of grey. Therefore, we used a standardised ratio that describes colour in terms of differences in the amount of shorter to longer wavelengths of light (an approach commonly used to calculate opponent channels): hue = ((LW+MW)–(SW+UV))/(LW+MW+SW+UV). In experiment 2, whereby we used backgrounds that were either red or blue, we predicted specific changes in coloration with fish moving more towards these two colour types. As such, we defined hue as (LW–SW)/(LW+SW). Higher values mean that an individual is relatively red in colour, whereas smaller values mean an individual is relatively blue. To derive a measure of achromatic change in appearance, we calculated luminance (perceived lightness) based on the double cone values, as in birds achromatic vision is widely thought to be driven by these receptors [37].

Finally, we calculated how changes in the appearance of fish equated to differences in their level of match to the experimental backgrounds. To do so we used a log form of a model of visual discrimination, the Vorobyev-Osorio model [38], which is based on differences in colour or luminance based on photo catch values, including estimates of neural noise and relative photoreceptor proportions. We used a Weber fraction value of 0.05 for the most abundant cone type [70, 90], and relative proportions of cone types in the retina of the peafowl (LW = 0.95, MW = 1.00, SW = 0.86, UV = 0.45; [30]). The model gives values of 'just noticeable differences' (JNDs), whereby differences of 1.00–3.00 mean that two stimuli are unlikely to be discriminated by an observer, and larger values above 3.00 are increasingly likely to equate to discriminable differences [39].

Statistics

We did not specifically expect an overall difference in appearance between fish on each background at all time points. Instead, our key prediction was that there should be no difference at the start of the experiment (time zero) when fish have been on the same intermediate grey background, whereas there should be differences as the experiment progresses. The exact time where differences arise should also depend on the speed of colour change. As such, we conduced a series of planned comparisons [40] between fish on each background type, separately at each time point. Data for all metrics except hue were non-normal and resistant to transformation and so we conduced Wilcoxon Mann-Whitney tests. For hue we conducted two-sample t-tests. Owing to the repeated testing for each experiment (one test per time point), we adjusted the critical p-values needed for significance by using a sequential Bonferroni [41]. For each experiment, p-values are ranked in order of significance and then compared to an adjusted critical value in turn, which becomes more stringent with each additional test. Critical thresholds for significance for each of the four statistical tests per experiment were therefore 0.050, 0.025, 0.016, and 0.012. To test for changes in the level of camouflage over time for both colour and brightness/luminance, we conducted Kruskal-Wallis tests.

Results

Experiment 1

Changes in Colour and Luminance. For luminance, there was no significant difference between fish on black or white backgrounds at time 0 (W = 414.0, n = 20, p = 0.925), but there were significant differences at one minute (W = 587.0, n = 20, p< 0.001), 10 minutes (W = 600.0, n = 20, p<0.001), and at 60 minutes (W = 610.0, n = 20, p<0.001), with fish on white backgrounds having higher luminance values (figure 1). Note, however, that the magnitude of differences is generally quite small with the largest difference in luminance values between time 0 and time 60 being 0.09 (with photon catch values on a scale of 0–1), and average differences being 0.03 (across both backgrounds).

In terms of colour change, for saturation, there was also no significant difference between fish on black or white backgrounds at time 0 (W = 406.0, n = 20, p = 0.925), but significant differences occurred at one minute (W = 305.0, n = 20, p = 0.005), 10 minutes (W = 268.0, n = 20, p<0.001), and 60 minutes (W = 308.0, n = 20, p = 0.006), with saturation values being higher on the black backgrounds (figure 1). The results for hue were very similar, again with no significant difference at time 0 (T = −0.92, df = 35, p = 0.363), but significant differences at one minute (T = −9.15, df = 36, p<0.001), 10 minutes (T = −12.24, df = 37, p<0.001), and 60 minutes (T = −10.49, df = 37, p<0.001), with hue values being higher (more LW and MW and less SW and UV in colour) for fish on white backgrounds (figure 1).

Differences in Camouflage Over Time. On the white background there was no significant reduction in JNDs over time (better match to the substrate) for colour (H = 0.46, df = 3, p< 0.927), but there was for luminance JNDs (H = 31.77, df = 3, p< 0.001; figure 1D). On the black background there was a significant difference in colour JNDs with time (H = 12.81, df = 3, p = 0.005). However, note that there was no decline in JNDs with time, but rather lower JND values at times 0 and 60 than at times 1 and 10, indicating that fish did not actually improve camouflage for colour over time intervals. There was no significant difference in luminance JNDs (H = 1.59, df = 3, p = 0.661). Therefore, fish improved in their achromatic match to the white background, but not to the black background.

Experiment 2

Changes in Colour and Luminance. There was no significant difference between fish on red or blue backgrounds for luminance at time 0 (W = 410.0, n = 20, p = 1.000), at one minute (W = 373.0, n = 20, p = 0.324), or at 60 minutes (W = 346.0, n = 20, p = 0.086), nor was there a significant difference at 10 minutes when controlling for multiple testing (W = 327.0, n = 20, p = 0.025); figure 2.

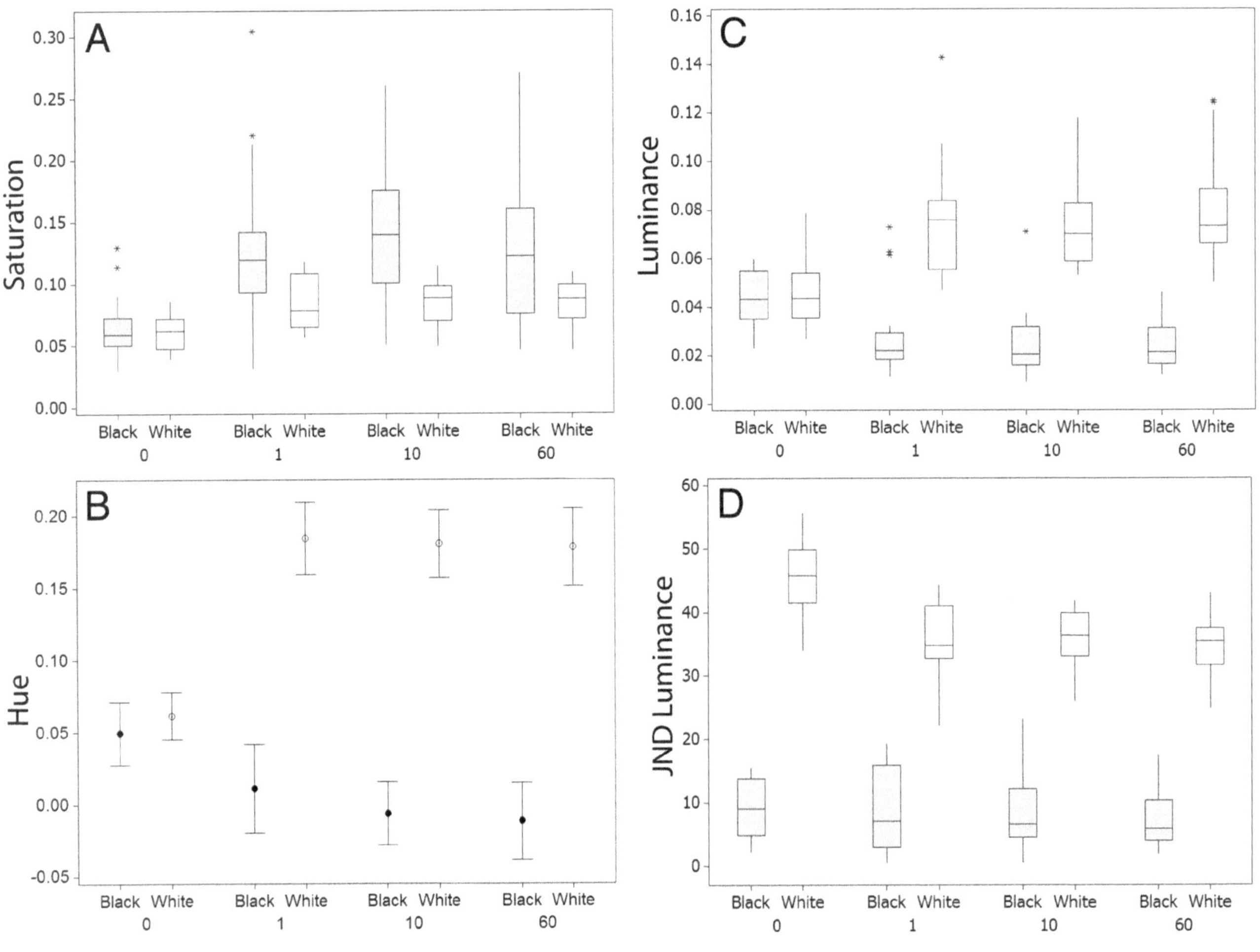

Figure 1. Changes in the different colour and achromatic metrics for fish on black and white backgrounds in experiment 1 for saturation (A), hue (B), and luminance (C) at the start (0 minutes) of the experiment and at 1, 10, and 60 minutes. Panel D shows the level of similarity for fish against the white and black backgrounds for JNDs in luminance (see main text) at 0, 1, 10, and 60 minutes. Graphs A, C, and D show medians plus inter-quartile range (IQR), whiskers are lowest and highest values that are within 1.5*IQR from the upper and lower quartiles, asterisks represent outliers. B shows means plus standard error.

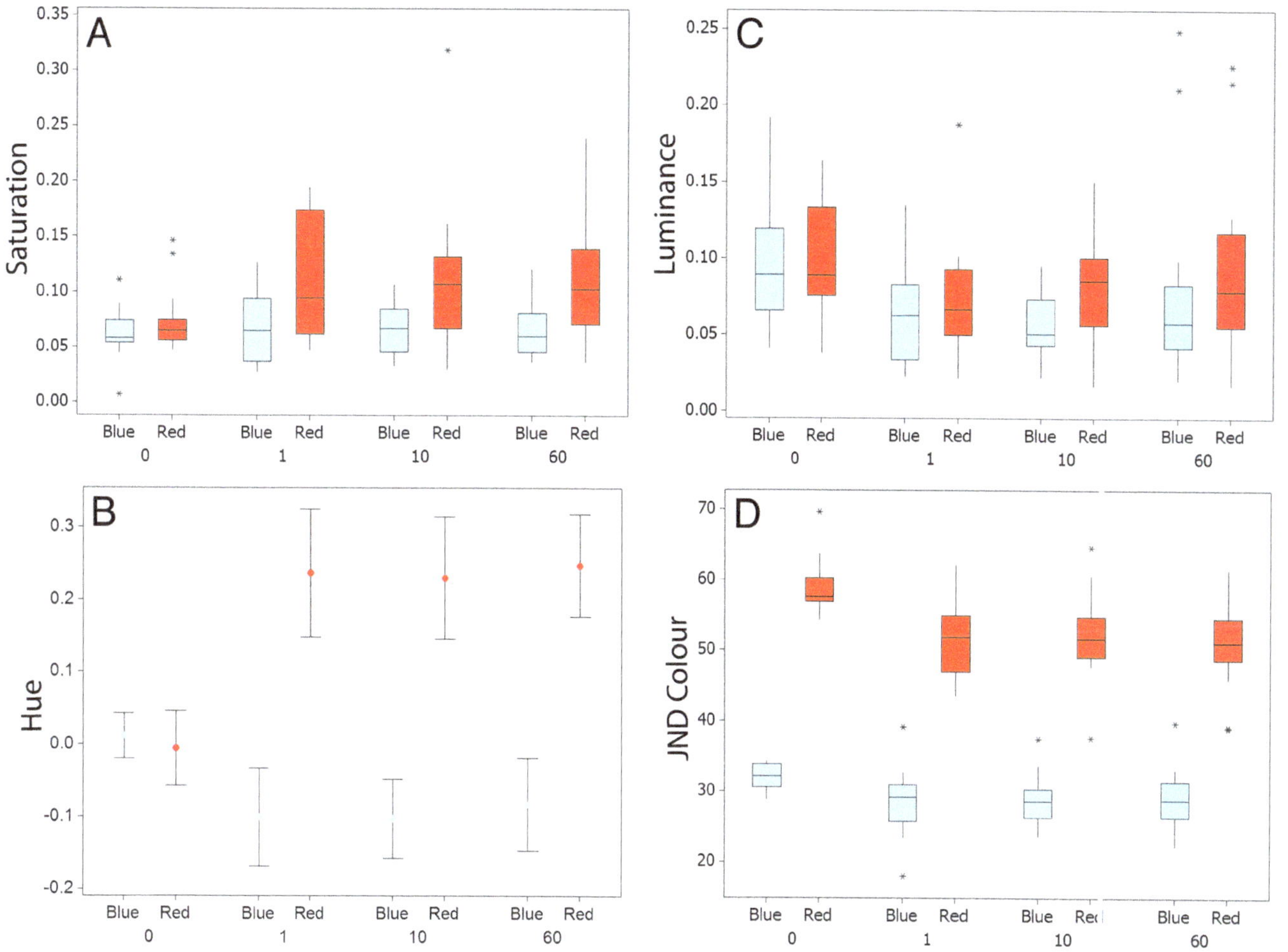

Figure 2. Changes in the different colour and achromatic metrics for fish on red and blue backgrounds in experiment 2 for saturation (A), hue (B), and luminance (C) at the start (0 minutes) of the experiment and at 1, 10, and 60 minutes. Panel D shows the level of similarity for fish against the red and blue backgrounds for JNDs in colour at 0, 1, 10, and 60 minutes. Graphs A, C, and D show medians plus inter-quartile range (IQR), whiskers are lowest and highest values that are within 1.5*IQR from the upper and lower quartiles, asterisks represent outliers. B shows means plus standard error.

Regarding colour, for saturation, there was no significant difference between fish on red or blue backgrounds at time 0 ($W = 378.0$, $n = 20$, $p = 0.394$), but there were significant differences at one minute ($W = 311.0$, $n = 20$, $p = 0.008$), 10 minutes ($W = 307.0$, $n = 20$, $p = 0.006$), and 60 minutes ($W = 289.0$, $n = 20$, $p = 0.001$), with fish being more saturated on the red background (figure 2). The results for hue were similar, with no significant difference at time 0 ($T = 0.60$, $df = 31$, $p = 0.552$), but significant differences at one minute ($T = -6.35$, $df = 35$, $p < 0.001$), 10 minutes ($T = -6.90$, $df = 32$, $p < 0.001$), and 60 minutes ($T = -7.24$, $df = 37$, $p < 0.001$). Fish on red backgrounds had higher hue values (more LW and less SW in coloration; figure 3).

Differences in Camouflage Over Time. On a red background, there was a significant reduction in JNDs for colour ($H = 25.31$, $df = 3$, $p < 0.001$), but not for luminance JNDs ($H = 3.10$, $df = 3$, $p = 0.376$); figure 2D. On a blue background, there was also a significant reduction in colour JNDs over time for colour ($H = 17.56$, $df = 3$, $p = 0.001$). Although there was a significant change in luminance JNDs, this was generally a decrease rather than improvement in luminance matching over time 0 ($H = 14.64$, $df = 3$, $p = 0.002$).

Discussion

Here, we tested whether rock gobies can change either their luminance (lightness) or colour depending on the background on which they are placed. As predicted, in experiment 1, fish changed in their overall luminance when put onto either a white or a black background, with individuals getting lighter or darker respectively. This led to changes in the level of similarity of fish to each background in terms of luminance, improving camouflage matching over time. In contrast, although there were some statistically significant changes in hue and saturation in this experiment too, these generally did not affect the overall match to the background, indicating that these changes were perceptually small and unlikely to be of significance in terms of camouflage, similar to other work [15].

In contrast, in experiment 2 where fish were placed onto either red or blue backgrounds, individuals underwent marked changes in colour with regards to both hue and saturation. At least some goby species have been shown to have three cone types, sensitive to relatively shorter and medium/longer parts of the spectrum [42], and so they should be able to distinguish between the blue and red backgrounds. In accordance with this, on a red background fish became more red in colour and more saturated,

Figure 3. Examples of changes in brightness of fish. Three individuals are shown on the left having been placed on a black background, and then the same individuals are shown on the right after being on a white background.

whereas on a blue background they became less red more grey in colour. These differences led to significant improvements in the level of colour match to the background. In contrast, there was little change in the luminance of fish on these backgrounds, demonstrating that fish can change their overall colour without changing their luminance. The exact mechanism of luminance perception in rock gobies is unknown, but this result suggests that they likely perceived the two background types as being of about the same brightness because in experiment 1, where the backgrounds were very different, fish did change in luminance. Overall, in both experiment 1 and 2 the changes were very rapid, with the majority of colour and brightness change occurring in the first minute.

The result that fish changed to become more red in coloration on the red background, yet that changes towards the blue colour were much smaller (they mostly become more grey in colour) is interesting. It suggests that some types of colour are easier for the fish to adopt than others. This fits with the background environment in the habitat where the fish were collected, whereby blue colours are rare, yet red encrusting algae and brown stones and seaweed are common. Past work has shown that different types of chromatophore control different colours, with black melanin being controlled by melanophores, yellow pteridine controlled by xantophores, red carotenoids controlled by erythrophores and more rarely blue cyanophores controlling a yet unknown cyan biochrome [25,26]. Colour responses may also be elicited by more than one type of chromatophore, and Fries [26] suggested that blue response in common gobies are due to a response of the erythrophores, xanohpores, and iridophores. However, more work is needed to test what cellular mechanisms cause changes in goby coloration, and whether other populations might be capable of greater changes in blue.

The levels of change in luminance were relatively small in this study even for the fish that changed the most. Thus the decrease in difference to the background, although significant, was not very large. However, this did equate to a decrease in discrimination thresholds of almost 10 JNDs on average for fish on the white backgrounds. In general across the experiment fish were quite dark, and in nature they are likely to be a better match to the general colour and brightness of the substrate in the rockpools (especially the dark rocks). Therefore, in such cases in the wild when fish are already well matched in appearance to the background even relatively small differences may equate to a valuable benefit in improved camouflage. Furthermore, to our eyes, changes in the brightness of fish are clearly perceptible (figure 3). One possibility to resolve this apparent discrepancy is that in this study we analysed the appearance of the entire body of each fish. In reality, gobies often have quite strong patterns that to us have key characteristics of disruptive coloration to break up the body shape against the background [1,4,5]. We often noted that the prominence of such patterns changed as fish change colour, and we think it quite possible that even when the overall brightness of an individual stays essentially the same that there can be pronounced changes in pattern. For example, on a uniform background fish may reduce the contrast and prominence of their markings and adopt a more uniform appearance, but this may be broadly similar in overall brightness to that of their starting appearance. Otherwise, better brightness match to the background may be brought about through longer-term changes, such as through morphological colour change that occurs over days and months and may be caused by changes in the overall density of chromatophores in the skin [14].

It is interesting that the ability of fish to change colour seems to be better than their ability to change brightness. Until we test fish on more natural coloured backgrounds we can only speculate as to why this may be. In the rockpool environment, the background is highly heterogeneous in terms of brightness, with stones and gravel of a range of shades occurring on a small scale (smaller than the size of the fish). Thus, overall changes in brightness may have a relatively small benefit. In contrast, some rockpools and larger backgrounds seem to have broadly different colours, meaning that colour change may be more valuable. This is likely to be especially the case with changes in shore height too, whereby there are changes in the amount of substrate types, especially greater brown and green algae cover lower down the shore.

Here, we have focussed on changes in colour and brightness using relatively artificial background appearances. Next, it will be important to test for colour change and camouflage ability on backgrounds that more closely resemble those in the environment where the individuals live. Moreover, given that gobies often have strongly contrasting patterns that appear disruptive, it would be important to test whether individuals have the capacity to change their markings on backgrounds of different marking sizes and contrasts. Previous studies have shown that flatfish are capable of impressive changes in pattern depending on the substrate appearance [20]. While gobies are unlikely to match the extent of this ability, the potential is there for them to change their patterns for concealment. In addition, the environment in which gobies live is both highly challenging (the intertidal) and heterogeneous, and so being able to adjust individual markings is likely to provide a strong advantage. A number of other species live in the same habitat, including several other common species of goby and blenny that have been suggested to change colour too. Thus, there also exists great potential for comparative studies of colour change within and among habitat types in intertidal fish species. It should also be noted that in this study we have not directly explored the level of individual colour change possible because this would require placing the same individual fish on different backgrounds in a repeated measures design and analysing their colour change abilities.

While in the present study we have focussed on colour change for camouflage a number of recent studies on colour change in gobies report change colour in response to breeding, with individuals becoming less camouflaged and more attractive to mates [43,44]. This change appears to be largely hormonally induced [44]. In gobies, breeding coloration change in other species is also influenced by predation pressure, with courtship colouration less intense under high predation risk [45,46]. When in aquaria devoid of predators, however, blennies from high predation sites showed full courtship colouration [46]. In other fish, arctic charr (*Salvelinus alpinus*) have shown differences in aggression when paired over white or black backgrounds [47,48]. Darker skin colour in salmonids may relate to subordination and be used to reduce aggressive interactions between conspecifics. When fish are paired within a tank and acclimatised to light coloured backgrounds, fish generally show increased aggression towards each other, whereas this aggression is not observed if fish have been acclimatised to dark backgrounds [47]. Many species of fish will also change colour under stressful conditions, such as when odour cues suggest a predator is in close proximity [49]. While some species of goby have been found to react to this predator cue, *Gobius paganellus* has not been found to display any defensive reaction to conspecific skin extract ([49] – see [50]). In *Gobius minutus*, physical handling also provoked rapid darkening and reddening in pale fish [26]. Thus, it would be valuable to investigate how capacity for colour change is affected by factors such as season and mating behaviour, dominance, and condition. Overall, colour change in fish and other species presents an excellent system to study camouflage, what makes this effective and how it is controlled, and other life history factors that may affect camouflage responses and tuning.

Acknowledgments

We thank Zeehan Jaafar and another anonymous referee for a range of helpful comments on the work and manuscript. We also thank Jolyon Troscianko for help by writing the Image J calibration programmes and assistance with camera calibration.

Author Contributions

Conceived and designed the experiments: MS AD AEL. Performed the experiments: AEL AD MS. Analyzed the data: MS. Contributed to the writing of the manuscript: MS AEL. Calibrated the images and measured them: AEL AD. Undertook visual modelling of the data: MS.

References

1. Cott HB (1940) Adaptive Coloration in Animals. London: Methuen & Co. Ltd.
2. Ruxton GD, Sherratt TN, Speed MP (2004) Avoiding Attack. Oxford: Oxford University Press.
3. Wallace AR (1889) Darwinism. An Exposition of the Theory of Natural Selection With Some of its Applications. London: Macmillan & Co.
4. Stevens M, Merilaita S (2009) Introduction. Animal camouflage: current issues and new perspectives. Phil Trans R Soc B 364: 423–427.
5. Cuthill IC, Stevens M, Sheppard J, Maddocks T, Párraga CA, et al. (2005) Disruptive coloration and background pattern matching. Nature 434: 72–74.
6. Rowland HM, Cuthill IC, Harvey IF, Speed MP, Ruxton GD (2008) Can't tell the caterpillars from the trees: countershading enhances survival in a woodland. Proc R Soc B 275: 2539–2545.
7. Skelhorn J, Rowland HM, Speed MP, Ruxton GD (2010) Masquerade: Camouflage without crypsis. Science 327: 51.
8. Merilaita S, Stevens M (2011) Chapter 2: Crypsis through background matching. In: M Stevens and S Merilaita, editors. Animal Camouflage. Cambridge: Cambridge University Press. 17–33.
9. Barbosa A, Mäthger LM, Buresch KC, Kelly J, Chubb C, et al. (2008) Cuttlefish camouflage: the effects of substrate contrast and size in evoking uniform, mottle or disruptive body patterns. Vis Res 48: 1242–1253.
10. Hultgren K, Stachowicz J (2011) Camouflage in decorator crabs: ntegrating ecological, behavioural and evolutionary approaches. In: M Stevens and S Merilaita, editors. Animal Camouflage: From Mechanisms to Function. Cambridge: Cambridge University Press.
11. Lovell PG, Ruxton GD, Langridge KV, Spencer KA (2013) Individual quail select egg-laying substrate providing optimal camouflage for their egg phenotype. Curr Biol 23: 260–264.
12. Rosenblum EB, Römpler H, Schöneberg T, Hoekstra HE (2010) Molecular and functional basis of phenotypic convergence in white lizards at White Sands. PNAS 107: 2113–2117.
13. Todd PA, Briers RA, Ladle RJ, Middleton F (2006) Phenotype-environment matching in the shore crab (*Carcinus maenas*). Mar Biol 148: 1357–1367.
14. Stuart-Fox D, Moussalli A (2009) Camouflage, communication and thermoregulation: lessons from colour changing organisms. Phil Trans R Soc B 364: 463–470.
15. Stevens M, Lown AE, Wood LE (2014) Colour change and camouflage in juvenile shore crabs *Carcinus maenas*. Front Ecol Evol 2: 14.
16. Zylinski S, Osorio D, Shohet A (2009) Perception of edges and visual texture in the camouflage of the common cuttlefish, *Sepia officinalis*. Phil Trans R Soc B 364: 439–448.
17. Bond AB, Kamil AC (2002) Visual predators select for crypticity and polymorphism in virtual prey. Nature 415: 609–613.
18. Merilaita S, Lind J (2005) Background-matching and disruptive coloration, and the evolution of cryptic coloration. Proc R Soc B 272: 665–670.
19. Webster RJ, Hassall C, Herdman CM, Sherratt TN (2013) Disruptive camouflage impairs object recognition. Biol Lett 9: 20130501.

20. Kelman E, Tiptus P, Osorio D (2006) Juvenile plaice (*Pleuronectes platessa*) produce camouflage by flexibly combining two separate patterns. J Exp Biol 209: 3288–3292.
21. Stevens M, Pei Rong C, Todd PA (2013) Colour change and camouflage in the horned ghost crab *Ocypode ceratophthalmus*. Biol J Linn Soc 109: 257–270.
22. Stuart-Fox D, Moussalli A, Whiting MJ (2008) Predator-specific camouflage in chameleons. Biol Lett 4: 326–329.
23. Marshall J, Johnsen S (2011) Camouflage in marine fish. In: M Stevens and S Merilaita, editors. Animal Camouflage: From Mechanisms to Function. Cambridge: Cambridge University Press.
24. Sumner FB (1940) Quantitative changes in pigmentation, resulting from visual stimuli in fishes and amphibia. Biol Rev 15: 351–374.
25. Sköld HN, Aspengren S, Wallin M (2013) Rapid color change in fish and amphibians - function, regulation, and emerging applications. Pigment Cell & Melanoma Res 26: 29–38.
26. Fries EFB (1942) Notes on colour change and pegmentary innervation in a goby, a wrasse and the plaice. Biol Bull 82: 273–283.
27. Svensson O, Kvarnemo C (2007) Parasitic spawning in sand gobies: an experimental assessment of nest-opening size, sneaker male cues, paternity, and filial cannibalism. Behav Ecol 18: 410–419.
28. Stevens M, Hopkins E, Hinde W, Adcock A, Connelly Y, et al. (2007) Field experiments on the effectiveness of 'eyespots' as predator deterrents. Anim Behav 74: 1215–1227.
29. Stevens M, Párraga CA, Cuthill IC, Partridge JC, Troscianko TS (2007) Using digital photography to study animal coloration. Biol J Linn Soc 90: 211–237.
30. Hart NS (2002) Vision in the peafowl (Aves: *Pavo cristatus*). J Exp Biol 205: 3925–3935.
31. Pike TW (2011) Using digital cameras to investigate animal colouration: estimating sensor sensitivity functions. Behav Ecol Sociobiol 65: 849–858.
32. Stevens M, Cuthill IC (2006) Disruptive coloration, crypsis and edge detection in early visual processing. Proc R Soc B 273: 2141–2147.
33. Ödeen A, Håstad O, Alstrom P (2009) Evolution of ultraviolet vision in shorebirds (Charadriiformes). Biol Lett 6: 370–374.
34. Endler JA, Mielke PW (2005) Comparing entire colour patterns as birds see them. Biol J Linn Soc 86: 405–431.
35. Komdeur JM, Oorebeek M, van Overveld T, Cuthill IC (2005) Mutual ornamentation, age, and reproductive performance in the European starling. Behav Ecol 16: 805–817.
36. Spottiswoode CN, Stevens M (2011) How to evade a coevolving brood parasite: egg discrimination versus egg variability as host defences. Proc R Soc B 278: 3566–3573.
37. Osorio D, Vorobyev M (2005) Photoreceptor spectral sensitivities in terrestrial animals: adaptations for luminance and colour vision. Proc R Soc B 272: 1745–1752.
38. Vorobyev M, Osorio D (1998) Receptor noise as a determinant of colour thresholds. Proc R Soc B 265: 351–358.
39. Siddiqi A, Cronin TW, Loew ER, Vorobyev M, Summers K (2004) Interspecific and intraspecific views of color signals in the strawberry poison frog *Dendrobates pumilio*. J Exp Biol 207: 2471–2485.
40. Ruxton GD, Beauchamp G (2008) Time for some a priori thinking about post hoc testing. Beh Ecol 19: 690–693.
41. Rice WR (1989) Analyzing tables of statistical tests. Evolution 43: 223–225.
42. Utne-Palm AC, Bowmaker JK (2006) Spectral sensitivity of the two-spotted goby *Gobiusculus flavescens* (Fabricius): a physiological and behavioural study. J Exp Biol 209: 2034–2041.
43. Immler S, Mazzoldi C, Rasotto MB (2004) From sneaker to parental male: change of reproductive traits in the black goby, *Gobuis niger* (Teleostei, Gobiidae). J Exp Zool 301: 177–185.
44. Sköld HN, Amundsen T, Svensson PA, Mayer I, Bjelvenmark J, et al. (2008) Hormonal regulation of female nuptial coloration in a fish. Hormones Behav 54: 549–556.
45. Forsgren E, Magnhagen C (1990) Behavioural responses to different types of predators by sand goby *Pomatoschistus minutus*: an experimental study. Mar Ecol Prog Ser 70: 11–16.
46. Hastings PA (1991) Flexible responses to predators in a marine fish. Ethol Ecol & Evol 3: 177–184.
47. Abbott JC, Dunbrack RL, Orr CD (1985) The interaction of size and experience in dominance relationships of juvenile steelhead trout (*Salmo gairdneri*). Behav 92: 241–253.
48. Höglund E, Blam PHM, Winberg S (2002) Behavioural and neuroendocrine effects of environmental background colour and social interaction in Arctic charr (*Salvelinus alpinus*) J Exp Biol 205: 2535–2543.
49. Smith RJF, Lawrence BJ (1992) The response of a bumblebee goby, *Brachygobius sabanus*, to chemical stimuli from injured conspecifics. Environ Biol Fishes 34: 103–108.
50. Pfeiffer W (1977) The distribution of fright reaction and alarm substance cells in fishes. Copeia 1977: 653–665.

10

Behavioural Stress Responses Predict Environmental Perception in European Sea Bass (*Dicentrarchus labrax*)

Sandie Millot[1]*, Marco Cerqueira[1], Maria-Filipa Castanheira[1], Øyvind Øverli[2], Rui F. Oliveira[3,4], Catarina I. M. Martins[1]

1 CCMAR-CIMAR L.A., Universidade do Algarve, Campus de Gambelas, Faro, Portugal, **2** Norwegian University of Life Sciences, Department of Animal and Aquacultural Sciences, Ås, Norway, **3** ISPA Unidade de Investigação em Eco-Etologia Integrative Behavioural Biology Group, Lisboa, Portugal, **4** Champalimaud Neuroscience Programme, Instituto Gulbenkian de Ciência, Oeiras, Portugal

Abstract

Individual variation in the response to environmental challenges depends partly on innate reaction norms, partly on experience-based cognitive/emotional evaluations that individuals make of the situation. The goal of this study was to investigate whether pre-existing differences in behaviour predict the outcome of such assessment of environmental cues, using a conditioned place preference/avoidance (CPP/CPA) paradigm. A comparative vertebrate model (European sea bass, *Dicentrarchus labrax*) was used, and ninety juvenile individuals were initially screened for behavioural reactivity using a net restraining test. Thereafter each individual was tested in a choice tank using net chasing as aversive stimulus or exposure to familiar conspecifics as appetitive stimulus in the preferred or non preferred side respectively (called hereafter stimulation side). Locomotor behaviour (*i.e.* time spent, distance travelled and swimming speed in each tank side) of each individual was recorded and analysed with video software. The results showed that fish which were previously exposed to appetitive stimulus increased significantly the time spent on the stimulation side, while aversive stimulus led to a strong decrease in time spent on the stimulation side. Moreover, this study showed clearly that proactive fish were characterised by a stronger preference for the social stimulus and when placed in a putative aversive environment showed a lower physiological stress responses than reactive fish. In conclusion, this study showed for the first time in sea bass, that the CPP/CPA paradigm can be used to assess the valence (positive *vs.* negative) that fish attribute to different stimuli and that individual behavioural traits is predictive of how stimuli are perceived and thus of the magnitude of preference or avoidance behaviour.

Editor: Claudia Mettke-Hofmann, Liverpool John Moores University, United Kingdom

Funding: This research project has been support by the European Commission under the 7th Framework Programme FP7–KBBE-2010-4 Contract n°: 265957 Copewell. The work of Sandie Millot is supported by the Portuguese Fundação para a Ciência e Tecnologia (FCT) grant n°: FRH/BPD/72952/2010. The PhD of Marco Cerqueira is supported by the FCT grant n°: SFRH/BD/80029/2011. The funders had no role in study design, data collection and analysis, decision to publish, or preparation of the manuscript.

Competing Interests: The authors have declared that no competing interests exist.

* Email: sandiemillot@yahoo.fr

Introduction

How and for what reasons individuals differ in the way they react to potential risks, handle novelty, or interact with conspecifics remain fascinating questions. Scherer [1] suggested that the individual evaluates the significance of an event according to a set of stimulus evaluation checks. These evaluate the relevance of the event according to various dimensions (*e.g.* novelty, pleasantness, and importance of the event for the individual), its implication in terms of the individual's needs, the possibility for the individual to cope with the event and the compatibility of the event with social or individual standards. A variety of related concepts have been used to describe individual differences in behaviour that are consistent over time and across situations (see Budaev and Brown [2]). Wilson et al. [3] proposed that the shy-bold continuum (propensity to take risk) is a fundamental axis of behavioural variation in many species. Another concept frequently used in the study of animal personality is behavioural syndrome: a suite of correlated behaviours that are expressed either within a given context or across context [4]. A third concept frequently used to investigate individual differences in behaviour is coping styles or strategies. Two alternative coping styles are frequently distinguished: proactive and reactive [5–7]. Proactive individuals are more active, aggressive, bold, tend to form inflexible routines and learn more slowly about small changes in the environment. Reactive individuals, in contrast, are shyer, non-aggressive and more sensitive to environmental changes. The existence such contrasting phenotypes seems to be a widespread phenomenon, with some aspects of this individual variation being reported in invertebrates (e.g. squids, *Euprymna tasmanica* [8]), lizards (*Anolis carolinensis* [9]) and in various species of fish (sticklebacks, *Gasterosteus aculeatus* [10–12], rainbow trout, *Oncorhynchus mykiss*, [13–15]; Nile tilapia, *Oreochromis niloticus*, [16,17]; Gilthead sea bream, *Sparus aurata*, [18,19]). Far from being stereotyped and invariant, differences in the behavioural repertoires, learning and memory abilities observed in both phenotypes suggest that fish are curiously plastic [20].

Recent reviews of fish cognition suggest fish show a rich array of sophisticated behaviours. For example they have functional long-term memories, develop complex traditions, show signs of Machiavellian intelligence, cooperate with and recognise one another and are even capable of tool use [21–23]. Emerging evidences also suggest that, despite appearances, the fish brain is also more similar to higher vertebrate one than previously thought [24–26]. Although this amount of knowledge, the way fish perceive stimuli and the affective value they attribute to them are still poorly known phenomena.

For most people, this is either linked to animal sentience or consciousness. Sentience is quite difficult to define or measure, and the meaning is constantly debated by scientists and philosophers alike, but it might be summed up in an ethical context as the ability to experience pleasure and pain (*i.e.* subjective perceptual experiences [27,28]). Being increasingly used in animal welfare evaluations and recognised as adaptive products of natural selection [29], affective states are not directly observable, and behavioural and physiological proxies have to be used in order to probe animal affective states. Preference/avoidance and motivation tests have been used for this purpose, based on the assumption that affective states are linked to motivation/preference and ultimately drive behaviour [30]. In these tests the animal is given some control over its environment, so that we can observe their choices in preference tests, or how much they are willing to work to access or avoid given resources or threats in operant motivation tests [30–33]. Thus, the ability of fish to express choice according to their preference or avoidance is a well established phenomenon; however, the extent of intraspecific variation and whether coping style influences this type of behaviour are still unknown.

In this study we investigated how European sea bass (*Dicentrarchus labrax*), one of the most important commercial species in Europe, evaluated putative appetitive (presence of social partners) or putative aversive (net chasing) stimuli in a conditioned place preference/avoidance test (CPP/CPA [34,35]). CPP/CPA is a behavioural paradigm in which a reward or a punishment is paired repeatedly with environmental cues so that the animal associates the cues with the appetitive or aversive stimulus and eventually develops preference or avoidance for the marked location even in the absence of the stimulus [36]. In order to validate this paradigm as a gauge of valence attributed by the fish to the putative appetitive/aversive stimuli, we used physiological (cortisol and glucose) and behavioural (distance travelled and swimming speed) measures. In addition, we investigated whether pre-existing differences in the behavioural response to acute stress (a putative indicator of stress coping style or animal personality in fish [18,37]) would predict individual variation in the response to putative appetitive and aversive stimuli.

Materials and Methods

The experiment described was conducted in accordance with the Guidelines of the European Union Council (86/609/EU) and Portuguese legislation for the use of laboratory animals, and approved by the ethics committee from the Veterinary Medicines Directorate, the Portuguese competent authority for the protection of animals, Ministry of Agriculture, Rural Development and Fisheries, Portugal. Permit number 0420/000/000-n.99-09/11/2009. The rules and regulations which protect experimental animals from unnecessary pain and suffering have been strictly followed during the experiment. In preparing the experiment, we have carefully considered the application of the 3R (Replacement, Reduction and Refinement, 2010/63/UE).

1. Experimental fish, housing and feeding

Fish were hatched and reared at the experimental research station of Ifremer in Palavas-les- Flots (France) until they weighted 0.1 g and then transported to Ramalhete Research Station (Faro, Portugal). Fish were housed in stock tanks (500 L) under sea bass standard rearing conditions [38] during 8 months before the start of the experimental procedures (rearing density from 0.3 kg m^{-3} (mean fish weight = 0.1 g) to 10 kg m^{-3} (mean fish weight = 45 g) which are considered as low rearing densities and reach all welfare demand). Fish were fed a commercial diet (Aquagold 3 mm, Aquasoja, Sorgal SA, Portugal; 44% crude protein, 14% crude fat, 8% ash, 2.5% crude fibres, 1.0% phosphorus) using automatic feeders (1.5% BW day^{-1}). Fish were reared in open water circuit tanks, with a temperature of 20±7°c, salinity of 35±2‰ and dissolved oxygen above 95%, and a 12L:12D photoperiod was employed with light on at 08:00. One month before the start of experimental procedures, 90 fish were randomly selected, anaesthetised with 2-phenoxyethanol (0.3‰, Sigma-Aldrich) and individually identified with a PIT-tag (Micro BE, France) injected in the flesh under the dorsal fin and with a visible implant elastomer tag (VIE; Northwest Marine Technology, USA) in the caudal fin. At the start of the experiment the body mass of the fish was 45±1.3 g (mean ± SE).

2. Set up and experimental procedures

2.1. Restraining test. Escape behaviour during restraining or confinement has been used to discriminate personality traits as well as physiological correlates of coping style in different fish species [16,37,39,40]. More recently, Castanheira et al. [18] showed in sea bream that escape behaviour during restraining was consistent over time and across contexts. Moreover, Ferrari S., Millot S., Leguay D., Chatain B., Bégout ML (unpubl. data) demonstrated in sea bass that escape attempts during restraining test were negatively correlated to plasma cortisol concentration. For these reasons, the net restraining test was performed only one time 15 days before the conditioned place preference/avoidance (CPP/CPA) tests. The restraining test consisted of holding each fish individually in an emerged net for three minutes [16–19,39,41]. The following behaviours were measured: latency to escape (time in seconds taken by each fish to show an escape attempt; escape attempt was defined as a elevation of the body from the net); number of escape attempts and total time spent on escape attempts (total time in seconds taken by each fish escaping since the first to the last escape attempts). Behaviours measured were collapsed into first principal component scores using Principal Components Analysis (PCA) in order to obtain a score allowing the characterisation of coping styles. PC1 explained 86% of the variation and the number of escapes was the variable that contributed the most for PC1 (Table 1).

Fish presenting a high latency to escape, small number of escape attempts and shorter total time to escape were characterised by a low score and identified as reactive fish. On contrary, fish presenting a lower latency to escape, high number of escape attempts and longer total time to escape were characterised by a high score and identified as proactive (based on Silva et al. [39]; Martins et al. [16,17] and Castanheira et al. [18,19]). No threshold was applied to separate subjectively the fish in two categories *i.e.* proactive and reactive. Instead these data (Coping Style, CS, score) were used as a continuous variable.

2.2. Conditioned place preference/avoidance test. Four days before the start of the CPP/CPA test, 3 groups of 12 fish each (6 focal fish with distinct VIE (Visible Implant Elastomer tag) and 6 familiar conspecifics, *i.e.* all fish coming from the same rearing tank) were placed in 3 different 100 L home tanks located in the

Table 1. Mean ± SEM, minimum (Min.) and maximum (Max.) values of behavioural variables obtained for the restraining test (N = 90) and PCA loading used to generate a principal component scores (PC1).

Behavioural variables	Mean ± SEM	Min.	Max.	Loadings for PC1	Eigenvalues	% variation explained
Latency escape (s)	90.9±0.75	2	180	−0.855	85.991	85.991
Number escape	5.3±0.06	0	18	0.962	12.481	
Total escape time (s)	1.6±0.02	0	6.8	0.96	1.528	

experimental room. This was done to acclimatize fish to their new environment. The photoperiod and the water temperature, salinity and oxygenation were the same as in rearing tanks. The fish were fed *ad libitum* each morning. This procedure was repeated 5 consecutive times in order to test 30 fish per treatment: appetitive (APP), aversive (AVER) and control (CONT).

The CPP/CPA test was performed in 6 glass aquaria of 80 L (70 cm length×40 cm width and 30 cm depth). Each aquarium was divided into two halves by a 10 cm wide grey central alley: one half marked by white walls without dots and the other half marked by white walls with black dots used as visual cues for fish (Figure 1). One infrared LED projector (IR-294S/60, Monacor) was placed beneath each aquarium.

The CPP/CPA test consisted of three experimental phases conducted over a 3 days period. An initial habituation phase was performed on the first day. During this phase, each fish was placed individually in the tank and allowed to swim freely throughout the whole tank for 40 min (this period was determined based on preliminary observations). For each individual the preference for white or dotted side (>50% time spent) was assessed through a 20 min additional recording. Then the fish was put back in the home tank. Animals that showed a very strong initial preference (>90% time spent) in either side or strong freezing behaviour (<500 cm distance moved) were excluded from the study because their initial position during the habituation phase could not be representative of their real preference for that side. Therefore, animals that showed an initial preference between 50.1% and 89.9% for either side and which had moved more than 500 cm were used for data analyses. The habituation phase was followed by a conditioning phase, during which treatments differed between the appetitive and aversive stimulus groups. For the aversive stimulus (AVER), fish was placed in the same aquarium as during the habituation phase but had only access to the initial preferred side for 20 min, hereafter termed the stimulation side (SS). Afterwards, the fish was chased with a net during 10 s each 4 min for a period of 20 min. For the appetitive stimulus (APP), fish was placed in the non initial preferred side (new SS) for 20 min and then 2 familiar conspecifics were added in the tank for a period of 20 min. The control fish (CONT) were handled exactly the same way as the tested animals (maintained in the preferred side or in the non preferred side) except that the stimulus was omitted during the training phase. After each treatment, fish was placed back in the home tank.

The test phase was performed on the last day of the experiment (the third day) and consisted exactly of the same procedure as the habituation phase in order to record any behavioural changes.

After this last phase, fish were immediately caught and euthanized with an overdose of 2-phenoxyethanol (1‰, Sigma-Aldrich). Blood was thus sampled one hour after fish were transferred to the experimental aquarium (based on Fanouraki et al. [42]). Blood was withdrawn within 3 min from the caudal vein using heparinised syringes and centrifuges at 2000× g for 25 minutes at room temperature. After centrifugation plasma was frozen in dry ice and stored at −80°c for glucose (QCA, Spain) kit analysis. Plasma cortisol levels were measured by means of a commercial ELISA kit RE52061 (IBL Hamburg, Germany),

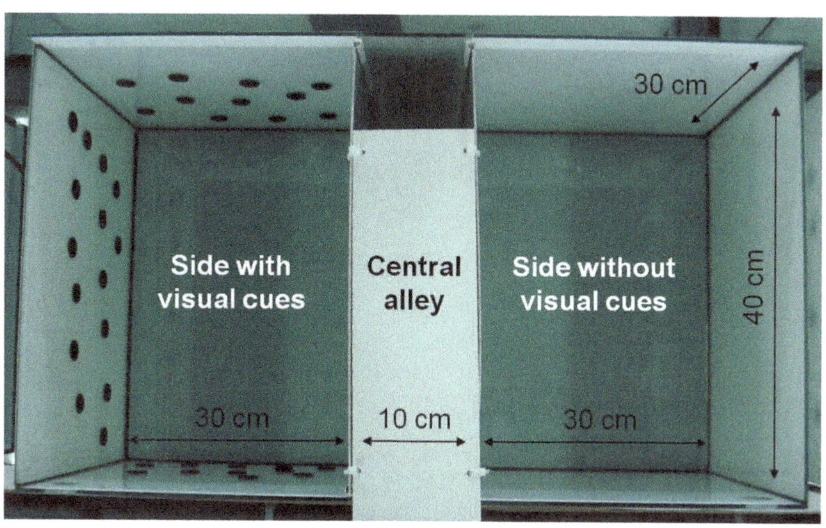

Figure 1. Experimental tank. CPP/CPA test glass aquarium (80L) divided into three compartments: one central alley with grey walls and two lateral compartments with white walls and with or without visual cues (black dots).

with a sensitivity of 2.5 ng/ml, and intra and inter-assay coefficients of variation (CV) of 2.9 and 3.5%, respectively. After blood sampling, fish were identified and measured for standard length (cm).

During each phase, individual behaviour was recorded by infrared sensitive video camera (TVCCD-623-COL, Monacor, Denmark) equipped with infrared filter (dark red, Schneider Optics, USA) and positioned 1 m above the tank. The videos were stored in AVI files on a hard drive and analysed afterwards with the Lolitrack 2.0 software (Loligo Systems, Denmark). Before each video analyses, the background image of each tank was divided into three arenas (Arena 1 = white side, Arena 2 = grey middle alley, Arena 3 = dotted side). For each tank the background was calibrated by marking the length of the Arena 2 in the image and entering its actual value (10 cm). The Lolitrack 2.0 software tracks the fish as a dark object on a light background. By using infrared light underneath the tank we avoid light reflexion on the water surface and optimise the fish tracking by the software. The following parameters were quantified by the software: time spent in each arena (min), distance travelled in each arena (m) and the swimming speed in each arena (cms^{-1}). In order to remove the influence of fish size in swimming speed data, these values were transformed in body length per second (BLs^{-1}). To evaluate the fish behavioural changes between the habituation and the test phase, percent change of time spent, distance travelled and swimming speed were calculated as: [(Test phase value – Habituation phase value)/Habituation phase value] ×100.

3. Statistics

Behaviours measured in restraining test were collapsed into first principal component scores (PC1) with orthogonal rotation (varimax) using Principal Components Analysis (PCA). The correlation matrix was used to check for multicollinearity (*i.e.* to identify variables that did not correlate with any other variable, or correlate very highly, r = 0.9, with one or more variables). Kaiser–Meyer–Olkin (KMO) test for sample adequacy was always greater than 0.5 and the Bartlett's test of sphericity was significant for all tests, indicated that correlations between items were sufficiently large for PCA. PCA analyses were performed using SPSS 18.0 for windows. The results are expressed as mean ± standard error of the mean (SEM).

All other statistical analyses were performed using Statistica 7 software (Statsoft, USA). The results were expressed as mean ± standard error of the mean (SEM).

A null model of side preference was tested by comparing the observed fish distribution to the theoretical homogeneous distribution in the side with or without dots (50% in each side) by a Kolmogorov–Smirnov test. Fish did not show a systematic initial preference for one side or the other one. Consequently, the stimulation was performed 36 times in the side without dots *vs* 42 times in the side with dots (d = 0.28; p>0.05).

One way ANOVA was used to analyse the differences in percent change of time spent on SS (after arcsine(sqrt (x/100)) transformation) by the experimental (APP or AVER) vs. control fish. Repeated-measures ANOVA's were used to analyse the differences in distance travelled (m) and swimming speed (BLs^{-1}) between experimental and control fish, experimental phases (*i.e.* before and after conditioning phase) and tank sides (4-levels repeated factor: before SS *vs.* before non-stimulation side (nSS) *vs.* after SS *vs.* after nSS; categorical variables: experimental *vs.* control fish). Newman & Keuls tests were subsequently used to test differences between the habituation and the test phase of each treatment and between the control and the experimental groups both at the habituation and at the test phase of each treatment.

One way ANOVA followed by Newman & Keuls tests were used to analyse the differences in plasma concentrations of cortisol ($ngml^{-1}$) and glucose ($mmol^{-1}$) between experimental and control fish.

For both APP and AVER treatments, Pearson correlations matrices between time spent on SS, distance moved in SS and nSS, percent change of time spent on SS, percent change of distance moved in SS and nSS, plasma concentration of glucose and cortisol with Coping Style (CS) score were calculated. The significance level of each correlation matrice was defined according to the table of critical values of Pearson correlation coefficient corrected by the individual number (n) in Scherrer p792 [43].

Results

From the 90 fish tested in this study, 12 fish did not comply with our CPP acceptance criteria and were thus removed from the analysis. This resulted in the following sample sizes: n = 28 for APP, n = 23 for AVER and n = 27 for CONT.

1. Restraining test

During the restraining test, fish waited on average 91 s before the first escape attempt and they performed a mean of 5 escape attempts for a total escape time of around 2 s (Table 1).

2. Time spent on the stimulation side

On average fish subjected to the appetitive stimulus showed a high increase (+163%) in the time spent on the SS during the test phase, whereas fish exposed to the aversive stimulus showed a significant decrease (−42%) in the time spent on the SS during the test phase (Fig. 2; One way ANOVA $F_{2,75} = 6.60$, $p<0.01$).

3. Distance travelled

The main effect of treatment (*i.e.* APP *vs.* AVER *vs.* CONT groups) on the distance travelled was not significant ($F_{2,75} = 0.54$, $p = 0.58$), but there were a significant main effect of the repeated measure (*i.e.* before SS *vs.* before nSS *vs.* after SS *vs.* after nSS; $F_{3,225} = 103.11$, $p<0.001$) and a significant interaction between treatment and the repeated measure ($F_{3,225} = 7.32$, $p<0.001$). Thus, regardless of treatment or tank side, fish significantly decreased the distance travelled between the habituation and the test phase (Fig. 3).

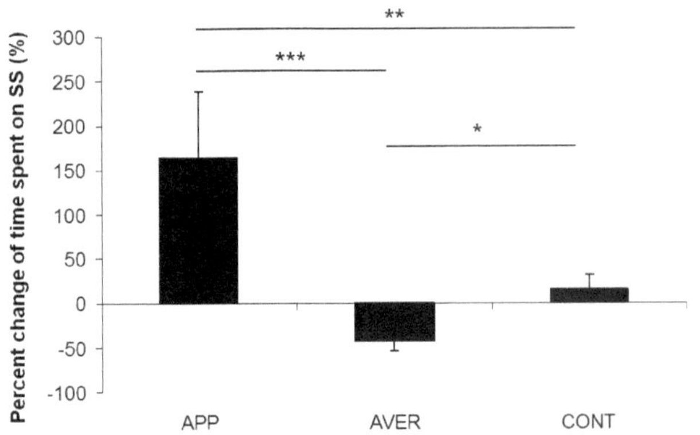

Figure 2. Time spent on the stimulation side. Percent change of time spent (Mean ± SEM; in %) by the fish on the stimulation side between the habituation and test phases for each treatment. One way ANOVA, * <0.05; ** <0.01; *** <0.001.

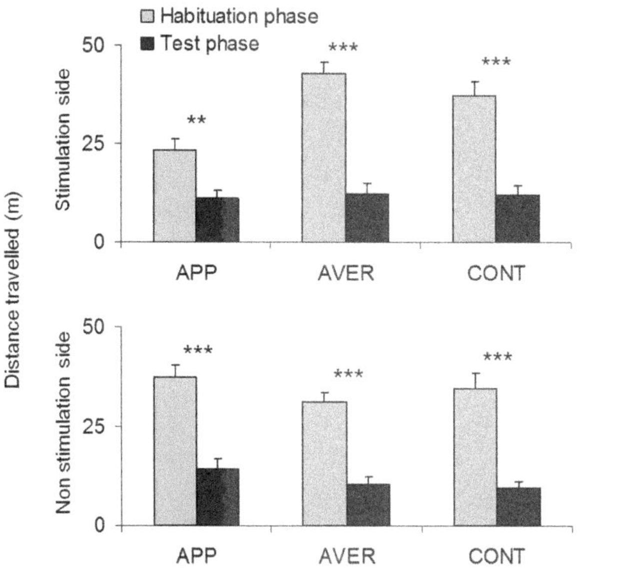

Figure 3. Distance travelled. Distance travelled (Mean ± SEM; in m) by the fish on the stimulation side and on the non stimulation side during the habituation and test phases for each treatment. Repeated ANOVA, ** <0.01; *** <0.001.

4. Swimming speed

The main effect of treatment (*i.e.* APP *vs.* AVER *vs.* CONT groups) on swimming speed was not significant ($F_{2,75} = 0.10$, $p = 0.90$), but there were a significant main effect of the repeated measure (*i.e.* before SS *vs.* before nSS *vs.* after SS *vs.* after nSS; $F_{3,225} = 50.47$, $p<0.001$) and a significant interaction between treatment and the repeated measure ($F_{3,225} = 2.94$, $p<0.01$).

Whatever the treatment and the tank side, fish significantly decrease the swimming speed during the test phase (data not shown).

5. Blood plasma analysis

There was no significant difference between APP, AVER and CONT in plasma concentration of cortisol (237±24; 212±25; 270±20 ng ml^{-1} respectively; $F_{2,71} = 1.11$, $p = 0.33$) and glucose (4.32±0.1; 4.33±0.1; 4.26±0.1 mmol l^{-1} respectively; $F_{2,67} = 0.14$, $p = 0.86$).

6. Correlations between coping style, behaviour and physiology parameters

The correlations matrice for the APP treatment showed that fish that spent the most time on SS during the test phase also had a higher CS score (Table 2, Fig. 4).

The correlations matrice for the AVER treatment showed that fish characterised by a low CS score increased the distance moved in SS and had higher plasma cortisol concentrations (Table 2, Fig. 4).

Discussion

This study demonstrated that a single exposure to an appetitive or an aversive stimulus significantly increased or decreased respectively, the time spent by sea bass on the stimulation side of a CPP/CPA setup. This behavioural change clearly suggests that the experimental fish attributed a positive valence to the presence of social partners, and a negative valence to net chasing. Further, it would appear that altered place preference observed during the final test phase (in the absence of stimulus) is due to associative learning of visual cues (black dots) coupled to expected appetitive or aversive stimulus.

Notably, both experimental and control fish showed a strong decrease in swimming activity (distance travelled and swimming speed) in both sides of the tank. This result could be explained either by the fact that fish became habituated to the aquarium and thus reduced exploration or by the experimental procedure which consisted of handling fish each day to transfer them from their home tank to the experimental tank and which could have induced stress (*i.e.* freezing behaviour). Thus, even if fish showed preference or avoidance for the tank zone associated with appetitive or aversive stimulus respectively, they did not express a swimming activity comparable to that observed during the habituation phase. This last interpretation is supported by the similar high plasma concentrations of cortisol and glucose observed across treatments. Thus, the stimulus effect on fish physiology was probably masked by the stress due to the handling procedure. To summarise, even if the experimental set up to assess the affective value that sea bass attribute to a stimulus was not

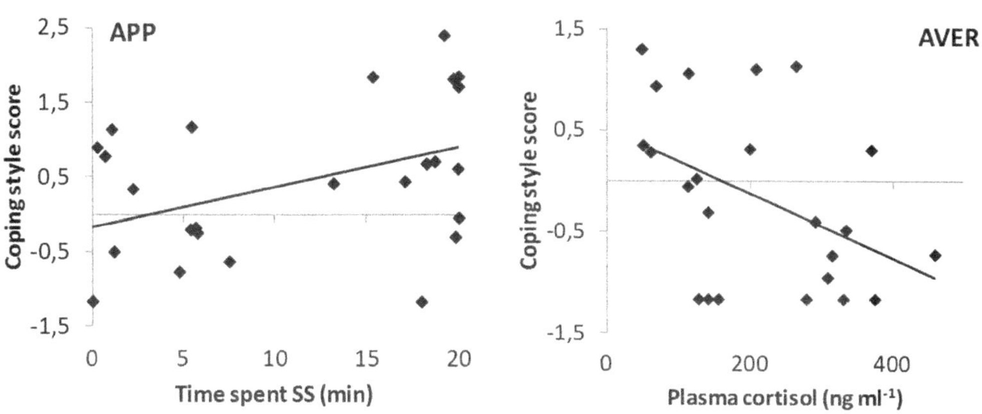

Figure 4. Correlations between coping style, behaviour and physiology parameters. Correlations between coping style score and time spent on SS (min) for appetitive treatment and coping style score and plasma cortisol concentration (ng ml^{-1}) for aversive treatment. The full black lines represent the linear regressions.

Table 2. Pearson correlations matrice and p value for appetitive (APP) and aversive (AVER) treatments between time spent on stimulation side (SS), distance moved in SS and on non stimulation side (nSS), percent change of time spent on SS, percent change of distance moved on SS and nSS and plasma concentration of cortisol and glucose with Coping Style score.

	APP		AVER	
	Coping style score		Coping style score	
Time spent SS (min)	**0.470**	**P = 0.019**	−0.225	P = 0.340
Distance moved SS (m)	−0.050	P = 0.796	−0.379	P = 0.099
Distance moved nSS (m)	−0.180	P = 0.384	−0.178	P = 0.452
Percent change time spent SS (%)	0.210	P = 0.030	−0.161	P = 0.498
Percent change distance moved SS (%)	−0.130	P = 0.539	**−0.441**	**P = 0.052**
Percent change distance moved nSS (%)	−0.240	P = 0.259	−0.169	P = 0.476
Cortisol (ng ml^{-1})	0.043	P = 0.841	**−0.495**	**P = 0.027**
Glucose (mmol l^{-1})	0.121	P = 0.573	0.009	P = 0.969

The significance level of each correlation matrice was defined according to the table of critical values of Pearson correlation coefficient corrected by the individual number (n) in Scherrer p792 [43] (APP: n = 28, |R|>0.374; AVER: n = 23, |R|>0.413). The significant results are in bold.

optimal due to the handling procedure which has masked part of the behavioural and physiological responses, the results showed nevertheless that fish can learn, remember and make decisions to avoid being exposed to aversive stimulus (net chase) or in contrary to seek for appetitive stimulus (familiar conspecifics).

The consistency of behaviours has not been tested in this study and should be done in the future in order to clearly define if the behavioural stress response measured during the restraining test could be considered as coping style, behavioural trait or personality. However, restraining test has been shown to be repeatable in sea bream [18] and correlated with physiological stress in sea bass (see method section) and thus the observed behaviour is likely to be a coping style. More interestingly, this study showed that behavioural stress response or coping style modulates the response (*i.e.* appraisal) of appetitive and aversive stimuli. When fish were subjected to an appetitive stimulus, proactive individuals expressed a higher preference (*e.g.* time spent) for the stimulation side than reactive fish. But when fish were submitted to an aversive treatment, reactive fish exhibited an increase of distance moved in the stimulation side (anxiety) and a higher plasma cortisol concentration than proactive fish. These behavioural and physiological changes showed that fish exhibiting proactive behaviour were more responsive to the appetitive stimulus while reactive phenotypes responded to the aversive one. These results suggested also that the proactive phenotype is less fearful when presented with a signal previously associated with an aversive stimulus, as compared to individuals of the reactive type. Previous studies have demonstrated in fish relationship between anxiety or fear behaviour and corticosterone response [44] and limbic neural systems [7,29] and also between fearfulness and coping style [16], but it is the first time that a study highlights the link between behavioural stress and fear responses and physiological patterns simultaneously.

Even if correspondence between boldness and sociability is not clearly established (Cote et al., 2010; Trompf and Brown, 2014), in our study, proactive fish seemed more responsive to social stimulus than the reactive ones. Pike et al. [45] showed in three-spined stickleback (*Gasterosteus aculeatus*) that bolder individuals had fewer overall interactions than shy fish, but tended to distribute their interactions more evenly across all group members. Thus, the fact that proactive fish were more attracted by the side where congeners were present did not necessarily means that they are more social but simply that they attribute a higher positive value to this stimulus than reactive fish.

In conclusion, this study showed for the first time in sea bass, that the CPP/CPA paradigm can be used (with some set up improvements in order to reduce the fish freeze behaviour due to the handling procedure) to assess the valence (positive *vs.* negative) that fish attribute to environmental stimuli and that the individual's behaviour under stress predicts how stimuli are perceived and thus the subsequent preference or avoidance behaviour.

Supporting Information

File S1 Supporting information file shows data of restraining test, CPP-CPA test and blood analysis.

Acknowledgments

The authors are grateful to João Reis and Miguel Viegas for their technical assistance. The authors are also very grateful to Béatrice Chatain to freely provide the sea bass and to Marie-Laure Bégout for device loan. We thank the editor and two referees for their helpful suggestions and advice.

Author Contributions

Conceived and designed the experiments: SM RFO CIMM. Performed the experiments: SM MC. Analyzed the data: SM. Contributed reagents/materials/analysis tools: SM MC M-FC. Wrote the paper: SM MC M-FC ØØ RFO CIMM.

References

1. Scherer KR (1987) Toward a dynamic theory of emotion: the component process model of affective states. Geneva Studies in Emotion and Communication 1: 1–98.
2. Budaev S, Brown C (2011) Personality traits and behaviour. In: Brown C, Laland K, Krause J, editors. Fish Cognition and Behaviour. Second Edition ed: Blackwell Publishing Ltd. pp. 135–165.
3. Wilson DS, Clark AB, Coleman K, Dearstyne T (1994) Shyness and boldness in humans and other animals. Trends in Ecology & Evolution 9: 442–446.

4. Sih A, Bell A, Johnson JC (2004) Behavioral syndromes: an ecological and evolutionary overview. Trends in Ecology & Evolution 19: 372–378.
5. Benus RF, Bohus B, Koolhaas JM, Vanoortmerssen GA (1991) Heritable variation for aggression as a reflection of individual coping strategies. Experientia 47: 1008–1019.
6. Koolhaas JM, Korte SM, De Boer SF, Van Der Vegt BJ, Van Reenen CG, et al. (1999) Coping styles in animals: current status in behavior and stress-physiology. Neuroscience & Biobehavioral Reviews 23: 925–935.
7. Øverli Ø, Sorensen C, Pulman KGT, Pottinger TG, Korzan W, et al. (2007) Evolutionary background for stress-coping styles: relationships between physiological, behavioral, and cognitive traits in non-mammalian vertebrates. Neuroscience and Biobehavioral Reviews 31: 396–412.
8. Sinn DL, Gosling SD, Moltschaniwskyj NA (2008) Development of shy/bold behaviour in squid: context-specific phenotypes associated with developmental plasticity. Animal Behaviour 75: 433–442.
9. Korzan W, Øverli Ø, Summers C (2006) Future social rank: forecasting status in the green anole (*Anolis carolinensis*). acta ethologica 9: 48–57.
10. Bell AM (2005) Behavioural differences between individuals and two populations of stickleback (*Gasterosteus aculeatus*). Journal of Evolutionary Biology 18: 464–473.
11. Huntingford F (1976) The relationship between anti-predator behaviour and aggression among conspecifics in the three-spined stickleback, *Gasterosteus aculeatus*. Animal Behaviour 24: 245–260.
12. Bell AM, Sih A (2007) Exposure to predation generates personality in threespined sticklebacks (*Gasterosteus aculeatus*). Ecology Letters 10: 828–834.
13. Sneddon LU (2003) The bold and the shy: individual differences in rainbow trout. Journal of Fish Biology 62: 971–975.
14. Øverli Ø, Winberg S, Pottinger TG (2005) Behavioral and neuroendocrine correlates of selection for stress responsiveness in rainbow trout—a review. Integrative and Comparative Biology 45: 463–474.
15. Winberg S, Schjolden J, Øverli Ø, Pottinger T (2007) Stress and stress coping in fish, behavioural correlates and neuroendocrine mechanisms. Comparaive Biochemistry and Physiology 146: S77.
16. Martins CIM, Silva PIM, Conceicao LEC, Costas B, Hoglund E, et al. (2011) Linking fearfulness and coping styles in fish. PloS ONE 6: e28084.
17. Martins CIM, Castanheira MF, Engrola S, Costa B, Conceição LEC (2011) Individual differences in metabolism predict coping styles in fish. Applied Animal Behaviour Science 130: 135–143.
18. Castanheira MF, Herrera M, Costas B, Conceição LEC, Martins CIM (2013) Can we predict personality in fish? Searching for consistency over time and across contexts. PloS ONE 8: e62037.
19. Castanheira MF, Herrera M, Costas B, Conceição LEC, Martins CIM (2013) Linking cortisol responsiveness and aggressive behaviour in gilthead seabream *Sparus aurata*: Indication of divergent coping styles. Applied Animal Behaviour Science 143: 75–81.
20. Laland KN, Brown C, Krause J (2003) Learning in fishes: from three-second memory to culture. Fish and Fisheries 4: 199–202.
21. Bshary R, Wickler W, Fricke H (2002) Fish cognition: a primate's eye view. Animal Cognition 5: 1–13.
22. Brown C, Laland KN, Krause J (2011) Fish cognition and behaviour 2e. Cambridge: Wiley-Blackwell.
23. Brown C (2014) Fish intelligence, sentience and ethics. Animal Cognition. DOI 10.1007/s10071-014-0761-0.
24. Broglio C, Gomez A, Duran E, Salas C, Rodriguez F (2011) Brain and cognition in teleost fish. In: Brown C, Krause J, Laland K, editors. Fish cognition and behavior. Oxford: Wiley. pp. 325–358.
25. Demski LS (2013) The pallium and mind/behavior relationships in teleost fishes. Brain, Behavior and Evolution 82: 31–44.
26. Rink E, Wullimann MF (2004) Connections of the ventral telencephalon (subpallium) in the zebrafish (Danio rerio). Brain Research 1011: 206–220.
27. Appleby MC, Sandøe PT (2002) Philosophical debate on the nature of well-being: implications for animal welfare. Animal Welfare 11: 283–294.
28. Dawkins MS (1998) Evolution and animal welfare. The Quarterly Review of Biology 73: 305–328.
29. Chandroo KP, Duncan IJH, Moccia RD (2004) Can fish suffer?: perspectives on sentience, pain, fear and stress. Applied Animal Behaviour Science 86: 225–250.
30. Kirkden RD, Pajor EA (2006) Using preference, motivation and aversion tests to ask scientific questions about animals' feelings. Applied Animal Behaviour Science 100: 29–47.
31. Yue S, Duncan IJH, Moccia RD (2008) Investigating fear in rainbow trout (*Oncorhynchus mykiss*) using the conditioned-suppression paradigm. Journal of Applied Animal Welfare Science 11: 14–27.
32. Endo M, Kumahara C, Yoshida T, Tabata M (2002) Reduced stress and increased immune responses in Nile tilapia kept under self-feeding conditions. Fisheries Science 68: 253–257.
33. Herrero MJ, Martinez FJ, Miguez JM, Madrid JA (2007) Response of plasma and gastrointestinal melatonin, plasma cortisol and activity rhytms of european sea bass (*Dicentrarchus labrax*) to dietary supplementation with tryptophan and melatonin. Journal of Comparative Physiology B 177: 319–326.
34. Tzschentke TM (2007) Measuring reward with the conditioned place preference (CPP) paradigm: update of the last decade. Addiction Biology 12: 227–462.
35. Prus AJ, James JR, Rosecras JA (2009) Conditioned place preference. In: Buccafusco JJe, editor. Methods of behavior analysis in neuroscience boca raton (FL): CRC Press. pp. Chapter 4.
36. Mathur P, Berberoglu MA, Guo S (2011) Preference for ethanol in zebrafish following a single exposure. Behavioural Brain Research 217: 128–133.
37. Backström T, Brännäs E, Nilsson J, Magnhagen C (2014) Behaviour, physiology and carotenoid pigmentation in Arctic charr *Salvelinus alpinus*. Journal of Fish Biology 84: 1–9.
38. Chatain B (1994) Abnormal swimbladder development and lordosis in sea bass (*Dicentrarchus labrax*) and sea bream (*Sparus auratus*). Aquaculture 119: 371–379.
39. Silva PIM, Martins CIM, Engrola S, Marino G, Øverli Ø, et al. (2010) Individual differences in cortisol levels and behaviour of Senegalese sole (*Solea senegalensis*) juveniles: evidence for coping styles. Applied Animal Behaviour Science 124: 75–81.
40. Øverli Ø, Sørensen C, Nilsson GE (2006) Behavioral indicators of stress-coping style in rainbow trout: do males and females react differently to novelty? Physiology & Behavior 87: 506–512.
41. Arends RJ, Mancera JM, Munoz JL, Wendelaar Bonga SE, Flik G (1999) The stress response of the gilthead sea bream (*Sparus aurata* L.) to air exposure and confinement. Journal of Endocrinology 163: 149–157.
42. Fanouraki E, Papandroulakis N, Ellis T, Mylonas CC, Scott AP, et al. (2008) Water cortisol is a reliable indicator of stress in European sea bass, *Dicentrarchus labrax*. Behaviour 145: 1267–1281.
43. Scherrer B (1984) Biostatistique. Chicoutimi, Canada: Gaëtan morin.
44. Egan RJ, Bergner CL, Hart PC, Cachat JM, Canavello PR, et al. (2009) Understanding behavioral and physiological phenotypes of stress and anxiety in zebrafish. Behavioural Brain Research 205: 38–44.
45. Pike TW, Samanta M, Lindström J, Royle NJ (2008) Behavioural phenotype affects social interactions in an animal network. Proceedings of the Royal Society B: Biological Sciences 275: 2515–2520.

11

Movements of Blue Sharks (*Prionace glauca*) across Their Life History

Frederic Vandeperre[1,2]*, Alexandre Aires-da-Silva[3], Jorge Fontes[1,2], Marco Santos[1,2], Ricardo Serrão Santos[1,2], Pedro Afonso[1,2]

1 Centre of IMAR of the University of the Azores; Department of Oceanography and Fisheries, Horta, Portugal, **2** LARSyS Associated Laboratory, Lisboa, Portugal, **3** Inter-American Tropical Tuna Commission, La Jolla, California, United States of America

Abstract

Spatial structuring and segregation by sex and size is considered to be an intrinsic attribute of shark populations. These spatial patterns remain poorly understood, particularly for oceanic species such as blue shark (*Prionace glauca*), despite its importance for the management and conservation of this highly migratory species. This study presents the results of a long-term electronic tagging experiment to investigate the migratory patterns of blue shark, to elucidate how these patterns change across its life history and to assess the existence of a nursery area in the central North Atlantic. Blue sharks belonging to different life stages (n = 34) were tracked for periods up to 952 days during which they moved extensively (up to an estimated 28.139 km), occupying large parts of the oceanic basin. Notwithstanding a large individual variability, there were pronounced differences in movements and space use across the species' life history. The study provides strong evidence for the existence of a discrete central North Atlantic nursery, where juveniles can reside for up to at least 2 years. In contrast with previously described nurseries of coastal and semi-pelagic sharks, this oceanic nursery is comparatively vast and open suggesting that shelter from predators is not its main function. Subsequently, male and female blue sharks spatially segregate. Females engage in seasonal latitudinal migrations until approaching maturity, when they undergo an ontogenic habitat shift towards tropical latitudes. In contrast, juvenile males generally expanded their range southward and apparently displayed a higher degree of behavioural polymorphism. These results provide important insights into the spatial ecology of pelagic sharks, with implications for the sustainable management of this heavily exploited shark, especially in the central North Atlantic where the presence of a nursery and the seasonal overlap and alternation of different life stages coincides with a high fishing mortality.

Editor: Athanassios C. Tsikliras, Aristotle University of Thessaloniki, Greece

Funding: This work was conducted in the framework of the EU FP7 project MADE, Mitigating adverse ecological impacts of open ocean fisheries (FP7 KBBE/2007/1/210496). FCT/MCTES-MEC provided individual funding to FV, JF and PA (SFRH/BD/46891/2008, SFRH/BPD/66532/2009, and Ciência 2008/POPH/QREN). IMAR-DOP/UAz is Research and Development Unit no. 531 and LARSyS-Associated Laboratory no. 9 funded by the Portuguese Foundation for Science and Technology (FCT) through FCT - Pest/OE/EEI/LA0009/2011–2014 and by the Azores Fund for Science and Technology (FRCT), funded by OE, COMPETE, QREN and ProConvergencia. The Open Access of this paper is funded by FCT - Pest/OE/EEI/LA0009/2014 (OE, COMPETE and QREN). The funders had no role in study design, data collection and analysis, decision to publish, or preparation of the manuscript.

Competing Interests: The authors have declared that no competing interests exist.

* Email: vandeperre@uac.pt

Introduction

Sharks are generally characterised by a complex spatial organisation of their populations resulting from trade-offs between components of their life history, social and environmental interactions [1–5]. This complex organisation is reflected by their sexual segregation and the existence of discrete locations for key events during their life history, like pupping, nursing and mating. However, the existence and delineation of boundaries within shark populations and in particular the movements of individual sharks within and across these boundaries remain largely unknown. In the context of declining shark populations (e.g., [6–8]) and their deleterious ecological effects [9], unravelling such spatial organisation and, with it, accurately identifying Essential Fish Habitats, is key to develop appropriate management plans for the protection of the most vulnerable life stages [10]. This need is most compelling considering our current limitations in understanding the impacts of heterogeneously distributed fishing pressure on spatially structured shark populations [5,11].

Blue shark is an oceanic predator with a global distribution [12] and a major constituent of the by-catch in pelagic longline fisheries [13–17]. The present status of blue shark stocks is the subject of much debate [18–21]. Blue sharks can be very abundant locally and have a productive life history strategy with a faster growth and a higher number of smaller offspring when compared to other pelagic sharks [21–24]. They are placental viviparous and females can give birth to up to 82 pups ([25]; mean litter size of 37 pups: [26]; up to 70 pups: [27]). Males and females mature at around 183 cm FL and 180–185 cm FL respectively [25,26] at an age of 5–6 years [25]. However, sub-adult females (145 to185 cm FL) can engage in copulation and store spermatozoa in oviducal glands for later insemination, even if they still possess underdeveloped reproductive organs [25].

Results from conventional tagging studies have supported the conclusion, and concomitant management strategy, of a single blue shark stock in the North Atlantic (NA) [28,29]. Yet, blue shark populations are known to segregate by sex and life stages and to make use of ecologically important areas [12]. Pupping areas are generally thought to be located off the western coasts of Portugal and North Africa [30,31], although alternative locations near the Mid-Atlantic ridge in the central NA [17] and the Gulf of Guinea [26] have also been proposed. Although juveniles can be found in large parts of the NA, juvenile females appear to dominate in the eastern NA [32,33] whereas juvenile males dominate in the western NA [18,25,34]. Mating has been recorded in the western NA [25], while observations of fresh bite marks indicating recent mating events [17,25,35] and the presence of dense male aggregations [36] suggests that mating also occurs in other areas.

These areas have all been identified based on analyses of fisheries dependent data and conventional tagging data. While these studies are fundamental in understanding the distribution across life stages, they tell us little about the dynamics and connectivity of these areas and are necessarily limited to those regions and seasons where fishing occurs. Furthermore, the definition of EFHs requires an evaluation of their ecological functionality. For example, nursery areas are generally considered to provide juveniles with an advantage, e.g. shelter or abundant food, thus increasing their possibilities to recruit to the adult population [37]. In practice, this is often difficult to evaluate [3,37], particularly for pelagic sharks. Heupel et al. [3] therefore proposed a practical definition based on three criteria: (1) young-of-the-year sharks are more abundant than in other areas, (2) they have the tendency to remain or return for extended periods and (3) the area is repeatedly used across years.

Shark nurseries are typically closed bays or sheltered coastal areas that mainly provide protection from predators (e.g., [38–41]). In addition, nursery areas can be considered as one component of a shark's life history that, as a consequence of trade-offs between these components, would particularly benefit species with low breeding frequency, small size-at-birth, small litter sizes and/or low juvenile growth rates [3,40]. It is therefore pertinent to test the practical definition proposed by Heupel et al. [3] on highly migratory, productive, oceanic species like blue shark [22–24].

Previous telemetry studies conducted on blue shark focussed on relatively short term (up to 210 days) horizontal and vertical movements [42–46]. In the Northwest NA, acoustic telemetry of mainly adult males showed consistent offshore and southerly movements between August and March and diel vertical niche expansion, extending hundreds of meters during daytime [42]. Juvenile and adult blue sharks of both sexes tagged in late summer and autumn off the eastern coast of Canada with PSAT tags were observed to expand their vertical niche as they entered the gulfstream and moved south and east to offshore overwintering grounds [46]. Juveniles of both sexes tagged in the English Channel and off southern Portugal during summer and autumn, displayed a general southward movement, with increased residence in frontal areas and behavioural plasticity in relation to their diel depth preference [44,45]. Yet, because of the design and limited duration of the deployments, these studies provide us with little information about the seasonal movements and residency of different life stages and the complex spatial organisation of the NA blue shark population. In particular, longer-term data series are needed to validate the nursery assumptions.

Analyses of pelagic longline fishery data have indicated that the Azores region, in the central NA, is seasonally and alternately visited by immature, male and female adult blue shark, that it is important as a nursery ground and, potentially, also a pupping and mating ground for this species [17,47]. The seasonal presence of all these life stages in a region with a central location in the NA offers an exceptional opportunity to investigate their movements, gain insights into the complex structuring of the NA blue shark population, and test the assumptions of the nursery concept in an oceanic shark.

This study addresses two objectives. The first was to investigate long-term migratory patterns of different life stages of blue shark, to study how their movements change across their life history, and to assess the connectivity between different areas they occupy. The second goal was to verify the existence of a nursery area for NA blue shark in the central NA, and to determine its boundaries and stability over time. A tagging experiment was set up in the Azores archipelago in which 37 blue sharks belonging to different life stages were tagged with satellite transmitters set for long-term deployments.

Materials and Methods

Ethics Statement

This study was performed according to national Portuguese laws for the use of vertebrates in research, and the work and tagging protocol approved by the Azorean Directorate of Sea Affairs of the Azores Autonomous region (SRAM 20.23.02/Of. 5322/2009), which oversees and issues permits for scientific activities. All procedures followed the guidelines for the use of fishes in research of the American Fisheries Society. The field studies did not involve endangered or protected species, no animals were sacrificed, and procedures for reduction, replacement and refinement were thoroughly adopted.

1. Tagging experiment

Blue sharks were captured onboard the R/V 'Arquipelago' (25 m) using a commercial style American longline (monofilament mainline and wire leaders) geared with 200 to 400 hooks, light sticks and baited with mackerel and squid. Fish selected for telemetry were then tagged off an auxiliary 7 m, low gunnel, fibreglass boat. Sharks were tagged at the surface after immobilization and induction of tonic immobility [48]. A total of 37 animals were tagged with different models of satellite tags, i.e. SPOT (Wildlife Computer SPOT5) and PSAT tags (Wildlife Computer MK10-PATs and Mini-PATs). SPOT tags were attached to the dorsal fin of male and female sharks measuring 127 to 211 cm FL through four nylon threaded rods fixed through stainless steel nuts and programmed to emit continuously on alternating days (one day on, one day off). MK10-PAT and Mini-PAT tags were attached intramuscularly under the first dorsal fin using a stainless steel tether and one of three types of anchors (small and large Wilton anchors, conventional titanium darts), and equipped with a guillotine to prevent descents bellow ca. 1800 m. MK10-PAT tags were deployed on males and females measuring 142 to 202 cm FL and programmed to release after 180 days. Mini-PATs were deployed on smaller individuals of 112 to 127 cm FL of both sexes and programmed to release after 180 or 270 days. Three individuals were double tagged with a SPOT and MK10-PAT tag.

The experimental design focussed on immature individuals in order to test the nursery hypothesis. However, it was also designed to address the movement patterns of different sexes and life stages of blue shark seasonally occurring in the area [17]. For this study, size limits of population segments, based on available knowledge of the species' biology [25] and seasonal occurrence [17], were as follows: (i) Small juveniles (SJ) up to 120–130 cm FL; (ii) Large juveniles (LJ) from 120 to 183 and 185 cm FL for males and females (i.e. including sub-adults females) respectively; and (iii) Adults (AD) above this latter thresholds. Four dedicated tagging cruises were performed during the four different seasons. Each cruise specific population segments were targeted for tagging, reflecting their seasonal availability (Table 1; Fig. S1): (i) SJ males and females (112–127 cm FL, n = 8) during summer and autumn; (ii) LJ females (127–178 cm FL, n = 9) during autumn and winter; (iii) LJ males (133–183 cm FL, n = 14) during summer and autumn; (iv) AD females (201–202 cm FL, n = 2) during winter and spring; and (v) AD males (200–211 cm FL, n = 4) during summer and autumn.

2. Analytical methods

Geographical positions of the SPOT tag transmissions and the popup locations of the PAT tags were obtained through the Argos system. Geographical positions from the PAT tags were reconstructed from archived light intensity curves transmitted by the tags after popup or retrieved after physical recovery of the tag. Geolocation (GLS, Global Location Sensing) was performed with the WC-GPE2 software, a program provided by the tag manufacturer (Wildlife Computers).

The raw tracks were post-processed using the IKNOS-WALK model [49]. In this approach, tracks are corrected and interpolated at fixed intervals through a non-state based random walk using a forward particle filter. In contrast with state-space models, no inferences are made about the unknown state of an animal in order to calculate subsequent positions [49]. Instead, locations are estimated from a cloud of weighted particles. The weighting can be manipulated to apply corrections based on known constraints (e.g. maximum speed) or available data. This flexible and intuitive approach has the additional advantage of being able to deal with both Argos and GLS positions and doing so taking into account the specific error distributions of each data type. In particular, longitude information is still preserved for position estimation during equinox, when latitude estimation is problematic for light based geolocation [50]. Another advantage over some methodologies is that, in case of data gaps, positions are interpolated in a straight line without over-fitting artefacts.

The tracks were interpolated to obtain one daily position after imposing a maximum speed and impeding tracks from crossing land. The speed constraint is important for modelling the GLS positions, which have very large error distributions. It was set to 3 km/h based on speed estimates calculated from quality 2 and 3 Argos positions separated by time intervals of 1 to 4 days (Fig. S2) and movement information from the literature [42,46,51]. The error around estimated positions is represented by the dispersion of 50 time-matching alternative positions (represented as a cloud in Fig. 1). Fifty percent of the distances between modelled locations and 'real' GPS locations were reported to be less then 20 km and 104.8 km for Argos and GLS locations respectively [49]. The accuracy of the modelled tracks is further dependent on the quality of the track [49]. Consequently, an additional filter was applied on the modelled tracks in which interpolated points were considered to belong to a data gap when the dataset contained no raw location estimates within 72 hours from its time stamp (represented as yellow dots in Fig. 1).

In order to quantify and visualise the seasonal space use of the blue shark population segments, we calculated home range parameters for each of four quarters: from February to March, April to June, July to October and November to January. Fixed quarterly Kernel Utilisation Distributions (KUDs) were estimated for the different life stages based on aggregated locations of individuals within a given segment with deployments exceeding 90 days. KUDs were only calculated from estimated positions that did not belong to a data gap and for quarters for which more than 100 aggregated positions were available. All KUD calculations were performed using the adehabitatHR package [52] on the R statistical platform [53]. The remotely sensed SST data that we used in the analysis were obtained from NOAA's CoastWatch program and were provided as a monthly blended product from microwave and infrared sensors carried on multiple platforms (Japan's Advanced Microwave Scanning Radiometer AMSR-E, NOAA Advanced Very High Resolution Radiometer AVHRR, the Imager on NOAA's GOES spacecrafts, and the Moderate Resolution Imaging Spectrometer MODIS aqua) with complete coverage and a spatial resolution of 0.1 degrees. Subsequently, the SST data were averaged to a resolution of 1° using the arithmetic mean.

Results

1. Tag performance

In total we tagged 37 blue sharks with 40 tags, of which only three MK10-PAT tags deployed on LJ males failed to report (Table 1). SPOT tags (n = 21) reported for periods between 31 and 952 days (median = 220 days), eight of which for over one year (> 365 days), resulting in a cumulative transmission period of 6429 days. While these SPOT transmission periods are long, some tracks showed considerable data gaps. The reporting MK10-PAT tags (n = 7) released after 3 to 180 days (median = 91 days). The number of light intensity curves received through the Argos system and the actual number of GLS positions that WC-GPE2 was able to calculate varied greatly between tags (2–41 GLS locations). Access to the full archive of one recaptured tag (Shark 32) allowed for the calculation of 100 GLS positions. Mini-PATs generally performed better than the MK10-PAT tags, resulting in longer deployments of 102 to 273 days (mean = 201 days). Transmission and data quality were also higher than for the MK10-PAT tags resulting in 20 to 220 GLS locations per tag.

2. Recaptures

At least 8 sharks (21.6%) were recaptured by commercial fishermen anywhere from 3 to 952 days after tagging (Table 1). Recaptures were either reported by the fishermen or confirmed by the detection of Argos tracks typical of fishing operations by commercial longliners. An additional two sharks were possibly recaptured, but this could not be verified.

3. Movements

In general, blue sharks tagged in this study displayed wide ranging movements (Fig. 1). The longest deployment was registered for a LJ female (Shark 13, 145 cm FL) that travelled an estimated 28139 km over a period of 952 days. All sharks were oceanic, with individuals exploring shelf break areas, but none apparently moving onto continental shelves. While the tracks revealed important individual variability (Fig. 1), the distance travelled as a function of time at liberty was not significantly different between individuals or life stages (Fig. 2A). The only outlier was shark 9, of which we lost track during 389 days.

3.1 Small Juveniles (SJ). SJ sharks of both sexes were tagged with Mini-PAT tags providing low resolution location data (GLS). In general, no differences are apparent between the quarterly movements of SJ males and females (Fig. 3). Both sexes used a relatively limited area, comprised between 25° and 45°N and 22°

Table 1. Summary data for the 34 reporting blue sharks tagged off the Azores with satellite tags.

Shark	Deployment	Size (cm FL)	Sex	Life Stage	Class	Tag	Programmed Release (Days)	Recorded Release (Days)	Track Duration (Days)	Track Length (Km)	Argos Locations (n)	GLS Locations (n)	Data gaps (Days)	Recapture
1	17-10-2009	116	F	Juvenile	SJ	Mini-PAT	180	179	179	5316	0	54	74	
2	20-08-2010	122	F	Juvenile	SJ	Mini-PAT	270	102	102	3958	0	74	36	
3	20-08-2010	120	F	Juvenile	SJ	Mini-PAT	270	235	235	9346	0	208	17	
4	17-10-2009	116	M	Juvenile	SJ	Mini-PAT	180	180	180	5379	0	64	53	
5	20-08-2010	112	M	Juvenile	SJ	Mini-PAT	270	174	174	6295	0	124	36	
6	20-08-2010	127	M	Juvenile	SJ	Mini-PAT	270	193	193	6348	0	20	35	Y
7	20-08-2010	120	M	Juvenile	SJ	Mini-PAT	270	273	273	8761	0	193	37	
8	21-08-2010	112	M	Juvenile	SJ	Mini-PAT	270	270	270	11533	0	220	38	
9	19-02-2009	127	F	Juvenile	LJ	SPOT			877	14494	236	0	389	
10	19-02-2009	139	F	Juvenile	LJ	SPOT			616	16907	779	0	21	
11	26-02-2009	175	F	Sub-adult	LJ	Double	180	91	226	8275	293	10	20	Y
12	26-02-2009	142	F	Juvenile	LJ	Double	180	30	30	911	40	26	3	(Y)
13	26-02-2009	145	F	Juvenile	LJ	Double	180	89	952	28139	699	22	82	Y
14	26-02-2009	148	F	Sub-adult	LJ	SPOT			90	2637	100	0	8	(Y)
15	06-03-2009	165	F	Sub-adult	LJ	MK10 PAT	180	3	3	76	0	2	3	Y
16	06-03-2009	156	F	Sub-adult	LJ	MK10 PAT	180	179	179	5307	0	41	31	
17	02-12-2009	178	F	Sub-adult	LJ	SPOT			579	15498	314	0	137	
18	16-10-2009	168	M	Juvenile	LJ	SPOT			36	1213	56	0	4	
19	16-10-2009	133	M	Juvenile	LJ	SPOT			206	5041	152	0	24	
20	16-10-2009	156	M	Juvenile	LJ	SPOT			82	2245	87	0	8	
21	02-12-2009	130	M	Juvenile	LJ	SPOT			42	973	41	0	6	
22	20-08-2010	164	M	Juvenile	LJ	SPOT			244	6892	196	0	22	
23	20-08-2010	180	M	Juvenile	LJ	SPOT			381	13066	387	0	16	
24	20-08-2010	140	M	Juvenile	LJ	SPOT			126	2364	189	0	6	Y
25	20-08-2010	183	M	Juvenile	LJ	SPOT			116	3846	160	0	9	
26	20-08-2010	183	M	Juvenile	LJ	SPOT			524	15066	612	0	32	Y
27	20-08-2010	159	M	Juvenile	LJ	SPOT			212	5994	332	0	10	
28	20-08-2010	172	M	Juvenile	LJ	SPOT			369	12451	610	0	12	Y
29	06-03-2009	202	F	Adult	AD	MK10 PAT	180	143	143	3577	0	2	89	
30	18-05-2012	201	F	Adult	AD	SPOT			33	1297	33	0	7	
31	16-10-2009	201	M	Adult	AD	SPOT			161	5108	133	0	22	
32	17-10-2009	200	M	Adult	AD	MK10 PAT	180	103	103	4338	0	100	13	Y
33	20-08-2010	207	M	Adult	AD	SPOT			228	6710	341	0	29	
34	20-08-2010	211	M	Adult	AD	SPOT			526	14500	929	0	24	

F – female; M – male; SJ – Small Juveniles; LJ – Large Juvenile, including sub-adults for females; AD – Adult; Double – Double tagged with SPOT and MK10 PAT tags; (Y) not confirmed recapture.

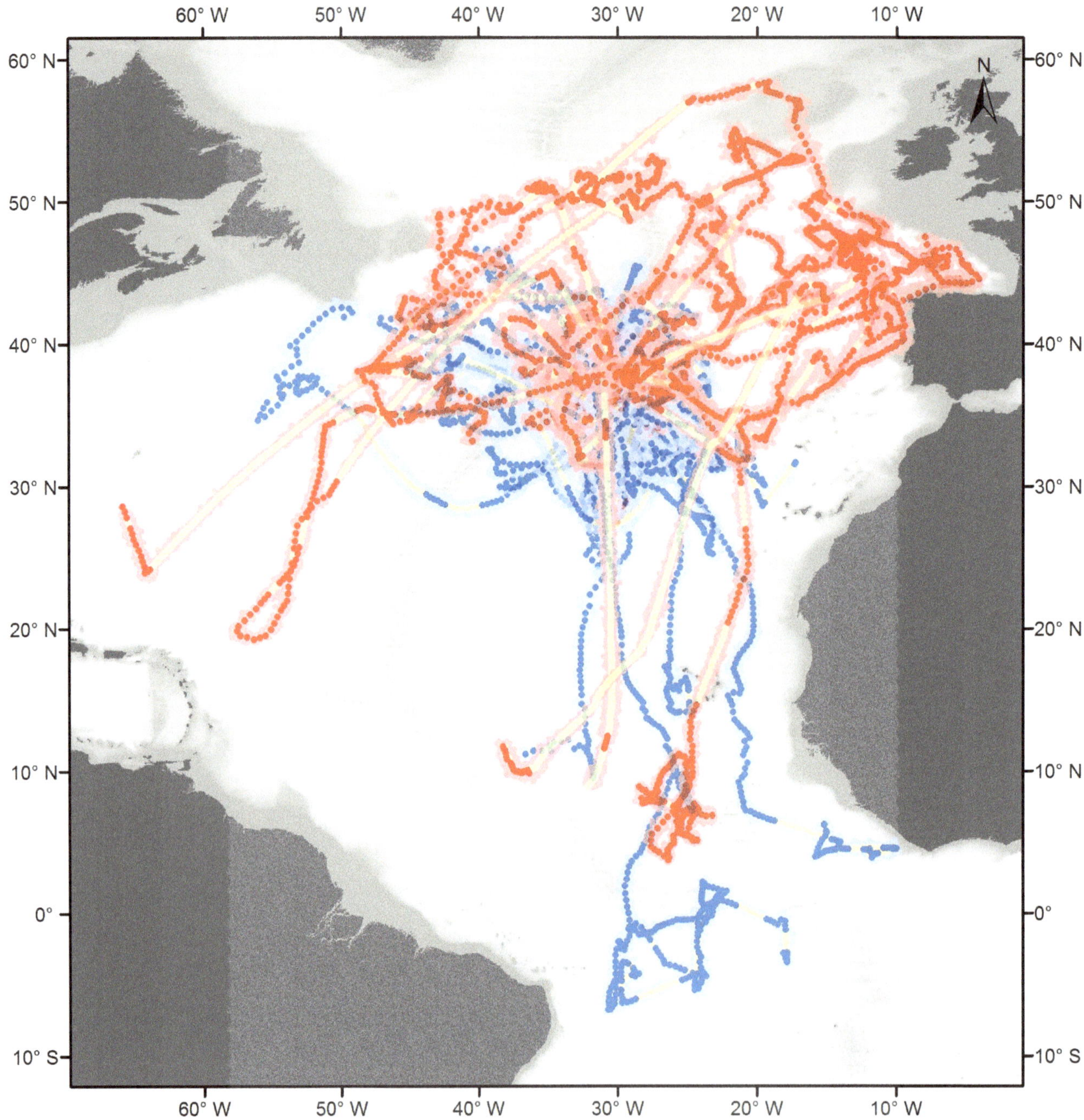

Figure 1. Reconstructed migratory pathways of blue sharks tagged in the Azores. Red and blue dots represent the most probable daily position estimates of respectively female and male sharks, with pink and light blue clouds representing the respective errors around the position estimates. Yellow dots indicate data gaps (i.e. days without position information within 72 h intervals).

and 56°W for up to 235 days. Two sharks (sharks 1 and 4), a male and a female tracked for a period of 6 months between October 2009 and May 2010 remained mainly in an area south-southwest of the Azores, near the MIDAR and the Atlantis – Great Meteor seamount complex. The six sharks, four males and two females tagged in August 2010 and tracked for periods between 101 and 270 days, spent most time in this same general area. Two males (sharks 7 and 8) initiated a westward movement during January. Shark 8 moved as far west as the eastern edges of the New England seamounts, before we lost its track near the southern slopes of the Grand Banks at the end of May. Shark 7 moved to an area close to 36.5°N 43°W, before returning to the same general area south-southeast of the Azores in May. Shark 5 moved to the Newfoundland Basin during autumn and returned by January.

3.2 Large Juveniles (LJ). Large juvenile females were tagged in autumn (2009, n = 1) and winter (2009, n = 8) with SPOT and/or MK10-PAT tags, providing high resolution location data (Argos data) for most individuals (sharks 9–15 and 17). In contrast with the SJ of both sexes, LJ females performed extensive latitudinal movements, utilising an area comprised between 31°N and 59°N and 3°W and 50°W (Fig. 4). LJ females displayed marked seasonal migrations, residing in the southern part of their range during

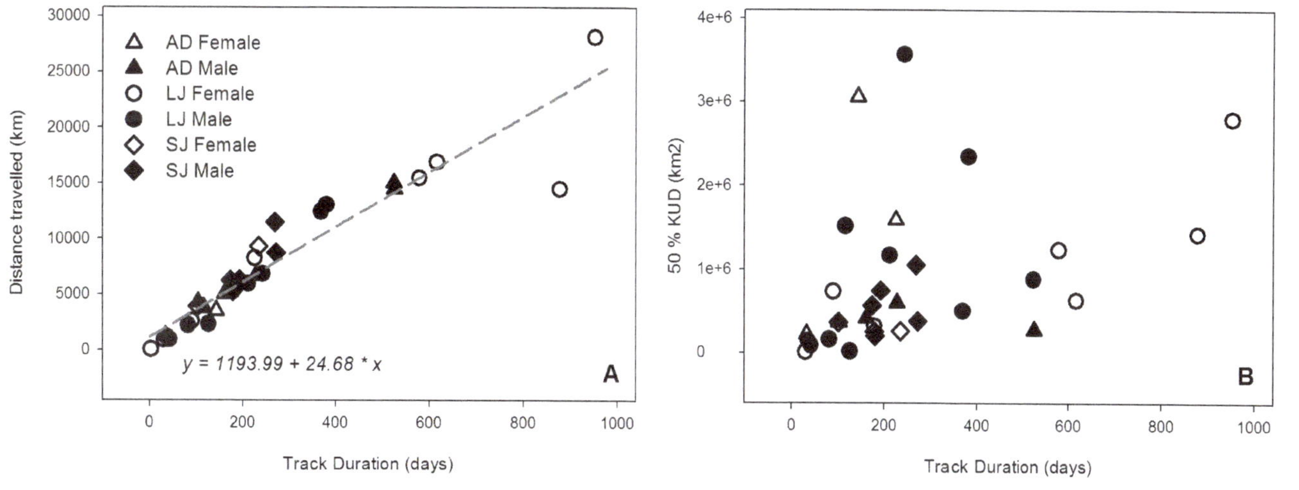

Figure 2. Distance travelled (A) and 50% fixed Kernel Utilisation Distributions (KUD) (B) in function of the track duration for the blue sharks belonging to different life stages. AD – Adults; LJ – Large juvenile; SJ – Small juvenile.

winter (i.e. between 31°N and 45°N) and moving northward during summer (i.e. between 45°N and 59°N). During the latter season, LJ females explored mainly seamount and slope areas in the north-eastern NA, from the Bay of Biscay to the Hatton bank (sharks 9, 10, 13 and 16), near the Charlie-Gibbs Fracture Zone (sharks 13 and 17) and east and south of the Flemish cap (sharks 13 and 17). Four individuals (three juveniles and one sub-adult, sharks 9, 10, 13, 17), were tracked for over a year allowing the observation of a full seasonal cycle (Fig. 5). All four sharks returned to the Azores approximately one year after release, despite having moved to different areas during summer. Whereas shark 10 moved to the eastern NA during two consecutive summers, in the same period shark 13 explored almost the entire longitudinal range between the Flemish cap and the Bay of Biscay. Sharks 12 and 15 were probably both recaptured close to the tagging site 30 and 3 days after release, respectively. Shark 14 moved to the Iberian coast.

LJ males were tagged with SPOT tags during summer (2010, n = 7) and autumn (2009, n = 4), providing high resolution location data for all sharks. This population segment also displayed extensive movements, but the seasonal nature of these movements appeared less pronounced as for LJ females (Fig. 6). Nonetheless, by December, 10 of 11 LJ males started moving south after spending up to 105 days (median = 37 days) in an area close to the tagging location. The core of this area, located south-southwest of the islands of Faial and Pico, is constituted by a complex of seamounts and offshore banks (e.g. Condor Seamount, Açores Bank and Princess Alice Bank) that joins the MIDAR further to the southwest. We will refer to this core area as the Princess Alice complex. Four sharks (sharks 21, 18, 20 and 24) were lost between 21 and 46 days after they initiated their southward movement. At that time these fish were located approximately at the latitude of Great Meteor Seamount, 8 degrees south of the tagging location. After the initial southward movement, four of the remaining individuals (sharks 22, 23, 25 and 27), dispersed further south to tropical latitudes with one shark eventually moving to the South Atlantic, where it remained 8 months. In contrast, two sharks that were tracked for 524 and 369 days (sharks 26 and 28, respectively), remained in the area south-southwest of the Azores after their initial southward movement in autumn. At the end of January, both sharks initiated a westward movement to areas in the north-western NA, where they remained for approximately 1.5 months before returning towards the Azores during May. The following autumn, shark 26 moved again to the area south-southwest of the Azores and initiated a new westward movement in December towards the Corner Rise Seamounts. In contrast with the other LJ

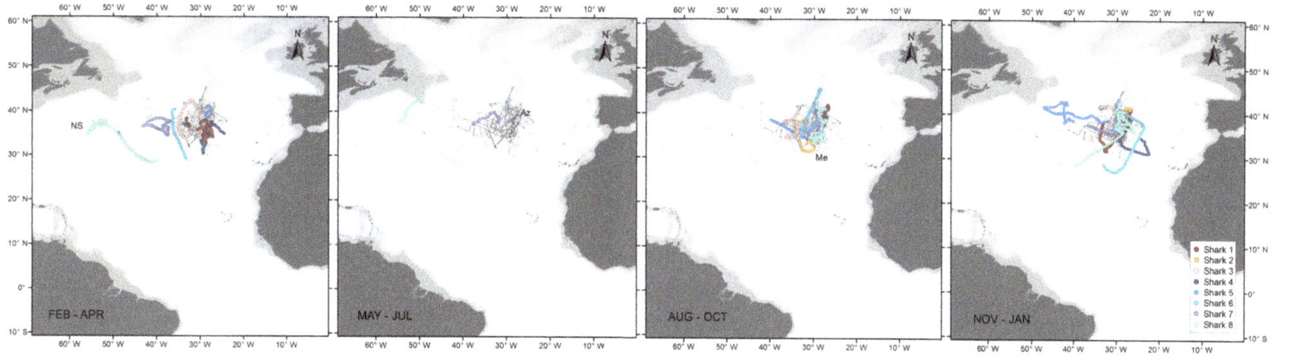

Figure 3. Reconstructed movements of small juvenile male (green – blue) and female (yellow – red) blue shark tagged in the Azores by quarter. Grey lines represent track segments of the individuals during other quarters. Az – Azores, Me – Great Meteor Seamount; NS – New England Seamounts.

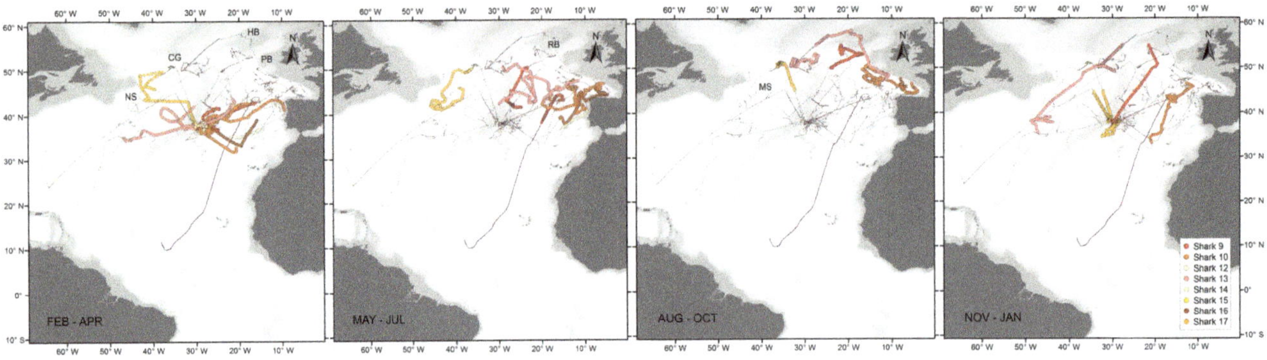

Figure 4. Reconstructed movements of large juvenile and sub-adult (LJ) female blue shark tagged in the Azores by quarter. Grey lines represent track segments of the individuals during other quarters. NS – Newfoundland Seamount; CG – Charlie-Gibbs Fracture Zone; HB – Hatton Bank; PB – Porcupine Bank; RB – Rockall Bank; MS – Milne Seamount.

males, shark 19 (140 cm FL) exhibited movements similar to SJ sharks, remaining in a smaller area close to the Archipelago during the entire duration of the deployment (206 days).

3.3 Adults. Two AD females (shark 29, 202 cm FL; shark 30, 201 cm FL) and one large LJ female (shark 11, 175 cm FL) were tagged in winter (2009, n = 2) and spring (2012, n = 1) with SPOT and/or MK10-PAT tags, providing high resolution location data for two individuals (sharks 11 and 30). Shark 30 released fully developed pups during capture, indicating late stage pregnancy at the time of tagging. All three individuals migrated south within 33 days after tagging (Fig. 7). After remaining 135 days (June to October) in tropical waters, in an area comprised between 3.5°N–11°N and 23°W–28.5°W, shark 11 was recaptured by a fisherman who reported that the shark was carrying embryos at the time. Similar southward migrations towards tropical and sub-tropical waters were also observed for 3 juvenile and sub-adult females that were tracked for over a year. After 830 days of tracking, shark 9 reappeared for a period of 29 days in tropical waters near 10°N 36°W in June (2011) after its signal had been lost for 389 days. Sharks 13 and 17 both migrated to the southern part of the Sargasso Sea by June 2011 after approximately 826 and 526 days, respectively. Shark 13 moved back north towards the Corner Rise Seamounts and then east towards the MIDAR near 33°N and 40.5°W, where we lost its track in October 2011. The southward migrations of sharks 9, 13, 17 and 29 were accompanied by large periods (137, 82, 389 and 89 days respectively) when no Argos locations or good GLS locations were obtained and reconstructed migratory pathways were difficult to produce.

The movements of the AD males were similar to those observed for LJ males (Fig. 7). Four adult males were tagged during summer (2010, n = 2) and autumn (2011, n = 2) with SPOT or MK10-PAT tags, providing high resolution location data for three sharks (sharks 31, 33 and 34). By December all four individuals started moving south. Before that time, three sharks (sharks 31, 33, 34) spent between 14 and 117 days within the Princes Alice complex. Shark 32 also remained in the vicinity of the archipelago. Thereafter, all four sharks stayed in the area south-southwest of the Azores, mainly near the MIDAR and the Atlantis – Great Meteor seamount complex, but moving as far south as 24°N, at least until the end of January. Shark 31 returned to the Archipelago by the middle of March, when we lost its signal. Shark 33 moved east until we lost its track between Madeira and the Canary Islands at the beginning of April. Shark 34 moved west towards the north-western NA during the month of February, reaching an area south of the Newfoundland Seamount in March. The second summer the shark moved again to the Azores, spending most time within the Princes Alice complex, initiating a new southward movement during November.

4. Space use

Distance travelled as a function of time at liberty was similar across individuals (Fig. 2A) but the concomitant variation on home range size changed dramatically (Fig. 2B). The 25% and 50% KUD areas were highly variable both between sexes and life stages (Fig. 8). The latitudinal and longitudinal displacements over time (Fig. 9) further illustrate the contrast in spatio-temporal dynamics across the species' life history. SJ of both sexes were contained within a narrow latitudinal range and mainly utilised an area delimited by the Azores archipelago to the North, the Atlantis – Great Meteor seamount complex to the South, and the MIDAR to the South-West. There was little difference between sexes, except that males extended their range more towards the West. In contrast, LJ of both sexes showed a wider range in both latitude and longitude. The segregation between both sexes becomes apparent during summer and autumn, when females extended their range to the North and East and males to the South. Conversely, AD females moved South and South-West while AD males used a more limited range, mainly extending their distribution South and West of the tagging location. The figures also clearly illustrate the intermediate geographical position of the tagging location as the distributions of all life stages spatially overlap in this area, albeit in different quarters.

5. Time at SST

The spatio-temporal segregation patterns are also reflected in the pooled time-at-SST profiles for the different population components (Fig. 10). No differences were apparent between SJ males and females, which show two preferred maxima, i.e. approximately 17°C and 25°C, corresponding to the most common temperatures within the narrow latitudinal range they occupied. The figure also shows a discrepancy between the SST recorded by the Mini-PATs and that obtained from remote sensing, which is probably due to heating of the surface layer as a consequence of increased sun exposure and low wind speeds during summer. In contrast, the temperature niche segregation between male and female LJ was evident, with males seemingly preferring warmer waters between 16°C and 30°C. This SST range was roughly identical as for AD males, whereas AD females used a wider temperature range (15°C to 30°C) and, apparently,

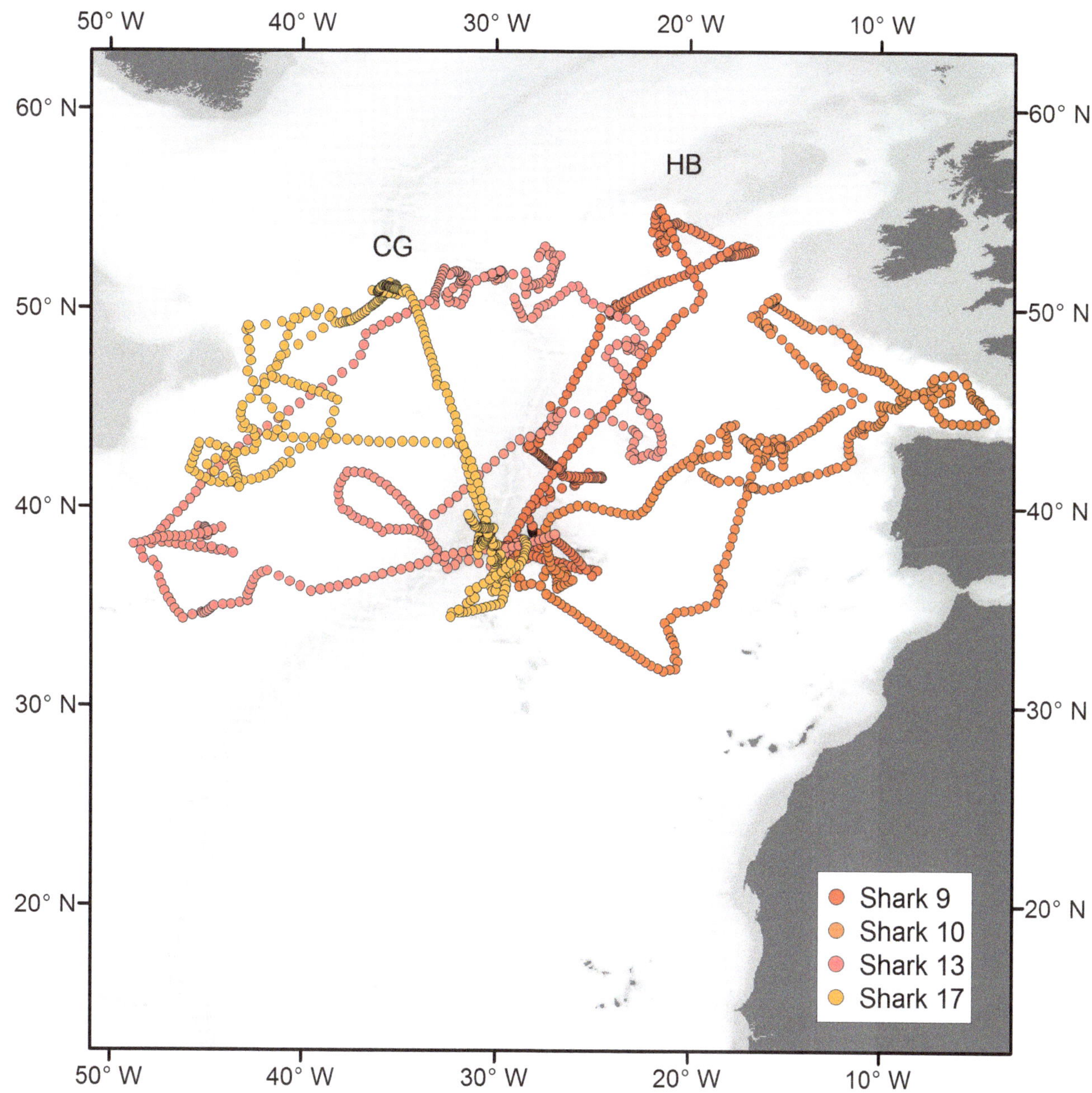

Figure 5. One year reconstructed tracks of four large juvenile and sub-adult (LJ) female blue shark tagged in the Azores. CG – Charlie-Gibbs Fracture Zone; HB – Hatton Bank.

preferred warmers waters within it (25°C to 30°C). However, this pattern is based on a smaller dataset and seasonal coverage.

Discussion

The blue sharks tagged in the Azores displayed extensive movements over vast areas of the North Atlantic, with one individual eventually moving into the southern hemisphere. The sharks appeared to be mainly oceanic and to explore shelf break areas, but did not venture onto the continental shelves, although many studies about blue shark catch data and conventional tagging are from shelf areas (e.g., [18,28,30,34,54]). This finding supports the conclusions of previous studies that pointed out the potential bias that release sites from conventional tagging studies can introduce in the interpretation of movements [12]. Furthermore, the sharks in this study exhibited considerable individual variability in their movements, both among individuals and life stages as well as individually across time.

1. Testing the nursery concept

Our data showed that both male and female SJ blue shark tend to remain for extended periods of time (up to 235 days) in a general area delimited by the Azores, the Atlantis – Great Meteor seamount complex and the Mid-Atlantic Ridge (MIDAR). Additionally, the movement patterns and SST preference did not show any indications of segregation between the two sexes at the spatial scale considered in this study. This pattern is in agreement with available data from fisheries-dependent conventional tagging programs, which suggest that juveniles of 100–

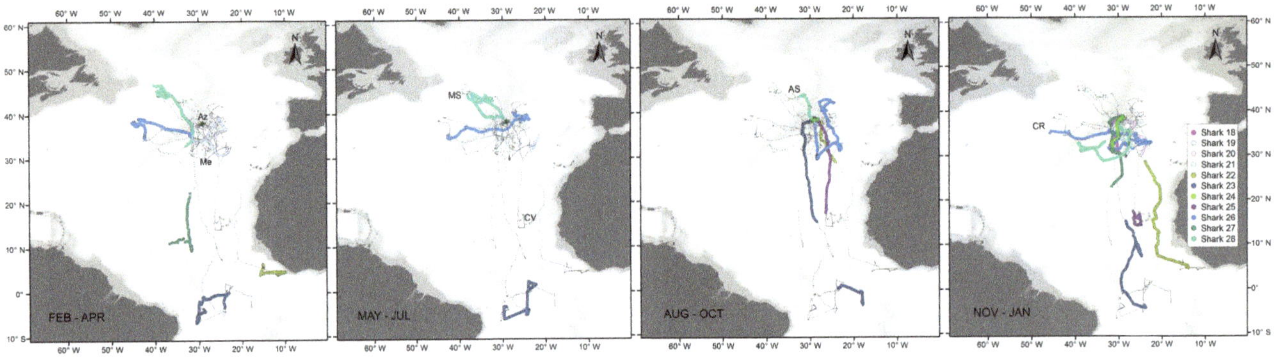

Figure 6. Reconstructed movements of large juvenile (LJ) male blue shark tagged in the Azores by quarter. Grey lines represent track segments of the individuals during other quarters. Az – Azores; Me – Great Meteor Seamount; MS – Milne Seamount; AS – Altair Seamount; CR – Corner Rise Seamounts.

130 cm FL do not participate in extensive latitudinal migrations [12,28,30,32,55]. Nonetheless, Queiroz et al. [44] found movements of juvenile blue shark closer to continental shelves to be more extensive, and Litvinov [36] concluded that male and female blue shark segregate during their first year, at lengths smaller than 70 cm FL. This is in disagreement with the results from the present study and a recent demographic analysis of fisheries data from the Azores region [17]. Both studies indicate that segregation does not occur before, at least, the second year, when blue sharks start to take part in larger scale migrations.

Our study does not present indications of connectivity between the central NA and other juvenile grounds, such as the continental shelves of the Iberian Peninsula and Northern Africa [12,32], reinforcing the hypothesis that parturition may also take place in this area of the NA [17]. Moreover, our study provides additional evidence of the existence of a nursery area for blue shark in the central NA. Not only are juvenile blue sharks locally abundant across years [17,47], but individual blue shark have the tendency to remain and return to the area for extended periods, meeting the three criteria proposed by Heupel et al. [3]. Still, in light of the classical notion of shark nursery areas [40], the classification of such a vast offshore area as a nursery could be disputed. Even so, oceanic nurseries have previously been suggested for some oceanic species, but solely based on observations of small individuals and/or pregnant females in offshore areas, e.g. salmon shark, *Lamna ditropis* [56,57], bigeye thresher, *Alopias superciliosus* [58], and oceanic whitetip, *Carcharhinus longimanus* [59]. Despite the consistent spatial patterns observed, our study was not designed to clarify whether the area provides juveniles with increased survival probability by providing them shelter from predators or some

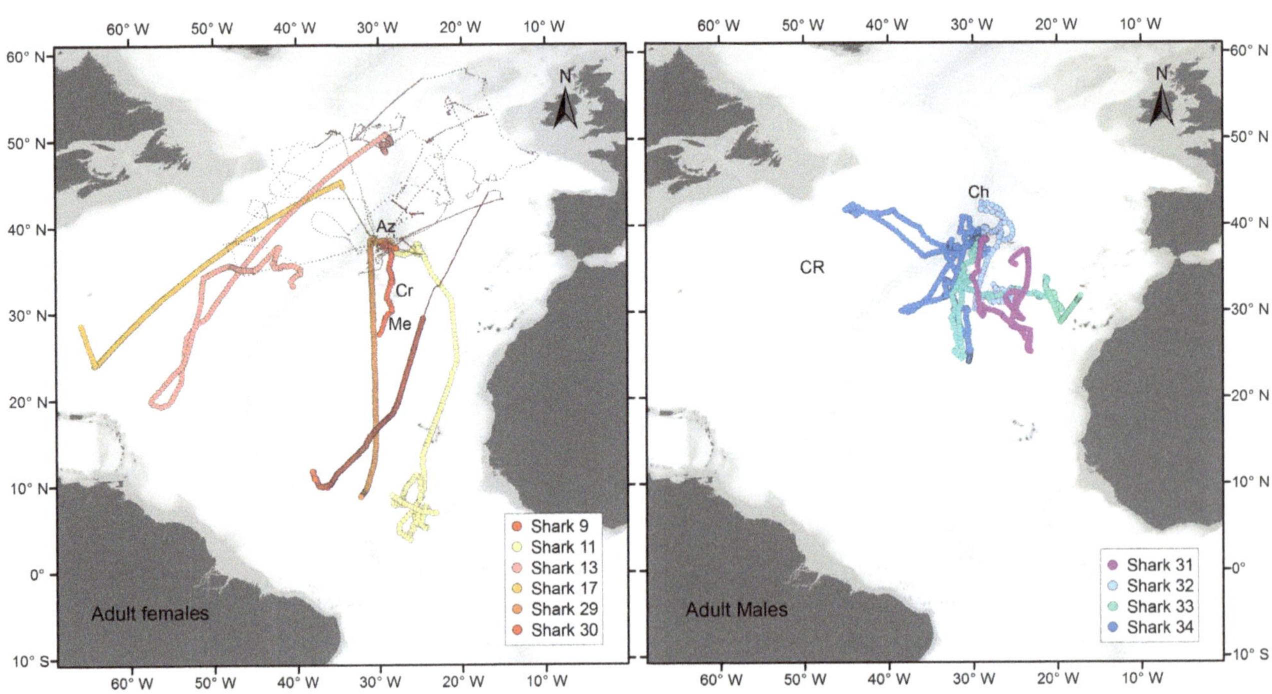

Figure 7. Reconstructed movements of adult (AD) male (right) and female (left) blue shark tagged in the Azores by quarter. Grey lines represent track segments of the individuals during other quarters. Az – Azores; Me – Great Meteor Seamount; Ch – Chaucer Seamount; Cr – Cruiser Seamount; CR – Corner Rise Seamounts.

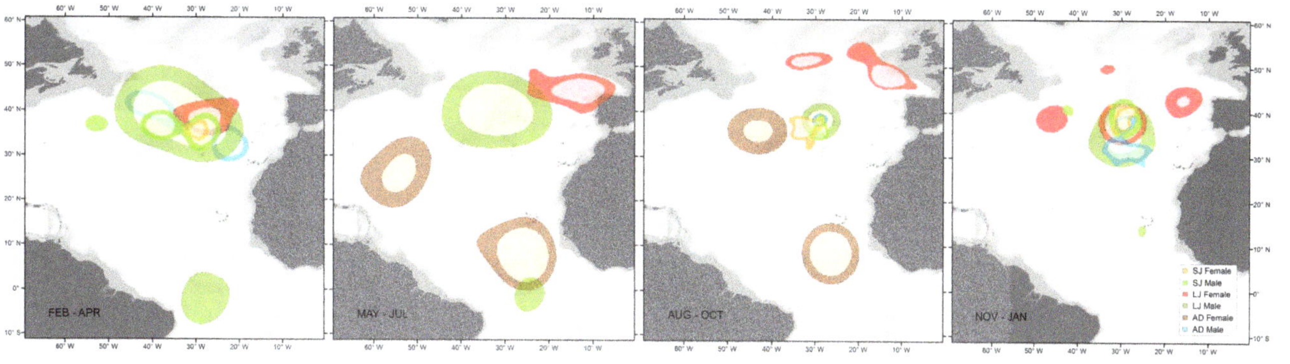

Figure 8. Quarterly 25% and 50% Kernel Utilisation Distributions (KUD) for the different life stages of blue sharks tagged in the Azores. Orange – Small Juvenile (SJ) females; Green – SJ males; Red – Large Juvenile and Sub-adult females; Dark green - Large Juvenile (LJ) males; Brown – Adult (AD) females; Blue – AD males.

other advantages [3,37]. Oceanic sharks have typically larger litters of smaller young than their coastal counterparts, which has been interpreted as an adaptive strategy of the offspring as a whole to thrive on patchy oceanic resources, benefiting from the supposedly lower predation pressure in the open ocean [60]. Lower predation risk in oceanic waters is also seen as an

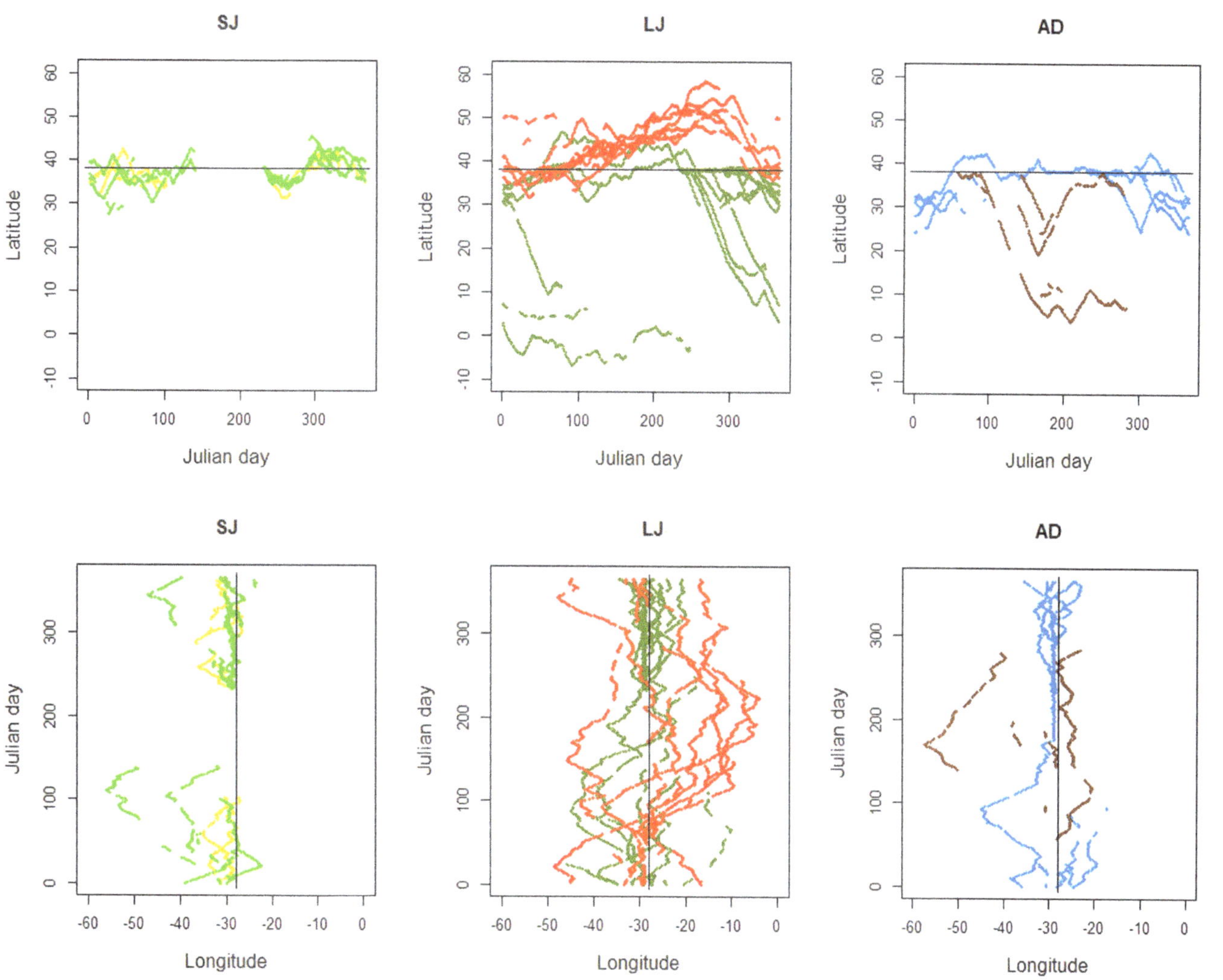

Figure 9. Latitudinal and longitudinal displacements by day of the year of different life stages of blue sharks tagged in the Azores. Orange – Small Juvenile (SJ) females; Green – SJ males; Red – Large Juvenile and Sub-adult females; Dark green - Large Juvenile (LJ) males; Brown – Adult (AD) females; Blue – AD males.

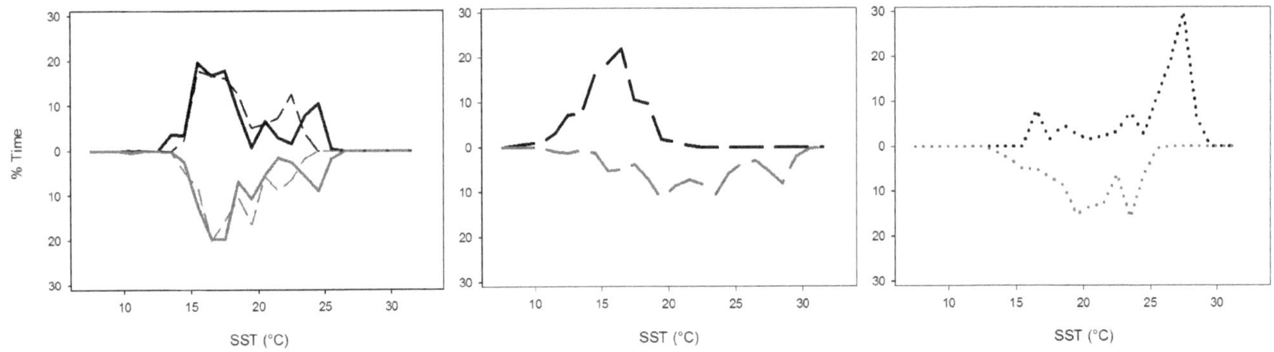

Figure 10. Aggregated long-term time at Sea Surface Temperature (SST, °C) profiles for male (grey) and female (black) blue sharks belonging to different life stages. Top panel: Small Juveniles (SJ); Centre panel: Large Juveniles and Sub-adults (LJ); Bottom panel: Adults (AD). SST values were obtained from averaged (1°) monthly SST images for NOAA's CoastWatch program, except for the dotted curves in the top panel that were obtained directly from the Mini-PAT tags.

evolutionary driver for the oceanic juvenile stage in some sea turtles [61,62]. A favourable habitat providing increased growth rates and thus a reduced time at vulnerable sizes [3] therefore appears to be a plausible advantage for an oceanic nursery habitat of blue shark.

Juvenile survival has been shown to have great impact on blue shark population growth [24]. Due to their small size-at-birth, juvenile blue sharks are probably highly vulnerable during their first year of life [22], which could be balanced through rapid juvenile growth [24,40]. While food is generally not considered a limiting factor for juvenile sharks in typically prey rich estuarine and coastal nursery areas [40], it has been shown to limit the survival of juvenile scalloped hammerheads in a Hawaiian nursery [41] and may play an important role in oceanic and oligotrophic environments. It can therefore be argued that the area provides juvenile blue shark with optimal growth conditions supported by ample food resources associated with the diversity of topographic features (seamounts and islands) [63] and localised oceanographic processes [64]. In addition, the intermediate geographical location of the wider Azores ensures a favourable temperature niche throughout the year. This combination of factors is probably what confers the region an important role for other highly migratory species, e.g. blue and fin whales [65] and juvenile loggerhead sea turtles [66,67]. Remarkably, the distance travelled by these SJ sharks as a function of time is similar to that of other life stages. This finding strongly suggests that their limited distribution when compared to that of larger life stages is unlikely to be merely a consequence of limited swimming capacities. In conclusion, our study provides strong support for the existence of an oceanic shark nursery, although further research is needed to clarify its primary drivers. The designation of the wider Azores as a nursery area also seems meaningful from a management perspective, considering the spatial scale at which the pelagic longline fishery operates.

2. Movements into adulthood

LJ females are known to undertake large scale latitudinal migrations, moving to northern latitudes during summer [12,30,32,33]. In the NA, they dominate the summer catches off south-west England and are present off the US and Canada [18,25,34] whereas in winter high abundances are mainly recorded off Portugal [55] and around the Azores [17,47]. This putative migratory pattern was directly observed in all of our tagged LJ females with periods at liberty exceeding 90 days (n = 6). Although the summer patterns revealed considerable individual variability as individuals explored almost the entire longitudinal range across the NA from the Flemish Cap to the Gulf of Biscay, LJ females displayed a strong connectivity with the eastern NA as indicated by conventional tagging [54]. Regardless of this variability, the four sharks that retained the transmitters for at least one full year were all tracked back to the Azores region during winter, indicating a high degree of site fidelity to the region. LJ females stayed within colder temperatures than younger or older sharks (10°–20°C) and showed a well marked preference for surface waters with an SST between 15° and 16°C throughout the year. The affinity of blue shark for certain water temperatures has been extensively described in the literature although these studies rarely made the same distinction between life stages as we do, making it difficult to compare findings. Nonetheless, highest abundances of juvenile and sub-adult females in the Northwest and central NA were recorded within the same range [17,18].

Tagging AD female blue shark was not the prime objective of the present study. Nonetheless, the few large sub-adult and adult female that we tagged all displayed directed southern movements to tropical latitudes in sharp contrast with LJ females. One AD female was later recaptured in tropical waters carrying embryos, indicating that these movements are related to the reproductive cycle. Additionally, we were able to observe a behavioural switch in three females up to 2.5 years after release, when they reached the theoretical adult size between 170 and 200 cm FL [68]. Up to that point, these sharks had remained in cold waters migrating between sub-tropical and temperate regions, after which all of them moved to southern tropical grounds in June and July utilising pathways on both sides of the ocean. Female blue sharks are thought to move offshore from coastal areas in the north-western NA after mating and to migrate towards the eastern NA to mature and deliver pups [18,25]. However, our results also show that a proportion of mature females move South towards the south-western NA, where their arrival in early summer coincides with the increasing seasonal catch rates of mature and pregnant females [27]. This finding supports the hypothesis of a southbound movement of mature blue sharks into the south-western NA as inferred from conventional tagging data [12,29]. In the eastern NA, adult and pregnant females are seasonally observed in different stages of the reproductive cycle. Mature and pregnant females are found during winter near the Canary Islands and African coast, with mating and pupping believed to occur off Portugal [12] and near the Azores during spring [17]. Castro and

Mejuto [26] found a large proportion of adult females in the Gulf of Guinea, many of which were pregnant. The present study indicates that small adult females also move to offshore areas further south (3°–12°N) into the south-eastern NA. This is in agreement with their presence in tropical latitudes [26,27], a behaviour that was hypothesised to facilitate fertilization and embryonic development [69].

LJ and AD males displayed similar movements and do not appear to segregate spatially. After spending up to 117 days within the Princes Alice seamount complex, both groups started moving south to an area south-southwest of the Azores during autumn. This autumn migration coincides with a decrease in abundance of both life stages in local catches [17]. Accordingly, Litvinov [70] observed high abundance of adult males above the seamounts south of the Azores during autumn and labelled these concentrations as "male clubs". In fact, three sharks (2 LJ and 1 AD) remained in that area until January before initiating a westward movement to offshore areas in the western Atlantic. By the end of spring these fish returned to the Azores, eventually resuming a new southward movement at the end of summer. Such a clear cyclic pattern of migration appears to be common throughout the life history of male blue sharks, as it was also observed in small juveniles. It demonstrates the strong connectivity between the north-western and central NA for males, as previously hypothesised [28,71].

In the northwest Atlantic, adult and sub-adult blue shark enter coastal waters during late spring and summer and engage in mating [25]. Thus, the subsequent migration of males from the central NA to waters off the north-western NA in late spring could well be related to such mating. The behavioural shift in spring of two tagged females in these off-shore areas appears to support this hypothesis, but their capacity to store sperm would not require such a temporal matching of events. In addition to the western migration cycle, LJ males also exhibited an important southward dispersal, which is in accordance with their dominance in tropical waters [26,27].

3. Underlying mechanisms of blue shark spatial ecology

The long-term tracking of blue shark in this study revealed the existence of a high degree of site fidelity with the central NA across almost all life stages, except for adult females for which no long-term tracks were available. Such site fidelity appears to be common in migratory shark species (e.g. great white shark, *Carcharodon carcharias* [72], tiger shark, *Galeocerdo cuvier* [48], broadnose sevengill shark, *Notorynchus cepedianus* [73]) and other pelagic predators like tuna [74] and cetaceans [75] and is generally associated with feeding, parturition or mating. The identification of these specific areas is considered an important step for the management of highly mobile species [76] and the fact that site fidelity towards the central NA is present throughout its life history emphasises the importance of the area for blue shark.

The cyclic migrations and dispersal to tropical latitudes of LJ and AD males suggest some degree of behavioural polymorphism within the male population supported by what seems to be a high tolerance and adaptation to habitat conditions. In fact, the wide geographical range occupied by these males was also reflected in their broad SST niche (12–30°C and 15–30°C, respectively), in agreement with fisheries catch analysis [17]. Such behavioural polymorphism, also observed in many fish (e.g., [77]) and sharks [48], may have ecological reasons, such as the existence of different feeding or reproductive strategies [48]. Given the segregation between AD and LJ females, one hypothesis to explain the behavioural polymorphism in male blue shark could therefore be the co-existence of different reproductive strategies aimed at mating with either of the two female life stages. Indeed, AD females are seldom observed at the mating areas in the north-western NA, where LJ females occur (see above) which are visited by males exhibiting the western migration cycle. Conversely, LJ females are virtually absent in tropical latitudes to where both AD females and (some) males appear to converge. This possibility casts a previously unknown complexity on the behavioural ecology of pelagic sharks and blue shark in particular.

The patterns of spatial segregation between sexes, which are considered a common trait in sharks [1,2,38,39], change over the course of the blue shark's life cycle indicating that the relative importance of the drivers behind the segregation changes along with it. This is particularly so for females, which undergo pronounced ontogenetic shifts in distribution and movement patterns during their juvenile phase as well as upon reaching maturity. Contrastingly, male blue sharks expand their range, mainly to warmer waters during their juvenile phase, but typical movement patterns are observed throughout their life history. As a result, LJ females segregate more clearly from other life stages, occupying colder waters and mainly occurring together with smaller sized individuals of both sexes [17,55]. Because this is a juvenile life stage and growth is similar for males and females, this pattern is likely a means for LJ females to avoid aggressive courtship behaviour until approaching sexual maturity [2,5], which is also why females develop an epidermis twice as thick as that of males [25]. The behavioural shift observed in females upon reaching maturity could then be related to the search of warmer thermal habitat to aid fertilisation and embryonic developments as proposed by Hazin et al. [69]. Other drivers in shaping the segregation patterns in blue shark can not be excluded (see [2,5] for reviews), and it is possible that certain patterns were not observed because they act at different spatial scales than could be detected in the present study.

Conclusion

The long term tracking of different life stages performed in this study demonstrates the complex structuring of the NA blue shark population. The study builds upon previous studies [17,47], providing strong evidence for the existence of a discrete central NA nursery roughly delimited by the Azores archipelago in the North, the Atlantis–Great Meteor seamount complex in the South and the MIDAR in the South-West. This oceanic nursery is used by juveniles of both sexes, at least during their first year, after which they start to show the typical segregation pattern that can be seen in the older juvenile component. Upon reaching maturity, females move to southern latitudes utilising migration pathways on both sides of the NA. Males extended their range, displaying a high connectivity with the western NA and dispersal to tropical latitudes. Moreover, there is a high degree of site fidelity with individuals of nearly all life stages returning to the central NA, emphasizing the importance of this region and their possible role as pupping grounds. Such complex spatial structuring and philopatric behaviour will have important consequences for the management and conservation of blue shark [11,78] in the Atlantic, especially considering that the seasonal overlap and alternation of different life stages in the central NA nursery coincides with high levels of fishing mortality [79].

Supporting Information

Figure S1 Overview of the tagging experiment.

Figure S2 Frequency distribution of blue shark speeds (km/h). Speed (km/h) was calculated as the displacement between quality 2 and 3 Argos locations separated by time intervals of 1–4 days.

Acknowledgments

We would like to thank the crew of the RV Arquipelago and the following individuals who were instrumental in helping with the fieldwork to tag the blue sharks: N. Serpa, M. Schmiing, C. Pham, G. Graça, R. Guedes and A. Morais. JA Musick and one anonymous reviewer provided comments on the manuscript.

Author Contributions

Conceived and designed the experiments: FV PA JF MS. Performed the experiments: FV PA JF MS. Analyzed the data: FV. Contributed reagents/materials/analysis tools: FV PA RSS. Contributed to the writing of the manuscript: FV PA JF AA.

References

1. Klimley AP (1987) The determinants of sexual segregation in the scalloped hammerhead shark, Sphyrna lewini. Env Biol Fish 18: 27–40.
2. Sims DW (2005) Differences in habitat selection and reproductive strategies of male and female sharks. In Ruckstuhl KE and Neuhaus P (Eds), Sexual Segregation in Vertebrates: Ecology of the Two Sexes. Cambridge University Press, Cambridge, UK, pp. 127–147.
3. Heupel MR, Carlson JK, Simpfendorfer CA (2007) Shark nursery areas: concepts, definition, characterization and assumptions. Mar Ecol Prog Ser 337: 287–297.
4. Robbins RL (2007) Environmental variables affecting the sexual segregation of great white sharks *Carcharodon carcharias* at the Neptune Islands South Australia. J Fish Biol 70: 1350–1364.
5. Wearmouth VJ, Sims DW (2008) Sexual segregation in marine fish, reptiles, birds and mammals: Behaviour patterns, mechanisms and conservation implications. Adv Mar Biol 54: 107–170.
6. Ward P, Myers RA (2005) Shifts in Open-Ocean Fish Communities Coinciding With the Commencement of Commercial Fishing. Ecology 86: 835–847.
7. Dulvy NK, Baum JK, Clarke S, Compagno LJV, Cortes E, et al. (2008). You can swim but you can't hide: the global status and conservation of oceanic pelagic sharks and rays. Aquatic Conserv: Mar Freshw Ecosyst 18: 459–482.
8. Ferretti F, Myers RA, Serena F, Lotze HK (2008) Loss of large predatory sharks from the Mediterranean Sea. Conserv Biol 22: 952–964.
9. Ferretti F, Worm B, Britten G, Heithaus MR, Lotze HK (2010) Patterns and ecosystem consequences of shark declines in the ocean. Ecol Lett 13: 1055–1071.
10. Grüss A, Kaplan DM, Guenette S, Roberts CM, Botsford LW (2011) Consequences of adult and juvenile movements for marine protected areas. Biol Conserv 144: 692–702.
11. Mucientes GR, Queiroz N, Sousa LL, Tarroso P, Sims DW (2009) Sexual segregation of pelagic sharks and the potential threat from fisheries. Biol Lett 5: 156–159.
12. Nakano H, Stevens J (2008) The biology and ecology of the blue shark *Prionace glauca*. In: In: Camhi M, Pikitch E, Babcock E (Eds), Sharks of the Open Ocean: Biology, Fisheries and Conservation. Blackwell Publishing, Oxford, UK, pp. 140–148.
13. Mejuto J (1985) Associated catches of sharks, *Prionace glauca*, *Isurus oxyrinchus*, and *Lamna nasus*, with NW and N Spanish swordfish fishery, in 1984. International Council for the Exploration of the Sea, Council Meeting 1985/H 42:16 pp.
14. Hurley PCF (1998) A review of the fishery for pelagic sharks in Atlantic Canada. Fish Res 39: 107–113.
15. Castro J, de la Serna JM, Macias D, Mejuto J (2000) Estimaciones cientificas preliminares de los desembarcos de especies asociados realizados por la flota Espanola de palangre de superficie en 1997 y 1998. ICCAT Coll Vol Sci Pap 51: 1882–1893.
16. Mejuto J, García-Cortés B, Ramos-Cartelle A, de la Serna JM (2009) Scientific estimations of by-catch landed by the Spanish surface longline fleet targeting swordfish (Xiphias gladius) in the Atlantic ocean with special reference to the years 2005 and 2006. ICCAT Coll Vol Sci Pap 64: 2455–2468.
17. Vandeperre F, Aires-da-Silva A, Santos M, Ferreira R, Bolten AB, et al. (2014) Demography and ecology of blue shark (*Prionace glauca*) in the central North Atlantic. Fish Res 153: 89–102.
18. Simpfendorfer CA, Hueter RE, Bergman U, Connett SMH (2002) Results of a Fishery-Independent Survey for Pelagic Sharks in the Western North Atlantic, 1977–1994. Fish Res 55: 175–192.
19. Burgess GH, Beerkircher LR, Cailliet GM, Carlson JK, Cortes E, et al. (2005). Is the Collapse of Shark Populations in the Northwest Atlantic Ocean and Gulf of Mexico Real? Fisheries 30: 19–26.
20. Aires-da-Silva A, Hoey JJ, Gallucci VF (2008b) A Historical Index of Abundance for the Blue Shark (*Prionace glauca*) in the Western North Atlantic. Fish Res 92: 41–52.
21. Anonymous (2009) Report of the 2008 shark stock assessment meeting. ICCAT Coll Vol Sci Pap 64: 1343–1491.
22. Cortés E (2000). Life history patterns and correlations in sharks. Reviews in Fisheries Science 8: 299–344.
23. Frisk MG, Miller TJ, Fogarty MJ (2001) Estimation and Analysis of Biological Parameters in Elasmobranch Fishes: a Comparative Life History Study. Can J Fish Aquat Sci 58: 969–981.
24. Aires-da-Silva A, Gallucci VF (2007) Demographic and Risk Analyses Applied to Management and Conservation of the Blue Shark (*Prionace glauca*) in the North Atlantic Ocean. Mar Freshwater Res 58: 570–580.
25. Pratt HL (1979) Reproduction in the Blue Shark, *Prionace glauca*. Fish Bull 77: 445–470.
26. Castro J, Mejuto J (1995) Reproductive parameters of blue shark *Prionace glauca*, and other sharks in the Gulf of Guinea. Mar Fresh Res 46: 967–973.
27. Tavares R, Ortiz M, Arocha F (2012) Population Structure, Distribution and Relative Abundance of the Blue Shark (*Prionace glauca*) in the Caribbean Sea and Adjacent Waters of the North Atlantic. Fish Res 129: 137–152.
28. Kohler N, Turner P, Hoey J, Natanson L, Briggs R (2002) Tag and recapture data for three pelagic shark species, blue shark (*Prionace glauca*), shortfin mako (Isurus oxyrinchus), and porbeagle (Lamna nasus) in the North Atlantic Ocean. ICCAT Coll Vol Sci Pap 54: 1231–1260.
29. Kohler N, Turner P (2008) Stock structure of the blue shark (*Prionace glauca*) in the North Atlantic Ocean based on tagging data. In: Camhi M, Pikitch E, Babcock E (Eds), Sharks of the Open Ocean: Biology, Fisheries and Conservation. Blackwell Publishing, Oxford, UK, pp. 339–348.
30. Stevens JD (1990) Further Results From a Tagging Study of Pelagic Sharks in the North-East Atlantic. J Mar Biol Assoc UK 70: 707–720.
31. Buencuerpo V, Rios S, Moron J (1998). Pelagic Sharks Associated With the Swordfish, Xiphias gladius, Fishery in the Eastern North Atlantic Ocean and the Strait of Gibraltar. Fish Bull 96: 667–685.
32. Stevens JD (1976) 1st Results of Shark Tagging in Northeast Atlantic, 1972–1975. J Mar Biol Assoc UK 56: 929–937.
33. Henderson AC, Flannery K, Dunne J (2001) Observations on the Biology and Ecology of the Blue Shark in the North-East Atlantic. J Fish Biol 58: 1347–1358.
34. Campana SE, Marks L, Joyce W, Kohler NE (2006) Effects of Recreational and Commercial Fishing on Blue Sharks (*Prionace glauca*) in Atlantic Canada, With Inferences on the North Atlantic Population. Can J Fish Aquat Sci 63: 670–682.
35. Stevens JD (1974) Occurrence and Significance of Tooth Cuts on Blue Shark (*Prionace glauca* L) From British Waters. J Mar Biol Assoc UK 54: 373–378.
36. Litvinov FF (2006) On the Role of Dense Aggregations of Males and Juveniles in the Functional tructure of the Range of the Blue Shark *Prionace glauca*. J Ichthyol 46: 613–624.
37. Beck MW, Heck KL, Able KW, Childers DL, Eggleston DB, et al. (2001) The identification, conservation, and management of estuarine and marine nurseries for fish and invertebrates. BioScience 51: 633–641.
38. Springer S (1967) Social organisation of shark populations. In Gilbert PW, Mathewson RF, Rall DP (Eds), Sharks, Skates and Rays. The Johns Hopkins Press, Baltimore, pp. 149–174.
39. Bass AJ (1978) Problems in studies of sharks in the Southwest Indian Ocean. In: Hodgson ES, Mathewson RF (Eds) Sensory biology of sharks, skates and rays. Office of Naval Research, Department of the Navy, Arlington, VA, pp. 545–594.
40. Branstetter S (1990) Early life-history implications of selected carcharhinoid and lamnoid sharks of the northwest Atlantic. In Pratt HL, Gruber SH and Taniuchi T (Eds), Elasmobranchs as Living Resources: Advances in the Biology, Ecology, Systematics, and the Status of the Fisheries. NOAA Technical Report NMFS 90. pp. 17–28. (US Department of Commerce: Washington, DC.)
41. Duncan KM, Holland KN (2006) Habitat use, growth rates and dispersal patterns of juvenile scalloped hammerhead sharks *Sphyrna lewini* in a nursery habitat. Mar Ecol Prog Ser 312: 211–221.
42. Carey FG, Scharold JV (1990) Movements of blue sharks (*Prionace glauca*) in depth and course. Mar Biol 106: 329–342.
43. Klimley AP, Beavers SC, Curtis TH, Jorgensen SJ (2002) Movements and swimming behaviour of three species of sharks in La Jolla. Env Biol Fish 63: 117–135.
44. Queiroz N, Humphries NE, Noble LR, Santos AM, Sims D (2010) Short-term movements and diving behaviour of satellite-tracked blue sharks *Prionace glauca* in the northeastern Atlantic Ocean. Mar Ecol Prog Ser 406: 265–279.
45. Queiroz N, Humphries NE, Noble LR, Santos AM, Sims DW (2012) Spatial Dynamics and Expanded Vertical Niche of Blue Sharks in Oceanographic Fronts Reveal Habitat Targets for Conservation. PloS ONE 7: e32374.
46. Campana SE, Dorey A, Fowler M, Joyce W, Wang Z, et al. (2011) Migration Pathways, Behavioural Thermoregulation and Overwintering Grounds of Blue Sharks in the Northwest Atlantic. PloS ONE 6: e16854.

47. Aires-da-Silva A, Ferreira R, Pereira J (2008a) Case study: blue shark catch patterns from Portuguese swordfish longline fishery in the Azores. In: Camhi M, Pikitch E, Babcock E (Eds), Sharks of the Open Ocean: Biology, Fisheries and Conservation. Blackwell Publishing, Oxford, UK, pp. 230–234.
48. Meyer CG, Papastamatiou YP, Holland KN (2010) A multiple instrument approach to quantifying the movement patterns and habitat use of tiger (*Galeocerdo cuvier*) and Galapagos sharks (*Carcharhinus galapagensis*) at French Frigate Shoals, Hawaii. Mar Biol 157: 1857–1868.
49. Tremblay Y, Robinson PW, Costa DP (2009) A Parsimonious Approach to Modeling Animal Movement Data. PLoS ONE 4, e4711.
50. Nielsen A, Bigelow KA, Musyl MK, Sibert JR (2006) Improving light-based geolocation by including sea surface temperature. Fish Oceanogr 15(4): 314–32.
51. Stevens JD, Bradford RW, West GJ (2010) Satellite tagging of blue sharks (*Prionace glauca*) and other pelagic sharks off eastern Australia: depth behaviour, temperature experience and movements. Mar Biol 157: 575–591.
52. Calenge C (2006). The package adehabitat for the R software: a tool for the analysis of space and habitat use by animals. Ecol Model 197: 516–519.
53. R Core Team (2013) R: A language and environment for statistical computing. R Foundation for Statistical Computing, Vienna, Austria. (http://www.R-project.org/).
54. Fitzmaurice P, Green P, Keirse G, Kenny M, Clarke M (2005) Stock discrimination of the blue shark, based on Irish tagging data. ICCAT Coll Vol Sci Pap 58: 1171–1178.
55. Queiroz N, Lima FP, Maia A, Ribeiro PA, Correia JP, et al. (2005) Movement of Blue Shark, *Prionace glauca*, in the North-East Atlantic Based on Mark - Recapture Data. J Mar Biol Assoc UK 85: 1107–1112.
56. Nakano H, Nagasawa K (1996) Distribution of Pelagic Elasmobranchs Caught by Salmon Research Gillnets in the North Pacific. Fish Sci 62(5): 860–865.
57. Goldman KJ, Musick JA (2008) The biology and ecology of the salmon shark, Lamna ditropis. In: Camhi M, Pikitch E, Babcock E (Eds), Sharks of the Open Ocean: Biology, Fisheries and Conservation. Blackwell Publishing, Oxford, UK, pp. 95–104.
58. Matsunaga H, Yokawa K (2013) Distribution and ecology of bigeye thresher Alopias superciliosus in the Pacific Ocean. Fish Sci 79:737–748.
59. Bonfil R, Clarke S, Nakano H (2008) The biology and ecology of the Oceanic whitetip shark, Carcharhhinus longimanus. In: Camhi M, Pikitch E, Babcock E (Eds), Sharks of the Open Ocean: Biology, Fisheries and Conservation. Blackwell Publishing, Oxford, UK, pp.128–139.
60. Snelson FF, Roman BL, Burgess GH (2008) The reproductive biology of pelagic elasmobranchs. In: Camhi M, Pikitch E, Babcock E (Eds), Sharks of the Open Ocean: Biology, Fisheries and Conservation. Blackwell Publishing, Oxford, UK, pp. 24–45.
61. Musick JA, Limpus CJ (1997) Habitat utilization and migration in juvenile sea turtles. In: Lutz PL, Musick JA (Eds) The biology of sea turtles. CRC Press, Boca Raton, Florida, USA, pp. 137–164.
62. Bolten AB (2003) Variation in sea turtle life history patterns: neritic vs. oceanic developmental stages. In: Lutz PL, Musick JA, Wyneken J (Eds) The biology of sea turtles, Vol II. CRC Press, Boca Raton, Florida, USA, pp. 243–257.
63. Morato T, Varkey DA, Damaso C, Machete M, Santos M, et al. (2008) Evidence of a seamount effect on aggregating visitors. Mar Ecol Prog Ser 357: 23–32.
64. Mailly T, Blayo E, Verron J (1997) Assessment of the ocean circulation in the Azores region as predicted by a numerical model assimilating altimeter data from Topex/Poseidon and ERS-1 satellites. Ann. Geophysicae 15: 1354–1368.
65. Silva MA, Prieto R, Jonsen I, Baumgartner MF, Santos RS (2013) North Atlantic Blue and Fin Whales Suspend Their Spring Migration to Forage in Middle Latitudes: Building up Energy Reserves for the Journey? PLoS ONE 8: e76507.
66. Bolten AB (2003) Active swimmers passive drifters: the oceanic juvenile stage of loggerheads in the Atlantic system. In: Bolten AB and Witherington BE (Eds). Loggerhead Sea Turtles. Smithsonian Institution Press, Washington, DC. p.63–78.
67. Santos MA, Bolton AB, Martins HR, Riewald B, Bjorndal KA (2007) Air-breathing visitors to seamounts: sea turtles. In: Pitcher TJ, Morato T, Hart PJB, Clarke MR, Haggan N, Santos RS (Eds), Seamounts: Ecology, Fisheries and Conservation. Blackwell Publishing, Oxford, UK, pp. 239–244.
68. Skomal GB, Natanson LJ (2003) Age and Growth of the Blue Shark (*Prionace glauca*) in the North Atlantic Ocean. Fish Bull 101: 627–639.
69. Hazin FHV, Couto AA, Kihara K, Otsuka K, Ishino M, et al. (1994) Reproduction of the blue shark *Prionace glauca* in the south-western equatorial Atlantic Ocean. Fisheries Sci 60: 487–491.
70. Litvinov FF (2007) Aggregations of large pelagic sharks above seamounts. In: Pitcher TJ, Morato T, Hart PJB, Clarke MR, Haggan N, Santos RS (Eds), Seamounts: Ecology, Fisheries and Conservation. Blackwell Publishing, Oxford, UK, pp. 202–206.
71. Mejuto J, García-Cortés B, Ramos-Cartelle A (2005) Tagging-recapture activities of large pelagic sharks carried out by Spain or in collaboration with the tagging programs of other countries. ICCAT Coll Vol Sci Pap 58: 974–1000.
72. Jorgensen SJ, Reeb CA, Chapple TK, Anderson S, Perle C, et al. (2009) Philopatry and migration of Pacific white sharks. Proc R Soc B 277: 679–688
73. Barnett A, Abrantes KG, Stevens JD, Semmens JM (2011) Site fidelity and sex-specific migration in a mobile apex predator: implications for conservation and ecosystem dynamics. Anim Behav 81: 1039–1048.
74. Walli A, Teo SLH, Boustany A, Farwell CJ, Williams T, et al. (2009) Seasonal movements, aggregations and diving behavior of Atlantic bluefin tuna (Thunnus thynnus) revealed with archival tags. PloS One 4: e6151.
75. Foote AD, Simila T, Vikingsson GA, Stevick PT (2010) Movement, site fidelity and connectivity in a top marine predator, the killer whale. Evol Ecol 24: 803–814.
76. Martin TG, Chades I, Arcese P, Marra PP, Possingham HP, et al. (2007) Optimal conservation of migratory species. PloS One 2: e751.
77. Afonso P, Fontes J, Holland KN, Santos RS (2009) Multi-scale patterns of habitat use in a highly mobile reef fish, the white trevally *Pseudocaranx dentex*, and their implications for marine reserve design. Mar Ecol Prog Ser 381: 273–286.
78. Hueter RE, Heupel MR, Heist EJ, Keeney DB (2005) Evidence of Philopatry in Sharks and Implications for the Management of Shark Fisheries. J Northw Atl Fish Sci 35: 239–247.
79. Aires-da-Silva A, Maunder NM, Gallucci VF, Kohler NE, Hoey JJ (2009) A spatially structured tagging model to estimate movement and fishing mortality rates for the blue shark (*Prionace glauca*) in the North Atlantic Ocean. Mar Fresh Res 60: 1029–1043.

12

Identification and Expression Profiling of MicroRNAs in the Brain, Liver and Gonads of Marine Medaka (*Oryzias melastigma*) and in Response to Hypoxia

Karen Lau[1,2]*, Keng Po Lai[1,2], Jessie Yun Juan Bao[3], Na Zhang[3], Anna Tse[1,2], Amy Tong[3¤], Jing Woei Li[4], Si Lok[3¤], Richard Yuen Chong Kong[2], Wing Yee Lui[1], Alice Wong[1], Rudolf Shiu Sun Wu[1,2]*

1 School of Biological Sciences, Kadoorie Biological Sciences Building, The University of Hong Kong, Hong Kong SAR, China, **2** The State Key Laboratory in Marine Pollution, Hong Kong, Hong Kong SAR, China, **3** Genome Research Centre, The Hong Kong Jockey Club Building for Interdisciplinary Research, The University of Hong Kong, Pokfulam, Hong Kong SAR, China, **4** School of Life Sciences, Hong Kong Bioinformatics Centre, The Chinese University of Hong Kong, Hong Kong SAR, China, **5** Department of Biology and Chemistry, City University of Hong Kong, Kowloon, Hong Kong SAR, China

Abstract

The marine medaka (*Oryzias melastigma*) has been increasingly used as a fish model for detecting environmental stresses and chemical contaminants in the marine environment. Recent mammalian studies have shown that environmental stresses can alter the expression profiles of microRNAs (miRNAs), leading to transgenerational effects. Here, we use high-throughput Illumina RNA sequencing (RNA-Seq) for miRNA transcriptome analysis of brain, liver, and gonads from sexually mature male and female marine medaka. A total of 128,883,806 filtered sequence reads were generated from six small RNA libraries, identifying a total of 2,125,663 non-redundant sequences. These sequences were aligned and annotated to known animal miRNAs (miRBase) using the BLAST method. A total of 223 distinct miRNA types were identified, with the greatest number expressed in brain tissue. Our data suggested that 55 miRNA types from 34 families are common to all tested tissues, while some of the miRNAs are tissue-enriched or sex-enriched. Quantitative real-time PCR analysis further demonstrated that let-7a, miR-122, and miR-9-3p were downregulated in hypoxic female medaka, while miR-2184 was specifically upregulated in the testis of hypoxic male fish. This is the first study to identify miRNAs in *O. melastigma* using small RNA deep sequencing technology. Because miRNA expression is highly conserved between marine medaka and other vertebrates, marine medaka may serve as a good model for studies on the functional roles of miRNAs in hypoxia stress response and signaling in marine fish.

Editor: Christophe Antoniewski, CNRS UMR7622 & University Paris 6 Pierre-et-Marie-Curie, France

Funding: These authors have no support or funding to report.

Competing Interests: The authors have declared that no competing interests exist.

* Email: rudolfwu@hku.hk (RSSW); karenytlau@hotmail.com (KL)

¤ Current address: The Chinese University of Hong Kong, Li Ka Shing Institute of Health Sciences, Prince of Wales Hospital, Shatin, New Territories, Hong Kong SAR, China

Introduction

MicroRNAs (miRNAs) are small non-coding RNA molecules (~22 nucleotides in length) encoded in the genome. They have emerged as one of the most important and abundant group of gene expression regulator in a wide range of species spanning from mammals to plants and viruses [1]. MiRNAs are first transcribed into primary miRNAs (pri-miRNAs) and then processed in the nucleus into precursor miRNAs (pre-miRNAs) with hairpin stem-loop structures. In animals, pre-miRNAs are then exported to the cytoplasm, where they are processed into mature miRNAs [2]. Mature miRNAs preferentially target 3′-untranslated regions (3′-UTRs) of mRNAs with which they share partial complementary sequence, leading to post-transcriptional gene silencing through translational repression. They can also lead to degradation of the target mRNA when complete sequence complementarity exists. Each miRNA may have multiple gene targets, and each gene target may also be regulated by more than one miRNA [3,4].

Because of the important biological functions of miRNA, considerable efforts have been made to establish miRNA databases such as miRBase (http://www.mirbase.org), which detail all published miRNA sequences, annotations, and predicted gene targets [5]. The number of miRBase entries has increased exponentially since it was established in [6]. The latest version of miRBase comprises 30,424 distinct mature miRNA products from 206 species, including over 2,578 mature human miRNAs. Considerably fewer mature miRNAs have been reported in fish: 255 have been reported from zebrafish (*Danio rerio*), 108 from pufferfish (*Fugu rubripes*) and 146 from the Japanese freshwater medaka (*Oryzias latipes*) [5,7–8]. Many of these fish miRNAs are conserved and have homologs in the human and mouse genomes [9–10].

Mammalian studies have shown that prenatal exposure to certain environmental stresses (e.g., hypoxia) and chemical contaminants (e.g., endocrine disrupting chemicals) can lead to alterations in phenotypes (e.g., delayed physical and sexual development, infertility, increased mortality and behavioral changes) in successive generations, despite the offspring have never been directly exposed to the stress and contaminants before [11,12]. Such transgenerational effects may not be caused by alterations in DNA sequence but by changes in the epigenome. For example, changes may occur in DNA methylation, covalent modification of histones, or the activation or silencing of genes by miRNAs [13]. MiRNAs have been shown to modulate cellular differentiation and development, and various hypoxia-responsive miRNAs (HRMs) have been reported in human breast and colon cancers [14–16]. Although the functional roles of several HRMs have been extensively studied in mammals [17,18], the biological roles of miRNAs in fish remain poorly understood. Small freshwater fish models (e.g. zebrafish and freshwater medaka) have been used extensively to identify and study biological responses to stresses in the freshwater environment [19,20]. The marine medaka (which is very closely related to *O. latipes*) has been developed as a small fish model for similar studies in the marine environment [21]. However, the miRNA profile of this species is unknown.

New technologies such as high throughput RNA sequencing (RNA-Seq) have rapidly increased the rate of miRNA discovery. The high sensitivity of RNA-Seq allows for both the detection of species-specific, conserved miRNAs and the detection of weakly expressed miRNAs. MiRNA discovery has often relied on the mapping of small RNA sequence reads to a reference genome. Because a reference genome is currently not available for *O. melastigma*, predicting miRNAs in this species is difficult. However, with the development of alignment-type methods such as BLAST [22] for identifying homologs of known miRNAs, it has become possible to predict candidate miRNAs in the marine medaka. This is the first report on the identification and expression profiles of miRNAs in male and female *O. melastigma,* and also the differential responses of selected miRNAs to hypoxic stress. The results presented here will be invaluable for future transgenerational studies and risk assessment using this model marine fish species.

Methods

Medaka maintenance and RNA Isolation

All animal research procedures were approved by the Committee on the Use of Live Animals in Teaching and Research (CULATR, #2714-12) at the University of Hong Kong. The stock of marine medaka used in our experiment was obtained from Interocean Industries (Taiwan) and has been reared in our laboratory for over 10 generations. Marine medaka were maintained under optimal growth and breeding conditions as described in Kong *et al*, (5.8 mg O_2 L^{-1}, 28±2°C, pH 7.2 in a 14-h light: 10-h dark cycle) [21]. The fish were fed with hormone-free Aquatox Feed (Zeigler Bros. Inc. Gardners, PA) and live artemia. For hypoxia study, fifty juvenile male and fifty female marine medaka were reared under normoxia (5.8+0.2 mg O_2/L) or hypoxia (1.5+0.2 mg O_2/L) for three months. At 120 days post fertilization, the fish were anesthetized on an ice bath. Then, brain, liver, ovary and testis tissue were dissected from randomly selected male ($n = 10$ from normoxia and $n = 10$ from hypoxia) and female ($n = 10$ from normoxia and $n = 10$ from hypoxia) fish. Total RNA from pooled tissue samples was extracted using the mirVana miRNA isolation kit (Applied Biosystems) and treated with DNase (Ambion) to remove contaminating genomic DNA. RNA quality was assessed using the Agilent 2100 Bioanalyzer system and samples with a RNA Integrity Number (RIN) greater than 9 were used for small RNA library construction.

Illumina small RNA library construction and sequencing

Six small RNA (cDNA) libraries were constructed (brain, liver and ovaries from female; brain, liver and testis from male), each prepared from 1 μg of total RNA. Small RNAs were gel purified to enrich for molecules less than 30 nucleotide in length, and Illumina adaptors (Integrated DNA technology) were ligated to their 5′ and 3′ ends to enable the construction of strand-specific cDNA libraries [23]. The ligation products were used as a template for cDNA synthesis using SuperScript II Reverse Transcriptase (Invitrogen), and an oligonucleotide primer with sequence complementary to the 3′ adapter was used to initiate cDNA synthesis (5′-CTCGGCATTCCTGCTGAACCGCTC-3′). cDNAs were amplified in 12 cycles with AccuPrime *Pfx* DNA Polymerase (Life Technologies) using PCR primer A (5′ - AATGATACGGCGACCACCGAGATCTACACTCTTTCCC-TACACGACGCTCTTCCGATCT 3′) and PCR primer B (5′-CAAGCAGAAGACGGCATACGAGATCGGTCTCGGCAT-TCCTGCTGAACCGCTCTTCCGATCT 3′), and products were purified on Novex 8% TBE polyacrylamide gels (Invitrogen). Amplified DNA of 140–190 nt was extracted from the gel and purified using the QiaQuick Gel Extraction Kit (Qiagen). Sequencing was performed on the Illumina Genome Analyzer GAIIx. Sequences were extracted from image files using the Illumina pipeline, set at default parameters to yield sequence reads of 58 base lengths. Low-quality sequences, homopolymers and adaptor sequences were removed. The filtered reads were processed into different read lengths for further analysis.

Illumina sequencing data analysis

Because the libraries were constructed strand-specifically, only reads in the sense orientation were considered in our analysis. The filtered reads were binned into non-redundant sets of unique small RNAs to tabulate the expression level profile for each library. The relative abundance of each unique miRNA was given as the number of tabulated reads per million filtered reads (RPM). Non-redundant miRNAs of between 20 and 23 bp having RPM≥6 were annotated using BLAST (Basic Local Alignment Search Tool) against miRBase version 17 [5], a database of known mature miRNA and mature star sequences. For unique species, only the highest-scoring alignments were considered for further analysis. Candidates were required to have at least 15 nucleotide matches to reference miRNAs with no more than 3 mismatches and to have an E-value lower than 0.06. In cases where a single reference miRNA from the database might match more than one sequence species from our *O. melastigma* small RNA library, the most abundant miRNA having the highest homology to the reference miRNA sequence was selected as the canonical candidate miRNA [24].

qPCR analysis and identification of hypoxia-responsive miRNAs

TaqMan MicroRNA Assays (primer and probe sets PN: 4427975 and 4440886, Applied Biosystems) were used to detect and quantify five representative miRNA genes (let-7a, miR-9-3p, miR-27a, miR-122 and miR-2184) from the RNA tissue samples collected from male ($n = 10$ from normoxia and $n = 10$ from hypoxia) and female ($n = 10$ from normoxia and $n = 10$ from hypoxia) fish, according to the manufacturer's protocol. U6 small nuclear RNA and 5S ribosomal RNA (5S rRNA) were used as an endogenous control for the analysis. All target sequences are shown in Table S1. cDNA was synthesized from total RNA using gene-specific primers. Reverse transcriptase reactions contained 10 ng of total RNA, 50 nM stem-loop RT primer, 1X RT buffer, 0.25 mM dNTPs, 3.3 U/µl MultiScribe reverse transcriptase, and 0.25 U/µl RNase inhibitor (TaqMan MicroRNA Reverse Transcription Kit, Applied Biosystems). The RT reactions were incubated in a C1000 Thermal Cycler (Bio-Rad) at 16°C for 30 min, then at 42°C for 30 min, at 85°C for 5 min and then held at 4°C. qRT-PCR was performed using the StepOnePlus Real-Time PCR system (Applied Biosystems). The 20 µl PCR included 1.3 µl RT product, 10 µl TaqMan Universal PCR Master Mix II (2X) without UNG (Applied Biosystems), 1 µl TaqMan Small RNA Assay, and 7.7 µl nuclease-free water. Reactions were incubated in a 96-well optical plate at 95°C for 10 min, followed by 40 cycles of 95°C for 15 sec and 60°C for 1 min. Reactions were run in triplicate and included a no-template control for each gene. Expression levels were determined based on the threshold cycle values (C_t) of each target miRNA normalized to the 5S rRNA and U6 snRNA. Statistical significance was assessed using Wilcoxon–Mann–Whitney test.

Availability of sequence data

The sequence data from this study have been submitted to the NCBI SRA (http://www.ncbi.nlm.nih.gov/sra) under the accession number SRP041922.

Results

Construction of small RNA libraries and Illumina sequencing

To identify miRNA from marine medaka (*O. melastigma*), six small RNA libraries were generated and sequenced using the Illumina GAIIx. A schematic description of the workflow for *O. melastigma* miRNA discovery is shown in Figure 1. A total of 128,883,806 high quality filtered sequence reads were generated, and 2,125,663 non-redundant RNAs were identified. Between 18 and 25 million usable pre-processed reads were obtained for each library. The number of unique species following removal of redundant sequences is shown in Table 1.

The length distributions of sequenced small RNAs in the six libraries were analyzed (Figure S1). In brain and liver, small RNAs of 21–23 nt in length were the most abundant species. However, in male and female gonads, the most abundant species are those between 26 and 28 nt. This distribution might reflect the presence of piRNAs in germ cells. It is consistent with the distribution that Houwing *et al* reported for zebrafish, in which ovary- and testis-specific piRNAs were detected [25]. The identification of piRNAs in ovaries and testis in the medaka will be assessed in future studies.

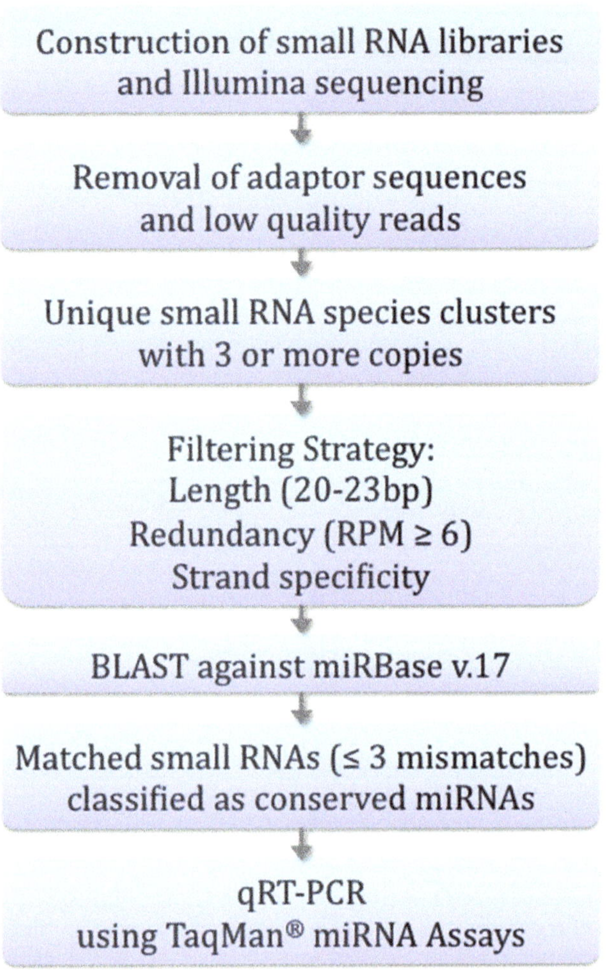

Figure 1. Schematic diagram of the workflow for *O. melastigma* miRNA discovery. Brain, liver and gonadal (ovary and testis) tissues of male and female marine medaka were submitted to small RNA libraries preparation and were sequenced using Illumina GAIIX. After the removal of low quality reads and filtering, high quality sequencing reads were blasted against miRBase v.17 to identify the conserved miRNAs of marine medaka.

A stringent screening process for the most promising miRNA candidates was implemented, using sequence length (20–23 bp),

Table 1. Statistics of pre-processed sequencing reads in six small RNA libraries of *O. melastigma*.

Small RNA Library	Number of Usable Reads	Total Number of Unique miRNA (All sizes)	Unique miRNA (20–23 bp)
Female brain	22,651,811	82,780	22,419
Female liver	18,898,991	54,580	8,763
Female ovary	22,890,085	745,265	55,506
Male brain	20,247,293	65,818	15,968
Male liver	18,660,890	68,024	10,463
Male testis	25,534,736	1,109,196	111,644

expression of RPM≥6, and strand specificity as screening criteria. A total of 3,750 miRNAs were found to satisfy the criteria for potential miRNA (Table 2). The highest complexity was found in brain tissue, with 1,150 and 1,127 screened miRNAs detected in female and male brains, respectively. In contrast, liver, ovary and testis samples each yielded between 300 and 400 miRNAs.

Identification of miRNAs in marine medaka

To identify miRNAs in marine medaka tissues, all screened miRNAs were mapped against the current release of miRBase at the time of analysis [5]. Approximately 20–30% of the screened unique sequences from each library could be annotated as miRNA candidates based on homology to miRBase entries. Some canonical sequences from our libraries differed by only one or two bases from miRBase entries, indicating they are the likely marine medaka orthologs of those miRNAs. The annotated miRNA sequences for the six individual libraries are shown in Tables S2–S7. We also observed potential miRNA paralogs in our libraries: sequence variants related to canonical miRNA sequences. At this time, we cannot rule out the possibility that these are sequencing or amplification errors and not true paralogs encoded by related miRNA genes or arising by differential processing. However, the great sequencing depth of our study ensures a high level of accuracy. To simplify our dataset, potential paralogs were assigned the same miRNA number pending independent analysis. Using BLAST searches to reveal matches to known sequences, a total of 223 known miRNA types from 109 miRNA families were identified (Table S8), 183 of which were known miRNAs and 40 of which were known miRNAs* (all previously named star forms have been re-assigned the −5p/−3p nomenclature).

Tissue- and sex-enriched miRNA expression in marine medaka

Tissue-enriched miRNAs in brain, ovary, testis and liver was elucidated. The highest numbers of genes were expressed in brain tissue (Table 3). In females, 198, 91, and 100 annotated miRNAs were identified in brain, liver and ovary, respectively. In males, 195, 103, and 83 annotated miRNAs were identified in brain, liver and testis, respectively. Comparison of expression profiles showed that 47.5% of identified miRNAs were tissue-enriched under the criteria used in this study: 39.9% of were brain-enriched, while only 3.6% were liver-enriched and 4% were gonad-enriched (Figure 2). The distribution of miRNAs across all three tissues in female and male medaka is shown in Figure 3; 69 miRNAs are found in female tissues and 60 miRNAs are found in male tissues, with only 55 miRNAs being common to all tissues in both males and females (24.7% of total miRNAs). A list of all commonly conserved miRNAs is shown in Table S9.

Further analysis showed that a small number of miRNA candidates are sex-enriched (Figure 4, Table 3). In the brain tissue, 89 miRNAs were identified, 9 of which (miR-20b, miR-83, miR-124-5p, miR-133, miR-184, miR-203a, miR-205a, miR-725, miR-962-3p) are only found in female brain and 3 (let-7a-2-3p, miR-17-3p, miR-124-3p) are enriched in male brain (Figure 4a). Of the 8 liver-enriched miRNAs, only 1 (miR-4448) is enriched in female liver and 2 (miR-749, miR-199b-3p) are only expressed in male liver (Figure 4b). We also observed 9 gonadal enriched miRNAs; 2 are only found in the female gonad (miR-27a, miR-1692) and 3 are enriched in the male gonad (miR-2184, miR-2895, and miR-4682) (Figure 4c).

Highly expressed miRNAs in marine medaka tissues

Among the 223 orthologous miRNA types, miR-9a-3p was the most highly expressed in brain tissue, with RPM values of 23,080 and 26,939 in female and male, respectively (Tables S2–S3). Other highly expressed miRNAs in brain included let-7a, miR-29b and miR-153, each having RPM values greater than 6,000. In liver tissues, the most highly expressed miRNAs were miR-122 with RPM>1000 and miR-194 with RPM>6000 (Tables S4–S5). In gonads, the most highly expressed miRNA was let-7a, with RPM values of 3,563 in ovary and 2,351 in testis (Tables S6–S7). However, the expression levels of let-7a in gonads were fairly low compared to those in brain tissue: 13,722 RPM in females and 10,299 RPM in males. Other orthologous miRNAs that were abundantly expressed in ovary were miR-21, miR-26a, miR-26b, miR-216b, miR-202-5p and miR-143, and in testis miR-21, miR-29b, miR-216b, miR-202-5p, miR-2476, and miR-143.

Interspecies comparison of miRNA expression in marine medaka tissues

The 223 identified miRNA types from 109 miRNA families were found to be homologous to miRNAs identified in 21 different species. Brain miRNAs showed a higher level of species conservation (21 species) than those identified in liver (15 species) and gonads (14 species). For a detailed species conservation analysis, *O. melastigma* miRNAs were compared with those identified in different model organisms including human (*Homo sapiens*), mouse (*Mus musculus*), nematode (*Caenorhabditis elegans*), fruit fly (*Drosophila melanogaster*) and four fish species: zebrafish (*Danio rerio*), tiger blowfish (*Fugu rubripes*), green-spotted pufferfish (*Tetraodon nigroviridis*) and Japanese freshwater medaka (*O. latipes*). BLAST searches revealed that the majority of identified *O. melastigma* miRNAs aligned well with human and zebrafish miRNA sequences (Table S10). A total of 144 of the identified *O. melastigma* miRNAs have previously been identified in zebrafish, and 92 have been identified in freshwater medaka.

Table 2. Annotation of canonical sequences based on known reference (miRBase).

Small RNA Library	Number of screened miRNA	Number of annotated canonical sequences	Number of annotated miRNA types
Female brain	1,150	243	198
Female liver	372	107	91
Female ovary	380	113	100
Male brain	1,127	241	195
Male liver	400	126	103
Male testis	321	89	83

The screened miRNAs satisfy four criteria: (1) sequencing length between 20 and 23 bp; (2) redundancy ≥6 RPM; and (3) strand specificity.

Table 3. Tissue- and sex-enriched miRNA expression in *O. melastigma*.

Tissue Type	Identified *O.melastigma* miRNA
Brain	let-7, let-7a-1-3p, let-7i-3p, miR-99-3p, miR-18a, miR-106, miR-10d, miR-124, miR-124a, miR-125b-1-3p, miR-129-1-3p, miR-129-3p, miR-129-5p, miR-130b, miR-132, miR-132-5p, miR-212, miR-135, miR-135b, miR-137, miR-137-5p, miR-137b, miR-138, miR-138-1-3p, miR-138-2-3p, miR-138b, miR-142-5p, miR-150, miR-153-5p, miR-153b, miR-22-5p, miR-22b, miR-23b-3p, miR-24-1-5p, miR-92a, miR-27b, miR-27b-5p, miR-365, miR-34a, miR-7-1-3p, miR-9a, miR-9a-3p, miR-9-3p, miR-181a-3p, miR-182, miR-183, miR-187, miR-190a-3p, miR-200b, miR-203b-5p, miR-216, miR-218, miR-218b, miR-219, miR-219-2-3p, miR-221-5p, miR-222a-5p, miR-301a, miR-301c, miR-375, miR-454, miR-455, miR-456, miR-458, miR-460b-3p, miR-489, miR-723, miR-724, miR-727, miR-727-5p, miR-728, miR-734, miR-737, miR-338, miR-2187, miR-2188-3p, miR-2188
Brain (Female only)	miR-20b, miR-124-5p, miR-133, miR-83, miR-184, miR-203a, miR-205a, miR-725, miR-962-3p
Brain (Male only)	let-7a-2-3p, miR-17-3p, miR-124-3p
Liver	miR-122, miR-122-3p, miR-148, miR-192, miR-199
Liver (Female only)	miR-4448
Liver (Male only)	miR-199b-3p, miR-749
Gonads	miR-196, miR-196a, miR-202, miR-202-5p
Ovary only	miR-27a, miR-1692
Testis only	miR-2895, miR-4682, miR-2184

Only 3 *O. melastigma* miRNAs (Let-7, miR-1 and miR-124) were highly conserved across all model animals, 52 were conserved in all of the vertebrate models, and 31 were found to be teleost-specific (Table 4).

qRT-PCR and stress response of marine medaka miRNAs

To validate miRNAs discovered during high-throughput sequencing, five representative miRNA candidates were selected for qRT-PCR analysis using RNA from the brain, liver and gonads of male and female *O. melastigma*. TaqMan MicroRNA Assays specific for mature miRNAs were performed with stem-loop RT primers. MiRNAs that were highly expressed in both sexes or that were tissue-enriched were chosen: let-7a (non-tissue enriched), miR-9-3p (brain-enriched), miR-122 (liver-enriched), miR-27a (ovary-enriched), and miR-2184 (testis-enriched). Consistent with our RNA sequencing data, qRT-PCR analysis (Figure S2) showed that let-7a was amplified in all of the male and female tissues examined, with the highest expression level observed in brain. Likewise, miR-9-3p was expressed only in the brain, while miR-122 was expressed only in the liver. As shown in our RNA-seq results, miR-2184 was most highly expressed in testis but was also found at extremely low levels in all other tissues. However, testicular expression of miR-2184 was significantly greater than that of all other tissues ($p<0.05$ for all). The highest level of miR-27a expression was observed in fish ovary. However, in contrast to the sequencing data, RT-PCR analysis indicated that miR-27a expression is not ovary-enriched. This discrepancy could be due to the presence of iso-miRNAs (multiple variants of miRNA genes) miR-27a, b, b-5p, c and e, which qRT-PCR may not be able to distinguish. Interestingly, RNA sequencing identified the presence of miR-27b and b-5p in fish brain only, while miR-27c expression was detected in all tissues and miR-27e in brain and gonads.

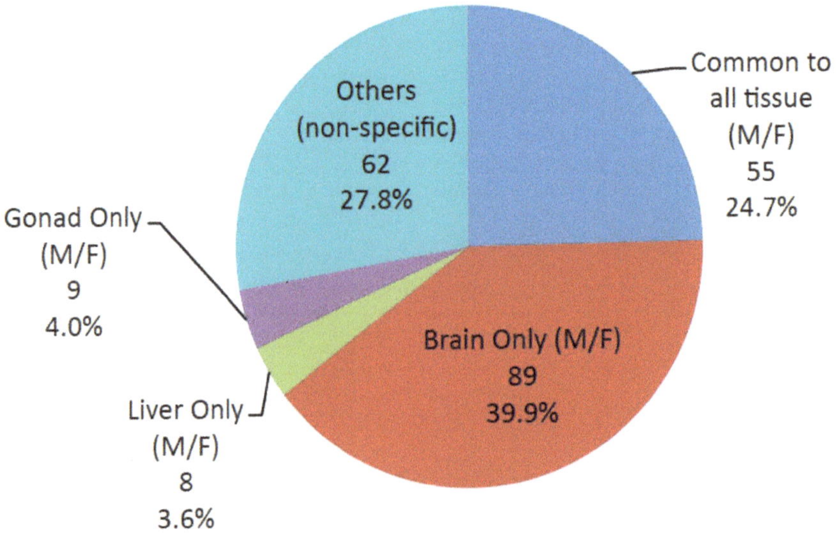

Figure 2. Pie chart of tissue-enriched miRNAs and conserved miRNAs common to all tissues in marine medaka. Diagram demonstrated brain-enriched, liver-enriched and gonad-enriched miRNAs of marine medaka. 'Others' are representative of miRNAs that are not tissue-enriched and not common to all male and female tissues.

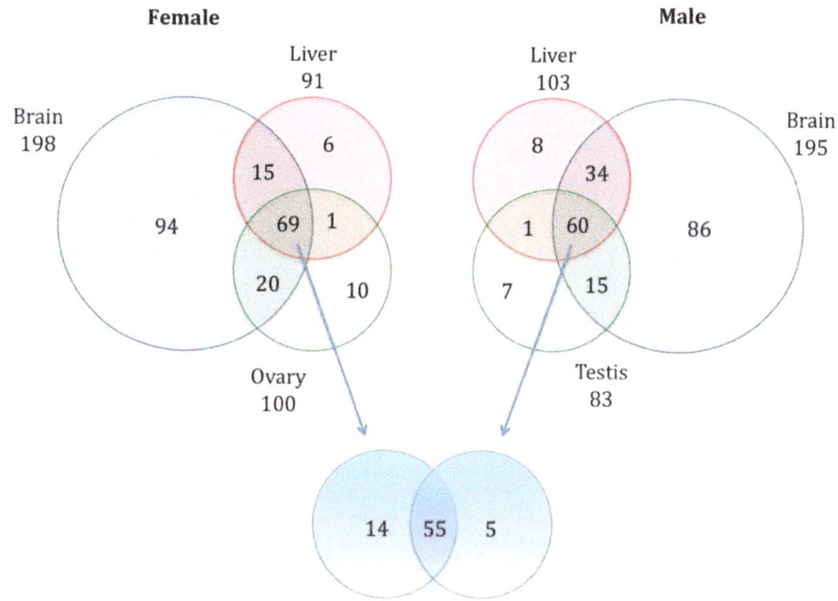

Figure 3. Venn diagrams of distribution of conserved miRNAs across tissues in female and male marine medaka. Diagram showed the distribution of conserved miRNAs in brain (grey), liver (red) and gonads (green) of marine medaka. Bottom diagram (blue) demonstrated the overlapped miRNAs between female and male medaka.

Globally, hypoxia is an important environmental stressor affecting every area in the marine environment [26]. Experiments were therefore further carried out to test the response of five selected miRNA (let-7a, miR-9-3p, miR-27a, miR-122 and miR-2184) to hypoxic stress. qRT-PCR analysis was performed on the brain, liver and gonadal tissues of male and female medaka fish that have been exposed to normoxia or hypoxia for 3 months. Our results demonstrated a significant suppression of let-7a in hypoxic liver and brain tissues of female marine medaka as compared to normoxic tissues, and a similar reduction of let-7a was also found in male liver (Figure 5A–C). However, there is no significant change of let-7a in gonadal tissues and male brain of marine medaka (data not shown). Moreover, a profound inhibition of miR-122 was observed in hypoxic female and male liver (Figure 5D–E). A significant reduction of mi-9-3p was also demonstrated in hypoxic female brain, but not in male brain tissue (Figure 5F and data not shown). In gonadal tissues, a 2-fold induction of miR-2184 was found in testis in response to hypoxia (Figure 5G). However, there is no significant change of miR-27a in ovary under hypoxic exposure (data not shown). Taken together, our data indicated that hypoxia would deregulate the expression profile of tissue-enriched miRNAs in marine medaka, but the biological implications of these changes have yet to be elucidated.

Discussion

The complete genome sequence of the *O. melastigma* has yet to be reported; thus, the established methods for novel miRNA discovery that involve mapping to the genome are not yet possible. The primary aim of the present study is to identify the known miRNAs in the organs along the brain-pituitary-gonad axis and the center of detoxification, liver in marine medaka. All of these organs are important for ecotoxicological studies. BLAST-based method was utilized to identify 223 distinct miRNA candidates in the brain, ovary, testis and liver of marine medaka based on homology to reported orthologous miRNAs. Using high-throughput sequencing, we identified unique miRNAs with expression levels ranging from 6 to nearly 27,000 RPM. Thus the methodology was highly sensitive and had sufficient dynamic range to detect miRNAs with both high and low expression levels. Our qRT-PCR analysis further confirmed that the identified *O. melastigma* miRNAs (let-7a, miR-9-3p, miR-122, miR-27a and miR-2184) are authentic miRNAs, which are differentially expressed by specific tissue and sex. The miRNAs were also highly tissue specific, with the majority brain-enriched. Many identified *O. melastigma* miRNAs in the present study are found to be highly conserved between species and taxa, and orthologous to miRNAs from as many as 21 other species, ranging from nematodes to humans. The majority of these miRNAs aligned well with miRNAs from humans and zebrafish and was also conserved in most vertebrates. Our analysis showed that 92% of *O. melastigma* miRNAs identified in the present study have orthologs previously found in zebrafish. Thus, species in the same taxonomic group appear to express similar miRNAs. The specific functions and gene targets of these teleost miRNAs are not well understood. Our qRT-PCR analysis further identified the hypoxia-responsive miRNAs: let-7a & miR-122 (female liver); let-7a & miR-9-3p (female brain) and miR2184 (testis) in hypoxia-exposed marine medaka.

In mammals, miRNAs are of great importance in the regulation of many fundamental biological processes such as cell proliferation, differentiation, apoptosis, signal transduction and organ development [27–29]. Several known classes of miRNAs have been predicted to regulate specific biological pathways. For example, the first known miRNAs, lin-4 and let-7, were found to play a major role in developmental timing [30,31]. Highly expressed miRNAs are likely to have essential and broad regulatory functions. For example, mature let-7 regulates cell proliferation and differentiation and is also highly conserved across animal species [32]. Let-7 is one of the most highly expressed across all tissues in our data. Thirteen members of the let-7 family were identified in *O. melastigma*, 8 of which were common to the

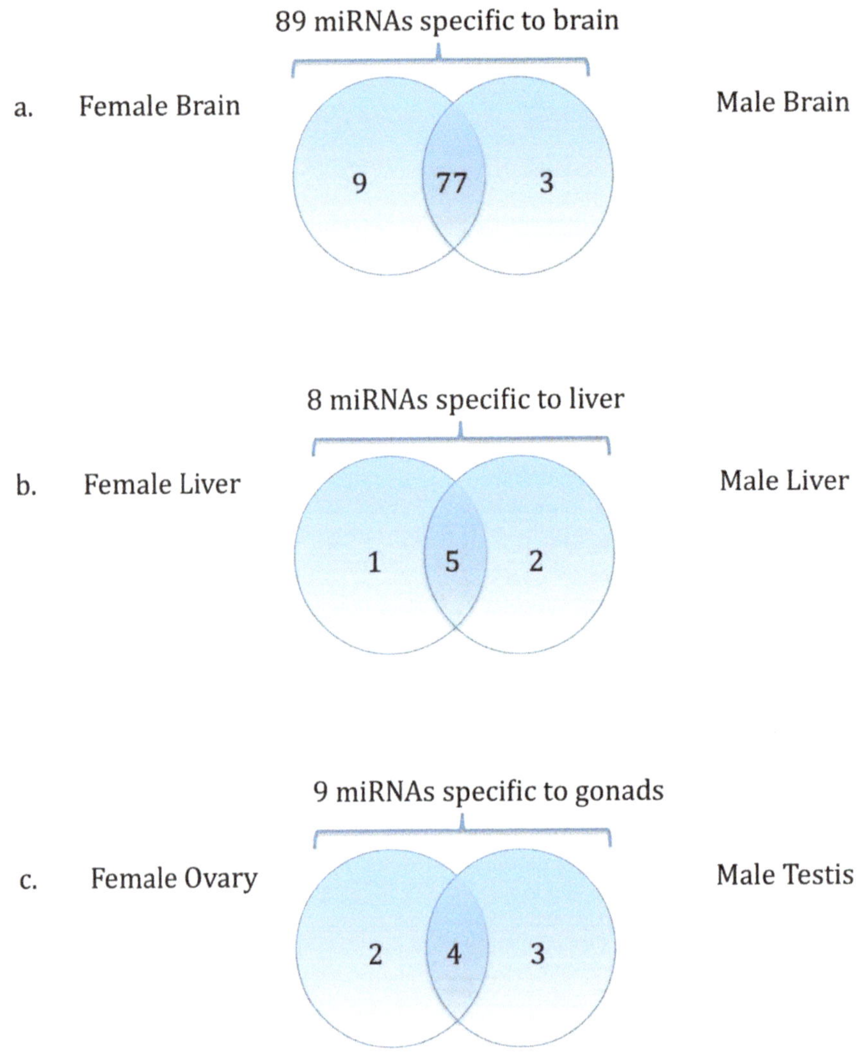

Figure 4. Venn diagrams of sex-dependent distribution of the tissue-enriched miRNAs. Diagram showed the overlapping of tissue-enriched miRNAs in a) brain, b) liver, and c) gonads of female and male medaka.

male and female tissue investigated (let-7a, let-7b, let-7d, let-7e, let-7g, let-7h, let-7i and let-7j). All let-7 members share a similar seed region which regulates the interaction between miRNA and its target genes, so it is generally believed that let-7 family imposes similar biological functions among different species. There is ample reports that demonstrated the importance of let-7 in different biological functions of brain, liver and gonads [33–35]. Sabrina *et al* reported that the introduction of let-7 caused the activation of Toll-like receptor 7, resulting in neurodegeneration in mouse's brain [33]. Also, let-7 was found to control glucose homeostasis and insulin sensitivity in liver [34]. Furthermore, let-7 regulated ageing of testis stem cell niche in drosophila [35].

Other highly expressed miRNAs identified in *O. melastigma* include miR-9, miR-21, miR-29, miR-122, miR-124, miR-143 and miR-202-5p. The orthologous miRNAs identified in the present study have diverse annotated regulatory functions. The brain-enriched miR-9 and miR-124 are both known to have a crucial role in neurogenesis and neuronal development, particularly in the regulation of neural differentiation, proliferation and cell migration [36–38]. The expression of miR-122 was found to be liver-enriched. This is consistent with the results of previous studies and is consistent with its known important role in cholesterol and fatty acid metabolism [39–41]. Recently, let-7, miR-122 and miR-143 have been shown to have tumor suppressive activity, and de-regulated let-7 expression has been associated with cancer [14,16,41]. Some miRNAs have been implicated in the regulation of apoptosis including miR-21 and miR-29, which we found to be expressed in all medaka tissues. MiR-21 is thought to be anti-apoptotic and oncogenic, while miR-29 is thought to be pro-apoptotic, suppressing tumor growth in mice [42,43]. We also observed gonad-enriched expression of miR-202-5p in medaka. This miRNA has been linked to the regulation of gonadal sex-differentiation and together with miR-21 appears to be estrogen-responsive [44,45]. Thus, there is potential for highly expressed medaka miRNAs to respond to endocrine disruptors. However, the effects of such alterations are completely unknown.

Our data also demonstrated the suppression of let-7a in liver and brain tissues under hypoxic condition. This response to oxygen depletion is concordant to the previous finding that hypoxia caused reduction of let-7a in nasopharyngeal carcinoma [46]. Moreover, let-7 transgenic mice study indicated an

Table 4. Fish-enriched miRNAs identified in *O. melastigma.*

O. melastigma miRNA	Sequence	Species Conservation
let-7h	UGAGGUAGUAAGUUGUGUUGUU	dre, fru, tni
let-7j	UGAGGUAGUUGUUUGUACAGUU	dre, fru, tni
miR-10c	UACCCUGUAGAUCCGGAUUUGU	dre, fru, tni
miR-10d	UACCCUGUAGAACCGAAUGUGU	dre, fru, tni, ola
miR-16b	UAGCAGCACGUAAAUAUUGGAG	dre
miR-457a	AAGCAGCACAUCAUUACUGGUA	dre
miR-19d	UGUGCAAACCCAUGCAAAACUG	dre, ola
miR-22b	AAGCUGCCAGUUGAAGAGCUGU	dre, fru, tni
miR-27c	UUCACAGUGGUUAAGUUCUGC	dre, fru, tni, ola
miR-27e	UUCACAGUGGCUAAGUUCAGU	dre, fru, tni
miR-130c	CAGUGCAAUAUUAAAAGGGCAUU	dre, ola
miR-135c	UAUGGCUUUCUAUUCCUAUGUG	dre
miR-301c	CAGUGCAAUAGUAUUGUCAUA	dre
miR-456	CAGGCUGGUUAGAUGGUUGUCU	dre
miR-458	AUAGCUCUUUAAAUGGUACU	dre, fru, tni, ola
miR-460	CCUGCAUUGUACACACUGUGC	dre, fru, tni, ola
miR-462	UAACGGAACCCAUAAUGCAGCUG	dre, ola
miR-722	UUUUGCAGAAACGUUUCAGAUU	dre
miR-723	AGACAUCAGAAAAAUCUGUGCU	dre
miR-724	UUAAAGGGAAUUUGCGACUGUU	dre
miR-725	UUCAGUCAUUGUUUCUGGUCGU	dre
miR-727	UUGAGGCGAGUUGAAGACUUCA	dre
miR-728	AUACUAAGUAUACUACGUUUAC	dre
miR-730	UCCUCAUUGUGCAUGCUGUGUG	dre
miR-731	AAUGACACGUUUUCUCCCGGAUU	dre, ola
miR-734	UAAAUGCUGCAGAAUUGUGC	dre
miR-737	AAAUCAAAGCCUAAAGAAAAUA	dre
miR-1388	AUCUCAGGUUCGUCAGCCCAUG	dre, ola
miR-2184	AACAGUAAGAGUUUAUGUGCUG	dre
miR-2187	UUACAGGCUAUGCUAAUCUGU	dre
miR-2188	AAGGUCCAACCUCACAUGUCCU	dre

dre: *Danio rerio*; fru: *Fugu rubripes*; tni: *Tetraodon nigrovirdis*; ola: *Oryzias latipes*.

impairment of glucose tolerance through the deregulation of insulin receptor (*Insr*) and insulin receptor substrate 2 (*Irs2*) in liver [34], suggesting that the hypoxia-mediated let-7a suppression may result in the dysfunction of medaka liver under hypoxia. But, it has been reported that let-7 can be induced by HIF-1α in response to hypoxia in vascular endothelial cell (HUVECs) [18]. Also, we found the suppression of miR-9-3p in brain tissue of hypoxic fish, which is contradictory to the findings in pulmonary artery smooth muscle cells where hypoxia caused the upregulation of miR-9 [47]. The opposite response of miRNAs to hypoxia treatment suggested that the differential expression of miRNAs in response to stresses may vary among different models. Moreover, we demonstrated the reduction of miR-122 in hypoxic liver tissues. It is similar to the finding in hypoxia-exposed Atlantic salmon [48]. The testis-enriched miRNA, miR-2184 is found to be upregulated in response to hypoxia. So far, there is no report to demonstrate the biological roles of miR-2184. But H-cadherin (CDH13), one of the predicted gene targets of miR-2184 reported in TargetScan-Fish, has been shown to be downregulated in testicular germ cell tumors and associated with differentiation of seminoma [49]. Thus, it would be interesting to study the functional roles of miR-2184-CDH13 pathway in testis in response to hypoxia.

Our previous report has demonstrated that hypoxia is an endocrine disruptor [50]. And, recent studies in mammals have suggested that endocrine-disrupting chemicals can modify the epigenome of the germ line via miRNAs, producing phenotypes in subsequent generations [13,51]. The miRNA profiling and identification of hypoxia-responsive miRNAs of *O. melastigma* in this study serves as an important platform, not only for understanding fundamental biological processes but also provides new insight into the functional roles of miRNAs in responses to environmental stresses such as hypoxia. Conceivably, changes in miRNA profile and specific miRNAs may be employed as biomarkers for transgenerational effects of environmental stress such as hypoxia, which is likely to be an important emerging environmental problem.

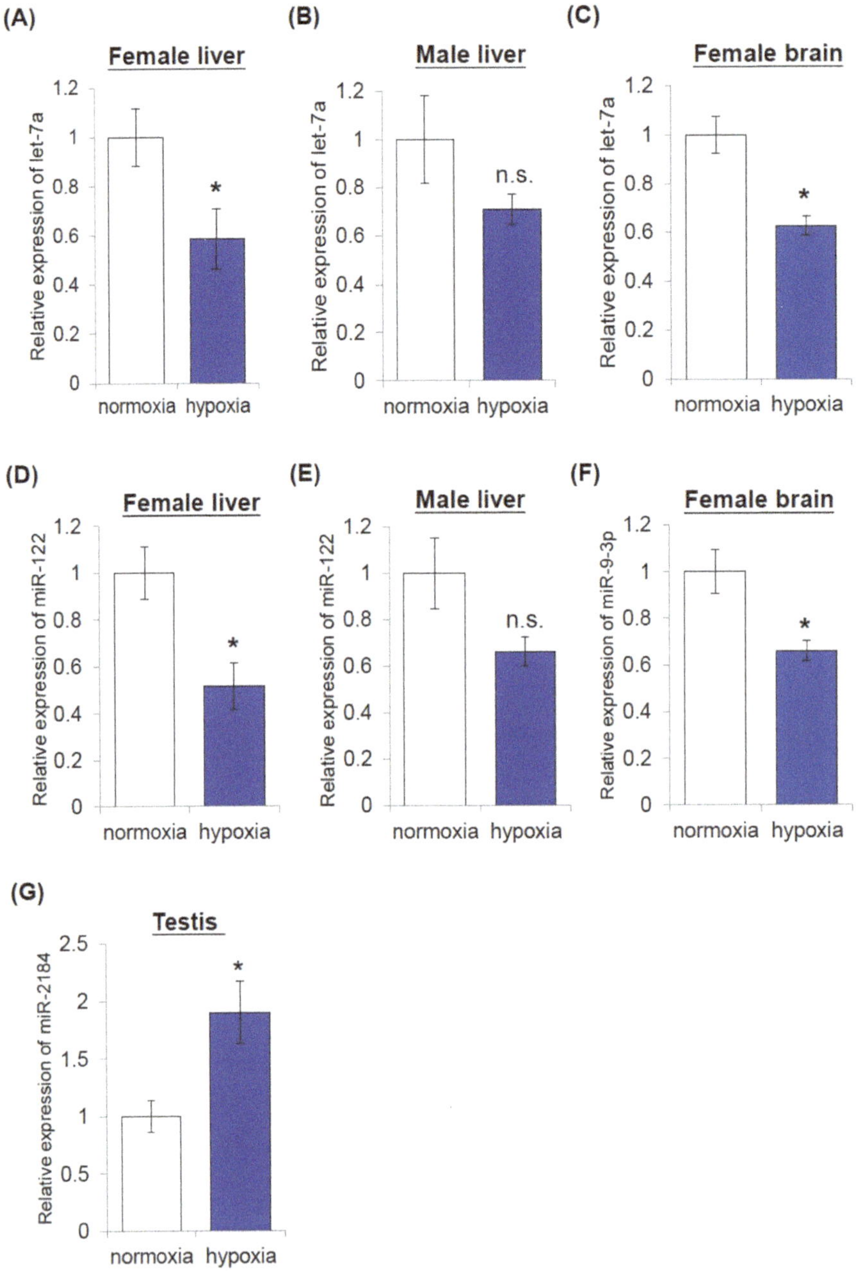

Figure 5. Identification of hypoxia-responsive miRNAs in marine medaka using qRT-PCR analysis. Female and male marine medaka were exposed to either hypoxia or normoxia for 3 months. The differentiation expressions of miRNAs in response to hypoxic stress were analyzed by quantitative real-time PCR. qRT-PCR analysis demonstrated the suppression of let-7a in female liver, male liver and female brain under hypoxia, respectively (A–C). Diagram showed the downregulation of miR-122 in female liver (D) and male liver (E). (F) Demonstrated the reduction of miR-9-3p in female brain. Diagram showed the upregulation of miR-2184 in testis of marine medaka (G). Data are presented as the means ± SEM (*$p<0.05$; n.s.: not significant).

Supporting Information

Figure S1 Length (nt) distribution of small RNAs in male and female brain, liver and gonads of *O. melastigma*. Diagram showed distributions of small RNAs in female brain (upper left), male brain (upper right), female liver (middle left), male liver (middle right), female ovary (bottom left) and male testis (bottom right) of marine medaka.

Figure S2 The qRT-PCR analysis of five representative miRNA candidates using TaqMan MicroRNA Assays. (A) let-7a was amplified in all of the male and female tissues, (B) miR-9-3p was expressed in brain only, (C) miR-122 was expressed in liver only, (D) miR-2184 was most highly expressed in testis, and (E) miR-27a was most highly expressed in the ovary.

Table S1 Sequence information for all TaqMan MicroRNA Assays.

Table S2 Annotated miRNA sequences and expression levels in female brain.

Table S3 Annotated miRNA sequences and expression levels in male brain.

Table S4 Annotated miRNA sequences and expression levels in female liver.

Table S5 Annotated miRNA sequences and expression levels in male liver.

Table S6 Annotated miRNA sequences and expression levels in female ovary.

Table S7 Annotated miRNA sequences and expression levels in male testis.

Table S8 Total conserved miRNAs identified in *O. melastigma*.

Table S9 MiRNAs common to all *O. melastigma* tissues.

Table S10 Highly conserved miRNAs identified in *O. melastigma*.

Author Contributions

Conceived and designed the experiments: KL RSSW. Performed the experiments: KPL A. Tse KL. Analyzed the data: AW WYL JWL. Contributed reagents/materials/analysis tools: JYJB NZ SL A. Tong. Contributed to the writing of the manuscript: KL RSSW KPL RYCK.

References

1. Bartel DP (2004) MicroRNAs: genomics, biogenesis, mechanism, and function. Cell 116: 281–297.
2. Zeng Y (2006) Principles of micro-RNA production and maturation. Oncogene 25: 6156–6162.
3. Lim LP, Lau NC, Garrett-Engele P, Grimson A, Schelter JM, et al. (2005) Microarray analysis shows that some microRNAs downregulate large numbers of target mRNAs. Nature 433: 769–773.
4. Rajewsky N (2006) microRNA target predictions in animals. Nat Genet 38: S8–S13.
5. miRBase website (2013) http://www.mirbase.org. Accessed 3 Dec 2013.
6. Kozomara A, Griffiths-Jones S 11(2011) miRBase: integrating microRNA annotation and deep-sequencing data. Nucleic Acids Res 39: D152–D157.
7. Lewis BP, Shih IH, Jones-Rhoades MW, Bartel DP, Burge CB (2003) Prediction of mammalian microRNA targets. Cell 115: 787–798.
8. Lagos-Quintana M, Rauhut R, Meyer J, Borkhardt A, Tuschl T (2003) New microRNAs from mouse and human. RNA 9: 175–179.
9. Li SC, Chan WC, Ho MR, Tsai KW, Hu LY, et al. (2010) Discovery and characterization of medaka miRNA genes by next generation sequencing platform. BMC Genomics 11(Suppl 4): S8.
10. Chen PY, Manninga H, Slanchev K, Chien M, Russo JJ, et al. (2005) The developmental miRNA profiles of zebrafish as determined by small RNA cloning. Genes Dev 19: 1288–1293.
11. Anway MD, Cupp AS, Uzumcu M, Skinner MK (2005) Epigenetic transgenerational actions of endocrine disruptors and male fertility. Science 308: 1466–1469.
12. Dunaeva TY, Trofimova LK, Graf AV, Maslova MV, Maklakova AS, et al. (2008) Transgeneration effects of antenatal acute hypoxia during early organogenesis. Bull Exp Biol Med 146: 385–387.
13. Skinner MK, Manikkam M, Guerrero-Bosagna C (2011) Epigenetic transgenerational actions of endocrine disruptors. Reprod Toxicol 31: 337–343.
14. Esquela-Kerscher A, Slack FJ (2006) Oncomirs - microRNAs with a role in cancer. Nat Rev Cancer 6: 259–269.
15. Kulshreshtha R, Ferracin M, Wojcik SE, Garzon R, Alder H, et al. (2007) A microRNA signature of hypoxia. Mol Cell Biol 27: 1859–1867.
16. Boyerinas B, Park SM, Hau A, Murmann AE, Peter ME (2010) The role of let-7 in cell differentiation and cancer. Endocr Relat Cancer 17: F19–F36.
17. Huang X, Ding L, Bennewith KL, Tong RT, Welford SM, et al. (2009) Hypoxia-inducible mir-210 regulates normoxic gene expression involved in tumor initiation. Mol Cell 35: 856–867.
18. Chen Z, Lai TC, Jan YH, Lin FM, Wang WC, et al. (2013) Hypoxia-responsive miRNAs target argonaute 1 to promote angiogenesis. J Clin Invest 123: 1057–1067.
19. Volz DC, Bencic DC, Hinton DE, Law JM, Kullman SW (2005) 2,3,7,8-Tetrachlorodibenzo-p-dioxin (TCDD) induces organ- specific differential gene expression in male Japanese medaka (*Oryzias latipes*). Toxicol Sci 85: 572–584.
20. Kissling GE, Bernheim NJ, Hawkins WE, Wolfe MJ, Jokinen MP, et al. (2006) The utility of the guppy (*Poecilia reticulata*) and medaka (*Oryzias latipes*) in evaluation of chemicals for carcinogenicity. Toxicol Sci 92: 143–156.
21. Kong RY, Giesy JP, Wu RS, Chen EX, Chiang MW, et al. (2008) Development of a marine fish model for studying *in vivo* molecular responses in ecotoxicology. Aquat Toxicol 86: 131–141.
22. Altschul SF, Gish W, Miller W, Myers EW, Lipman DJ (1990) Basic local alignment search tool. J Mol Biol 215: 403–410.
23. Hafner M, Landgraf P, Ludwig J, Rice A, Ojo T, et al. (2008) Identification of microRNAs and other small regulatory RNAs using cDNA library sequencing. Methods 44: 3–12.
24. Bar M, Wyman SK, Fritz BR, Qi J, Garg KS, et al. (2008) MicroRNA discovery and profiling in human embryonic stem cells by deep sequencing of small RNA Libraries. Stem Cells 26: 2496–2505.
25. Houwing S, Kamminga LM, Berezikov E, Cronembold D, Girard A, et al. (2007) A role for Piwi and piRNAs in germ cell maintenance and transposon silencing in zebrafish. Cell 129: 69–82.
26. Wu RS (2002) Hypoxia: from molecular responses to ecosystem responses. Mar Pollut Bull 45: 35–45.
27. Karp X, Ambros V (2005) Developmental biology. Encountering microRNAs in cell fate signaling. Science 310: 1288–1289.
28. Miska EA (2005) How microRNAs control cell division, differentiation and death. Curr Opin Genet Dev 15: 563–568.
29. Kloosterman WP, Plasterk RH (2006) The diverse functions of microRNAs in animal development and disease. Dev Cell 11: 441–450.
30. Reinhart BJ, Slack FJ, Basson M, Pasquinelli AE, Bettinger JC, et al. (2000) The 21-nucleotide let-7 RNA regulates developmental timing in *Caenorhabditis elegans*. Nature 403: 901–906.
31. Lee RC, Feinbaum RL, Ambros V (1993) The *C. elegans* heterochronic gene lin-4 encodes small RNAs with antisense complementarity to lin-14. Cell 75: 843–854.
32. Roush S, Slack FJ (2008) The let-7 family of microRNAs. Trends Cell Biol 18: 505–516.
33. Lehmann SM, Krüger C, Park B, Derkow K, Rosenberger K, et al. (2012) An unconventional role for miRNA: let-7 activates Toll-like receptor 7 and causes neurodegeneration. Nat Neurosci 15: 827–835.
34. Frost RJ, Olson EN (2011) Control of glucose homeostasis and insulin sensitivity by the Let-7 family of microRNAs. Proc Natl Acad Sci U S A 108: 21075–21080.
35. Toledano H, D'Alterio C, Czech B, Levine E, Jones DL (2012) The let-7-Imp axis regulates ageing of the Drosophila testis stem-cell niche. Nature 485: 605–610.
36. Delaloy C, Liu L, Lee JA, Su H, Shen F, et al. (2010) MicroRNA-9 coordinates proliferation and migration of human embryonic stem cell–derived neural progenitors. Cell Stem Cell 6: 323–335.
37. Gao FB (2010) Context-dependent functions of specific microRNAs in neuronal development. Neural Dev 5: 25.
38. Kawahara H, Imai T, Okano H (2012) MicroRNAs in neural stem cells and neurogenesis. Front Neurosci 6: 30.
39. Wienholds E, Kloosterman WP, Miska E, Alvarez-Saavedra E, Berezikov E, et al. (2005) MicroRNA expression in zebrafish embryonic development. Science 309: 310–311.
40. Esau C, Davis S, Murray SF, Yu XX, Pandey SK, et al. (2006) miR-122 regulation of lipid metabolism revealed by *in vivo* antisense targeting. Cell Metab 3: 87–98.
41. Jopling C (2012) Liver-specific microRNA-122: Biogenesis and function. RNA Biol 9: 137–142.
42. Chan JA, Krichevsky AM, Kosik KS (2005) MicroRNA-21 is an anti-apoptotic factor in human glioblastoma cells. Cancer Res 65: 6029–6033.
43. Xiong Y, Fang JH, Yun JP, Yang J, Zhang Y, et al. (2010) Effects of microRNA-29 on apoptosis, tumorigenicity, and prognosis of hepatocellular carcinoma. Hepatology 51: 836–845.
44. Wickramasinghe NS, Manavalan TT, Dougherty SM, Riggs KA, Li Y, et al. (2009) Estradiol downregulates miR-21 expression and increases miR-21 target gene expression in MCF-7 breast cancer cells. Nucleic Acids Res 37: 2584–2595.
45. Bannister SC, Smith CA, Roeszler KN, Doran TJ, Sinclair AH, et al. (2011) Manipulation of estrogen synthesis alters MIR202* expression in embryonic chicken gonads. Biol Reprod 85: 22–30.

46. Hua Z, Lv Q, Ye W, Wong CK, Cai G, et al. (2006) MiRNA-directed regulation of VEGF and other angiogenic factors under hypoxia. PLoS One 1: e116.
47. Shan F, Li J, Huang QY (2014) HIF-1 Alpha-Induced Up-Regulation of miR-9 Contributes to Phenotypic Modulation in Pulmonary Artery Smooth Muscle Cells During Hypoxia. J Cell Physiol 229(10): 1511–1520.
48. Kure EH, Sæbø M, Stangeland AM, Hamfjord J, Hytterød S, et al. (2013) Molecular responses to toxicological stressors: profiling microRNAs in wild Atlantic salmon (Salmo salar) exposed to acidic aluminum-rich water. Aquat Toxicol 138–139: 98–104.
49. Lind GE, Skotheim RI, Fraga MF, Abeler VM, Esteller M, et al. (2006) Novel epigenetically deregulated genes in testicular cancer include homeobox genes and SCGB3A1 (HIN-1). J Pathol 210: 441–449.
50. Wu RS, Zhou BS, Randall DJ, Woo NY, Lam PK (2003) Aquatic hypoxia is an endocrine disruptor and impairs fish reproduction. Environ Sci Technol 37: 1137–1141.
51. Brevik A, Lindeman B, Brunborg G, Duale N (2012) Paternal benzo[a]pyrene exposure modulates microRNA expression patterns in the developing mouse embryo. Int J Cell Biol 2012: 407431.

13

Characteristics of Settling Coral Reef Fish Are Related to Recruitment Timing and Success

Tauna L. Rankin[1,2], Su Sponaugle[1,3]*

1 Division of Marine Biology and Fisheries, Rosenstiel School of Marine and Atmospheric Science, University of Miami, Miami, Florida, United States of America, 2 NOAA National Marine Fisheries Service, Office of Habitat Conservation, Coral Reef Conservation Program, Silver Spring, Maryland, United States of America, 3 Department of Integrative Biology, Oregon State University, Hatfield Marine Science Center, Newport, Oregon, United States of America

Abstract

Many marine populations exhibit high variability in the recruitment of young into the population. While environmental cycles and oceanography explain some patterns of replenishment, the role of other growth-related processes in influencing settlement and recruitment is less clear. Examination of a 65-mo. time series of recruitment of a common coral reef fish, *Stegastes partitus*, to the reefs of the upper Florida Keys revealed that during peak recruitment months, settlement stage larvae arriving during dark lunar phases grew faster as larvae and were larger at settlement compared to those settling during the light lunar phases. However, the strength and direction of early trait-mediated selective mortality also varied by settlement lunar phase such that the early life history traits of 2–4 week old recruit survivors that settled across the lunar cycle converged to more similar values. Similarly, within peak settlement periods, early life history traits of settling larvae and selective mortality of recruits varied by the magnitude of the settlement event: larvae settling in larger events had longer PLDs and consequently were larger at settlement than those settling in smaller pulses. Traits also varied by recruitment habitat: recruits surviving in live coral habitat (vs rubble) or areas with higher densities of adult conspecifics were those that were larger at settlement. Reef habitats, especially those with high densities of territorial conspecifics, are more challenging habitats for young fish to occupy and small settlers (due to lower larval growth and/or shorter PLDs) to these habitats have a lower chance of survival than they do in rubble habitats. Settling reef fish are not all equal and the time and location of settlement influences the likelihood that individuals will survive to contribute to the population.

Editor: Sebastian C. A. Ferse, Leibniz Center for Tropical Marine Ecology, Germany

Funding: This study was made possible through National Science Foundation grants OCE-9986359 to S. Sponaugle and OCE 0550732 to R.K. Cowen, S. Sponaugle, C. Paris and V. Kourafalou. T. Rankin was also supported by a University of Miami Maytag Fellowship and Harding Michel Biological Oceanography Fellowship. During the preparation of the manuscript, S. Sponaugle received support from National Oceanic and Atmospheric Administration Center for Sponsored Coastal Ocean Research under award NA11NOS4780045 to the University of Miami. The funders had no role in the study design, data collection and analysis, decision to publish, or preparation of the manuscript.

Competing Interests: The authors have declared that no competing interests exist.

* Email: su.sponaugle@oregonstate.edu

Introduction

Understanding the mechanisms that drive variability in population dynamics is a fundamental goal of ecology. Most marine organisms have complex life cycles with multiple ontogenetic shifts, and processes occurring during the early life stages can influence population demography of later stages. For taxa such as coral reef fishes, which are demersal as juveniles and adults but possess a pelagic larval phase, population replenishment can be highly variable in space and time (reviewed in [1]) and patterns in settlement and post-settlement processes can strongly influence population size and distribution [2–4].

Common causes of temporal and spatial variability in settlement and recruitment range from environmental processes (e.g., reproductive seasonality related to water temperature [5,6]; spawning and settlement synchrony to lunar and tidal amplitude cycles [7,8]), and transport of larvae by climatological or oceanographic processes [8-11], to location and quality of settlement habitat [12], and distribution and composition of reef residents (e.g., predators, competitors [13]). Recently, more attention has focused on the characteristics of settlers and how variation in the quality of settlers may contribute to observed patterns in settlement and recruitment.

Early life history traits, such as larval growth and size-at-settlement, vary in space and time and have been linked to settlement magnitude [14–17]. Larval encounter with particular water temperatures, water masses, or oceanographic features can influence the early life history traits of settlers [18–21]. Further, these traits can carry over to affect post-settlement survival and recruitment success [22–24]. For many reef fish species, faster larval growth, shorter larval duration, larger size-at settlement, or higher condition-at-settlement [20,22–30] confers a survival advantage to new recruits. Such trait-mediated selective mortality influences the composition of young entering the benthic population.

The magnitude of benthic mortality also likely varies with timing and location of settlement. For instance, many species of reef fish settle during the darkest half of the lunar cycle [7,8,31,32], potentially due to reduced predation by visual predators during this time [33]. While the presence of conspecifics can enhance post-settlement survival in some species [34], high densities of conspecifics also can increase the mortality of new recruits [28]. Habitats also vary in the abundance and composition of predators [35,36] and refuges from such predation [37]. Thus, spatially and temporally variable predation pressure may interact with the variable traits of incoming settlers [38,39] and substantially complicate patterns of selective mortality of recruits.

Here, we examined spatial and temporal variation in settlement and recruitment for a common coral reef fish and tested whether this variation was associated with certain growth-related early life history traits. Specifically, we explored whether the early life history traits of settling larvae varied with the lunar timing of settlement and recruitment habitat, including density of conspecifics, and how the trait composition of juveniles changed over time.

Materials and Methods

Ethics Statement

All fish were collected and handled in accordance with the University of Miami Animal Care and Use Committee regulations as approved under UM-ACUC Permits 01–056ad and renewal, 05–019ad, 07–068ad01 and renewal02–056. Field collections were authorized under permits 00S–524 and 02R–524A, 04SR–524, 07SR–1032A from the Florida Fish and Wildlife Conservation Commission, and permits 2001–004, 2002–025A, 2004–024 from the Florida Keys National Marine Sanctuary.

Study site and physical data sampling

The seasonally variable Florida Current dominates the circulation of the Florida Keys. Cyclonic eddies propagate along the frontal boundary of the Florida Current and can transport larvae to nearshore reefs [10]. Mean water temperature in the Florida Keys varies seasonally by approximately 10°C and can significantly influence variation in reef fish early life history traits and recruitment strength [19,24]. Thus, to characterize the inshore environment of the Florida Keys and control for temperature effects, daily water temperatures were obtained from the National Underwater Research Center, which continuously records water temperature at 21 m depth at Conch Reef in the upper Florida Keys (24°59′N, 80°25′W).

Biological sampling

The bicolor damselfish *Stegastes partitus* is common throughout the Caribbean. Adults are highly territorial and spawn demersal eggs on a monthly basis throughout the year [31,40], after which, the males guard the embryos for ~4 d until hatching [31]. Larvae spend a mean of 30 d in the plankton before settling to the reef and metamorphosing into juveniles overnight [32]. Juveniles and adults maintain benthic territories on spur and groove reefs and are found in highest densities in areas of high current flow and within dead coral rubble piles [37,41]. The timing of settlement appears to be synchronized with lunar phase, with pulses occurring during the third quarter and/or new moon [8,31,32]. Seasonal peaks in *S. partitus* settlement to the Florida Keys typically occur during summer months [8,42]. *Stegastes partitus* is a useful model species for studying processes affecting the early life history of coral reef fishes because it is widely distributed, is integral to the trophic dynamics of the reef community [43,44], is easy to observe and collect at most life stages, and has otoliths (ear stones) that provide a daily record of events occurring during early life.

Newly recruited *S. partitus* juveniles were surveyed monthly on six different reefs within the Florida Keys National Marine Sanctuary offshore of Key Largo, Florida, for 65 mo. over 6 yrs from April 2003 until August 2008. To ensure that each survey captured all of the settlement for a given month, surveys were timed to coincide with the full moon, which is known to be a period of low settlement. Four of the reefs are part of the bank reef tract ~10 km offshore: French Reef (FR; 25°02.06′N, 80°21.00′W), Sand Island Reef (SI; 25°01.09′N, 80°22.08′W), Molasses Reef (MO; 25°00.74′N, 80°22.40′W), and Pickles Reef (PI; 24°59.23′N, 80°24.88′ W), two of which are sanctuary protected areas (MO, and FR; see Fig. S1 for map). The other two sites, Triangles (TR; 25°01.153′N, 80°26.272′W) and White Banks (WB; 25°02.609′N, 80°22.133′W), are inshore patch reefs. Recruit density was estimated for two habitats within the bank reef sites, the reef itself and dead coral rubble at the base of the reef, by counting the number of newly recruited juveniles within 15 haphazardly placed 5×1 m transects in each habitat. Within the two inshore patch reefs (TR and WB), only reef habitat was surveyed due to the lack of coral rubble habitats surrounding the reef. At SI, PI, TR, and WB, all censused recruits were collected by divers using hand nets and the anesthetic Quinaldine. Because FR and MO are Sanctuary Protected Areas, benthic sampling was not permitted. During 2007 and 2008, after each survey, additional collections were made off of the transects at the non-sanctuary protected sites to supplement sample sizes. All collected juveniles were immediately stored in 95% EtOH following collection to preserve their otoliths.

We also examined settlement-stage *S. partitus* larvae collected in larval light traps at FR, SI, MO, and PI from June 2001 to January 2004 as part of two separate studies [8,10] to compare differences in growth-related traits and magnitude of supply with the data from recruits. Three replicate light traps were deployed every other night at each of 2–3 sites for ~3 summer months in 2001, 6 summer months in each of 2002 and 2004, and twice a month for 6 winter months in each of 2003 and 2004.

Otolith analysis

Prior to dissection, the standard length (SL) of each specimen (n = 1291; 465 larvae and 826 juveniles) was measured digitally to the nearest 0.01 mm. The otoliths were extracted using standard dissecting techniques and placed in a drop of medium viscosity immersion oil on a microscope slide to clear for a minimum of 30 d. Based on ease of reading, only the lapilli were examined [32]. The clearest lapillus was chosen from each individual and viewed under 400× magnification through a Leica DMLB microscope equipped with a polarized filter between the first stage and light source. The image was captured by a Dage MTI video camera and analyzed using Image Pro Plus 4.5 software (Media Cybernetics). Each otolith was read once blind (i.e., without sample information available) and saved as a digital file. Every 5th individual was measured a second time (sensu [45]). The images of the remaining 1029 fish were examined to determine if there was any ambiguity in the placement of the increments and an additional 161 fish were aged a second time, resulting in a total of 418 fish aged twice. Otoliths were rejected where the difference between the first and second reads was >5%, resulting in six exclusions. Otolith analysis was utilized to determine the following early life history traits: post-settlement age (number of concentric increments after the settlement mark), pelagic larval duration (PLD; number of concentric increments from the primordium to the settlement mark), larval and juvenile growth rates (widths between consecutive increments), and size-at-age (otolith radius-at-each age, including settlement).

Data analysis

Interannual and monthly variation in settlement and recruitment magnitude. Otolith analysis was used to back-calculate date of hatching (collection date minus PLD and post-settlement age) and date of settlement (collection date minus post-settlement age). Spawning dates were calculated by subtracting 4 d from the hatch dates, for average egg development [31]. Daily settlement magnitude was estimated from the monthly measures of recruit density coupled with the settlement dates of the collected fish and adjusted for estimated mortality with age (sensu [46]). Mortality was estimated by fitting a function (equation: $y = 0.0671x^2 - 5.0165x + 89.763$; where x = juvenile age in days, and y = number of recruits) to the age frequency distribution of all 9+ d old recruits (Fig S2). For this calculation only these older recruits were used because the youngest fish were underrepresented in the samples due to the timing of sampling (during the low settlement period around the full moon) and the fact that older recruits do not shelter as much. The mortality function was then used to adjust the number of settlers per day based on the age of the recruits that settled on that day. Note that where recruitment densities from surveys were used in an analysis directly, we use the term "recruitment"; where recruit densities were adjusted for post-settlement mortality, we use the term "settlement." For larvae collected in light traps, we use the phrase "larval supply."

Lunar and tidal periodicity in settlement and spawning. A lunar day of settlement was assigned for each aged fish (day 1 corresponding to the new moon; day 8 corresponding to the first quarter; day 15 corresponding to the full moon; day 23 corresponding to the third quarter). Each fish was also assigned a tidal amplitude day (day 1 corresponding to the maximum tidal amplitude closest to the new moon). These data were collapsed into a single lunar and maximum tidal amplitude cycle. Rayleigh tests were utilized to determine if light trap catches or settlement magnitude were uniformly distributed across the lunar or tidal amplitude cycles [47]. When data were non-randomly distributed, the mean lunar or tidal amplitude day about which the data were centered was calculated.

Variation in early life history traits by lunar phase of settlement and over time. Aged fish were divided into three age groups: larvae (all settlement-stage larvae collected from 2001–2004 light traps), newly settled recruits (1–10 d old), and juveniles (11–28 d old). To determine if the composition of early life history traits varied with lunar timing of settlement, otolith derived traits were compared among lunar phases by age group using Kruskal-Wallis procedures (SYSTAT version 11.0; [48]), followed by post-hoc non-parametric multiple comparison tests [47]. For this and all subsequent analyses, individuals were classified into four groups by lunar phase, which included each lunar phase and 3–4 d on either side of the phase (i.e., first quarter moon: days 5–11, full moon: days 12–18, third quarter moon: days 19–26, new moon: 27–4). To determine whether patterns of selective mortality varied with lunar phase, we compared the traits among age groups by lunar phase in the same manner. Directional selection on a given trait was identified where there was a shift in the distribution of that trait in the older juveniles compared to younger recruits or larvae, and between recruits and larvae. Previous results demonstrated that temperature influences the intensity and direction of selective mortality acting on some *S. partitus* early life history traits (i.e., larval growth and PLD; [24]). In other species, such traits can influence recruitment magnitude [14,15]. To control for the effect of temperature we divided monthly cohorts into three categories by temperature (<25°C, 25 to <28°C, and >28°), which also corresponded roughly with season. Over 65% of the recruits collected were from the warmest temperature group and therefore comparisons of traits with other processes (i.e., settlement magnitude, lunar phase of settlement, location of settlement, and age group) were confined to individuals that settled during the warmest months.

Variation in early life history traits with settlement magnitude. Early life history traits of juveniles were more variable during smaller settlement events than larger ones, which precluded a significant linear relationship between monthly settlement magnitude and any of the individual traits. Thus, to determine whether the composition of growth-related traits varied with settlement magnitude and changed over time, the estimated total number of settlers was divided into two groups: those that settled in small events where monthly settlement was ≤10% of the seasonal total, and those that settled in large events where monthly settlement was >10% of the seasonal total. The 10% criterion was selected a posteriori based on the distribution of the data. Early life history traits were compared between large and small events using the Mann-Whitney U test (SYSTAT version 11.0; [48]). To control for the effects of temperature and lunar phase of settlement, comparisons were limited to individuals that settled during the darker half of the lunar cycle, from cohorts that encountered mean water temperatures of >28°C. Comparisons examining mean juvenile growth over the first two d post-settlement were limited to recruits that were at least 2 d old (2–10 d old versus 11–28 d old).

Variation in early life history traits by settlement habitat. Variable habitat quality (i.e., food abundance and quality, substrate suitability) and habitat-specific differential predation pressure may lead to preferential habitat selection by larvae and/or differential survival of recruits [49]. To examine whether early life history traits varied by microhabitat choice, traits were examined for recruits in reef and rubble habitats using the Mann-Whitney U test (SYSTAT version 11.0; [48]). To control for the influence of settlement magnitude on early life history traits, we compared the traits between reef crest and rubble habitats only for the warmest temperatures when recruitment pulses were relatively large.

Influence of older conspecifics on recruitment magnitude and early life history traits. The density of conspecifics may make an area more or less favorable for larval settlement or juvenile survival; therefore recruit density was also regressed against juvenile (20–30 mm SL), intermediate (30–50 mm SL), and adult (>50 mm SL) conspecific densities to determine if recruitment magnitude was related to densities of conspecifics. Growth-related traits were also regressed against conspecific density to determine if composition of traits was related to the presence of juvenile, intermediate, or adult conspecifics.

Results

Interannual and monthly variation in settlement and recruitment magnitude

Over six years, *Stegastes partitus* settlement varied on both an interannual and monthly basis (Fig. 1). Annual *S. partitus* settlement was similar for most years, with the exception of 2004, which was three-fold to over six-fold less than the other five years, and 2008, which had the highest magnitude. The highest mean recruitment across years occurred during July and August (Table 1). The coefficients of variation (CV) were relatively high for all months, indicating high interannual variability, but July and August were on the lower end of the spectrum (Table 1). February had the lowest mean recruitment and largest CV across years.

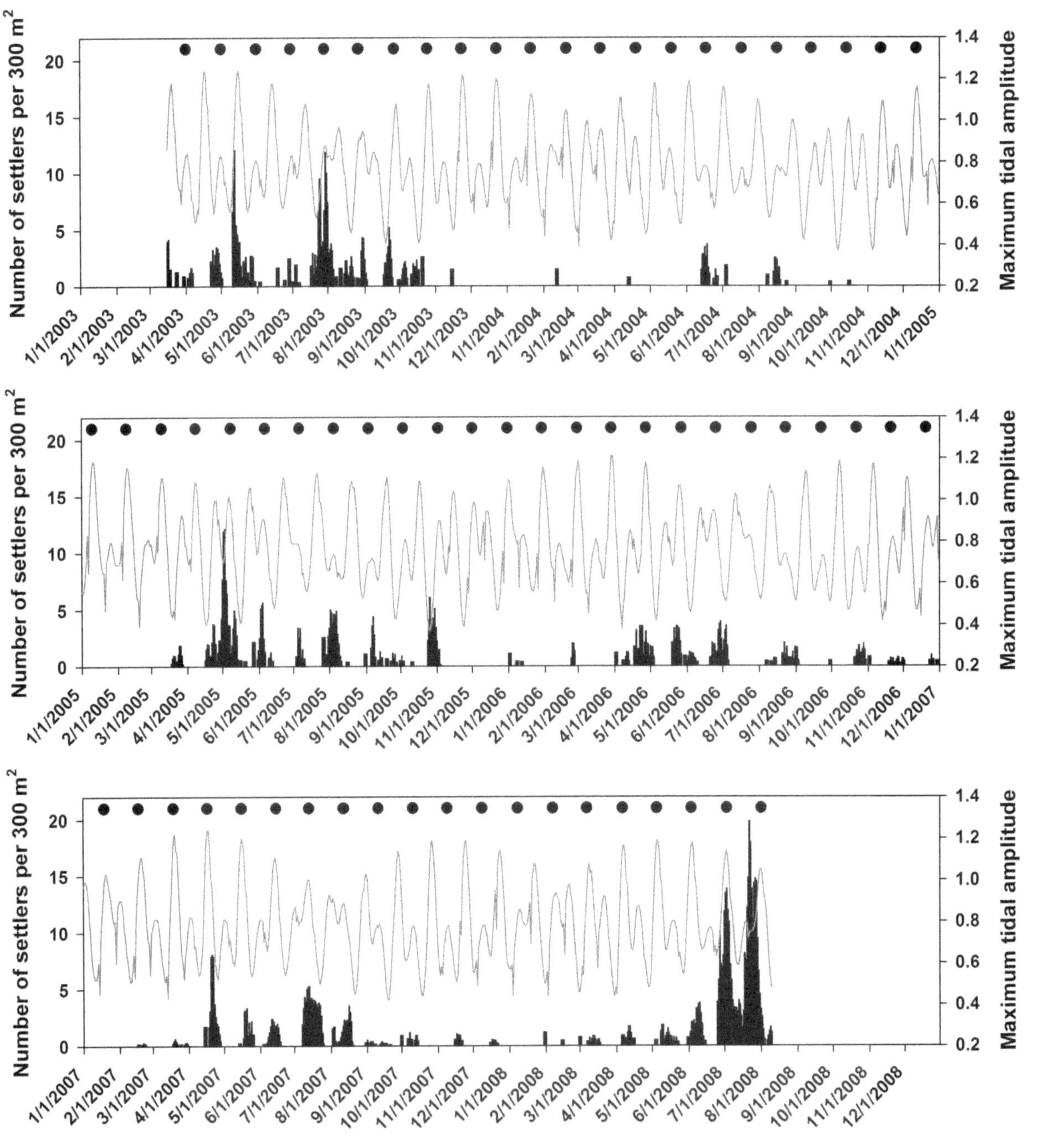

Figure 1. Daily back-calculated settlement of *Stegastes partitus* in reef and rubble habitat at Pickles and Sand Island Reefs combined (bars). Timing of settlement of collected recruits was adjusted for monthly density estimates and estimated post-settlement mortality. Maximum tidal amplitude is indicated by a line, and new moons by solid circles. Asterisks indicate months when sampling did not take place.

Spatial patterns in recruitment magnitude

Recruitment of *S. partitus* to the offshore bank reefs was almost 12-fold higher than to the inshore patch reefs. When recruitment to the rubble habitats around the base of the reefs was included, the difference increased to 29-fold, because there was high recruitment to the rubble surrounding the offshore reefs, but no recruitment to the seagrass surrounding the inshore patch reefs. Within the offshore reefs, recruitment was three-fold greater in the rubble habitat compared the reef habitat.

Lunar and tidal periodicity in settlement and spawning

Within months, back-calculated settlement was strongly periodic (Fig. 1). For the aged individuals, there was a significant peak in settlement on day 2 of the lunar cycle (one day after the new moon; Table 2). After adjusting for recruitment magnitude (density) and post-settlement mortality, the peak in settlement shifted to day 29 (one day before the new moon) as there was greater representation of juveniles that settled around the third quarter moon (days 19–26; Fig. S3). This pattern corresponded closely to light trap catches of settlement-stage larvae from 2003–2004, where the mean peak in settlement was calculated to be on

Table 1. Monthly variability in recruitment of *Stegastes partitus* to reefs in the upper Florida Keys.

Month	Mean number of recruits (#300 m^{-2})	SD	CV
January	0.31	0.79	2.53
February	0.17	0.73	4.23
March	0.29	0.80	2.72
April	1.70	3.11	1.83
May	3.39	7.00	2.07
June	4.55	6.91	1.52
July	5.39	8.58	1.59
August	6.57	9.73	1.48
September	2.64	3.16	1.20
October	1.19	2.66	2.23
November	2.02	3.21	1.59
December	0.25	0.61	2.43

New recruits were surveyed monthly by divers for 65 months from April 2003 to August 2008.

day 27 (Table 2). The maximum tidal amplitude cycle is tightly coupled with the lunar cycle in Florida, but specific timing of maximum amplitude tides with lunar phase shifts seasonally (Fig. 1). Settlement of juveniles corrected for post-settlement mortality peaked on day 1 of the maximum tidal amplitude cycle (Table 2, Fig. S3).

Back-calculation of spawning day for aged juveniles revealed a peak in successful spawning on day 1 of the lunar cycle, corresponding to the new moon (Table 2, Fig. S4). Using the adjusted values of settlement intensity as above, the distribution remained non-random with a mean peak in spawning on day 26, several days closer to the spawning distribution back-calculated from the light-trap larvae which peaked on day 20 (Table 2, Fig. S4).

Variation in traits by lunar phase of settlement

We have previously shown that early life history traits in *S. partitus* are closely linked. PLD is inversely related to larval growth and both of these traits determine size-at-settlement, the most important trait influencing juvenile survivorship in *S. partitus* [24]. Fish that settle at large sizes have higher survivorship but also spend more time in territorial defense and thus experience slower early juvenile growth [24]. All of these trait relationships obtained from tracking cohorts over time are evident in the patterns obtained in the present study; only here the focus is on variation over the lunar cycle. Early life history traits varied among fish settling on different lunar phases for all three age groups of *S. partitus*. Larvae settling to the reef on the third quarter and new moons experienced significantly faster larval growth and were significantly larger at settlement than larvae arriving on the first and full moons (Table 3, Fig. 2). Furthermore, the new moon settlers spent significantly less time in the plankton than larvae

Table 2. Rayleigh statistics for lunar and tidal amplitude synchrony in settlement and successful spawning of *Stegastes partitus* in the upper Florida Keys.

	n	Z	Day	s
Lunar settlement				
Aged recruits	786	134.5	2	5.0
Recruits adj. for mortality	1429	142.0	29	5.4
Settlement-stage larvae	415	64.1	27	5.1
Tidal amplitude settlement				
Recruits adj. for mortality	1383	161.6	1	5.3
Lunar spawning				
Aged recruits	788	89.0	1	5.3
Recruits adj. for mortality	1433	86.4	26	5.7
Settlement-stage larvae	196	21.7	21	5.3

Recruits were obtained from monthly surveys; settlement-stage larvae were obtained from light traps. Otolith microstructure of recruits was used to back-calculate settlement timing and recruit densities were adjusted for mortality to provide a more precise estimate of settlement magnitude. All tests were significant at $p<0.001$. Z is the Rayleigh test statistic; Day is the mean peak in lunar or maximum tidal amplitude calculated from mean vector angle; s is the mean angular deviation; Day 1 is the new moon.

settling on the other lunar phases (Table 3, Fig. 2). At 1–10 d post-settlement, recruits had fairly similar traits among the different lunar phases of settlement, with the exception that new moon settlers had significantly faster larval growth than third quarter and full moon settlers. By 11–28 d post-settlement, juveniles that settled on the third quarter and new moons had significantly longer PLDs than first quarter moon settlers, and new moon settlers had significantly faster early post-settlement growth than first, full, and third quarter moon settlers (Table 3, Fig. 2).

Trait-related selective mortality across age groups was evident for most settlers, but the strength and direction of selection varied by lunar phase. First and full moon settlers experienced selection for faster larval growth, lower PLDs, larger settlement sizes, and slower early juvenile growth (Table 3, Fig. 2). For third quarter moon settlers, there was no selective mortality related to larval growth or PLD, but selection for larger settlement sizes and slower early juvenile growth. For new moon settlers, directional selection on larval growth and PLD was opposite to that for first and full moon settlers. While the direction of selective mortality related to the critical trait of size-at-settlement was consistent across all lunar phases, first quarter/full moon settlers experienced stronger selective mortality than third quarter/new moon settlers (Table 3, Fig. 2).

Variation in traits with settlement magnitude

During peak settlement during the dark half of the lunar cycle between May-August, late stage larvae that settled to the reef in large events had significantly slower larval growth, longer PLDs, and larger sizes-at-settlement that those that arrived in small events (Table 4; Fig. 3). Selective mortality was evident for fish settling in both small and large events: there was selection for slower larval growth, shorter PLDs, and larger sizes-at-settlement (Table 4; Fig. 3). However, there were larger differences in traits between larvae and juveniles for fish that settled in small events, indicating stronger selective mortality pressure. Consequently, for older juveniles, there were few differences in traits between fish that settled in large vs. small events ($p>0.05$). However, juveniles that settled in large events had significantly faster early juvenile growth ($p=0.013$; Fig. 3).

Variation in early life history traits by settlement habitat & conspecific density

Comparisons of mean traits among microhabitats for the two age groups of *S. partitus* revealed that there were generally few differences in traits among habitats. However, the trait most directly related to mortality, size-at-settlement, varied significantly between habitats in the oldest age group. While size-at-settlement did not differ between habitats for young recruits, over time, survivors (11–28 d old) in reef habitats were those that settled at larger sizes than those settling to rubble habitats ($p=0.018$), suggesting that selective mortality is stronger in reef habitats. The only other trait varying between habitats was larval growth, which was significantly higher in young recruits (1–10 d old) in reef habitats ($p=0.030$), but evened out between habitats over time (Fig. 4). General patterns of trait-based selective mortality between the age groups echoed those reported above and previously [24], however, this analysis by habitat reveals a larger shift in both PLD and size-at-settlement over time in the reef habitat, reflecting stronger selective pressures in that habitat.

Considering the whole dataset, monthly recruit density was positively related to the density of >1 mo. old juvenile conspecifics (20–30 mm SL; $R^2=0.83$, $p<0.001$). This positive relationship was maintained when the analysis was constrained to only cohorts settling during the warmest water temperatures ($R^2=0.45$,

Table 3. Results of Kruskal-Wallis tests comparing *Stegastes partitus* early life history traits among different phases of lunar settlement (1 = first quarter; F = full moon; 3 = third quarter; N = new moon); and among three age groups (L = larvae; R = recruits 1–10 d post-settlement; J = juveniles 11–28 d post-settlement).

	Trait comparison across lunar phases						Trait comparison by age group							
	LARVAE		RECRUITS (1–10 d)		JUVENILES (11–28 d)		1ST		FULL		3RD		NEW	
TRAIT	*P*	post-hoc	*P*	post-hoc	*P*	post-hoc	*P*	post-hoc	*P*	post-hoc	*P*	post-hoc	*P*	post-hoc
Larval growth	<0.001	1, F <3, N	0.008	3<1, N; F <N	0.113		<0.001	L <R, J	0.015	L <R, J	0.044	L, J> R	<0.001	L> R, J
PLD	<0.001	1, F, 3> N	0.962		0.003	1<3, N	0.016	L> R, J	0.045	L <R	0.125		<0.001	L <R <J
Size-at-settlement	0.001	1, F <3, N	0.533		0.334		<0.001	L <R, J	0.016	L <R, J	<0.001	L <R, J	<0.001	L <R, J
Juvenile growth 1–2 d			0.844		0.001	1, F, 3 <N	0.009	R> J	0.025	R> J	0.004	R> J	0.001	R> J

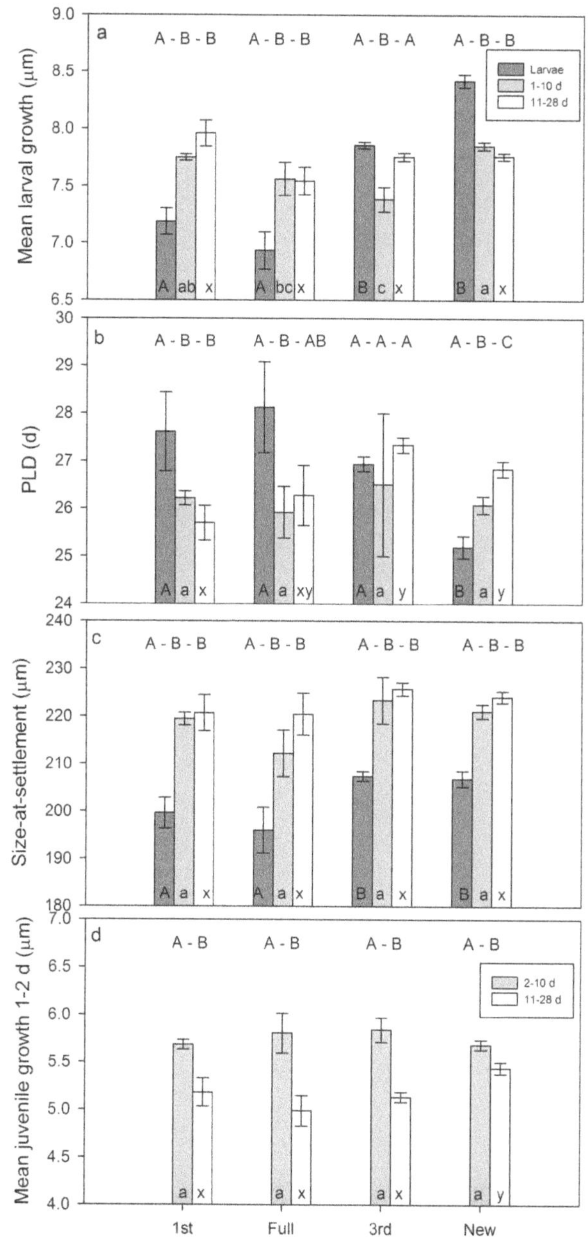

Figure 2. Mean otolith-derived (a) larval growth, (b) pelagic larval duration (PLD), (c) size-at-settlement, and (d) early (2 d) juvenile growth of *Stegastes partitus* settlement-stage larvae, recruits (1–10 d post settlement), and juveniles (11–28 d) by lunar phase of settlement. Effect of temperature was minimized by including only 25 monthly cohorts collected from May 2003 to August 2008 with mean water temperatures of >28°C. Error bars represent SE. Different letters above the bars indicate significant relationships among age groups, within lunar phases. Different letters within the bars indicate significant relationships among lunar phases, within age groups (capital A, B denote differences within larvae, lowercase a, b denote differences within 1–10 d old recruits, lowercase x, y denote differences within 11–28 d old juveniles). Settlement-stage larvae are not included in the last panel because they had not yet settled so did not exhibit juvenile growth.

$p = 0.025$). Recruit density was also positively related to the density of intermediate age conspecifics (30–50 mm SL; $R^2 = 0.042$, $p = 0.034$), but there was no significant relationship between recruit density and adult conspecific density ($p = 0.120$). However, larval growth and size-at-settlement of recruits were positively related to adult density (Fig. 5). PLD and early juvenile growth of recruits were not significantly related to adult conspecific density ($p>0.05$).

Discussion

Results of this nearly six year time series indicate that there is interannual, monthly (seasonal), and lunar periodicity in the settlement and recruitment of the common reef fish, *Stegastes partitus*, to the upper Florida Keys. Similar patterns have been demonstrated for *S. partitus* settling to other locations [7,31,32], as well as in a shorter study of larval supply in the Florida Keys [8]. In Panama, large pulses of *S. partitus* settlement were associated with faster early growth [14] and larger sizes-at-settlement [15]. Such traits not only influence recruitment magnitude, but can also affect juvenile survival of *S. partitus* [24,28]. Here we tested whether the composition of young entering the population varies with the specific timing and location of settlement and whether trait-based selective mortality similarly varies over time and space. Results demonstrate that early life history traits differ with lunar timing of settlement, magnitude of settlement, and settlement habitat, including density of conspecifics. Further, selective mortality interacts with the timing and location of settlement to influence the suite of traits of surviving juveniles.

Early life history traits vary by lunar phase of settlement

Monthly settlement of *S. partitus* consistently peaked during the summer months, especially in July and August, but there was also substantial interannual variability. These settlement patterns obtained by examining the otoliths of new recruits from monthly surveys and adjusting for mortality paralleled patterns in the supply of settlement stage *S. partitus* to Florida Keys reefs as measured in light traps [8]. For *S. partitus,* there is a linear relationship between the supply of settlement-stage larvae and recruitment in both the Florida Keys [50] and in Barbados [32]. Our 65-mo long settlement record revealed lunar cyclic settlement peaking during the third quarter and new moons, but sufficient settlement also occurred during other phases to enable a comparison of traits across settlement periods. Growth-related traits varied among lunar phases of settlement: Larvae settling to the reef during the dark portion of the lunar cycle, around the third quarter and new moons, were significantly larger and faster growing, and spent less time in the plankton than larvae that settled during the brighter portion of the lunar cycle, around the first quarter and full moons. Thus larvae that settle during the dark half of the lunar cycle are not only potentially less susceptible to visual predation [33], but also arrive with more "optimal" traits [51]. This may be a simple consequence of differential larval growth following spawning. Spawning in *S. partitus* peaks during the dark half of the lunar cycle (third quarter-new moon). Larvae that grow sufficiently rapidly reach a suitable settlement size and are able to settle during the following dark lunar phase (third quarter-new moon). Larvae that do not find sufficient food sources and thus grow more slowly take longer to develop and reach a suitable settlement size, are not ready to settle during the subsequent dark lunar phase, and consequently, settle during the following less optimal light phase (first quarter-full moon). Thus, for larvae settling during the non-optimal light half of the lunar cycle, larval growth is lower, PLDs longer, and settlement sizes smaller. This timing of larval growth and development is likely related to synchrony with larval food sources. In other subtropical environments, zooplankton abundance in the epipelagic layer

Table 4. Results of Mann-Whitney U tests comparing *Stegastes partitus* early life history traits between small and large settlement events for larvae, recruits (1–10 d post-settlement), and juveniles (1–28 d post-settlement); and among age groups (L = larvae; R = recruits; J = juveniles).

	Trait comparison by size of settlement event						Trait comparison by age group			
	LARVAE		RECRUITS (1–10 d)		JUVENILES (11–28 d)		SMALL		LARGE	
TRAIT	*P*	Relationship	*P*	Relationship	*P*	Relationship	*P*	post-hoc	*P*	post-hoc
Larval growth	<0.001	Small > large	0.550		0.857		0.023	L> J	<0.001	L> R> J
PLD	<0.001	small <large	0.210		0.220		0.038	L <J	<0.001	L, J> R
Size-at-settlement	<0.001	small <large	0.128		0.190		<0.001	L <R, J	<0.001	L <R <J
Juvenile growth 1-2 d			0.413		0.010	small <large	0.040	R> J	0.006	R> J

peaks just before the new moon [52], coinciding with the timing of hatching of numerous species [53], including *S. partitus* [54]. Such an increase in prey items around the new moon could be beneficial for larvae at first feeding as well as just prior to settlement. However, prey are typically patchy and larvae will have differential success in finding food. As noted above, larvae that experience poor feeding conditions and grow more slowly will have to spend longer in the plankton and settle during less optimal times. Protracted delays in settlement have been shown to result in striking declines in larval condition in invertebrate taxa with non-feeding larval stages or specific diet requirements [55,56], and reduced growth in other fish larvae [57,58].

Not only did larvae settling during different lunar phases exhibit contrasting traits, but the selective loss of these traits over time also varied among lunar phases. Fish settling during the light lunar phases generally experienced stronger selective loss of individuals with less optimum traits. The primary trait driving reef-based mortality is size-at-settlement [24]. Fish settling during the light half of the lunar cycle were generally smaller and there was stronger selective loss of the smallest settlers relative to those settling during the dark lunar phase. Size-at-settlement is a function of both larval growth and PLD, which are typically inversely related to one another. Thus, light phase settlers experienced selective mortality favoring faster larval growth (associated with larger sizes-at-settlement) and shorter PLDs. Third quarter moon settlers experienced no clear selective mortality related to either of these two traits and selection favored slower larval growth and longer PLDs in new moon settlers. Although fish that are larger at settlement typically have higher survivorship, they spend more time in aggressive territorial interactions and consequently have slower early juvenile growth [24]. Thus settlers during all lunar phases experienced selection against fast early juvenile growth. As a result of different patterns of selective loss of traits across lunar phases, traits of the oldest surviving juveniles (11–28 d old) were generally similar across lunar phases of settlement, except that fish settling during the new moon had significantly faster early juvenile growth. Together these results indicate that the initial composition of settlers interacts with potential lunar-cyclic differences in predation pressure to influence the population composition. In most cases, traits shifted sharply between the larval stage and young recruit stage, suggesting rapid selective loss. Indeed, most post-settlement mortality is thought to occur within the first 2 d of settlement [59]. The relative brightness of the nocturnal reef environment during and immediately following settlement likely influences overall mortality due to predation by visually oriented predators. Nighttime illumination varies predictably over the lunar cycle: regardless of tidal state, the number of moonless hours at night is higher during the period between the third quarter and new moon [8]. Stronger selective loss of early life history traits during the light phases of the lunar cycle is consistent with higher predation pressure by visually oriented predators during and immediately following settlement.

Early life history traits and relationship to settlement magnitude

While examination of *S. partitus* traits in relation to settlement magnitude across seasonal temperatures indicates that individuals from larger settlement events exhibited faster larval growth than those that settled during small events, this likely simply reflects the general trend of higher recruitment during warm seasons. Controlling for the effect of temperature on growth by examining this relationship *within* the peak settlement season during the warmest months, and limiting the analysis to individuals settling during the darker half of the lunar phase, revealed the opposite

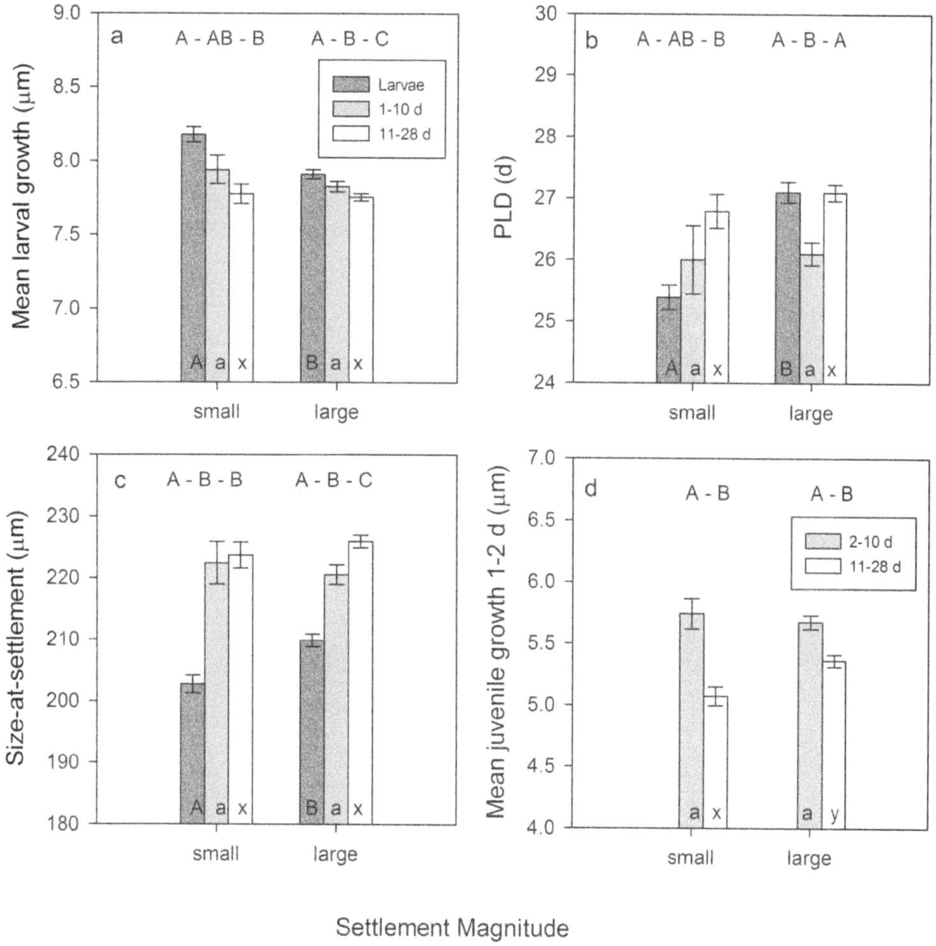

Figure 3. Mean otolith-derived (a) larval growth, (b) pelagic larval duration (PLD), (c) size-at-settlement, and (d) early (2 d) juvenile growth of *Stegastes partitus* settlement-stage larvae (dark gray bars), recruits (1–10 d post-settlement; light gray bars), and juveniles (11–28 d; white bars) by size of light trap pulse (larvae) and size of settlement event (recruits/juveniles). Effects of temperature and lunar phase were minimized by including only individuals from cohorts that encountered mean water temperatures of >28°C and settled during the darker half of the lunar cycle. Error bars are SE. Different letters above the bars indicate significant differences among age groups, within events of a given size. Different letters within the bars indicate significant differences between small and large events, within age groups (capital A, B denote differences within larvae, lowercase a, b denote differences within 1–10 d old recruits, lowercase x, y denote differences within 11–28 d old juveniles). Settlement-stage larvae are not included in the last panel because they had not yet settled so did not exhibit juvenile growth.

pattern. Within the peak settlement season, individuals arriving in the largest pulses during the dark lunar phases had slower larval growth, longer PLDs, and were larger at settlement. As noted earlier, size-at-settlement is the most important trait influencing early survivorship of *S. partitus* [24]. In this case, within the overall warm season when larval growth was already relatively high, the largest sizes-at-settlement were achieved by longer PLDs. Since PLD is inversely related to larval growth [24], somewhat reduced larval growth may simply be a consequence of a significantly longer PLD. The magnitude of *S. partitus* settlement across seasons in San Blas, Panama, was positively related to larval growth [15], but because temperature-related influences on growth were not controlled for, it is hard to resolve whether fast growth was directly related to successful settlement or simply correlated with seasonal patterns of recruitment.

Once settled, *S. partitus* recruits in the Florida Keys did not maintain these differences in traits. Fish that settled in small pulses experienced more trait-based selective loss (i.e., larger difference in traits of larvae and juveniles) than fish that settled in large pulses. Strong selection favoring slower larval growth, longer PLDs, and larger sizes at settlement for fish settling in small events resulted in surviving juveniles from all events exhibiting similar ETHTs. The only exception was that stronger selection for slow early juvenile growth in fish from small pulses resulted in these fish having significantly slower early juvenile growth than juveniles from large events. Thus, overall, fishes settling in small pulses arrived with less optimal traits and consequently, suffered stronger selective mortality which resulted in survivors with a similar range of early life history traits.

It is important to note that we used a cross-sectional approach to compare traits of fishes of different ages to detect patterns in selective mortality. These fish were not from the same cohort, but were from many cohorts integrated over time. There are disadvantages to this method, as neither time of hatching nor settlement was controlled for and individuals may have been exposed to different environmental features that were unmeasured, but contributed to differences in traits with age. However, such cross-sectional approaches have been used before to examine selective mortality processes [60,61], and in the present study we integrated over a large number of cohorts spread over 6 years,

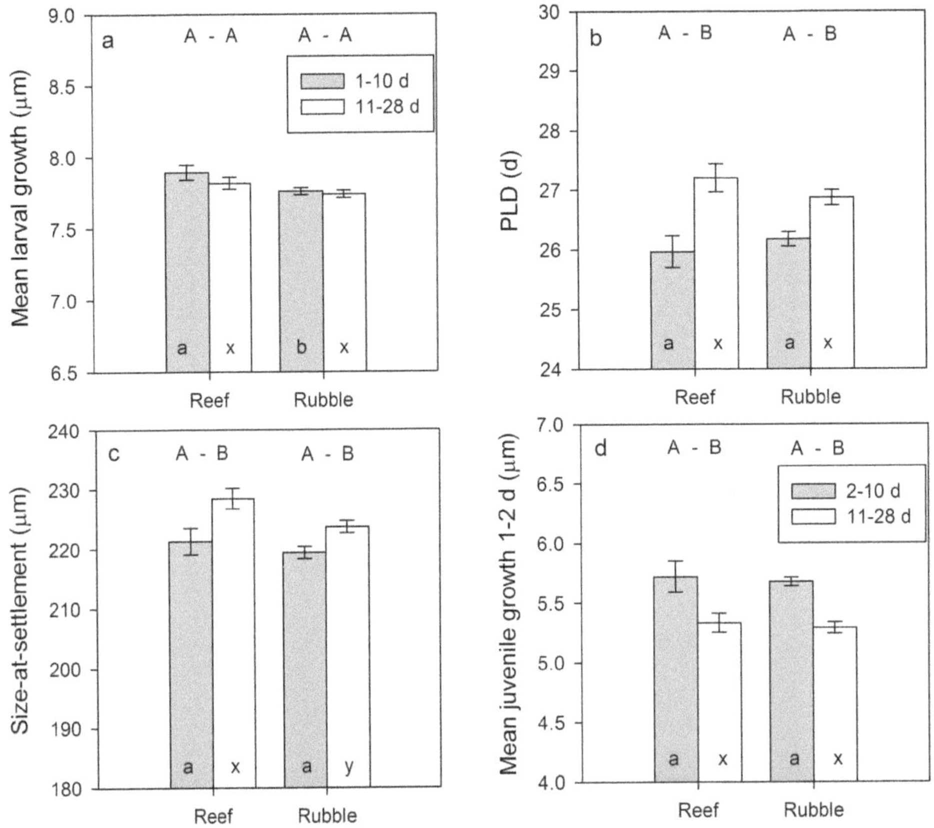

Figure 4. Mean otolith-derived (a) larval growth, (b) pelagic larval duration (PLD), (c) size-at-settlement, and (d) early (2 d) juvenile growth for two different age groups of *Stegastes partitus* recruits by habitat type (i.e., reef crest and rubble). Effect of temperature was minimized by including only 25 monthly cohorts collected from May 2003 to August 2008 that encountered mean water temperatures of >28°C. Error bars are SE. Different letters above the bars indicate significant relationships between habitats for a given age group. Different letters within the bars indicate significant relationships between age groups (selective mortality) by habitat (a, b denote differences between recruits (1–10 d post-settlement) and juveniles (11–28 d) in reef habitats, x, y denote differences between recruits and juveniles in rubble habitats.

potentially averaging out the noise associated with individual cohort variation. Importantly, the overall patterns of selective mortality we obtained from this approach are consistent with those we obtained from tracking 16 individual cohorts over time [24].

Spatial patterns in recruitment and early life history traits

Recruitment to offshore bank reefs of the upper Florida Keys was 12-fold higher than to the inshore patch reefs. This may be due, in part, to differences in physical processes that influence larval transport, and depletion of larvae as they cross the reef crest [8], however, there is also evidence that habitat in the inshore patch reefs is less optimal than along the offshore bank reefs for this species [62]. There was a three-fold greater abundance of recruits in coral rubble than on the reef crest and *S. partitus* juveniles have been shown to suffer higher mortality on boulder coral than in coral rubble piles [37,62]. While there were not enough recruits from the patch reef sites to compare traits of recruits to those habitats with those that settled along the fore reef, we compared growth-related early life history traits among individuals that settled on the reef crest versus surrounding rubble on the offshore bank reefs. Individuals that settled on the reef crest and survived 11–28 d were larger-at-settlement than those in rubble habitats. While 1–10 d old recruits in reef habitats also had higher larval growth rates, these differences disappeared over time. Because habitat-specific traits of settling larvae could not be determined due to the sampling method (light traps), we cannot fully disentangle the relative importance of pre- and post-settlement processes in creating these spatial patterns. For example, differences in larval growth of recruits could have been due to differences in the traits of settling larvae or due to selective processes acting immediately post-settlement (i.e., before the recruits were collected; [63]). However, we can conclude that the magnitude of selective mortality based on settlement size and PLD was larger in the reef habitat. Consistent with habitat-specific abundances and previous mortality estimates [37], selective mortality appears to be stronger in reef habitats.

Density-dependent effects on recruitment and early life history traits

For the warmest months from 2007–2008, monthly *S. partitus* recruit density was strongly positively correlated with the density of >1 mo. old juvenile conspecifics (2–3 cm SL). This is consistent with a pattern of monthly pulses in settlement following one another, but could also indicate enhanced recruitment as a result of the presence of conspecifics. Further evidence that the presence of conspecifics can enhance recruitment is the positive relationship between recruit density and density of intermediate-sized conspecifics (30–50 mm SL), which should be at least several months old and could have settled during a different season. Presence of conspecifics can enhance reef fish settlement [64,65] or inhibit it [34,66]. Interestingly, the relationship between recruit density and

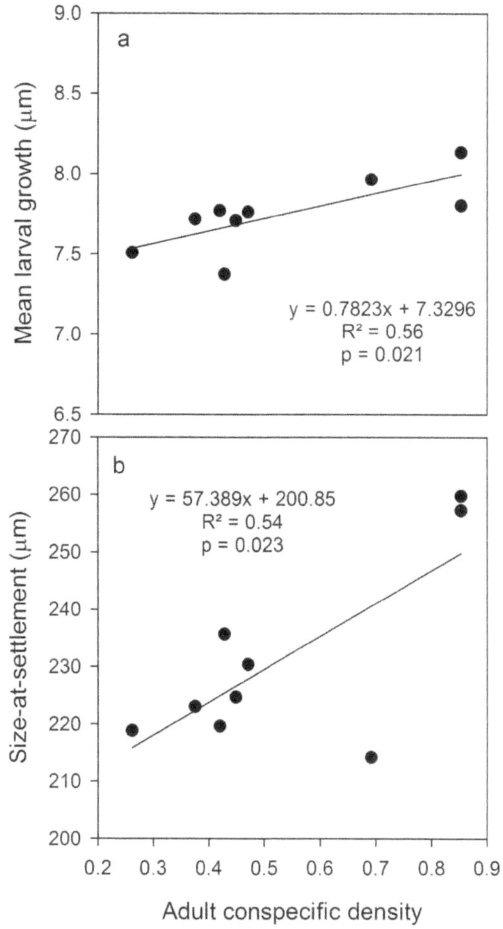

Figure 5. Relationship between mean otolith-derived (a) larval growth and (b) size-at-settlement of *Stegastes partitus* recruits and adult conspecific density from June 2007 to August 2008.

adult conspecific density was not significant, and if anything, appeared negative. The presence of conspecifics may signal high quality habitat, but *S. partitus* is a highly territorial species and reproductively mature adults may aggressively deter settlement of potential competitors. Younger conspecifics may not be as disturbed by new settlers or as effective in removing them. While recruitment magnitude was not significantly correlated with adult conspecific density, some early life history traits of recruits were. Larval growth and size-at-settlement were greater for recruits surrounded by higher densities of adult conspecifics. Since these traits are related to enhanced survival [24], it is likely that only the best performers can survive high densities of adult conspecifics (high densities of conspecifics have been associated with elevated mortality of juvenile *S. partitus*; [28,67]. The presence of conspecifics may initially serve as a cue for settlement, but the aggressive nature of adult *S. partitus* may also prevent all but the strongest recruits from remaining within high densities of adults.

Conclusions

Results of this study demonstrate that early life history traits interact with both temporal and spatial patterns of settlement to influence reef fish juvenile demography. The distribution of early life history traits of settlers is related to both timing and location of settlement. Post-settlement processes tend to obscure pre-settlement patterns, and in some cases, create new patterns of variation. Individuals settling during optimal times (i.e., dark portion of lunar cycle, larger events) arrive with better traits, such as larger sizes-at-settlement. Large settlement sizes can be achieved by faster larval growth or longer PLD and it appears that both scenarios are used by *S. partitus*. Fish settling during the dark as opposed to the light half of the lunar cycle achieved larger settlement sizes through faster larval growth, but within peak settlement periods, larvae settling during large events achieved larger sizes by spending more days growing in the plankton. Regardless of the mechanism, most of the variation in traits among individuals that settled during different lunar phases and settlement magnitudes was not maintained in the juveniles over time. Selective mortality processes varied in strength and direction depending on lunar phase, resulting in juveniles with rather similar traits.

Early life history traits also varied as a function of settlement habitat. Recruits in reef habitats or within high densities of adult conspecifics were larger at settlement compared to those in rubble habitat or within low densities of adults. Differential selective mortality pressures acting across variable settlement habitats resulted in spatial differences in the distribution of early life history traits. In sum, differences in early life history traits of settling larvae due to timing or magnitude of settlement are further modified over spatial scales depending on habitat and strength of selective mortality processes. Therefore, for *S. partitus*, and probably other species with similar life histories, the specific timing and location of settlement define the composition and demography of the juvenile population.

Supporting Information

Figure S1 Map of the upper Florida Keys with sampling sites for light trap deployment and recruit surveys and collection.

Figure S2 Age distribution of *Stegastes partitus* recruits in the upper Florida Keys (bars) with a function ($y = 0.0671x^2 - 5.0165x + 89.763$) fitted to the slope of all recruits >9 d old post-settlement (line) to estimate mortality.

Figure S3 Back-calculated settlement of *Stegastes partitus* recruits collected in the upper Florida Keys from April 2003 to August 2008, adjusted for recruitment magnitude (density) and mortality, plotted over (upper) a single lunar cycle and (lower) a single maximum tidal amplitude cycle. Two sample maximum tidal amplitude cycles plotted for representative days when maximum occurred in conjunction with the new moon (solid line) and the full moon (dashed line). New moon indicated by a solid circle and full moon by an open circle. Arrow indicates mean day about which settlement peaked.

Figure S4 Back-calculated timing of successful spawning by *Stegastes partitus* for (a) recruits collected from April 2003 to August 2008, and (b) adjusted for recruitment magnitude (density) and mortality. (c) Back-calculated timing of successful spawning of settlement-stage larvae collected in light traps from May 2003 to January 2004 plotted over a single lunar cycle.

Acknowledgments

Thanks to the many divers who contributed to the fish surveys over the years – the primary divers E D'Alessandro, K Huebert, K Grorud-Colvert, M Paddack, D Meier, K Shulzitski, K Walter, K Denit, S Bignami, J Boulay, L Havel, L Jones, as well as many other substitute divers. Diving is not possible without boat tenders and we thank our volunteer tenders and in particular, A Exum, L Gundlach. Thanks also to those who assisted with measuring and dissecting fish: especially N Holocombe, L Glade, F Graham, A Hogarth, and C Gioia. Early drafts benefitted from the comments of RK Cowen, M Schmale, D Williams, and T Kellison.

Author Contributions

Conceived and designed the experiments: SS TR. Performed the experiments: TR SS. Analyzed the data: TR SS. Contributed reagents/materials/analysis tools: SS TR. Wrote the paper: TR SS.

References

1. Doherty PJ (2002) Variable replenishment and the dynamics of reef fish populations. In: Sale PF, editor. Coral reef fishes: Dynamics and diversity in a complex ecosystem. New York: Academic Press. pp. 327–355.
2. Jones GP (1990) The importance of recruitment to the dynamics of a coral reef fish population. Ecology 71: 1691–1698.
3. Caley MJ, Carr MH, Hixon MA, Hughes TP, Jones GP, et al. (1996) Recruitment and the local dynamics of open marine populations. Annu Rev Ecol Syst 27: 477–500.
4. Hixon MA (2012) 60 years of coral reef fish ecology: past, present, future. Bull Mar Sci 87: 727–765.
5. Robertson DR (1990) Differences in the seasonalities of spawning and recruitment of some small neotropical reef fishes. J Exp Mar Biol Ecol 144: 49–62.
6. Bergenius MAJ, McCormick MI, Meekan MG, Robertson DR (2005) Environmental influences on larval duration, growth and magnitude of settlement of a coral reef fish. Mar Biol 147: 291–300.
7. Robertson DR (1992) Patterns of lunar settlement and early recruitment in Caribbean reef fishes at Panama. Mar Biol 114: 527–537.
8. D'Alessandro E, Sponaugle S, Lee T (2007) Patterns and processes of larval fish supply to the coral reefs of the upper Florida Keys. Mar Ecol Prog Ser 331: 85–100.
9. Raventos N, Macpherson E (2005) Environmental influences on temporal patterns of settlement in two littoral labrid fishes in the Mediterranean Sea. Estuar Coast Shelf Sci 63: 479–487.
10. Sponaugle S, Lee T, Kourafalou V, Pinkard D (2005) Florida Current frontal eddies and the settlement of coral reef fishes. Limnol Oceanogr 50: 1033–1048.
11. Ben-Tzvi O, Kiflawi M, Gildor H, Abelson A (2007) Possible effects of downwelling on the recruitment of coral reef fishes to the Eilat (Red Sea) coral reefs. Limnol Oceanogr 52: 2618–2628.
12. Ohman MC, Munday PL, Jones GP, Caley MJ (1998) Settlement strategies and distribution patterns of coral-reef fishes. J Exp Mar Biol Ecol 225: 219–238.
13. Almany GR (2004) Priority effects in coral reef fish communities of the Great Barrier Reef. Ecology 85: 2872–2880.
14. Bergenius MAJ, Meekan MG, Robertson DR, McCormick MI (2002) Larval growth predicts the recruitment success of a coral reef fish. Oecologia 131: 521–525.
15. Wilson DT, Meekan MG (2002) Growth-related advantages for survival to the point of replenishment in the coral reef fish *Stegastes partitus* (Pomacentridae). Mar Ecol Prog Ser 231: 247–260.
16. Macpherson E, Raventos N (2005) Settlement patterns and post-settlement survival in two Mediterranean littoral fishes: influences of early-life traits and environmental variables. Mar Biol 148: 167–177.
17. Lemberget T, McCormick MI (2009) Replenishment success linked to fluctuating asymmetry in larval fish. Oecologia 159: 83–93.
18. Sponaugle S, Pinkard DR (2004) Impact of variable pelagic environments on natural larval growth and recruitment of the reef fish *Thalassoma bifasciatum*. J Fish Biol 64: 34–54.
19. Sponaugle S, Grorud-Colvert K, Pinkard D (2006) Temperature-mediated variation in early life history traits and recruitment success of the coral reef fish *Thalassoma bifasciatum* in the Florida Keys. Mar Ecol Prog Ser 308: 1–15.
20. Hamilton SL, Regetz J, Warner RR (2008) Postsettlement survival linked to larval life in a marine fish. Proc Natl Aca Sci US 105: 1561–1566.
21. Shima JS, Swearer SE (2009) Larval quality is shaped by matrix effects: implications for connectivity in a marine metapopulation. Ecology 90: 1255–1267.
22. Searcy SP, Sponaugle S (2001) Selective mortality during the larval-juvenile transition in two coral reef fishes. Ecology 82: 2452–2470.
23. McCormick MI, Hoey AS (2004) Larval growth history determines juvenile growth and survival in a tropical marine fish. Oikos 106: 225–242.
24. Rankin TL, Sponaugle S (2011) Temperature influences selective mortality during the early life stages of a coral reef fish. PLoS One 6: e16814.
25. Vigliola L, Meekan MG (2002) Size at hatching and planktonic growth determine post-settlement survivorship of a coral reef fish. Oecologia 131: 89–93.
26. Hoey A, McCormick MI (2004) Selection for low body condition at the larval-juvenile transition of a tropical marine fish. Oecologia 139: 23–29.
27. Sponaugle S, Grorud-Colvert K (2006) Environmental variability, early life-history traits, and survival of new coral reef fish recruits. Integr Comp Biol 46: 623–633.
28. Johnson DW (2008) Combined effects of condition and density on post-settlement survival and growth of a marine fish. Oecologia 155: 43–52.
29. Samhouri JF, Steel MA, Forrester GE (2009) Inter-cohort competition drives density dependence and selective mortality in a marine fish. Ecology 90: 1009–1020.
30. Johnson DW, Grorud-Colvert K, Rankin TL, Sponaugle S (2012) Measuring selective mortality from otoliths: a practical guide for describing multivariate selection from cross-sectional data. Mar Ecol Prog Ser 471: 151–163.
31. Robertson DR, Green DG, Victor BC (1988) Temporal coupling of production and recruitment of larvae of a Caribbean reef fish. Ecology 69: 370–381.
32. Sponaugle S, Cowen RK (1996) Larval supply and patterns of recruitment for two Caribbean reef fishes, *Stegastes partitus* and *Acanthurus bahianus*. Mar Freshw Res 47: 433–447.
33. Johannes R (1978) Reproductive strategies of coastal marine fishes in the tropics. Environ Biol Fish 3: 65–84.
34. Almany GR (2003) Priority effects in coral reef fish communities. Ecology 84: 1920–1935.
35. Caley MJ (1993) Predation, recruitment and the dynamics of communities of coral-reef fishes. Mar Biol 117: 33–43.
36. Holbrook SJ, Schmitt RJ (2003) Spatial and temporal variation in mortality of newly settled damselfish: patterns, causes and co-variation with settlement. Oecologia 135: 532–541.
37. Nemeth RS (1998) The effect of natural variation in substrate architecture on the survival of juvenile bicolor damselfish. Environ Biol Fishes 53: 129–141.
38. Holmes TH, McCormick MI (2009) Influence of prey body charateristics and performance on predator selection. Oecologia 159: 401–413.
39. Holmes TH, McCormick MI (2010) Location influences size-selective predation on newly settled reef fish. Mar Ecol Prog Ser 317: 203–209.
40. Schmale MC (1981) Sexual selection and reproductive success in males of the bicolor damselfish, *Eupomacentrus partitus* (Pisces, Pomacentridae). Anim Behav 29: 1172–1184.
41. Emery AR (1973) Atlantic bicolor damselfish (Pomacentridae) – Taxonomic question. Copeia: 590–594.
42. Sponaugle S, Walter KD, Grorud-Colvert KA, Paddack MJ (2012) Influence of marine reserves on fish recruitment in the upper Florida Keys. Coral Reefs 31: 641–652.
43. Randall JE (1967) Food habits of reef fishes of the West Indies. Stud Trop Oceanogr 5: 665–847.
44. Emery AR (1968) Comparative ecology of damselfishes at Alligator Reef, Florida Keys. PhD thesis, University of Miami, Coral Gables.
45. Baumann H, Pepin P, Davidson FJM, Mowbray F, Schnack D, et al. (2003) Reconstruction of environmental histories to investigate patterns of larval radiated shanny (*Ulvaria subbifurcata*) growth and selective survival in a large bay of Newfoundland. ICES J Mar Sci 60: 243–258.
46. Houde ED (2002) Mortality. In: Fuiman LA, Werner RG, editors. Fishery Science: The unique contributions of early life stages. Oxford: Blackwell Science. pp. 64–87.
47. Zar JH (1999) Biostatistical analysis. Upper Saddle River: Prentice Hall. 662 p.
48. Wilkinson L (1992) SYSTAT: the system for statistics. Evanston: SYSTAT, Inc.
49. Shima JS, Osenberg CW (2003) Cryptic density dependence: Effects of covariation between density and site quality in reef fish. Ecology 84: 46–52.
50. Grorud-Colvert K, Sponaugle S (2009) Larval supply and juvenile recruitment of coral reef fishes to reserves and non-reserves in the upper Florida Keys. Mar Biol 156: 277–288.
51. Anderson JT (1988) A review of size dependent survival during prerecruit stages of fishes in relation to recruitment. J Northwest Atl Fish 8: 55–66.
52. Hernandez-Leon S, Almeida C, Yebra L, Aristegui J (2002) Lunar cycle of zooplankton biomass in subtropical waters: biogeochemical implications. J Plankton Res 24: 935–939.
53. Hernandez-Leon S (2008) Natural variability of fisheries and lunar illumination. Fish Fish 9: 138–154.
54. Dorsey SE (1999) The role of oceanographic and seasonal environmental processes in the early life history of two coral reef fish (Pomacentridae). PhD thesis, State University of New York at Stony Brook, Stony Brook.
55. Marshall DJ, MJ Keough (2003) Variation in the dispersal potential of non-feeding invertebrate larvae: the desperate larva hypothesis and larval size. Mar Ecol Prog Ser 255: 145–153.
56. Botello G, PJ Krug (2006) 'Desperate larvae' revisited: age, energy and experience affect sensitivity to settlement cues in larvae of the gastropod *Alderia* sp. Mar Ecol Prog Ser 312: 149–159.

57. Victor BC (1986) Larval settlement and juvenile mortality in a recruitment-limited coral-reef fish population. Ecol Monogr 56: 145–160.
58. Cowen RK (1991) Variation in the planktonic larval duration of the temperate wrasse *Semicossyphus pulcher*. Mar Ecol Prog Ser 69: 9–15.
59. Almany GR, Webster MS (2006) The predation gauntlet: early post-settlement mortality in reef fishes. Coral Reefs 25: 19–22.
60. Hawn AT, Martin GB, Sandin SA, Hare JA (2005) Early juvenile mortality in the coral reef fish *Chromis cyanea* (Pomacentridae): The growth-mortality hypothesis revisited. Bull Mar Sci 77: 309–318.
61. Sponaugle S, Boulay JN, Rankin TL (2011) Growth- and size-selective mortality in pelagic larvae of a common coral reef fish. Aquat Biol 13: 263–273.
62. Nemeth RS (1997) Spatial patterns of bicolor damselfish populations in Jamaica and St. Croix are determined by similar post-settlement processes. Proc 8th Int Coral Reef Symp 1: 1017–1022.
63. Bonin MC, Srinivasan M, Almany GR, Jones GP (2009) Interactive effects of interspecific competition and microhabitat on early post-settlement survival in a coral reef fish. Coral Reefs 28: 265–274.
64. Sweatman HPA (1983) Influence of conspecifics on choice of settlement sites by larvae of two pomacentrid fishes (*Dascyllus aruanus* and *Dascyllus reticulatus*) on Coral Reefs. Mar Biol 75: 225–229.
65. Lecchini D, Planes S, Galzin R (2007) The influence of habitat characteristics and conspecifics on attraction and survival of coral reef fish juveniles. J Exp Mar Biol Ecol 341: 85–90.
66. Steele MA, Forrester GE, Almany GR (1998) Influences of predators and conspecifics on recruitment of a tropical and a temperate reef fish. Mar Ecol Progr Ser 172: 115–125.
67. Tupper M, Hunte W (1994) Recruitment dynamics of coral-reef fishes in Barbados. Mar Ecol Prog Ser 108: 225–235.

14

Environmental Quality of Italian Marine Water by Means of Marine Strategy Framework Directive (MSFD) Descriptor 9

Chiara Maggi*, Serena Lomiri, Bianca Di Lorenzo, Marco d'Antona, Maria Teresa Berducci

ISPRA, Institute for Environmental Protection and Research, Roma, Italy

Abstract

ISPRA, on behalf of the Italian Ministry of Environment, carried out the initial assessment of environmental quality status of the 3 Italian subregions (Mediterranean Sea Region) on Descriptor 9. The approach adopted to define the GES started to verify that contaminants in fish and other seafood for human consumption did not exceed levels established by Community legislation (Reg. 1881/2006 and further updates). As the Marine Strategy Framework Directive (MSFD) requires to use health tools to assess the environment, Italy decided to adopt a statistical range of acceptance of thresholds identified by national (D.Lgs. 152/2006 concerning water quality required for mussel farms) and international legislation (Reg. 1881/2006 and further updates), which allowed to use the health results and to employ them for the assessment of environmental quality. Italy proposed that Good Environmental Status (GES) is achieved when concentrations are lower than statistical range of acceptance, estimated on samples of fish and fishery products coming from only national waters. GIS-based approach a to perform different integration levels for station, cell's grid and years, was used; the elaborations allowed to judge the environmental quality good.

Editor: Silvia Mazzuca, Università della Calabria, Italy

Funding: This work has been carried out by the Institute for Environmental Protection and Research (ISPRA) on behalf of the Italian Ministry of Environment (http://www.strategiamarina.isprambiente.it/). The funders had no role in study design, data collection and analysis, decision to publish, or preparation of the manuscript.

Competing Interests: The authors have declared that no competing interests exist.

* Email: chiara.maggi@isprambiente.it

Introduction

Marine waters have traditionally been used by population for different activities (e.g. fishing, aquaculture, shipping, tourism, discharges from agriculture and urban areas). Currently, new activities are being developed or increasing (e.g. renewable energies, extraction of minerals, etc.), and thus competing for space with traditional uses, causing many spatial conflicts and increasing human impacts on marine ecosystems [1,2,3]. On the other hand, numerous anthropogenic sources such as land-based industrial activity, urban discharge, pesticide use, nuclear accidents and discharges, aquaculture, heavy shipping lines might be responsible for elevated levels of contaminants in sediment and biota, including species for human consumption.

To face this status the European Parliament and the Council of the European Union on 17 June 2008 have issued the Framework Directive 2008/56/EC (MSFD, [4]) on the Strategy for the Marine Environment, subsequently implemented in Italy with the Legislative Decree n. 190 of 13 October 2010 [5].

The aim of the Marine Strategy is to protect more effectively the marine environment across Europe. The MSFD requires that Member States take measures to achieve or maintain Good Environmental Status (GES) by 2020. The goal of the MSFD is in line with the objectives of the Water Framework Directive (WFD–2000/60/EC Directive [6]), achieving good ecological status by 2015, for surface water including marine-coastal waters and in case of chemical status, also territorial waters, and providing for the first review of the River Basin Management Plans in 2020 [7,8]. MSFD Annex I includes a set of 11 descriptors, on the basis of which GES should be determined. Some descriptors deal with general status of the environment, others with specific habitat integrity or specific pressures. The Commission Decision of 1 September 2010 ([9] COM DEC 2010/477/EU) listed 29 criteria and 56 indicators, based on the scientific and technical assessment prepared by the Task Groups set up by the Joint Research Centre and the International Council on the Exploration of the Seas, to assess GES for each descriptor of the MSFD Annex I. These criteria and indicators include a factors' combination in relation to state, impacts and pressures.

Member States had to make an initial assessment of own marine waters, taking account of existing data, where available, specifying the essential features and characteristics and current environmental status of the waters. The initial assessment was provided by 2012, together with the definition of targets and monitoring indicators. For this evaluation, it was necessary to point out predominant pressures and impacts existing in each region or subregion (MSFD art. 8, [4]). By 2015, Member States will have to develop coherent and coordinated programmes of measures. To ensure clean, healthy and productive marine waters, Member

States strategies will have to be coordinated, consistent and properly integrated with regulations provided by existing Community legislation (such as transport, fisheries, tourism, infrastructure, research) and international agreements.

This work describes the initial assessment by means of criteria and indicators defined on Descriptor 9 ("*Contaminants in fish and other seafood for human consumption do not exceed levels established by Community legislation or other relevant standards*"), carried out by the Institute for Environmental Protection and Research (ISPRA) on behalf of the Italian Ministry of Environment.

Focus on Descriptor 9: Contaminants in fish and other seafood for human consumption

MSFD Descriptor 9 requires to use health tools to assess the marine environment; it starts from ensuring that contaminants in fish and other seafood for human consumption, detected in the sea, do not exceed levels established by Community legislation or other relevant standards in order to protect the public health and, as indirect effect, to avoid negative influence on the sustainable use of marine resources. Thus, Good Environmental Status should be achieved when contaminants are below the levels fixed for human consumption, although the absence of human health effects may however involve environmental pollution effects. Descriptor 9 necessarily requires a balance between health information and environmental assessment; so, it becomes necessary to identify possible relations between contaminant levels in seafood tissues and the status of marine environment. In this way, starting from health information, it should be possible to reconstruct a general description of the environmental status of the area where the analyzed organisms live.

Decision 2010/477/EU [9] defines the need to monitor the presence of contaminants and hazardous substances in edible tissues of fisheries species fit for human consumption, by the use of two indicators:

- 9.1.1: actual levels of contaminants that have been detected and number of contaminants which have exceeded maximum regulatory levels;
- 9.1.2: frequency of regulatory levels being exceeded.

So, the goal of the Descriptor 9 is to check the public health. Nevertheless the aim of the Marine Strategy Directive is providing to the achievement of the Good Environmental Status, taking into account all descriptors together.

Indeed, also regarding data retrieval, we observed that existing national health monitoring programs often do not cover the information needs for compliance to GES, because monitoring programs on fish and seafood for public health reasons focus on estimating consumer exposure rather than assessing environmental status. Sampling procedures include all sizes of fish sold for human consumption rather than focusing on a standardized sample (offering greater possibilities in comparing degrees of contamination in the marine environment), without considering biological factors [10] that can influence concentrations of contaminants in fish, such as seasonal variation, age and sex [11]. However, the standardized sampling shuld produced a loss of representativeness compared to the great variability of samples available for human consumption.

Anyway, GES for Descriptor 9 should be judged in view of the monitoring of Descriptor 8 ("*Concentrations of contaminants are at levels not giving rise to pollution effects*"), also dealing with contaminants in marine environment. Descriptors 8 and 9 both measure contaminants in the marine environment, but focus on different state indicators. The importance of pay serious attention to the public health comes from the increasing recognition that the body burdens of environmental contaminants in the European human population is a growing cause for concern and may be associated with alterations in the frequency of occurrence of a range of diseases including diabetes, heart attacks, strokes, neurological disorders and cancers [12,13]. Otherwise, the global per-capita demand for seafood has reached an all-time high, and is likely to continue to increase [14].

It was noted that almost all the contaminants thought to be of concern for marine vertebrates are also of concern for human health, due to the transfer along the food web. Thus, the relationship between the degree of chemical contamination and health of marine vertebrates is of great relevance to contaminant-related human diseases [15]. This is likely to be an even greater area of interest in the coming years as the human demographic ageing proceeds across Europe, allowing longer-term accumulation of contaminants in human tissues.

Principles for assessing Italian marine waters

Concerning to Mediterranean area, there were some studies, most of them conducted in non-Italian waters, stale and with a very local profile (i.e. [16,17,18,19,20,21]). Among them, European Environmental Agency published data (1985–2002) relating to concentrations of selected metals and organic contaminants in Mediterranean Sea mussels, highlighting a moderate variability and values non-exceeding regulatory levels. That being so, a more recent data collection was needed.

Italy used both results from health and environmental monitoring, so as to allow a proper evaluation and to cover the information needs for monitoring compliance to GES under Descriptor 9.

Human health programs monitored fish from different sizes and ages; so, higher levels can not automatically be interpreted as a negative status or evolution of the marine environment. In fact contamination appears to change according to the trophic level considered, with estimates higher in larger species, i.e. pelagic fish, containing more fat and occupying an higher food web position [22].

That being so, the "one out, all out" (OOAO) approach, based on the practice that the status of the worst element determines the final status, doesn't seem applicable although it may be regarded as a precautionary approach [3,7]. In case one species shows concentrations exceeding the regulatory levels, the (sub) region should be assessed as bad status for Descriptor 9; this judgment could easily be interpreted that consumption of all fish and seafood originating in that (sub) region would be dangerous.

Otherwise the Marine Strategy does not give specific rules, allowing Member States to adopt different criteria in order to use correctly health data for assessing environmental quality.

In Italy, a statistical range of acceptance was applied in order to avoid losing information, as reported below. Regarding the assessment of contaminants and the number of pollutants that exceeded law levels, Italy proposed that Good Environmental Status is achieved when concentrations are statistically lower than the thresholds identified by national and international legislation (Reg. 1881/2006 [23] and further updates [24,25,26], D.Lgs. 152/2006 [27]), estimated on samples of wild caught fish and aquaculture products (mussels) coming from national waters. Special attention was paid to assure traceability, selecting samples directly linked to the Italian (sub) regions laid down in the MSFD. The need arose since seafood is widely transported, its provenance may not be identifiable and human exposure monitoring programs often lack the necessary data to link the

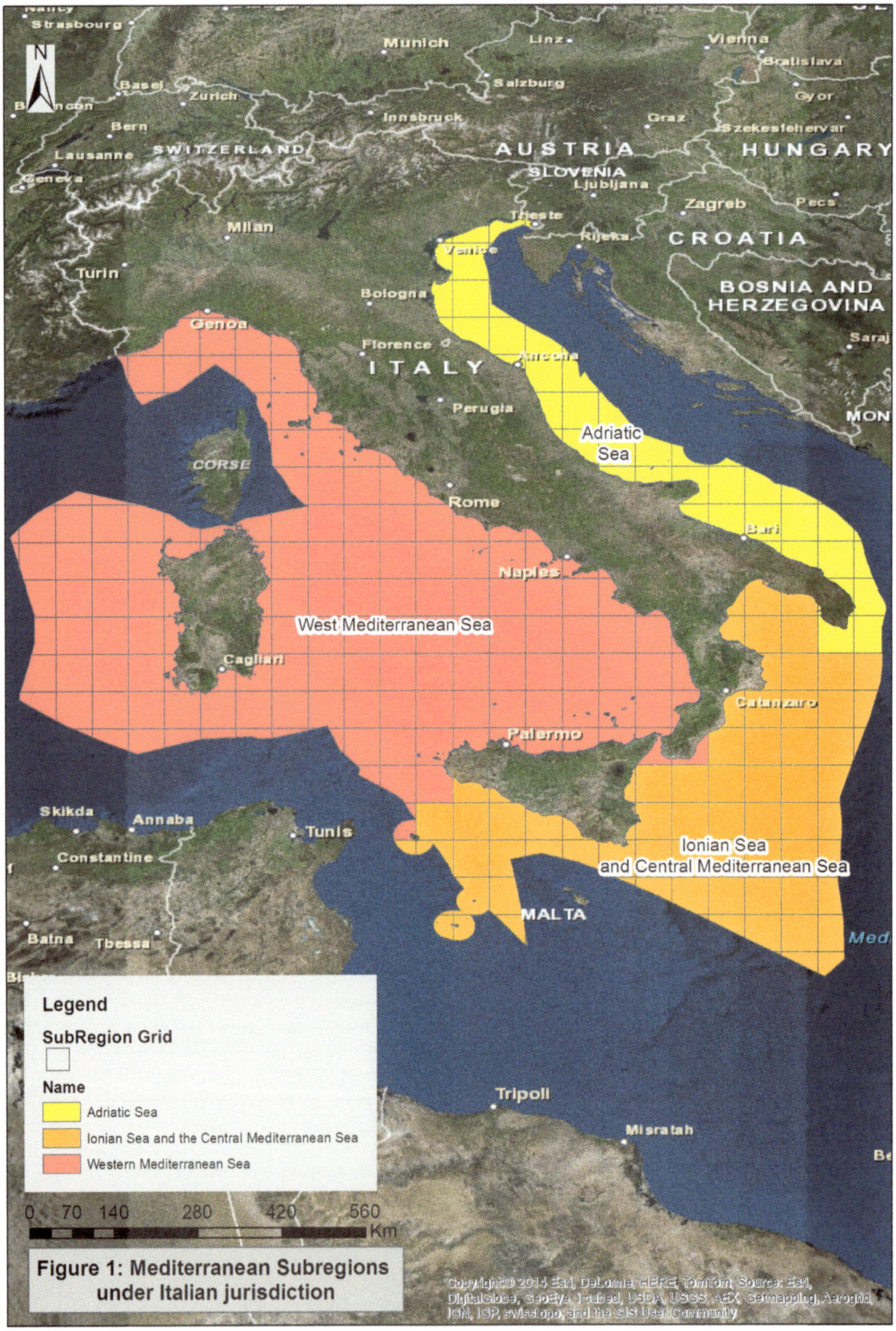

Figure 1. Mediterranean Subregions under Italian jurisdiction.

samples to specific (sub) regions. In addition, Italy, as Task Group 9 suggests, took care to make a selection of species, choosing those more prone to biomagnify/bio-accumulate regulatory classes of contaminants, species representative of the different trophic levels or habitats, species representative for different subregions.

Table 1. Regulatory levels, reference legislation, code and foodstuff categories.

Category code	Legislation	Foodstuff	Regulatory levels
Cd 3.2.5	Reg.1881/2006/CE	Muscle meat of fish (footnote 24)	0,05 mg/kg w.w.
Cd 3.2.6	Reg.1881/2006/CE	Muscle meat of listened fish	0,10 mg/kg w.w.
Cd 3.2.8	Reg.1881/2006/CE	Crustaceans	0,50 mg/kg w.w.
Cd 3.2.9	Reg.1881/2006/CE	Bivalve molluscs	1,0 mg/kg w.w.
Cd 3.2.10	Reg.1881/2006/CE	Cephalopods	1,0 mg/kg w.w.
Hg 3.3.1	Reg.1881/2006/CE	Fishery products and muscle meat of fish (footnotes 24, 25, 26)	0,50 mg/kg w.w.
Hg 3.3.2	Reg.1881/2006/CE	Muscle meat of listened fish	1,0 mg/kg w.w.
Pb 3.1.5	Reg.1881/2006/CE	Muscle meat of fish (footnote 24)	0,3 mg/kg w.w.
Pb 3.1.6	Reg.1881/2006/CE	Crustaceans	0,50 mg/kg w.w.
Pb 3.1.7	Reg.1881/2006/CE	Bivalve molluscs	1,5 mg/kg w.w.
Pb 3.1.8	Reg.1881/2006/CE	Cephalopods	1,0 mg/kg w.w.
Dioxins 5.3	Reg.1259/2011/CE	Muscle meat of fish and Bivalve molluscs	3,5 pg/g w.w.
Sum dioxins and dioxin like PCBs 5.3	Reg.1259/2011/CE	Muscle meat fish and Bivalve molluscs	6,5 pg/g w.w.
Benzo(a)pyrene 6.1.4	Reg.1881/2006/CE	Muscle meat of fish (footnote 24)	2,0 μg/kg w.w.
Benzo(a)pyrene 6.1.5	Reg.1881/2006/CE	Crustaceans and Cephalopods	5,0 μg/kg w.w.
Benzo(a)pyrene 6.1.6	Reg.835/2011/CE	Bivalve molluscs	5 μg/kg w.w.
Sum PAH 6.1.6	Reg.835/2011/CE	Bivalve molluscs	30 μg/kg w.w.

Methodology

Marine Strategy Directive divided European marine waters in 4 regions: Baltic, North East Atlantic Ocean, Mediterranean Sea and Black Sea. Some of them, as Mediterranean Sea, had been subdivided in subregions and Commission Decision of 1 September 2010 [9] established that Member States should performed initial assessment in each subregions of jurisdiction.

Italian waters fall into Mediterranean region and Western Mediterranean Sea (WMS), Adriatic Sea (AS) and Ionian Sea and the Central Mediterranean Sea (ISCMS) subregions are under Italian authority (Fig. 1).

The regulatory framework is essentially composed of Regulation (EC) No 1881/2006 [23] and further updates [24,25,26] on contaminants levels, set taking into account their toxicity as well as their prevalence in the food chain, and of Regulations (EC) No 882/2004 [28], No 1883/2006 [29] and No 333/2007 [30] (and respective further updates) regarding sampling procedures, methods of analysis and analytical quality. The Regulations provide for consumers protection from chemical substances in food and set maximum levels for certain contaminants in foodstuffs in order to keep contaminants at levels which are toxicologically acceptable. Each regulated contaminant has a different threshold level, laid down according to the specific tissue analyzed. Descriptor 9 refers to these regulations, but also to other relevant national standards (in Italy i.e. D.Lgs. 152/2006 fixes mercury and lead levels in mussel tissues, concerning to water quality required for mussel farms).

Chemical contaminants monitored were those for which regulatory levels have been laid down for public health reasons: heavy metals (lead, cadmium, and mercury), polycyclic aromatic hydrocarbons (benzo(a)pyrene, benz(a)anthracene, benzo(b)fluoranthene and chrysene) and dioxins (including dioxin-like PCBs). Each contaminant was identified by category code reported in Reg. 1881/2006 [23], Reg. 835/2011[24] and Reg. 1259/2011 only for dioxins and dioxin-like PCBs [26], providing for specific values of edibility, relative to different species and analyzed tissues. In table 1, regulatory values, reference legislation, code and foodstuff categories are summarized.

It was noted that almost all the contaminants thought to be of concern for marine vertebrates are also of concern for human health, due to the transfer along the food web. Thus, the relationship between the degree of chemical contamination and health of marine vertebrates is of great relevance to contaminant-related human diseases [15]. This is likely to be an even greater area of interest in the coming years as the human demographic ageing proceeds across Europe, allowing longer-term accumulation of contaminants in human tissues.

For each contaminant, the number of excesses and their frequency on the total number of observed data were indicated.

Calculation of statistical range of acceptance

Based on the thresholds identified in the relevant national and international laws, statistical ranges of acceptance were calculated for the concentrations of regulated contaminants.

To calculate these ranges, it was utilized R software. The assumption used is that contaminants' concentrations follow lognormal distribution $Y \sim LN(\mu, \sigma)$ [31], [32], with parameter μ equal to the geometric mean of collected contaminant's data and σ equal to 20% of this value, as required for environmental data [33]. We have also tested this assumption, with Shapiro Normality test on logarithm of actually used data. In order to identify statistical ranges, it was assumed that threshold level (TL) is the 90th percentile of theoretical distribution and that 90th percentile of empirical data distribution must not exceed this level.

GIS elaborations

Useful data for the assessment were georeferenced in GIS (Software ArcGis 10.1) and exported into Feature Classes, taking

Table 2. Statistical outputs for Metals.

AS* (total cells 51)	Cd 3.2.5	Cd 3.2.6	Cd 3.2.8	Cd 3.2.9	Cd 3.2.10	Hg 3.3.1	Hg 3.3.2	Pb 3.1.5	Pb 3.1.7	Pb 3.1.8	Pb 3.1.6
% of spatial coverage	**27.5**	**11.8**	**5.9**	**47.1**	**9.8**	**82.4**	**11.8**	**33.3**	**49.0**	**9.8**	**5.9**
% of data within the limits	100	100	100	99.4	100.0	99.4	93.8	92.9	99.7	100	100
% of data outside the limits				0.6		0.6	6.2	7.1	0.3		
% of station within the limits	100	100	100	100	100	97.3	85.7	94.7	100	100	100
% of station outside the limits						2.7	14.3	5.3			
% of cells within the limits	100	100	100	100	100	95.2	83.3	94.1	100	100	100
% of cells outside the limits						4.8	16.7	5.9			
WMS (total cells 166)**	**Cd 3.2.5**	**Cd 3.2.6**	**Cd 3.2.8**	**Cd 3.2.9**	**Cd 3.2.10**	**Hg 3.3.1**	**Hg 3.3.2**	**Pb 3.1.5**	**Pb 3.1.7**	**Pb 3.1.8**	**Pb 3.1.6**
% of spatial coverage	**3.6**	**2.4**	**-**	**10.8**	**-**	**15.1**	**3.6**	**7.8**	**11.4**	**-**	**-**
% of data within the limits	85	75	-	100	-	88.3	87.5	87.2	96	-	-
% of data outside the limits	15	25	-		-	11.7	12.5	12.8	4	-	-
% of station within the limits	85.7	75	-	100	-	92.3	85.7	86.7	97.7	-	-
% of station outside the limits	14.3	25	-		-	7.7	14.3	13.3	2.3	-	-
% of cells within the limits	83.3	75	-	100	-	96	83.3	92.3	94.7	-	-
% of cells outside the limits	16.7	25	-		-	4	16.7	7.7	5.3	-	-
ISCMS* (total cells 77)**	**Cd 3.2.5**	**Cd 3.2.6**	**Cd 3.2.8**	**Cd 3.2.9**	**Cd 3.2.10**	**Hg 3.3.1**	**Hg 3.3.2**	**Pb 3.1.5**	**Pb 3.1.7**	**Pb 3.1.8**	**Pb 3.1.6**
% of spatial coverage	**5.2**	**5.2**	**-**	**5.2**	**-**	**7.8**	**9.1**	**7.8**	**3.9**	**-**	**-**
% of data within the limits	100	75	-	100	-	100	100	100	99.8	-	-
% of data outside the limits		25	-		-				0.2	-	-
% of station within the limits	100	60	-	100	-	100	100	100	100	-	-
% of station outside the limits		40	-		-					-	-
% of cells within the limits	100	50	-	100	-	100	100	100	100	-	-
% of cells outside the limits		50	-		-					-	-

* Adriatic Sea SubRegion;
** West Mediterranean Sea SubRegion;
*** Ionian Sea and Central Mediterranean Sea SubRegion.

Table 3. Statistical outputs for PAHs (Polycyclic Aromatic Hydrocarbons).

AS* (total cells 51)	**Benzo(a)pyrene 6.1.4**	**Benzo(a)pyrene 6.1.5**	**Benzo(a)pyrene 6.1.6**	**Sum of PAHs 6.1.6**
% of spatial coverage	**2**	**-**	**29.4**	**29.4**
% of data within the limits	100	-	100	100
% of data outside the limits		-		
% of station within the limits	100	-	100	100
% of station outside the limits		-		
% of cells within the limits	100	-	100	100
% of cells outside the limits		-		
WMS (total cells 166)**	**Benzo(a)pyrene 6.1.4**	**Benzo(a)pyrene 6.1.5**	**Benzo(a)pyrene 6.1.6**	**Sum of PAHs 6.1.6**
% of spatial coverage	**3.6**	**1.2**	**3.6**	**4.8**
% of data within the limits	100	100	94.3	98.7
% of data outside the limits			5.7	1.3
% of station within the limits	100	100	89.7	96.3
% of station outside the limits			10.3	
% of cells within the limits	100	100	66.7	88.9
% of cells outside the limits			33.3	
ISCMS* (total cells 77)**	**Benzo(a)pyrene 6.1.4**	**Benzo(a)pyrene 6.1.5**	**Benzo(a)pyrene 6.1.6**	**Sum of PAHs 6.1.6**
% of spatial coverage	-	-	**2.6**	**2.6**
% of data within the limits	-	-	85.7	94.4
% of data outside the limits	-	-	14.3	5.6
% of station within the limits	-	-	100	100
% of station outside the limits	-	-		
% of cells within the limits	-	-	100	100
% of cells outside the limits	-	-		

* Adriatic Sea SubRegion;
** West Mediterranean Sea SubRegion;
*** Ionian Sea and Central Mediterranean Sea SubRegion.

as reference system ETRS89. The Feature Classes have been created on the basis of regulatory codes. The data were processed both for a single year, in order to see the concentration trend in the subregion, and considering the whole period (2006–2012), as requested by MSFD. Polygonal grids were created to cover the entire subregion.

A critical aspect of the cartographic layers production was the definition of the reference grid (fishnet) by selecting:

- The starting point (lower left corner);
- the cell size(s);
- the number of rows and columns.

As in 2009 ISPRA was involved in the EuSeaMap project (Mapping European seabed habitats, [34]), in which a broad-scale modelled habitat map for the Western Mediterranean was produced together with the Spanish and French partners, the fishnet was created according to the grid used in this project.

GIS layers were created in geographic coordinates (ETRS 89), using angular measures in order to obtain a 1:1 cell size ratio. The chosen mesh was 50 km per side, coherently with the ecological differences (such as mobility, migratory patterns and feeding habits) of functional groups and habitats considered.

Value of "code" 1 and green colour's cell were assigned to data not exceeding the threshold levels. Value of "code" 2 and red colour's cell were assigned to data above reference thresholds.

The stations results which fall in the same cell and/or which have the same coordinates (e.g. data relating to the same station, but sampled in different years or related to differences species belonging to the same regulatory category) were integrated, as follows:

- if stations number with code 1 and 2 is different, the cell has code with the highest frequency.
- if stations number with code 1 is equal to stations number with code 2, the cell reaches code 2, in compliance with the precautionary principle [7];

So, in the cartographic representation, colour green had been assigned to cells with majority of station with code 1, while colour red to the cells with equal or majority of station with code 2.

Table 4. Statistical outputs for Dioxins and Dioxins like PCBs.

AS*(total cells 51)	Dioxins 5.3	Sum Dioxins and dioxin like 5.3
% of spatial coverage	-	**56.9**
% of data within the limits	-	100
% of data outside the limits	-	
% of station within the limits	-	100
% of station outside the limits	-	
% of cells within the limits	-	100
% of cells outside the limits	-	
WMS (total cells 166)**	**Dioxins 5.3**	**Sum Dioxins and dioxin like 5.3**
% of spatial coverage	**1.2**	**4.2**
% of data within the limits	100	92.9
% of data outside the limits		7.1
% of station within the limits	100	100
% of station outside the limits		
% of cells within the limits	100	100
% of cells outside the limits		
ISCMS* (total cells 77)**	**Dioxins 5.3**	**Sum Dioxins and dioxin like 5.3**
% of spatial coverage	**1.3**	**1.3**
% of data within the limits	87.5	100
% of data outside the limits	12.5	
% of station within the limits	85.7	100
% of station outside the limits	14.3	
% of cells within the limits	100	100
% of cells outside the limits		

* Adriatic Sea SubRegion;
** West Mediterranean Sea SubRegion;
*** Ionian Sea and Central Mediterranean Sea SubRegion.

For the three subregions, the percentage of spatial coverage for individual contaminants and the percentage of stations that exceed the limit (frequency of exceedances) were determined.

No review from the ethic committee was required as our research work did not involve any direct manipulation or disturbance of animals. Blood samples were collected by and under the permit of the Israel Nature and Parks Authority. Access to all sampling locations was not restricted and so no specific permissions were required.

Results and Discussion

Initial Assessment

The initial assessment examined substances for which maximum levels are established in fish, shellfish and other marine products destined for human consumption, defined a statistical range of acceptance, calculated the number and frequency of contaminants exceeding maximum regulatory levels and handed over these information to environmental quality assessment.

In the Mediterranean Sea approximately 15000 data were collected, concerning Metals (Hg, Cd, Pb), PAHs (benzo(a)pyrene and sum of benzo(a)pyrene, benz(a)anthracene, benzo(b)fluoranthene and chrysene) and dioxins (including dioxin-like PCBs).

Data related to Adriatic Sea were 7119 of which 5801 for Metals, 1086 for PAHs and 232 for HOCs.

In West Mediterranean Sea 1900 data were examined, of which 1344 for Metals, 513 for PAHs and 43 for HOCs. For Ionian Sea and Central Mediterranean Sea the number of analyzed data was equal to 5983, of which 5865 for Metals, 92 for PAHs and 26 for HOCs.

Data were elaborated and those unfitting in analytical quality (Annex III of Commission Regulation (EC) No 882/2004 [28]), in MSFD criteria and in GIS elaborations standards, were removed. Collected data concerned edible portion of different target species both from environmental and commercial value point of view, but also representing a coverage of the entire subregion as better as possible. Target species were principally coastal fish, tuna, mullet, mussels, shark, crustaceans, cephalopods, selected using Annex II of Task Group 9 Report [35]. For PAHs', bivalve molluscs were used as preferred species for initial assessment, being sedentary benthic species with a large lipid content [36,37].

Tables 2–4 show statistical outputs for each Subregion. Percentage of spatial coverage and percentage of data, stations and cells, within or outside regulatory limits, are reported. Regulatory levels (Tab. 1) are referred to categories established by Reg. 1881/2006 [23]. Limits of the sum of PAHs and benzo(a)pyrene in bivalve molluscs related to Reg. 835/2011/CE [24] while limits for s and dioxin like PCBs for muscle meat of fish and bivalve molluscs referred to Reg. 1259/2011/CE [26].

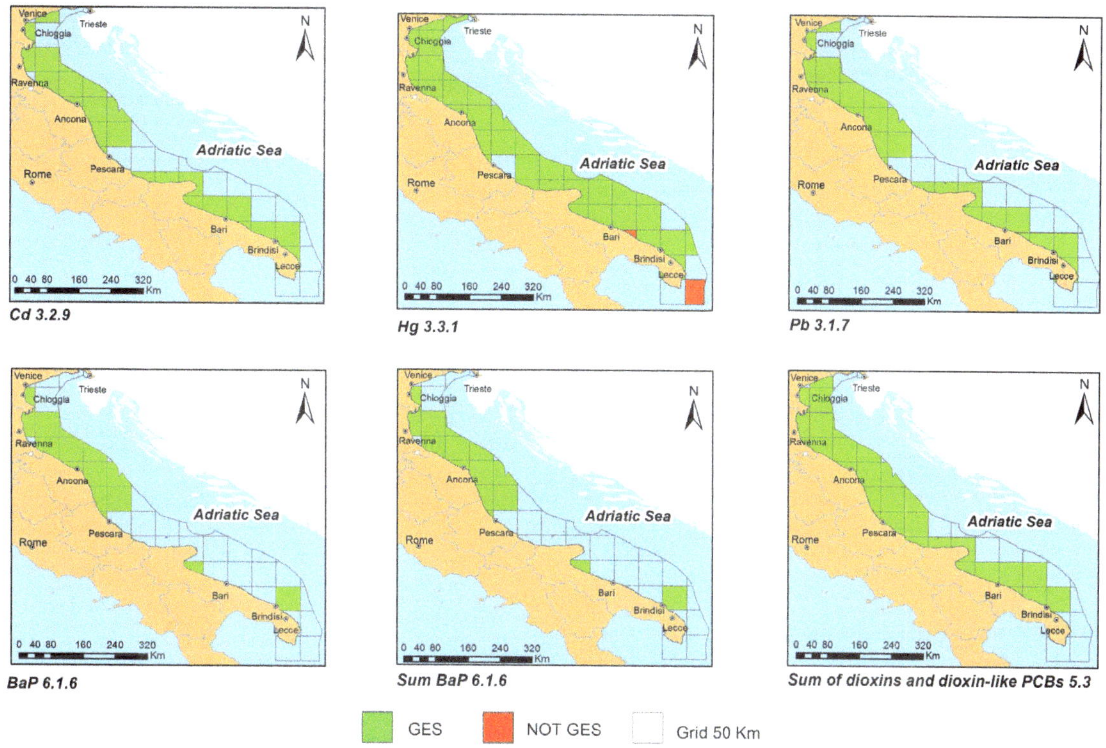

Figure 2. Results on Metals, PAH and Dioxins/Dioxin-like PCBs in Adriatic Sea Subregion (AS).

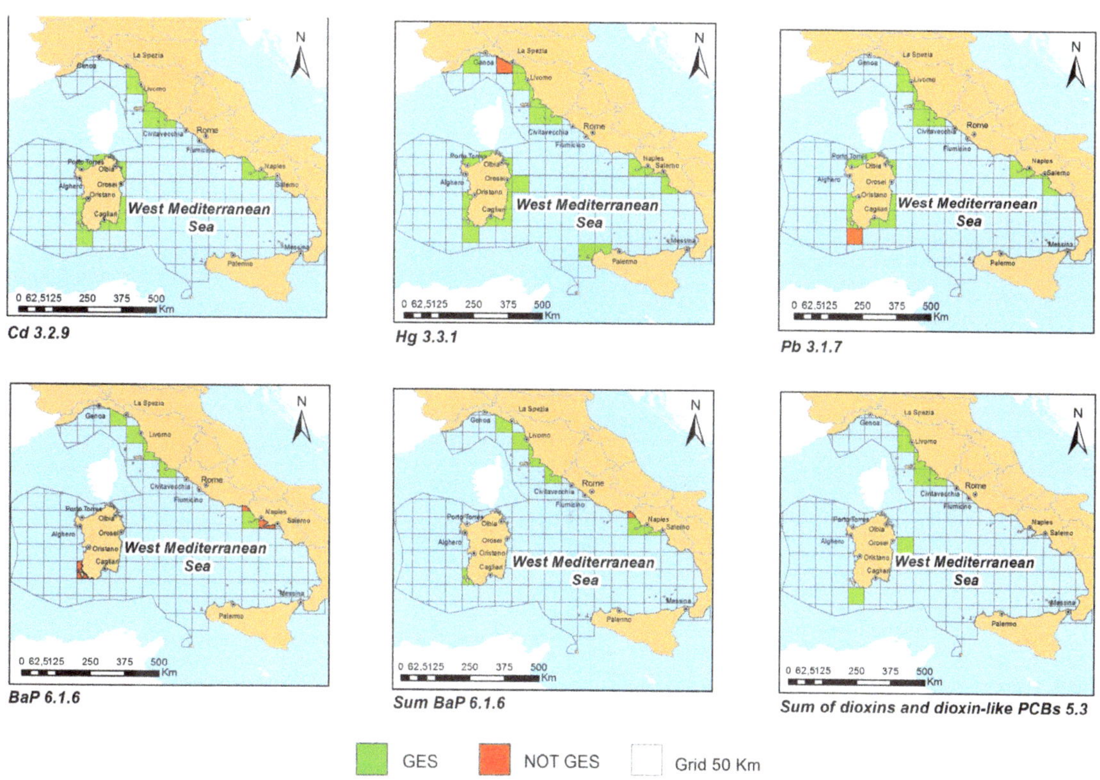

Figure 3. Results on Metals, PAH and Dioxins/Dioxin-like PCBs in Western Mediterranean (WMS).

Italy tried to hand over health information in terms of environmental assessment, without judging the outcome of a single sample, as requested by human consumption legislations, but integrating the results of all stations included into the same cell. For each regulatory category, different integration levels were performed: data within the same station; several stations falling in the same grid cell; same grid cell in different years.

The analyzed data showed that inorganic elements (metals) are usually more investigated than organic contaminants; indeed for metals the spatial coverage reaches 82% in AS subregion, while the maximum PAHs and sum Dioxins-dioxin like PCBs percentages are 29 and 57 respectively, still in AS subregion.

These data showed that the favourite organisms to assess the health risk are generally bivalve molluscs, as demonstrated by spatial coverage percentage of Cd 3.2.9, Hg 3.3.1(chiefly populated by molluscs), Pb 3.1.7 and BaP 6.1.6. Indeed, the mussel *Mytilus galloprovincialis* has all the desirable characteristics of a biological indicator and is the main species recommended to perform bio-monitoring activities; moreover, both the biology and ecology of the species are documented [38]. For these reasons mussel watch programs are widely diffused to evaluate environmental quality of marine water [38,39,40]. On the other hand, sedentary benthic species, having a high lipid content, are suggested as preferred species for PAHs' monitoring activities on human consumption [37,41].

Adriatic Sea (AS) subregion data referred to bivalve mollusc organisms, fishes (coastal fish, tuna, mullet, shark), crustaceans and cephalopods. In the Ionian Sea and Central Mediterranean Sea (ISCMS) subregion data distribution included mullet, nectobentonic fishes, bivalve molluscs, anchovies, sardines and seabream. Concerning Western Mediterranean (WMS) subregion, bivalve molluscs, crustaceans, mullet, seabream, seabass and flatfishes were studied.

Results on Metals, PAH and Dioxins/Dioxin-like compounds concerning to the seafood categories with greater spatial coverage are reported in figures 2 and 3. ISCMS results are not shown as the spatial coverage of data was very low compared to the number of subregion total cells.

Data were collected especially in coastal zones, where the main fishing activities take place; so the spatial coverage was often lacking in the open sea.

After integration, in all subregions, the environmental quality by metals was good, although some stations exceeded regulatory levels. In the AS subregion, Pb 3.1.5, and Hg 3.3.2 exceeded in muscle of twaite shad (*Alosa fallax*) and albacore tuna (*Thunnus alalunga*) while Hg 3.3.1 exceeded in blackbelly rosefish (*Helicolenus dactylopterus*), in thornback ray (*Raja clavata*) and in shellfish tissues. In the WMS subregion several stations exceeded regulatory levels concerning all codified categories of metals, except Cd 3.2.9. Indeed Pb 3.1.5–3.1.7 passed limits in fish muscle of seabream (*Dicentrarchus labrax*) and shells, Hg 3.3.1 in bivalves and Hg 3.3.2 in Atlantic bluefin tuna (*Thunnus thynnus*) and finally Cd 3.2.5–3.2.6 in fish muscle of seabream and in European anchovy (*Engraulis encrasicolus*). In ISCM subregion all data were below regulatory levels except Cd 3.2.6 in European anchovy, that covered only about 5% of the whole subregion.

These exceedances confirm the literature data regarding the different metal behaviour due to propensity to undergo biomagnification and to trophic level and foraging method/location of organisms. Thus, it is not surprising that tuna, that is at high trophic levels, had some excesses of mercury and that bottom feeding fishes had the excesses too. Animals living in close association with sediments, in which they bury and from where they mainly feed, are, in fact, more exposed to any sediment-associated contamination than other fish [42,43,44]. Another relevant point when evaluating the variability in data is the fish size. There is, in fact, a large body of data showing that mercury tissue concentrations increase with age/size of marine fauna [45,46,47]. Concerning Cd and Pb, the concentration ranges were quite narrow and similar, with exception of data from industrial sites in the coastal areas of ISCM and WMS subregions, suggesting that there is no trend in tissue residues of these metals based on taxonomic position or trophic status [48].

However, independently from the interaction complexity leading to a different toxic load among species, collected metals concentrations were in agreement with those reported in seafood from either the same environment [43,44,49] or other marine areas [50].

Concerning dioxins and dioxin-like compounds, in the AS and WMS subregions data were below the regulatory thresholds, while in the ISCMS only data referred to mussels (*Mytilus galloprovincialis*) showed some excesses; anyway the grid cells were all within the limits. The spatial cover for dioxins and dioxin-like compounds was higher in Adriatic subregion (56,9%), where the most part of mussel farming sites are located [51], while in the others subregions the coverage didn't reach 5% overall.

The monitoring of benzo(a)pyrene and the others congeners (benz(a)anthracene, benzo(b)fluoranthene and chrysene) was carried out in a systematic way in Adriatic Sea subregion, where data collected covered about 30% of the subregion and the environmental quality was good. Despite of that, in Western Mediterranean Sea and Ionian Sea and Central Mediterranean subregions the coverage percentage didn't go beyond 5% and some data in bivalve molluscs exceeded the regulatory thresholds for benzo(a)pyrene. It is well known that seasonal differences in biotic and abiotic factors regulate metabolic mechanisms and PAHs bioaccumulation and that bivalve molluscs are not able to metabolize PAHs, accumulating them in their organisms [49]. Nevertheless, several works described annual cyclic variation in the PAHs content of seafood, so that data from *Mytilus galloprovincialis* collected in winter exceeded regulatory levels [41,52,53].

Despite some exceedances, environmental quality of Italian subregions was good overall, regarding Metals, PAHs, Dioxins and Dioxin like compounds, taking into account coverage percentages too. Indeed, concentrations over thresholds were linked not only to the presence of the pollutant, but also to a whole range of biological (species, age, growth degree) and environmental (temperature, geochemical anomalies, salinity) factors which influence trace elements bioavailability [49,54,55,56]. Human health programmes effectively do not contemplate seasonal variation, age and sex which influence contaminants' bioavailability in fish and seafood. So higher levels of contaminants concentrations should not automatically be interpreted as a negative status.

Conclusion

The analysis of the collected data allowed to carry out the initial assessment on environmental quality of the 3 Italian subregions of the Mediterranean Sea. Results showed that the three subregions were overall in good status, since, for each substance or category, the greater part of the investigated cells had a concentration statistically lower than the acceptance range calculated assuming the regulatory levels as the 90 percentile of data distribution. However, it is needed to underline that data spatial coverage is limited, especially for synthetic compounds and, for this reason, results could have some limitations.

For this initial assessment Italy used data from health and environmental monitoring plans because existing national health monitoring programs didn't not cover all information needs for monitoring compliance to GES under Descriptor 9. It would be desirable improve monitoring activities considering Descriptor 9 in conjunction with requirements for Descriptor 8. Moreover it would be interesting to identify possible relations between contaminant levels in sediment and biota's, in order to explain the transfer of contaminants from sediment to biota and in order to establish a well-defined quantitative link between levels in fish and other seafood and levels of contaminants in marine environment. That might help to clearly demonstrate the transfer of them from marine environment to fish/fishery species and, at the end, to man.

So that researchers could best benefit from future monitoring programs (art.11 MSFD) and from information obtained, major adaptations would be needed regarding sampling plans' design, sampling procedures, matrix on which to perform the analysis as well as traceability to the location of catching or harvesting. In fact monitoring programs on human exposure use regulatory levels set for public health, but they often lack the necessary information to link the samples and results to specific subregions. In addition, these monitoring programs check out contaminants for which regulatory levels have been set, while other contaminants of relevance in fish and other seafood, distinguishing of a specific area, activity or input source, should be investigated to evaluate both spatial and temporal trends [35]. It would be opportune to set a limit also for total PAHs, organochlorine pesticides and organotin compounds. Infact, to date, the legislative Regulation 835/2011/CE [24] provides only a limit for the sum of 4 PAHs. Among toxic metals, only Pb, Cd and Hg have a limit set by E.C. Reg. 1881/2006 [23], while other metals also might become toxic when elevated doses are ingested [37,57].

Nevertheless there is still the need to improve the flow of information coming from institutions working in the health field, in order to validate the applied methodology and confirm the possibility to extend the use of health information arising from chemical contamination of seafood for the evaluation of the quality of the marine environment.

This shrewdness will be functional to fill part of the knowledge gaps that emerged from the initial assessment and nevertheless indispensable for a better definition of GES and Targets.

Acknowledgments

We are grateful to National Institute of Health, to Ministry of Health, to Experimental Zooprophylactic Institutes for provided data and to Prof. M.C. Fossi and coworkers for their expert advice and contribution that helped the completion of this work.

Author Contributions

Conceived and designed the experiments: CM. Performed the experiments: CM SL MTB. Analyzed the data: CM SL MTB MdA BDL. Contributed to the writing of the manuscript: CM SL MTB.

References

1. Ban N, Alder J (2008) How wild is the ocean? Assessing the intensity of anthropogenic marine activities in British Columbia, Canada. Aquatic Conserv.: Mar Freshw Ecosyst 18: 5–85.
2. Halpern BS, Walbridge S, Selkoe KA, Kappel CV, Micheli F (2008) A global map of human impact on marine ecosystems. Science 319: 948–952.
3. Borja A, Elliott M, Andersen JH, Cardoso A C, Carstensen J, et al. (2013) Good Environmental Status of marine ecosystems: What is it and how do we know when we have attained it? Mar Pollut Bull 76(1–2): 16–27.
4. Framework Directive 2008/56/EC (MSFD) on the Strategy for the Marine Environment.
5. Legislative Decree n. 190 of 13 October 2010.
6. Water Framework Directive (WFD) 2000/60/EC Directive.
7. Borja A, Elliott M, Carstensen J, Heiskanen A, van de Bund W (2010) Marine management – Towards an integrated implementation of the European Marine Strategy Framework and the Water Framework Directives. Mar Pollut Bull 60: 2175–2186.
8. http://ec.europa.eu/environment
9. Commission Decision 2010/477/EU of 1 September 2010.
10. Jakimska A, Konieczka P, Skóra K, Namieśnik J (2011) Bioaccumulation of metals in tissues of marine animals, part II: metal concentrations in animal tissues. Pol J Environ Stud 20(5): 1127–1146.
11. Cox GW (1997) Conservation Biology - 2nd ed. WCB.
12. Lang IA, Galloway TS, Scarlett A, Henley WE, Depledge M, et al. (2008) Association of urinary bisphenol A concentration with medical disorders and laboratory abnormalities in adults. JAMA 300(11): 1303–1310.
13. Melzer D, Rice NE, Lewis C, Henley WE, Galloway TS (2010) Association of urinary bisphenol a concentration with heart disease: evidence from NHANES 2003/06. PLoS One 5(1): 8673.
14. Gutierrez NL, Valencia SR, Branch TA, Agnew DJ, Baum JK, Bianchi PL, Cornejo-Donoso J, Costello C, Defeo O, Essington TE, Hilborn R, Hoggarth DD, LarsenAE, Ninnes C, Sainsbury K, Selden RL, Sistla S, Smith ADM, Stern-Pirlot A, Teck SJ, Thorson JT, Williams NE (2012) Eco-Label Conveys Reliable Information on Fish Stock Health to Seafood Consumers. PLoS One 7(8): e43765.
15. Fleming LE, Broad K, Clement A, Dewailly E, Elmira S, Knap A, Pomponi S A, Smith S, Solo Gabriele H, Walsh P (2006) Oceans and human health: Emerging public health risks in the marine environment. Mar Pollut Bull. 2006; 53(10–12): 545–560.
16. Kalay M, Ay Ö. and Canli M. (1999). Heavy metal concentrations in fish Tissues from the Northeast mediterrenean Sea. Bulletin of Environmental Contamination and Toxicology. 63: 673–681.
17. Gómez-Gutiérrez AI, Jover E, Bodineau L, Albaigés J, Bayona JM. (2006) Organic contaminant loads into the Western Mediterranean Sea: estimate of Ebro River inputs Chemosphere 65(2): 224–36.
18. El-Serehy HA, Aboulela H, Al-Misned F, Kaiser M, Al-Rasheid K, El-Din H E, (2012) Heavy metals contamination of a Mediterranean Coastal Ecosystem, Eastern Nile Delta, Egypt. Turk. J. Fish. Aquat. Sci. 12: 751–760.
19. Baumard P, Budzinski H, Garrigues P (May1998a) Polycyclic aromatic hydrocarbons in sediments and mussels of the western Mediterranean sea. Environ Toxicol Chem 17(5) 765–776.
20. Baumard P, Budzinski H, Michon Q, Garrigues P, Burgeot T, Bellocq J (July 1998b) Origin and Bioavailability of PAHs in the Mediterranean Sea from Mussel and Sediment Records. Estuar Coast Shelf S 47(1) 77–90.
21. Copat C, Bella F, Marine Castaing, Fallico R, Sciacca S, FerranteM, (2012) Heavy Metals Concentrations in Fish from Sicily (Mediterranean Sea) and Evaluation of Possible Health Risks to Consumers Bull Environ Contam Toxicol. 88(1): 78–83.
22. Miniero R, De Felip E, Magliulo M, Ferri F, Di Domenico A (2005) Selected persistent organic pollutants (POPs) in the Italian environment. Ann Ist Super Sanita 41(4): 487–492.
23. Commission Regulation (EU) No 1881/2006 of 19 December 2006.
24. Commission Regulation (EU) No 835/2011 of 19 August 2011.
25. Commission Regulation (EU) No 629/2008 of 2 July 2008.
26. Commission Regulation (EU) No 1259/2011 of 2 December 2011.
27. Legislative Decree 152/2006 Environment Regulation of April 29, 2006.
28. Commission Regulation (EU) No 882/2004 of 29 April 2004.
29. Commission Regulation (EU) No 1883/2006 of 19 December 2006.
30. Commission Regulation (EU) No 333/2007 of 28 March 2007.
31. Esman NA, Hammad YY (1977) Log-normality of environmental sampling data, Journal of Environmental Science and Health, Part A: Environmental Science and Engineering 12(1–2): 29–41.
32. Zou GY, Huo CY, Taleban J (2009) Simple confidence intervals for lognormal means and their differences with environmental applications - Environmetrics - Wiley Online Library.
33. Decreto Ministeriale Ambiente 14 aprile 2009, n. 56.
34. Cameron A, Askew N, (2011) EUSeaMap - Preparatory Action for development and assessment of a European broad-scale seabed habitat map final report. Available: http://jncc.gov.uk/euseamap
35. Swartenbroux F, Albajedo B, Angelidis M, Aulne M, Bartkevics V, et al. (2010) Marine Strategy Framework Directive – Task Group 9 Report Contaminants in fish and other seafood. EUR 24339 EN–2010. http://publications.jrc.ec.europa.eu/repository/handle/111111111/13669.
36. Perugini M, Visciano P, Giammarino A, Manera M, Di Nardo W, et al. (2007b) Polycyclic aromatic hydrocarbons in marine organisms from the Adriatic Sea, Italy. Chemosphere 66: 1904–1910.
37. Conti GO, Copat C, Ledda C, Fiore M, Fallico R, et al. (2012) Evaluation of heavy metals and Polycyclic Aromatic Hydrocarbons (PAHs) in *Mullus barbatus* from Sicily Channel and risk-based consumption limits. Bull Environ Contam Toxicol 88: 946–950.

38. Galgani F, Martínez-Gómez C, Giovanardi F, Romanelli G, Caixach J, et al. (2010) Assessment of polycyclic aromatic hydrocarbon concentrations in mussels (*Mytilus galloprovincialis*) from the Western basin of the Mediterranean Sea. Environ Monit Assess 172(1–4): 301–317.
39. Andral B, Galgani F, Tomasino C, Bouchoucha M, Blottiere C, et al. (2011) Chemical contamination baseline in the Western basin of the Mediterranean sea based on transplanted mussels. Arch Environ Contam Toxicol, 61(2): 261–271.
40. Scarpato A, Romanelli G, Galgani F, Andral B, Amici M, et al. (2010) Western Mediterranean coastal waters–Monitoring PCBs and pesticides accumulation in *Mytilus galloprovincialis* by active mussel watching: the Mytilos project. J Environ Monit 12(4): 924–935.
41. Perugini M, Visciano P, Manera M, Turno G, Lucisano A, et al. (2007a) Polycyclic aromatic hydrocarbons in marine organisms from the Gulf of Naples, Tyrrhenian Sea. J Agric Food Chem 55: 2049–2054.
42. Storelli MM, Giacominelli Stuffler R, Marcotrigiano GO (1998) Total mercury in muscle of benthic and pelagic fish from the South Adriatic Sea (Italy). Food Addit Contam 15: 876–883.
43. Storelli MM, Giacominelli Stuffler R, Storelli A, D'Addabbo R, Palermo C, et al. (2003) Survey of total mercury and methylmercury levels in edible fish from the Adriatic Sea. Food Addit Contam 20: 1114–1119.
44. Storelli MM, Giacominelli Stuffler R, Storelli A, Marcotrigiano GO, (2006) Cadmium and mercury in cephalopod molluscs: estimated weekly intake. Food Addit Contam 23: 25–30.
45. Joiris CR, Holsbeek L, Moatemri NL (1999) Total and methylmercury in sardines Sardinella aurita and Sardina pilchardus from Tunisia. Mar Pollut Bull 38: 188–192.
46. Storelli MM, Marcotrigiano GO (2000) Fish for human consumption: risk of contamination by mercury. Food Addit Contam 17: 1007–1011.
47. Storelli MM, Barone G, Piscitelli G, Marcotrigiano GO (2007) Mercury in fish: concentration versus fish-size and estimates intake dose. Food Addit Contam 6: 1–5.
48. Franca S, Vinagre C, Cacador I, Cabral HN (2005) Heavy metal concentrations in sediment, benthic invertebrates and fish in three salt marsh areas subjected to different pollution loads in the Tagus Estuary (Portugal). Mar Pollut Bull 50: 998–1003.
49. Martí-Cid R, Bocio A, Llobet JM, Domingo JL (2007) Intake of chemical contaminants through fish and seafood consumption by children of Catatonia, Spain: health risks. Food Chem Toxicol 45: 1968–1974.
50. Sivaperumal P, Sankar TV, Viswanathan Nair PG (2007) Heavy metal concentrations in fish, shellfish and fish products from internal markets of India vis-a-vis international standards. Food Chem 102: 612–620.
51. Prioli G, (2001) Censimento Nazionale sulla Molluschicoltura del Consorzio Unimar. 114 pp.
52. Marrone R, Mercogliano R, Palma G, Chirollo C, Smaldone G, et al. (2011) Presenza di idrocarburi policiclici aromatici in prodotti della piccola pesca del golfo di Napoli. AIVI online 1: 61–65.
53. Tornero V, d'Alcalà MR (2014) Contamination by hazardous substances in the Gulf of Naples nearby coastal areas: a review of sources, environmental levels and potential impacts in the MSFD perspective. Sci Total Environ 466–467: 820–840.
54. Jureša D, Blanuša M (2003) Mercury, arsenic, lead and cadmium in fish and shellfish from the Adritic Sea. Food Addit Contam 20(3): 241–246.
55. Storelli MM. (2008) Potential human health risks from metals (Hg, Cd, and Pb) and polychlorinated biphenyls (PCBs) via seafood consumption: Estimation of target hazard quotients (THQs) and toxic equivalents (TEQs). Food Chem. Toxicol. 46: 2782–2788.
56. Pastorelli AA, Baldini M, Stacchini P, Baldini G, Morelli Set al. (2012) Human exposure to lead, cadmium and mercury through fish and seafood product consumption in Italy: a pilot evaluation. Food Addit Contam Part A 29(12): 1913–1921.
57. Menzie CA, Ziccardi LM, Lowney YM, Fairbrother A, Shock SS, et al. (2009) Importance of considering the framework principles in risk assessment for metals. Environ Sci Technol 43: 8478–8482.

15

Spatial Scale, Means and Gradients of Hydrographic Variables Define Pelagic Seascapes of Bluefin and Bullet Tuna Spawning Distribution

Diego Alvarez-Berastegui[1]*, Lorenzo Ciannelli[2], Alberto Aparicio-Gonzalez[3], Patricia Reglero[3], Manuel Hidalgo[3], Jose Luis López-Jurado[3], Joaquín Tintoré[1,4], Francisco Alemany[3]

1 Balearic Islands Coastal Observing and Forecasting System (SOCIB), Palma de Mallorca, Balearic Islands, Spain, **2** College of Earth, Ocean, and Atmospheric Sciences (CEOAS), Oregon State University, Corvallis, Oregon, United States of America, **3** Instituto Español de Oceanografía - Centre Oceanogràfic de les Balears (COB-IEO), Palma de Mallorca, Balearic Islands, Spain, **4** Institute for Advanced Studies (IMEDEA), Consejo Superior de Investigaciones Científicas (CSIC) and the University of the Balearic Islands (UIB). Esporles, Spain

Abstract

Seascape ecology is an emerging discipline focused on understanding how features of the marine habitat influence the spatial distribution of marine species. However, there is still a gap in the development of concepts and techniques for its application in the marine pelagic realm, where there are no clear boundaries delimitating habitats. Here we demonstrate that pelagic seascape metrics defined as a combination of hydrographic variables and their spatial gradients calculated at an appropriate spatial scale, improve our ability to model pelagic fish distribution. We apply the analysis to study the spawning locations of two tuna species: Atlantic bluefin and bullet tuna. These two species represent a gradient in life history strategies. Bluefin tuna has a large body size and is a long-distant migrant, while bullet tuna has a small body size and lives year-round in coastal waters within the Mediterranean Sea. The results show that the models performance incorporating the proposed seascape metrics increases significantly when compared with models that do not consider these metrics. This improvement is more important for Atlantic bluefin, whose spawning ecology is dependent on the local oceanographic scenario, than it is for bullet tuna, which is less influenced by the hydrographic conditions. Our study advances our understanding of how species perceive their habitat and confirms that the spatial scale at which the seascape metrics provide information is related to the spawning ecology and life history strategy of each species.

Editor: Athanassios C. Tsikliras, Aristotle University of Thessaloniki, Greece

Funding: The development of this study was supported by the Project "PERSEUS" financed by the Seventh Framework Program for Research (FP7) under theme "Oceans Tomorrow", grant Agreement No. 287600 (http://www.perseus-fp7.eu/), the Spanish Foundation for Science and Technology, under the project "BALEARES" Grant number: CTM2009-07944/MAR (http://www.fecyt.es/fecyt/home.do) and the Science, Engineering and Education for Sustainability NSF-Wide Investment (SEES) Research Coordination Network, Grant 1140207 "Sustainability of Marine Renewable Resources in Subarctic Systems Under Incumbent Environmental Variability and Human Exploitation" (http://grantome.com/grant/NSF/OCE-1140207). Hydrographic data used in the study were provided by IBAMAR (http://www.ba.ieo.es/ibamar) a public data base founded by the Council for Economy and Competitiveness of the Balearic Islands Government (http://www.caib.es/govern/organigrama/area.do?coduo=6&lang=es), grant number AAEE0022/08. The research was developed under the framework of the BLUEFIN TUNA Project driven by the Balearic Island Observing and Forecasting System (www.socib.es) and the Spanish Institute of Oceanography (www.ieo.es). The funders had no role in study design, data collection and analysis, decision to publish, or preparation of the manuscript.

Competing Interests: The authors have declared that no competing interests exist.

* Email: dalvarez@socib.es

Introduction

Seascape ecology represents an emerging field in the study of how the habitat structure shapes the spatial distribution of marine species and influences key ecological processes [1],[2]. This discipline initiated applying techniques and metrics from the traditional landscape ecology to characterize and quantify spatial structure of benthic habitats, observed as a mosaic of patches of different habitat classes [1],[3],[4], [5],[6]. Nevertheless, there is still a gap in the development of concepts and techniques providing metrics to characterize the spatial structure of the seascape in the pelagic environments, where there are no clear boundaries delimitating different habitats [1],[2],[6]. In the framework of landscape ecology spatial gradients have been recently proposed as more appropriated metric than traditional categorical patch mosaic based metrics to characterize continuous habitats [7]. Accordingly, a location in a pelagic seascape would be better characterized by the combination of the value of a particular hydrographic variable and its spatial gradient.

Several studies have applied gradients of hydrographic parameters to characterize the spatial distribution of marine species during various life history stages, as nursery, foraging or spawning [8],[9],[10],[11],[12]. It is likely that the scale at which an individual perceive a change in the environment (i.e., a gradient) varies according to life history, ontogeny and to the hydrographic variable in exam. For instance, while large-scale gradients associated with the North Pacific transition zone drive the location of many pelagic predators including albacore tuna (*Thunnus alalunga*) during their feeding migratory stages [13], once off the west coast of the US, albacore tuna distribution is associated with smaller scale features linked to upwelling fronts [14]. In the Mediterranean Sea during spawning, bluefin tuna distribution is regulated by oceanographic variables that can change at relatively small scales [15],[16]. In spite of the expected importance of gradient scales, to our knowledge there are no studies that have evaluated the effect of changing the spatial scales at which environmental gradients are calculated to model the spatial distribution of fish. The goal of our study is to examine the distribution of large pelagic predators during spawning by explicitly considering mean, gradients and scale of gradients of hydrographic variables.

Atlantic bluefin tuna (*Thunnus thynnus*) and bullet tuna (*Auxis rochei rochei*) are two species of pelagic predators showing different spawning strategies. We target these species in the Balearic Sea (Figure 1), known as a recurrent spawning area for large pelagic species in the Western Mediterranean [17]. Bluefin tuna is a highly migratory species; the Eastern population enters in the Mediterranean Sea from the North Atlantic at the end of spring and early summer [18],[19]. Their spawning activity at the Balearic Sea is linked to the regional oceanography with spawning grounds located in the vicinity of frontal structures formed when the recent Atlantic water mass encounters the more saline resident surface Atlantic waters [15]. The area is characterized by highly dynamic processes that promote a seascape shaped by filaments, fronts and eddies whose location varies between years [20],[21]. Bullet tuna, by contrast, is smaller and more frequent in near coastal areas [22]. The spawning of bullet tuna is associated with the geography. Young larvae are found mainly in coastal areas and with little influence of the local oceanography in comparison with bluefin [16].

We expect that, when selecting spawning locations, a large-bodied and long-distance migratory pelagic fish, such as the bluefin tuna explores its environment at larger spatial scales than bullet tuna, a small-bodied and non-migratory pelagic fish. Therefore, we expect that pelagic seascape metrics based on the combination of hydrographic parameter values and their gradients calculated at appropriate spatial scales provide relevant information for bluefin tuna but not for bullet tuna, where we expect a greater reliance on geographic and hydrographic parameters, calculated at comparatively small-scale.

In this work we analyze the influence of the pelagic environment by depicting the spatial scales at which gradients of hydrographic variables are linked to the spawning ecology of these tuna species. We investigate the two most relevant hydrographic variables describing their spawning spatial distribution: salinity and geostrophic currents velocity [16], already determined in previous studies. Our analytical approach has two steps. Firstly, we identify the scale at which each hydrographic variable (Figure 2) maximizes the performance of a model fitted to larval distribution. Second, we evaluate whether components of the seascape (i.e., mean and gradients of oceanographic variables) are interactively affecting the spatial distribution of tuna larvae. By performing this analysis on two species we evaluate how fish with contrasting life history strategies perceive their environments when deciding for spawning locations.

Materials and Methods

Data acquisition

Bluefin and bullet tuna larvae were collected during ichthyoplankton surveys using Bongo nets from 2001 to 2005. The surveys were conducted by the Instituto Español de Oceanografía (www.ieo.es), a Spanish Government marine research institution. The sampling scheme was communicated and approved by the Spanish Directorate of Fisheries before the sampling was conducted. No specific ethical approval was required and the survey of biological data was conducted using Bongo nets,which are accepted standard

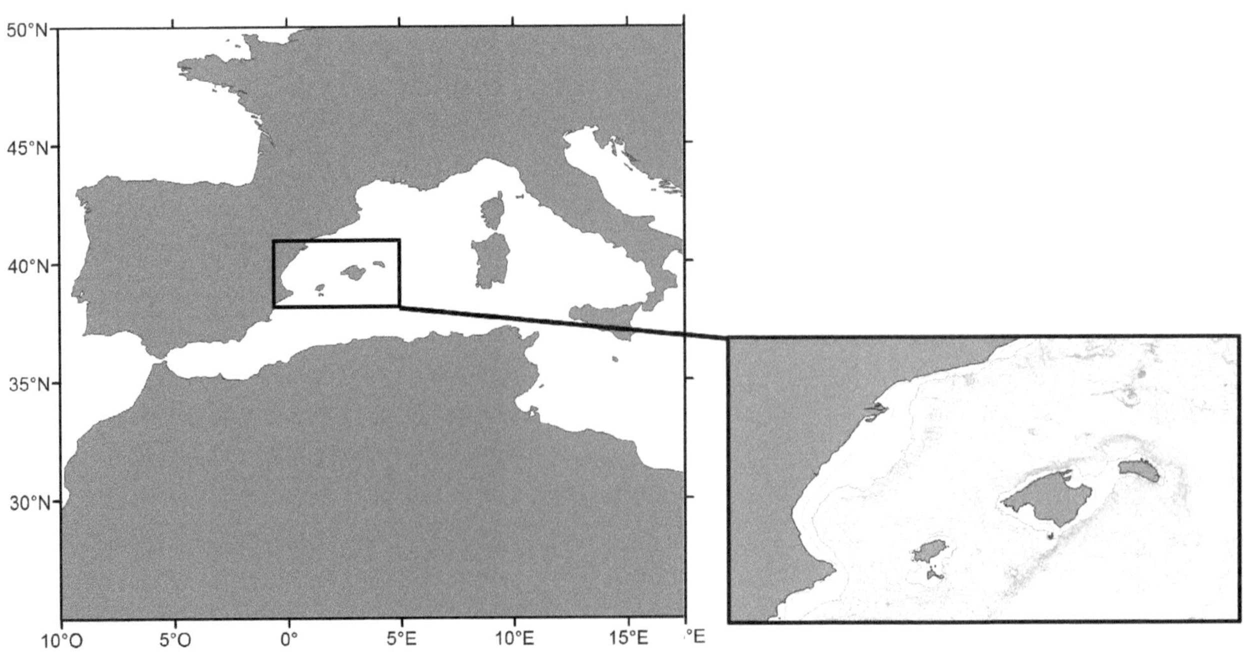

Figure 1. Location of the Balearic Islands, Western Mediterranean.

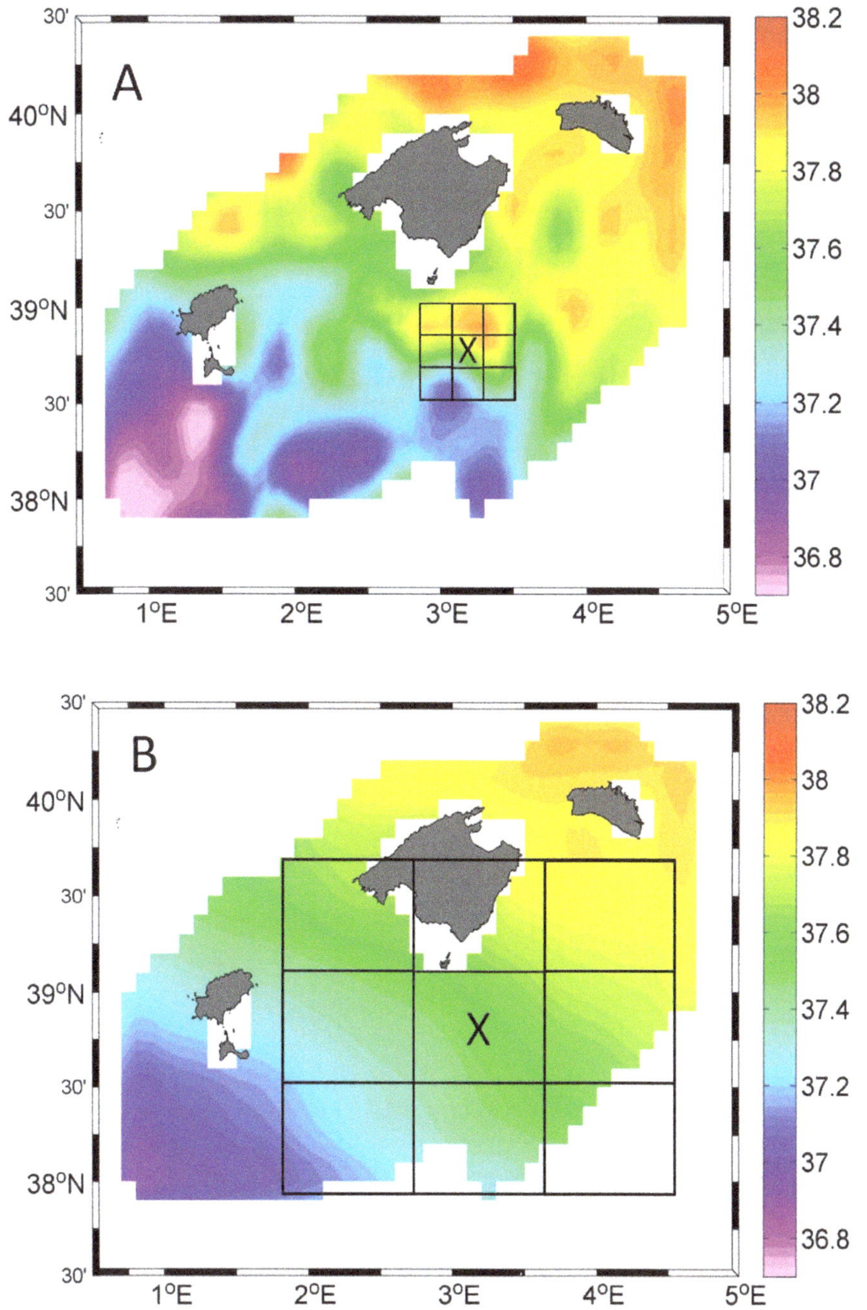

Figure 2. Sea surface salinity field in 2003. Spatial means of sea surface salinity processed at two different scales A) 0.15 degrees and B) 0.75 degrees. Spatial means were interpolated following an objective analysis onto a regular grid by using minimum error variance methods (Bretherton, 1976). The squares in each figure show the polygons used for the calculation of the spatial gradients at the two scales at station X.

techniques for this type of surveys, used worldwide for the collection of plankton samples, including billfish and tuna larvae [15],[17], [23], [24]. The nets were towed at low speeds, around 2 knots, during 8–10 minutes, and plankton samples were immediately preserved with 4% formalin buffered with borax onboard.

Around 200 stations were sampled every year, in a regular sampling grid of 10×10 nm located between 37.85°–40.35° N and 0.77°–4.91°E, covering a total area of 101360 km2 (= 280×362 km) around the Balearic archipelago. The sampling was conducted during June–July coinciding with the spawning period of bullet and bluefin tuna (see [15] for details of the sampling procedures). Tuna larvae were identified to the species level and measured in standard length. All larvae identified as yolk sac and preflexion stages (<4.5 mm) were classified as "stage 1".

Mean and maximum age of larvae under this size are 6 days and 11 days old respectively, accounting for growth [25] and hatching time [26]. Displacement from spawning areas during these intervals in the study area are below 25 kilometers for the mean ages, which are in the range of 1.4 sampling stations, and 46 kilometers for the maximum ages, what is in the range of 2.6 sampling stations. These values have been calculated following methods in [27]. Considering these values, abundances of stage 1 larvae have been defined to get a proxy of spawning locations as in previous research in the study area [28].

Vertical profiles of conductivity, temperature and pressure data were recorded at all stations, by means of Sbe911 CTD. Sea surface salinity at each station (SAL) was calculated as the mean salinity over the mixed layer depth. Geostrophic velocities (GVEL) were calculated at sea surface from the first-derivative of the sea surface height between adjacent points, which was obtained by vertical integration of the specific volume, using 600 m as the level of no motion [21].

These two variables (SAL and GVEL) were selected, since they have been demonstrated to be the two most relevant environmental variables describing the spawning spatial distribution of tuna [16]. Sea surface temperature was also included in the models since in this area it is a secondary but relevant variable mainly related to the phenology of the spawning process [29]. However, the spatial gradient was not explored because sea surface temperature during the summer changes relatively fast due to solar irradiance [21].

Processing of spatial gradients along continuous spatial scales

Spatial gradients from the sea surface salinity field (gSAL) and geostrophic velocity field (gGVEL) within the sampled region were calculated at six spatial scales, from 0.15° to 0.90° with a spatial increment of 0.15°. The minimum (0.15°) and the maximum scale (0.90°) were in the range of the smallest (from 0.13° to 0.27°) and largest (up to 0.92°) mesoscale oceanographic structures in the area [21]. For the computation of the gradients, nine square polygons at every scale were delimitated around each sampling position (see examples for scale 0.15° and 0.75° in figure 2A, 2B respectively). The gradient was then computed as the maximum absolute difference between the mean hydrographic variable at the center polygon and each of the eight surrounding polygons standardized to distance [9]. The software for the spatial processing was developed in R language [30].

Identification of spatial scales

Comparison of how models perform along scales allowed identifying the spatial scales at which information provided by gradients is maximized. The effect of gGVEL and gSAL at each scale on the abundance of bullet and bluefin tuna larvae was assessed using nonparametric regression statistical models (generalized additive models, GAMs, [31]). A base model was formulated to describe inter-annual variability (variable YEAR), sampling location (latitude and longitude variables), and the hour of the day on the catch of tuna larvae. Over-dispersed Poisson distribution family and a natural-log link were selected to model larval data. The volume of water filtered was included as an offset (after natural log transformation), to account for the effort expanded in catching the sample (Equation 1). The effects of these variables on the base model have been already analyzed in previous studies [16]. Here, the base model represented the null hypothesis of no gradient effect on tuna larvae distribution, against which all other more complex formulations will be compared.

Equation 1: Base model

$$Larvae\ abundance = offset(log(m3)) + factor(year) + sm_1(long, lat) + sm_2(hour)$$

m3 = volume filtered by the bongo nets (m3); long = longitude; lat = latitude; hour = hour of the day expressed from 0 to 1, sm_1 and sm_2 the smoothing functions.

At each spatial scale a GAM model was processed including the gradient of one hydrographic variable (gSAL, gGVEL) as a new additive term (s3) in the standardization model. The number of knots for the new smoother was always set to a maximum of three (i.e. two degrees of freedom) in order to avoid over fitting in the responses.

The identification of characteristic spatial scales (cgSAL, cgGVEL) was assessed with scalogram where the scale of the covariate is plotted against a measure of the model goodness of fit, which in our case were represented by the adjusted R-squared (Rsq, the higher the better), and the Generalized Cross Validation (GCV, the lower the better) [31]. We selected the scale that maximize Rsq and minimize the GCV. Results of the base model (when a seascape covariate was not included) were presented in the same graphics. Note that due to the greater complexity of the gradient model higher Rsq values do not necessarily imply an improvement in relation to the base model, while they do represent a better performance when compared to other gradient models.

Significant differences of Rsq values between models, or GCVs, were obtained from t-test of these parameters obtained from 500 iterations where 10% of the data was excluded. For all cases, alternative hypothesis (difference in means is not equal to 0) was accepted only if the P value was lower than 0.001, with a confidence level of 0.99. When one variable presented similar Rsq and GCV values at various scales, selection was assessed by inspection of the plot showing the response of the abundance in relation to the gradient processed at those scales.

Once the characteristic scale of the gradients was identified, we tackled the questions of whether the information provided by the gradients is different and complementary to the information provided by the hydrographic variables from which they were calculated, and in that case, how the information from these two variables (spatial mean and gradient) should be combined to maximize the goodness of fit of the models and the ecological information they provide. To assess these questions we analyzed the performance of models with different complexity:

i) The base model from equation 1.

ii) Hydrographic models combining the sea surface salinity, geostrophic velocity and sea surface temperature at the sampling station (stSAL, stGVEL, stSST).

iii) Seascape models combining the gradients at characteristic scales (cgSAL, cgGVEL) and the hydrographic variable at the sampling location (stSAL, stGVEL, stSST). Different seascape models were constructed including the two components of the seascape (values at stations and gradients) as additive and interactive terms. An interaction may be ecologically meaningful when a species is selecting its spawning habitat on a specific side of a frontal region, for example. In such case, it is the combination of both the gradient and the mean that provide the suitable conditions for spawning. The performance of different model configurations for each species was assessed by the delta AIC (ΔAIC), calculated as the difference between model AICs and the base model AIC. The AIC in this case is best suited for model comparisons because each model had different number of variables [32]. Rsq, GCV and explained deviances were used to compare how models perform between the two species, as AIC values among models with different dependent variables are not comparable.

Results

Identification of characteristic spatial scales

In all years considered, the recent Atlantic water masses encountered the more saline resident water masses forming an oceanic frontal zone inside the study area. The size of such frontal zone was bigger than other oceanographic phenomena as small eddies and meanders derived from the instabilities along the haline front and the effect of strong bathymetric changes (Figures S3, S4).

The scalogram of gGVEL for bluefin showed that Rsq values gradually improve as the spatial scales increased to a maximum at 0.6° (Rsq = 0.44, Figure 3A), which was chosen as the geostrophic velocity gradient characteristic scale for bluefin tuna. Values of GCV showed a similar pattern of model improvement, being significantly better than the base model at 0.6 degrees (Figure 3B). At this characteristic scale the response of the larvae abundance is positively related to the gradient of geostrophic velocity (Figure 3C).

The scalograms of gSAL for bluefin tuna showed also an increment of Rsq with higher values at 0.6° and 0.75° that also coincide with lower values of GCV (Figure 3D, 3E). Differences of R-sq between these two scales (0.6° and 0.75°) were not significant. The characteristic scale for gSAL was chosen at 0.6° as the model response at this scale presented a less ambiguous effect on larval abundance (Figure 3F). The gSAL at 0.75° spatial gradients displays a dome-shaped response with a less clear ecological interpretation (Figure S1). At this scale GCV was lower than the base model.

In contrast to the results obtained for salinity, the gradients of geostrophic velocities (gGVEL) did not show any single scale that maximizes R-square and minimizes GCV for bullet tuna (Figure 4A, 4B). The Rsq scalogram showed a flat trend with the highest value at 0.15 degrees. The Rsq value at this scale (= 0.18) showed similar values than other scales (values between 0.170 and 0.173) or when compared to the base model (= 0.166). On the contrary the GCV scalogram showed significantly lower values than the base model at 0.45 and 0.6 degrees, scales at which Rsq were not even significantly higher than the base model. Therefore, the contradictory response of the model performance indicators, the flat trend of Rsq scalogram and their very low values (despite the higher complexity of the gradient models in relation to the base model), may indicate that the spatial gradient of geostrophic velocity is not a valid seascape metric for the spawning locations of bullet tuna. Consequently, gGVEL was excluded from further analysis in relation to this species.

The gSAL scalogram of bullet tuna showed a moderate effect of the scale at which gradients were calculated. The shape of the scalogram did not show a peak scale at which model performs better (Figure 4C). Scales from 0.45°, 0.6° and 0.75° for gSAL seemed to maximize Rsq, but not being different from each other. In this case GCV-scalograms showed the lowest at 0.75° (Figure 4D), being significantly lower than the base model, which

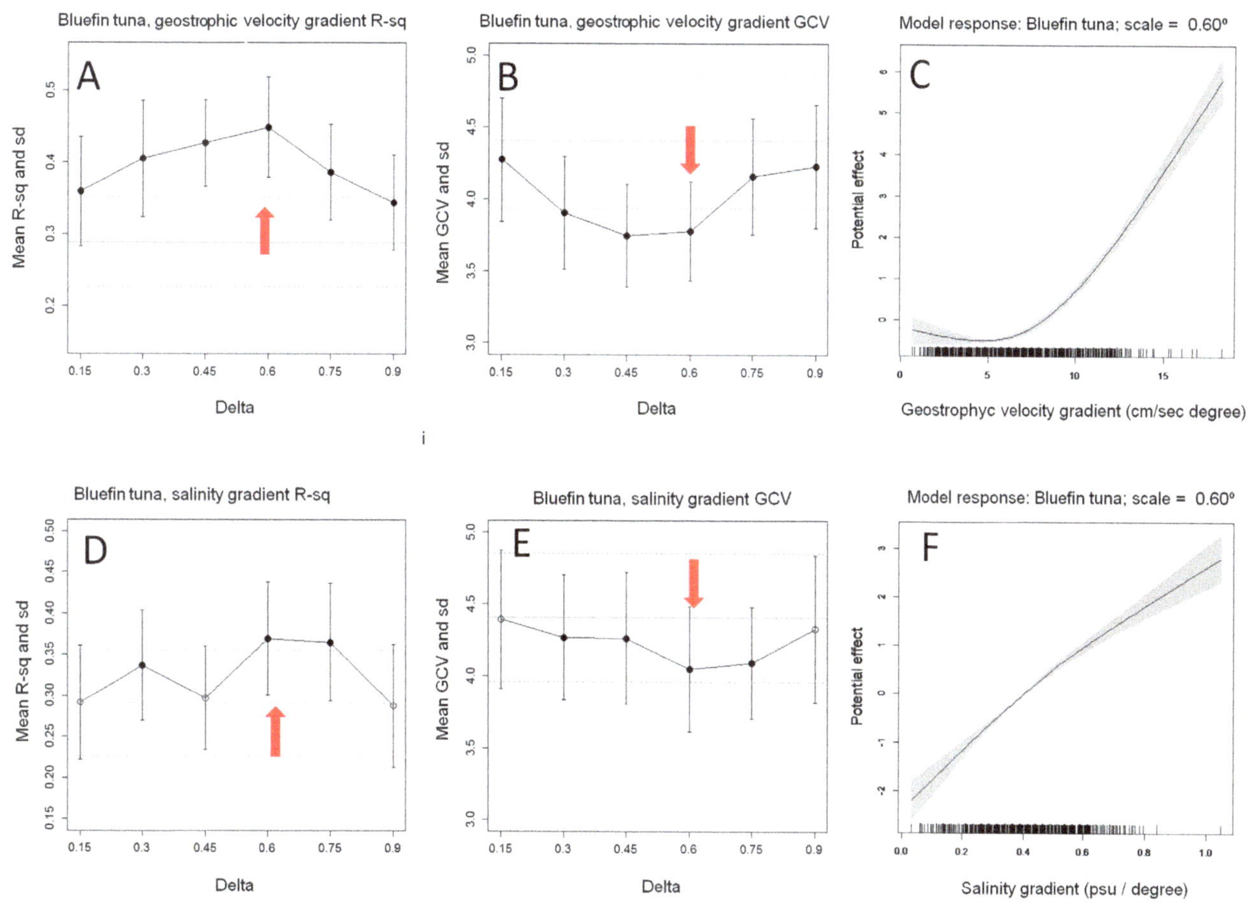

Figure 3. Rsq and GCV Scalograms of bluefin tuna larva abundance models along spatial scales, standard deviations. Horizontal grey lines indicate statistics from the base model (Straight line = mean, dashed line = Sd). Black dots show scales at which values are significantly different from the standardization model. Red arrows indicate the selected characteristic scale.

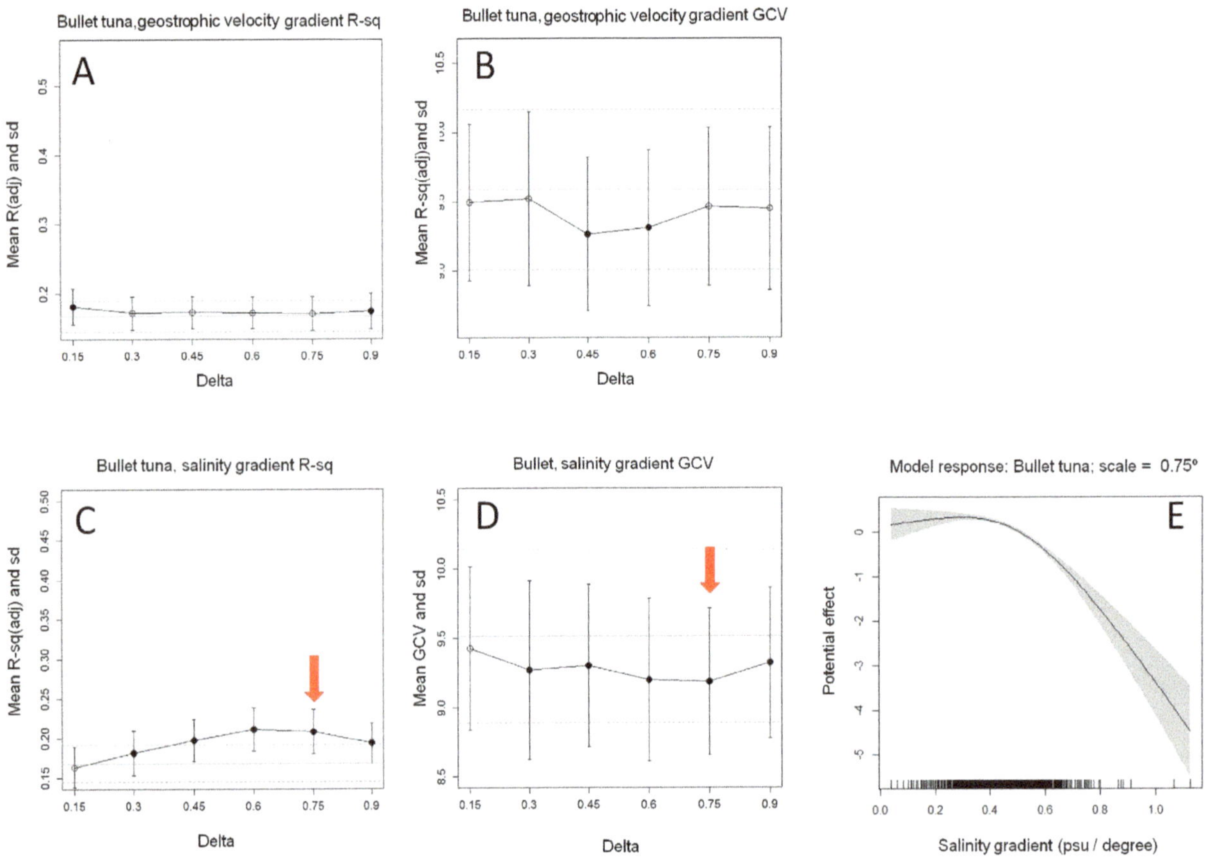

Figure 4. Rsq and GCV scalograms of bullet tuna larva abundance models along spatial scales, standard deviations. Horizontal grey lines indicate statistics from the standardization model (Strait line = mean, dashed line = Sd). Black dots show scales at which values are significantly different from the standardization model. Red arrows indicate selected characteristic scale.

was selected as characteristic scale. The Rsq at this scale was higher (= 0.21) than that of the base model (Rsq = 0.16). At this scale, gradients displayed a negative effect on the bullet tuna larvae abundance (Figure 4E) showing that bullet tuna spawning locations are found with more probability in areas where salinity is spatially homogeneous– a result that contrasted to that obtained for bluefin tuna.

Species-specific seascape characterization

The best model for each species included a gradient and a mean term (Tables 1 and 2). Note however that the hydrographic model already represents an improvement respect to the standardization model (Tables 1 and 2). For bluefin tuna the best seascape model had an improvement of 186% of the Rsq when compared to the base model (Rsq base mode = 0.23; Rsq best seascape model = 0.66, Table 1). The improvement for bullet was 61%, considerably lower compared to bluefin (Rsq base mode = 0.16; Rsq best seascape model = 0.25, Table 2).

Pearson correlation coefficients between hydrographical variables and their gradients at characteristic scales were in all cases below 0.50 and pair plots showed no clear tendencies on the correlations (Figure S2), indicating that selected gradients provided complementary information to that of the hydrographical variable. Models showed a better performance (i.e. lower GCV and higher ΔAIC; Tables 1 and 2) in all the cases when the gradient and the hydrographic value were considered as an interaction term, suggesting dependence in their effect on larvae abundance rather than an additive response. However, larvae abundance of each species responded differently to the interaction of seascape components (Figures 5A, 5B, 5C).

For Bluefin tuna, higher probability of spawning was associated to higher values of geostrophic velocity gradients, but where velocities at station may present either high or low values (Figure 5A), (Figures S7, S8). Considering that a gradient is characterized by an area with high current speed near an area of low current speed, this result indicates that spawning locations were not associated to a particular side of the gradient, but in an area around the location where maximum velocities occurs. The extension of this area would be around a circle of 0.6 degrees of radius (aprox. 65 km in the study area), the characteristic scale at which the gradients were more relevant. In contrast, the interaction of the salinity seascape variables showed high larvae abundances in areas with high salinity gradients and intermediate-high salinity levels (Figure 5B), indicating an effect of the location of the main haline front and a preference for spawning at the high salinity values of that front (Figures S5, S6). The characteristic spatial correlation scale of local oceanographic structures in the area is around 18 nmi (aprox. 0.15 meridian degrees) [21] and therefore surface oceanographic structures at smaller spatial scales are ephemeral. The oceanographic structures at larger spatial scales, relevant for bluefin tuna, are linked to the Med-Atlantic salinity front and last longer.

The functional form of the interaction terms was different for bullet tuna. The relation between the bullet tuna larvae

Table 1. Summary of GAM models of larvae abundance for Atlantic bluefin tuna (*Thunnus thynnus*).

Model group	Model variables	R2	Dev(%)	GCV	AIC	delta AIC
Base model	(latitude, longitude) + filtered volume + hour	0.232	40.8	4,596	3985,54	0
One additive variable models	base model + stGVEL0.15	0,271	43,4	4,412	3827,79	157,75
	base model + gGVEL0.6	0,39	49,4	3,947	3465,03	520,50
	base model + stSAL0.15	0,222	41,8	4,534	3924,70	60,84
	base model + gSAL0.6	0,301	45,1	4,275	3723,53	262,01
Hydrographic model	base model + stGVEL + stSAL + stTEMP	0,472	51,6	3,814	3338,62	646,92
GVEL seascape models	Hydrographic model + stGVEL + gGVEL0.6	0,53	55,8	3,500	3087,14	898,40
	Hydrographic model + (stGVEL,gGVEL0.6)	0,666	59,3	3,251	2881,29	1104,25
SAL seascape models	Hydrographic model + stSAL + gSAL0.6	0,506	54,8	3,571	3145,80	839,74
	Hydrographic model + (stSAL,gSAL0.6)	0,533	57,1	3,417	3009,58	975,96

Interaction terms included in parenthesis.

abundance and the interaction of spatial distribution of sea surface salinity and its gradients is presented in Figure 5C. Spawning locations were associated to areas where the salinity at the station were lower and gradients presented intermediate values, but the interaction plot revealed that spawning also appears in areas of higher salinities associated to very low gradients. Areas defined by this twofold combination were located at both sides of the front avoiding more mixed waters. This spatial distribution was more evident in years 2001, 2003 and 2005 (Figures S9, S10)

Discussion

We have found that the combination of sea surface current velocities, salinities and their gradients calculated at characteristic spatial scales are relevant for the parameterization of the pelagic seascape affecting a key ecological process of bluefin tuna. For bullet tuna only salinity and their gradient provided a valid seascape metric not being relevant the gradients of sea surface current velocities. In agreement with our expectations, the importance of these metrics was much higher for bluefin, a large-bodied, long distance migratory and more dependent on local oceanography than were for bullet tuna a smaller coastal species with shorter migration distance.

Previous studies have documented the links between the frontal activity and the spawning of bluefin tuna [15],[16],[33]. In this study we add to these results by examining the effect of gradients and their interactions with hydrographic mean. These metrics improved our understanding of the conditions for bluefin and bullet tuna spawning when compared to models using just the hydrographical values but not the gradients. Furthermore the identification of characteristic scales of gradients provided a new source of information for the interpretation of how local oceanography determines the selection of the site to spawn.

For bluefin tuna larvae, spatial salinity and geostrophic velocity gradients maximize spatial model performance when calculated at 0.6 degrees. The higher abundance of bluefin tuna larvae in areas with intermediate to high salinities and with high gradients of velocity is consistent among years. Higher abundance occurs around the location of the main frontal area, at the side of higher salinity of the front and where current speed presents high values. Higher salinity water likely has higher resident time near the islands than the less saline water, which may run along the front towards east getting farther from the archipelago. Spawning at the higher salinity side may favor spatial overlap with other larval species that are also located in this water mass [17],[34].

Results for bullet tuna showed that pelagic seascape metrics are not as relevant to explain the spawning distribution as they are for bluefin. In the western Mediterranean, bullet tuna spawning has been associated to near coastal areas [22], being less influenced by the local oceanography than bluefin tuna [16], which is consistent with our results.

Table 2. Summary of GAM models of larvae abundance for bullet tuna (*Auxis rochei rochei*). Interaction terms included in parenthesis.

Model group	Model variables	R2	Dev(%)	GCV	AIC	delta AIC
Base model	(latitude, longitude) + filtered volume + hour	0,158	32,8	9,615	8857,17	0
One additive variable models	base model + GVEL st0.15	0,177	35,3	9,281	8572,91	284,26
	base model + stSAL0.15	0,16	36,5	9,130	8440,06	417,11
	base model + gSAL0.75	0,207	39,6	8,741	8098,28	758,89
Hydrographic model	base model + stGVEL + stSAL + stTEMP	0,192	38,8	8,847	8183,39	673,78
SAL seascape models	Hydrographic model + stSAL + gSAL0.75	0,215	40,6	8,617	7984,18	872,99
	Hydrographic model + (stSAL,gSAL0.75)	0,255	43,1	8,327	7708,82	1148,35

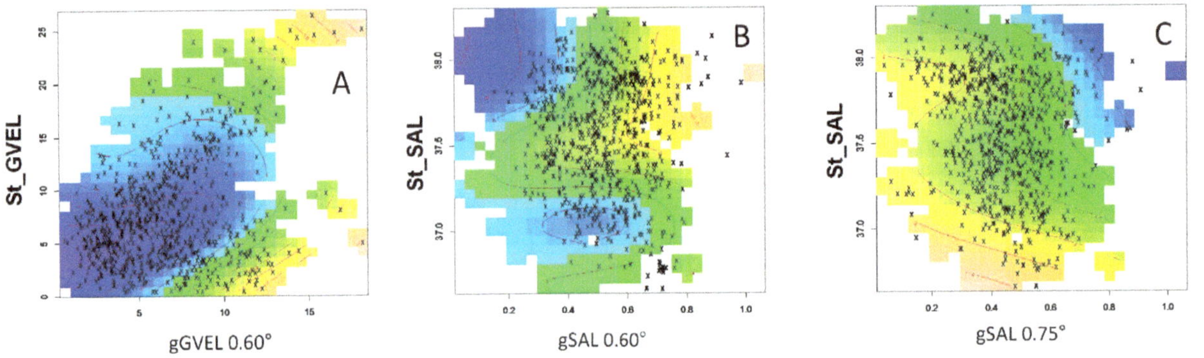

Figure 5. The effect of the interactions of the seascape components on the larval abundance as estimated from the seascape generalized additive model. The effects are shown for bluefin tuna (A–B) and bullet tuna (C). For bluefin tuna: A) the effect of the gGVEL and st_GVEL interaction. B) the effect f the gSAL and st_SAL interaction. For bullet tuna C) the effect of the gSAL and st_SAL interaction. Isolines indicate larval abundances predicted by the model. Peak of abundances are indicated in pink-yellow. Low and very low abundances are indicated in green and blue, respectively.

Despite the lower importance seascape metrics in bullet tuna, the inclusion of salinity gradients provided additional information for the identification of spawning sites. The analysis indicated that bullet tuna spawning areas are mostly found in areas where salinity gradients are low. Bullet tuna was found at both sides of the front but avoiding more mixed waters, located closer to the front. This was verified when observing the spatial distribution of larvae in relation to the salinity seascapes among the different years. For instance, in 2001, 2003 and 2005 high larvae abundances were observed North of the archipelago (high salinity waters with very low gradients), but intermediate abundances, indicating spawning, also occurs in Southern areas (low salinities and intermediate gradients). In 2002 and 2004 higher abundances were linked to low salinity and intermediate gradients shown in the south of the archipelago (Fig. S9 and S10). These results reinforce the theory of bullet tuna spawning occurs in widespread geographic areas, and not only close to the coast and suggest that the location of the main haline front negatively affects the spawning of this species.

Overall results related to bullet tuna point to the fact that, besides the avoidance of areas near strong surface haline gradients, other factors not considered in this study may also be relevant for spawning site selection in this species. It is also relevant that the spatial pattern in relation to the salinity is the opposite to that shown by bluefin tuna, located in areas near the front, suggesting possible avoidance of predators by bullet spawners [35].

The application of seascape metrics derived from salinity and geostrophic currents to characterize the spawning habitat provides new descriptors for environmental variables that improve model quality and predictions. This improvement allows a more precise identification of the relationships between the spatial location of the spawning grounds and the local oceanographic processes. Moreover, our study demonstrates that seascapes must be characterized at specific spatial scales to provide useful information as proposed in previous studies [36] and supporting results on terrestrial landscapes [37],[38] and bottom seascapes [39]. Therefore, the relations between the location of spawning sites and the mesoscale oceanographic processes may prove to be non significant if seascape metrics are not processed at the right spatial scales.

Seascape ecology is an emerging field generally being applied for the analysis of how benthic habitats pattern in coastal areas drives different aspects of marine species ecology [39]. Techniques are applied following categorical approach where the seascape is composed by a number of patches of different type of habitats [40],[7]. However, very little attention has been paid to the techniques and concepts to investigate pelagic seascape ecology due to the complex spatiotemporal dynamics of this system [2]. Thus, the work presented here sheds new light to modeling spatial distribution and investigating key ecological processes of species highly dependent on the variability of the pelagic environment, like spawning ecology of many of the big tuna species are [41]. In areas as the Balearic Sea, for which new operational oceanography platforms provide near real time data of hydrography [42] and also in combination with remote sensing data (e.g. altimetry [43]) and modeling [44],[45] these metrics will improve the species spatial distribution forecast that has proved effective for management [46].

In contrast to seascapes, landscape metrics have a long history in terrestrial ecology, and have improved over time. For instance, the effect caused on the habitat analysis derived from the spatial definition of the input habitat maps or the extent of the study area are common studied topics, [47],[48]. Likewise, calculation of seascape metrics and the final results from their application in ecological studies may be affected by different issues, like the different ways of computing the hydrographic variables and their gradients, or the origin of the input data source like from in situ measurements, remote sensing or hydrodynamic models, each with different sources of uncertainty. A relevant question is how seascapes can provide information for other type of species and ecological processes. Addressing all these challenges and developing comparative studies between different data sources, processing methods, species and ecological processes will allow advancing towards the understanding of how seascape metrics can provide information about how ecological processes and oceanography are linked together.

In summary, pelagic seascapes based on gradients and characteristic scales allow improving spatial distribution models and the identification of essential fish habitat of pelagic species. They also provide a tool for analyzing the links between particular ecological processes and local oceanography going far beyond than stochastic models based on just hydrographic parameters as salinity, temperature or geostrophic velocities. As a consequence these metrics will provide an improvement in all the management approaches and tools pending on the capability of models to identify essential habitats as near real-time spatial management based on habitat predictions [46],[10], pelagic species distribution from deterministic models [49] or the standardization of larvae indices to assess adult stock, [50],[51],[52].

Supporting Information

Figure S1 Model response of bluefin tuna in relation to salinity gradient processed at 0.75 degrees. Fitted line (solid line) and 95% confidence intervals (grey shaded areas) are shown. Whiskers on the x-axis show the locations of measurements.

Figure S2 Correlation between the gradients at the characteristic scales and the hydrographical variables at the sampled station. A) Current velocity and B) salinity for Atlantic Bluefin tuna. C) Salinity for Bullet tuna.

Figure S3 Sea surface salinities in the area during the five years analyzed (2001 to 2005).

Figure S4 Sea surface geostrophic currents in the area during the five years analyzed (2001–2005).

Figure S5 Spatial distribution of bluefin tuna (*Thunnus thynnus*) larvae in relation to the salinity mean calculated at its characteristic scale (0.6 degrees). Relative stage-1 larval abundances are shown in the maps such as dots.

Figure S6 Spatial distribution of bluefin tuna (*Thunnus thynnus*) larvae in relation to the salinity gradient calculated at 0.6 degrees. Relative stage-1 larval abundances are shown in the maps such as dots.

Figure S7 Spatial distribution of bluefin tuna (*Thunnus thynnus*) larvae in relation to the geostrophic velocity mean calculated at its characteristic scale (0.6 degrees). Relative stage-1 larval abundances are shown in the maps such as dots.

Figure S8 Spatial distribution of bluefin tuna (*Thunnus thynnus*) larvae in relation to the geostrophic velocity gradient calculated at the characteristic scale (0.6 degrees). Relative stage-1 larval abundances are shown in the maps such as dots.

Figure S9 Spatial distribution of bullet tuna (*Auxis rochei rochei*) in relation to the salinity mean calculated at 0.75 degrees. Relative stage-1 larval abundances are shown in the maps such as dots.

Figure S10 Spatial distribution of bullet tuna (*Auxis rochei rochei*) in relation to the salinity gradient calculated at 0.75 degrees. Relative stage-1 larval abundances are shown in the maps such as dots.

Acknowledgments

We thank the scientists and crew participating in the TUNIBAL cruises. We also thank P. Tugores and L. Rueda for their help in software development, and R. Balbin for their comments about methodological approaches.

Author Contributions

Conceived and designed the experiments: DAB LC AAG PR FA MH JLLJ JT. Performed the experiments: DAB LC FA. Analyzed the data: DAB LC AAG PR FA MH. Contributed reagents/materials/analysis tools: DAB LC AAG PR FA MH JLLJ JT. Wrote the paper: DAB LC AAG PR FA MH JLLJ JT. Sample collection: FA JLLJ PR. Laboratory sample processing: FA JLLJ AAG. Software development: DAB LC AAG. Numerical modelling design: DAB LC PR. Environmental variables calculation: DAB LC AAG PR FA MH JLLJ JT. Relevant literature review in different fields: DAB LC AAG PR FA MH JLLJ JT. Final manuscript edition: DAB LC AAG PR FA MH JLLJ JT.

References

1. Hinchey EK, Nicholson MC, Zajac RN, Irlandi EA (2008) Preface: marine and coastal applications in landscape ecology. Landscape Ecology 23: 1–5.
2. Pittman SJ, Kneib RT, Simenstad CA (2011) Practicing coastal seascape ecology. Marine Ecology-Progress Series 427: 187–190.
3. Turner MG (1989) Landscape Ecology - the Effect of Pattern on Process. Annual Review of Ecology and Systematics 20: 171–197.
4. Pittman SJ, McAlpine CA, Pittman KM (2004) Linking fish and prawns to their environment: a hierarchical landscape approach. Marine Ecology-Progress Series 283: 233–254.
5. Boström C, Pittman SJ, Simenstad C, Kneib RT (2011) Seascape ecology of coastal biogenic habitats: advances, gaps, and challenges. Marine Ecology Progress Series 427: 191–217.
6. Wedding LM, Lepczyk CA, Pittman SJ, Friedlander AM, Jorgensen S (2011) Quantifying seascape structure: extending terrestrial spatial pattern metrics to the marine realm. Marine Ecology-Progress Series 427: 219–232.
7. Cushman SA, Gutzweiler K, Evans JS, McGarigal K (2010) The gradient paradigm: a conceptual and analytical framework for landscape ecology. In: Huettmann F, Cushman S, editors. Spatial complexity, informatics and wildlife conservation. Tokyo: Springer. pp. 83–110.
8. Mannocci L, Laran S, Monestiez P, Dorémus G, Van Canneyt O, et al. (2013) Predicting top predator habitats in the Southwest Indian Ocean. Ecography 37: 261–278.
9. Worm B, Sandow M, Oschlies A, Lotze HK, Myers RA (2005) Global patterns of predator diversity in the open oceans. Science 309: 1365–1369.
10. Druon JN, Fromentin JM, Aulanier F, Heikkonen J (2011) Potential feeding and spawning habitats of Atlantic bluefin tuna in the Mediterranean Sea. Marine Ecology-Progress Series 439: 223–240.
11. Louzao M, Pinaud D, Peron C, Delord K, Wiegand T, et al. (2011) Conserving pelagic habitats: seascape modelling of an oceanic top predator. Journal of Applied Ecology 48: 121–132.
12. Hidalgo M, Gusdal Y, Dingsor GE, Hjermann D, Ottersen G, Stige LC, et al. (2012) A combination of hydrodynamical and statistical modelling reveals non-stationary climate effects on fish larvae distributions. Proceedings of the Royal Society B: Biological Sciences 279: 275–283.
13. Block BA, Jonsen ID, Jorgensen SJ, Winship AJ, Shaffer SA, et al. (2011) Tracking apex marine predator movements in a dynamic ocean. Nature 475: 86–90.
14. Phillips AJ, Ciannelli L, Brodeur RD, Pearcy WG, Childers J (2014) Spatio-temporal associations of albacore CPUEs in the Northeastern Pacific with regional SST and climate environmental variables. ICES, Journal of Marine Science. In press.
15. Alemany F, Quintanilla L, Velez-Belchi P, Garcia A, Cortés D, et al. (2010) Characterization of the spawning habitat of Atlantic bluefin tuna and related species in the Balearic Sea (western Mediterranean). Progress in Oceanography 86: 21–38.
16. Reglero P, Ciannelli L, Álvarez-Berastegui D, Balbín R, López-Jurado JL, et al. (2012) Geographically and environmentally driven spawning distributions of tuna species in the western Mediterranean Sea. Marine Ecology Progress Series 463: 273–284.
17. Torres AP, Reglero P, Balbin R, Urtizberea A, Alemany F (2011) Coexistence of larvae of tuna species and other fish in the surface mixed layer in the NW Mediterranean. Journal of plankton research 33: 1793–1812.
18. Block BA, Teo SL, Walli A, Boustany A, Stokesbury MJ, et al. (2005) Electronic tagging and population structure of Atlantic bluefin tuna. Nature 434: 1121–1127.
19. Rooker JR, Secor DH, De Metrio G, Schloesser R, Block BA, et al. (2008) Natal homing and connectivity in Atlantic bluefin tuna populations. Science 322: 742–744.
20. La Violette PE, Tintoré J, Font J (1990) The surface circulation of the Balearic Sea. Journal of Geophysical Research 95: 1559–1568.
21. Balbín R, López-Jurado JL, Flexas MM, Reglero P, Vélez-Velchí P, et al. (2013) Interannual variability of the early summer circulation around the Balearic Islands: driving factors and potential effects on the marine ecosystem. Journal of Marine Systems In Press, DOI: 10.1016/j.jmarsys.2013.12.007.

22. Sabatés A, Recasens L (2001) Seasonal distribution and spawning of small tunas (*Auxis rochei* and *Sarda sarda*) in the northwestern Mediterranean. Scientia Marina 65: 95–100.
23. Muhling BA, Lamkin JT, Roffer MA (2010) Predicting the occurrence of Atlantic bluefin tuna (*Thunnus thynnus*) larvae in the northern Gulf of Mexico: building a classification model from archival data. Fisheries Oceanography 19: 526–539.
24. Rooker JR, Simms JR, Wells RD, Holt SA, Holt GJ, et al. (2012) Distribution and habitat associations of billfish and swordfish larvae across mesoscale features in the Gulf of Mexico. PloS one 7: e34180.
25. de la Gándara F, Ortega A, Blanco E, Viguri FJ, Reglero P (2013) La flexión de la notocorda en larvas de atún rojo,*Thunnus thynnus* (L, 1758) cultivadas a diferentes temperaturas. In: Sociedad Española de Acuicultura, editors. XIV Congreso Nacional de Acuicultura. Gijón: Universidad Laboral de Gijón. pp. 181–182. http://www.repositorio.ieo.es/e-ieo/bitstream/handle/10508/1559/5905A.pdf?sequence=2&isAllowed=y. Accessed 23 July 2014.
26. Gordoa A, Carreras G (2014) Determination of Temporal Spawning Patterns and Hatching Time in Response to Temperature of Atlantic Bluefin Tuna (*Thunnus thynnus*) in the Western Mediterranean. PloS one 9: e90691.
27. Reglero P, Balbín R, Ortega A, Alvarez-Berastegui D, Gordoa A, et al. (2013) First attempt to assess the viability of bluefin tuna spawning events in offshore cages located in an a priori favourable larval habitat. Scientia Marina 77: 585–594.
28. Reglero P, Urtizberea A, Torres AP, Alemany F, Fiksen Ö (2011) Cannibalism among size classes of larvae may be a substantial mortality component in tuna. Marine Ecology Progress Series 433: 205–219.
29. Blank JM, Morrissette JM, Landeira-Fernandez AM, Blackwell SB, Williams TD, et al. (2004) In situ cardiac performance of Pacific bluefin tuna hearts in response to acute temperature change. Journal of Experimental Biology 207: 881–890.
30. RDevelopment CORE (2011) TEAM. 2008. R: A language and environment for statistical computing. R Foundation for Statistical Computing, Vienna, Austria. ISBN 3-900051.
31. Wood, Simon N. (2006) Generalized additive models: an introduction with R. CRC, Boca Raton, Forida: Chapman & Hall.
32. Burnham, Kenneth P. and Anderson, David R. (2002) Model selection and multimodel inference: a practical information-theoretic approach. New York: Springer-Verlag. 488 p.
33. Muhling BA, Reglero P, Ciannelli L, Alvarez-Berastegui D, Alemany F, et al. (2013) Comparison between environmental characteristics of larval bluefin tuna *Thunnus thynnus* habitat in the Gulf of Mexico and western Mediterranean Sea. Marine Ecology Progress Series 486: 257–276.
34. Rodriguez JM, Alvarez I, Lopez-Jurado JL, Garcia A, Balbín R, et al. (2013) Environmental forcing and the larval fish community associated to the Atlantic bluefin tuna spawning habitat of the Balearic region (Western Mediterranean), in early summer 2005. Deep Sea Research Part I: Oceanographic Research Papers 77: 11–22.
35. Bakun A (2013) Ocean eddies, predator pits and bluefin tuna: implications of an inferred 'low risk-limited payoff' reproductive scheme of a (former) archetypical top predator test. Fish and Fisheries 14: 424–438.
36. Steele JH (1989) The ocean landscape. Landscape Ecology 3: 185–192.
37. Wiens JA (1989) Spatial Scaling in Ecology. Functional ecology 3: 385–397.
38. Wu JG, Li HB (2006) Perspectives and methods of scaling. In: Jianguo Wu, K.. Bruce ones, Harbin Li, Orie L.. Loucks, editors. Scaling and Uncertainty Analysis in Ecology. Netherlands: Springer. pp. 17–44.
39. Bostrom C, Pittman SJ, Simenstad C, Kneib RT (2011) Seascape ecology of coastal biogenic habitats: advances, gaps, and challenges. Marine Ecology-Progress Series 427: 191–217.
40. Forman, R. T T. (1995) Land mosaics: the ecology of landscapes and regions. Cambridge, UK: Cambridge University Press.
41. Reglero P., Tittensor D.P., Álvarez-Berastegui D., Aparicio-González A., Worm B. (2014) Worldwide distributions of tuna larvae: revisiting hypotheses on environmental requirements for spawning habitats. Marine Ecology Progress Series 501: 207–224.
42. Tintore J, Vizoso G, Casas B, Heslop E, Pascual A, et al. (2013) SOCIB: The Balearic Islands Coastal Ocean Observing and Forecasting System, Responding to Science, Technology and Society Needs. Marine Technology Society Journal 47: 101–117.
43. Pascual A, Bouffard J, Ruiz S, Buongiorno Nardelli B, et al. (2013) Recent improvements in mesoscale characterization of the western Mediterranean Sea: synergy between satellite altimetry and other observational approaches. Scientia Marina 77: 19–36.
44. Juzza M, Mourre B, Renault L, Tintoré J (2014) Assessment and intercomparison of numerical simulations in the Western Mediterranean Sea. Geophysical Research Abstracts, Vol 16, EGU2014-6893, 2014, 27 April, 2014 Vienna, Austria. http://adsabs.harvard.edu/abs/2014EGUGA..16.6893J. Accessed 23 July 2014
45. Juza M, Renault L, Ruiz S, Tintoré J (2013) Origin and pathways of Winter Intermediate Water in the Northwestern Mediterranean Sea using observations and numerical simulation. Journal of Geophysical Research: Oceans 118: 6621–6633.
46. Hobday AJ, Hartmann K (2006) Near real-time spatial management based on habitat predictions for a longline bycatch species. Fisheries Management and Ecology 13: 365–380.
47. Cushman SA, McGarigal K, Neel MC (2008) Parsimony in landscape metrics: strength, universality, and consistency. Ecological Indicators 8: 691–703.
48. Wu JJ (2013) Landscape ecology. In: Rik Leemans, editors. Ecological Systems. New York: Springer. pp. 179–200.
49. Lehodey P, Senina I, Murtugudde R (2008) A spatial ecosystem and populations dynamics model (SEAPODYM)-Modeling of tuna and tuna-like populations. Progress in Oceanography 78: 304–318.
50. Ingram Jr GW, Richards WJ, Porch CE, Restrepo V, Lamkin JT, et al. (2008) Annual indices of bluefin tuna (*Thunnus thynnus*) spawning biomass in the Gulf of Mexico developed using delta-lognormal and multivariate models. ICCAT Working Document SCRS/2008/086.
51. Ingram Jr GW, Alemany F, Alvarez-Berastegui D, García A (2013) Development of indices of larval bluefin tuna (*Thunnus thynnus*) in the Western Mediterranean Sea. ICCAT,Collect Vol Sci Pap 69: 1057–1076.
52. Muhling BA, Lee SK, Lamkin JT, Liu Y (2011) Predicting the effects of climate change on bluefin tuna (*Thunnus thynnus*) spawning habitat in the Gulf of Mexico. ICES Journal of Marine Science: Journal du Conseil 68: 1051–1062.

16

Screening for Viral Hemorrhagic Septicemia Virus in Marine Fish along the Norwegian Coastal Line

Nina Sandlund[1]*, Britt Gjerset[3], Øivind Bergh[2], Ingebjørg Modahl[3], Niels Jørgen Olesen[4], Renate Johansen[3]

1 Research group Disease and Pathogen transmission, Institute of Marine Research, Bergen, Norway, **2** Research group Oceanography and climate, Institute of Marine Research, Bergen, Norway, **3** Section of Virology, National Veterinary Institute, Oslo, Norway, **4** Section of Virology, Technical University of Denmark, Frederiksberg C, Denmark

Abstract

Viral hemorrhagic septicemia virus (VHSV) infects a wide range of marine fish species. To study the occurrence of VHSV in wild marine fish populations in Norwegian coastal waters and fjord systems a total of 1927 fish from 39 different species were sampled through 5 research cruises conducted in 2009 to 2011. In total, VHSV was detected by rRT-PCR in twelve samples originating from Atlantic herring (*Clupea harengus*), haddock (*Melanogrammus aeglefinus*), whiting (*Merlangius merlangus*) and silvery pout (*Gadiculus argenteus*). All fish tested positive in gills while four herring and one silvery pout also tested positive in internal organs. Successful virus isolation in cell culture was only obtained from one pooled Atlantic herring sample which shows that today's PCR methodology have a much higher sensitivity than cell culture for detection of VHSV. Sequencing revealed that the positive samples belonged to VHSV genotype Ib and phylogenetic analysis shows that the isolate from Atlantic herring and silvery pout are closely related. All positive fish were sampled in the same area in the northern county of Finnmark. This is the first detection of VHSV in Atlantic herring this far north, and to our knowledge the first detection of VHSV in silvery pout. However, low prevalence of VHSV genotype Ib in Atlantic herring and other wild marine fish are well known in other parts of Europe. Earlier there have been a few reports of disease outbreaks in farmed rainbow trout with VHSV of genotype Ib, and our results show that there is a possibility of transfer of VHSV from wild to farmed fish along the Norwegian coast line. The impact of VHSV on wild fish is not well documented.

Editor: Oliver Schildgen, Kliniken der Stadt Köln gGmbH, Germany

Funding: Norwegian Research Council, grants no. 190245 and 224931 (http://www.forskningsradet.no/en/Home_page/1177315753906) RJ. The funders had no role in study design, data collection and analysis, decision to publish, or preparation of the manuscript.

Competing Interests: The authors have declared that no competing interests exist.

* Email: nina.sandlund@imr.no

Introduction

Viral hemorrhagic septicemia (VHS) is a severe virus infection causing great losses in farming of rainbow trout *Oncorhynchus mykiss*. Mortality rates are variable depending on the age of the fish with up to 100% in fry. It is often less in older fish, typically between 30–70% [1,2]. Hence, economical losses could be substantial. VHS causes clinical signs such as haemorrhages in internal organs, pale gills, exophthalmia and darkening of the body [2].

The causative agent is the viral haemorrhagic septicaemia virus (VHSV), an enveloped negative single-stranded RNA virus belonging to the family of Rhabdoviridae and genus *Novirhabdovirus*. VHSV was first thought of as a virus affecting freshwater fish species of continental Europe. The first VHSV isolation from wild fish in the marine environment was in 1979 from Atlantic cod, *Gadhus morhua* [3,4]. Since then the virus has proven to be both widely spread in the northern hemisphere and occurring in more than 80 marine and fresh water fish species [5]. This emphasizes the ability this virus has to adapt to new host species.

The four main genotypes of VHSV (I–IV) and the subtypes (a–e) are geographically distributed; genotype I, II and III are found in Europe while genotype IV occurs in North American and Northern Pacific waters [2,6]. Subtype Ia is highly virulent for rainbow trout and this subtype is found in VHS infected fresh water farms in continental Europe. Subtype Ib has been isolated from several marine species in the Baltic and North Sea (reviewed in [1]) and has caused two outbreaks of VHS in rainbow trout in Sweden [7,8]. Subtype Ic consists of older isolates from Denmark. Subtype Id represent another group of older isolates from the 1960s in Denmark and Norway as well as isolates from the more recent VHS outbreaks in rainbow trout farmed in sea cages in the Åland archipelago in the Baltic Sea [9]. Group Ie is only present in the Black Sea area. The Baltic Sea is also the main reservoir for the genotype II isolates. Genotype III has been isolated from wild marine fish in the North Sea and Skagerrak and has caused disease outbreaks in farmed turbot *Scophthalmus maximus* in Scotland and Ireland, in farmed rainbow trout in Norway [10] and recently also in wild caught farmed wrasse *Labridae* spp [11]. VHSV genotype IVa is widespread in wild marine fishes on both the American and the Asian side of the Northern pacific and has caused disease outbreaks in farmed Atlantic salmon in British Columbia, Canada [12]. It is still debated to what extent VHSV play a role in stock and population variations of Pacific herring

Clupea pallasii ([13], reviewed in [14]). Over the past decade VHSV genotype IVb has become an emerging problem in the Great Lakes region in North America causing high mortality in several wild fish species [13,15–18]. Subtype IVc has been isolated from brackish fishes during mortality events in Atlantic coastal regions of North America including mummichog *Fundulus heteroclitus*, stickleback Gasterosteidae, striped bass *Morone saxatilis* and brown trout *Salmo trutta* [12,19].

Screening surveys of VHSV in wild marine fish have been performed at several separate locations in European waters; the Barents Sea, North Sea, Norwegian Sea, Skagerrak, Kattegat and Baltic Sea [20–24]. VHSV has been detected from the North Sea, coastal areas around Scotland, Skagerrak, Kattegat, the Baltic Sea and Flemish Cap, in the North Atlantic Ocean near Newfoundland and the virus is assumed endemic in these waters (reviewed in [1]). VHSV prevalence of up to 16.7% was found in Atlantic herring *Clupea harrengus* [21,24,25]. But most screening surveys performed on both Atlantic herring and Pacific herring [14,26,27] reports relatively few positive detections, in spite of high number of sampled fish (reviewed in [1]). Recently high prevalence of VHSV was reported in Atlantic herring from the southern part of Norway during the spawning season, however no disease symptoms associated with the findings were reported [28].

Although VHS outbreaks mainly occur in freshwater rainbow trout farms there have been a few reports of outbreaks of VHS in sea farmed rainbow trout in Sweden, Finland, France, Denmark and Norway [1,7,8,10]. The VHS outbreak in Norway occurred in farmed rainbow trout in Storfjorden in 2007 [10]. This was the first detection of VHS in Norway since the eradication of the disease in 1974 and the source of the infection is still unknown. The outbreak in Norway was unexpectedly caused by a marine genotype III VHSV strain and challenge trials confirmed the marine virus-strain as virulent to rainbow trout [10,29]. Before this VHSV genotype III was considered low pathogenic to rainbow trout and other salmonid fish species [30,31]. This makes the outbreak unique and one of the main interests of the current screening was therefore to determine a possible marine reservoir in wild marine fish populations.

The Norwegian fish farming industry, especially the production of Atlantic salmon *Salmo salar*, has grown rapidly over the last decades. The annual production of Atlantic salmon and rainbow trout is now approx 1.2 mill and 70.000 tons respectively according to Statistics, Directorate of Fisheries in Norway (http://www.fiskeridir.no/). The possibility of pathogen transmission between farmed and wild fish is therefore increasing and a major concern. VHSV has so far not been isolated from Atlantic salmon in northern European waters, but VHSV genotype IVa has been isolated from farmed Atlantic salmon on the east coast of Canada (British Columbia) [15]. In general, Atlantic salmon has shown limited susceptibility to VHSV in immersion trails, but using intra peritoneal (i.p.) injection as challenge model has resulted in up to 78% mortality [1,32]. In challenge experiments exposing Atlantic salmon to the genotype III VHSV isolate from Norway, mortality was only experienced in the i.p. injected fish group and not in the immersion trail groups [10]. However, recent work demonstrated that Atlantic salmon were susceptible to VHSV genotype IVa and developed clinical signs after i.p. injection, immersion and when cohabited with VHSV diseased Pacific herring [33]. Transmission of VHSV from infected Atlantic salmon to sympatric Pacific herring was also demonstrated. This shows that VHSV has the ability to adapt to and infect Atlantic salmon. It should be added that Atlantic salmon seems less susceptible to VHSV genotype IVb than IVa [34].

With more than 80 susceptible species in the marine and freshwater environment, VHSV shows high ability of host adaptation and evolution of new strains with increased virulence. VHSV is thereby a potential risk to farming of susceptible species. Evidence support the theory that both wild and farmed fish can function as a reservoir and transmitter of VHSV [11,15,33,35], however more knowledge regarding specific species is needed. The presence of fish species that are asymptomatic carriers of VHSV also needs to be considered. Recently the Shetland Isles experienced an outbreak of VHS in wild caught wrasse held at a marine farm before use as cleaner fish in Atlantic salmon farming. The origin of this VHSV genotype III was likely from the marine environment, as several marine species in and around the locality tested positive for this virus [11]. As use of wrasse as cleaner fish is increasing, this points to another potential route of pathogen transmission to Atlantic salmon.

Screening of wild fish in northern parts of the Atlantic Ocean has earlier mainly been done in the open waters away from the aquaculture facilities. The aim of the present study was to investigate the occurrence of VHSV in marine wild fish in the coastal and fjord areas of Norway by analysing organ samples using both virus isolation and PCR.

Material and Methods

Field collection of fish

Five different research cruises in the fjords and coastal areas in Norway were carried out during 2009–2011. All cruises were part of the Institute of Marine Research annual coastal surveys and the fish samples used for VHSV analysis were randomly selected among the fish from the trawl haul. The Institute of Marine Research is a governmental research institute with given permission to perform research cruises including fish samplings by the Norwegian Government. Hence no additional permits were needed to sample organs from already deceased fish. All sampling and handling of fish were performed by experienced personnel.

Cruise one took place in September/October 2009 targeting the northern coast area. Cruise two was carried out in December 2009 targeting the coast of mid Norway including the Storfjorden fjord system were a VHSV outbreak occurred in farmed rainbow trout in 2007. Cruise three took place on the west coast in May 2010. During cruise four fish was sampled from the coast of north and mid Norway in October/November 2010. Cruise 5 took place in October/November 2011 in mostly the same areas as cruise one. The sampling area and geographical positions of each trawling station are shown in Figure 1.

During cruises one, two, four and five fishing was performed as either bottom or pelagic trawling, with trawling time of approximately 30 minutes. During cruise three only pelagic trawling was performed and it lasted up to 4 hours. Most fish died from suffocation in the trawling process or during handling on deck. All fish were kept in a cooled 4°C room until sampling and maximum time from fishing to sampling was 7–8 hours. During post mortem examination all fish were measured, weighed and any external signs of disease were recorded if present.

Sampling of organs

All tissue samples were processed on board the research vessels. The same sampling procedure was used during all cruises. Organ specimens of spleen, kidney and brain from maximum 5 fish was pooled and diluted 1:10 in transport medium (Eagle's Minimum Essential Medium, pH 7.6, supplemented with 10% newborn bovine serum and 100 µg ml^{-1} gentamicin). The samples were immediately transferred to a −80°C freezer for storage. Gonads

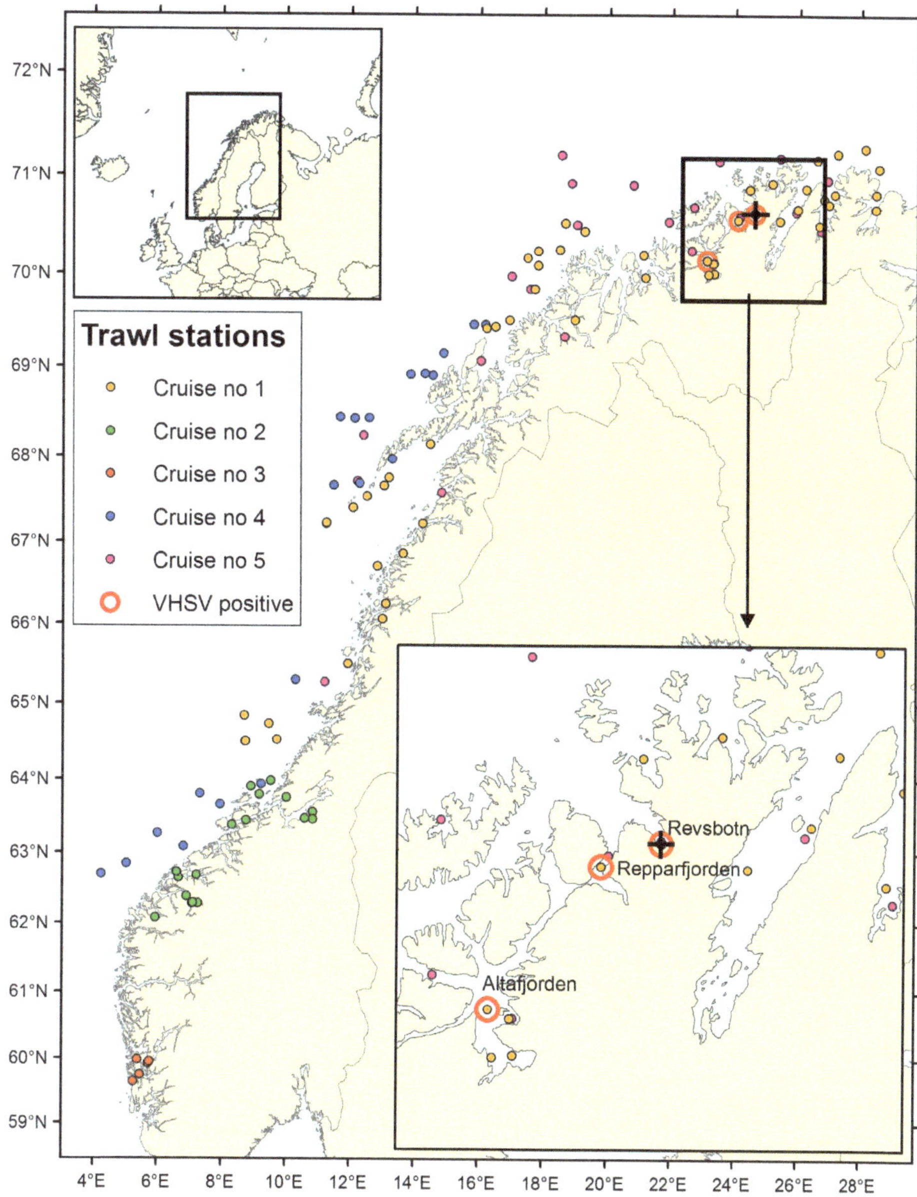

Figure 1. Map showing sampling locations (trawling stations) from cruises 1–5 indicated in colour. The various cruises are presented with different colour codes. Trawling stations with VHSV positive rRT-PCR samples are shown with red circles. The respective latitude/longitude for the locations are; Repparfjorden/Sammelsundet (70.5600/24.0800), Revsbotn (70.62667/24.6150) and Altafjorden (70.14667/23.0950). The black cross shows origin of the one pool sample testing positive by cell culture isolation.

(testes/ovaries) were sampled from sexually mature fish and kept in separate transport medium tubes. In addition individual samples of gills, heart, spleen, kidney, brain and gonads were collected in RNAlater (Sigma, USA), stored at 4°C for 24 hours prior to long term storage at −20°C. In cruise one the individual samples were randomly selected from three of the five pooled fish due to storage and sampling capacity. In cruise two-five individual samples were taken from all fish. Hearts were not individually sampled during cruise two, three and four due to storage capacity. Spleen and brain were diagonally sectioned and one half collected in transport medium and RNAlater for virus isolation and real-time Reverse Transcriptase Polymerase Chain Reaction (rRT-PCR) examination, respectively. The organs were aseptically sampled in this specific order; gills, heart, spleen, kidney and brain. If sexually matured gonads were present, these were sampled after the spleen. Equipment used to collect samples, were cleaned with alcohol, flamed between uses on each organ, and changed between each fish. Disposable gloves and tissue paper were changed between each fish. Equipment was washed and disinfected on daily basis to ensure sterile conditions; using Virkon S (Lilleborg, Norway) for minimum 30 minutes, rinsed in fresh water and boiled in fresh water for 15 minutes.

Following completion of the cruises, all samples were transported to the laboratory on dry ice and stored at either −20°C or −80°C prior to testing.

Virus isolation

Tissue samples pooled in transport medium were homogenized and cleared by low-speed centrifugation, and supernatants inoculated onto subconfluent monolayers of BF-2-cells (ECACC,

Salisbury, UK) in 1:10 and 1:100 dilutions in 24-well tissue culture plates according to the OIE procedure [5]. Inoculated cultures were incubated at 15°C and inspected after 1 week for cytopathic effect (CPE). Culture medium was collected from all wells and passed to new cell cultures. After a further week of incubation, the cultures were again inspected, and supernatant from wells with evident CPE in the second passage was collected, RNA extracted and tested for VHS-virus by rRT-PCR. Pooled organ samples that later were found VHSV positive by rRT-PCR were additionally incubated in BF-2 and EPC-cells three times for 14 days.

RNA extraction and real-time RT-PCR (rRT-PCR)

RNA extraction was performed on homogenized tissue from organ pools (100 µl) and individual organ samples (10–20 mg), and from 150 µl virus supernatant using the automated easyMAG protocol (Biomérieux) or the RNeasy Mini kit (Qiagen). Extracted RNA was measured using NanoDrop ND-1000 (NanoDrop Technologies). The rRT-PCR assay was conducted using 500–1000 ng template RNA with a QIAGEN OneStep RT-PCR kit (QIAGEN Nordic) and nucleoprotein (N) gene primers and probe described by Duesund et al. [29]. The assay was performed with 0.5 µM of each primer and 0.3 µM probe in a 20 µl reaction, with cDNA synthesis at 52°C for 30 min followed by 15 min at 95°C, then 45 cycles of 95°C for 15 s and 60°C for 1 min using the Mx3005p real time PCR system (Stratagene). During the study the laboratory changed the standard VHSV rRT-PCR method to a validated assay with higher analytical and diagnostic sensitivity for all VHSV genotypes [36]. The assay was conducted using the same reaction conditions, but cycling conditions: 30 min a 50°C and 15 min at 95°C, followed by 45 cycles of 94°C for 30 s and 60°C for 1 min. Required positive and negative controls were included in all runs. Samples with specific cycle treshold (Ct) value ≤40 were considered positive.

Sequencing and phylogenetic analysis

VHSV rRT-PCR positive samples were confirmed by sequence analysis of the viral G- and N-protein gene. Partial and full length glycoprotein (G) gene sequence was generated from overlapping sequences using primer sets V2, GB and Gseq [10,37]. Three primers were used to obtain a 1217 bp nucleoprotein (N) gene sequence: N-G1F 5′-GCT CAC AGA CAT GGG CTT CA-3′, N-G2R 5′-TGG ATT GGG CTT CTT CTT-3′, N-G3F 5′-GGC TCA ACG GGA CAG GAA-3′. The RT-PCR was performed using 5 µl extracted RNA, 0.5 µM primer concentration in a 50 µl QIAGEN OneStep reaction, with cDNA synthesis at 50°C for 30 min followed by 15 min at 95°C, then 40 cycles of 95°C for 1 min, 55°C 1 min and 72°C for 90 s. The RT-PCR products were visualised on an agarose gel and purified using the ExoSAP-IT protocol (Usb) prior to sequencing with BigDye Terminator v3.1 Cycle Sequencing kit (Applied Biosystems). Sequences derived were aligned and compared to related VHSV sequences using Vector NTI Advance 11 (Invitrogen). A maximum-likelihood (ML) phylogenetic analysis was conducted using MEGA version 5.0 [38] on the complete G-gene alignment employing the GTR+G model. The obtained G- and N-gene sequences were deposited in GenBank and given accession numbers HM632035–HM632036 and KJ768664–KJ768665 for the isolate from Atlantic herring and silvery pout, respectively.

Results

Fish sampling

During the five cruises, a total of 1927 fish representing 39 different species were caught and sampled at 121 different haul stations. A total overview of the various fish species sampled during the separate cruises, the number of pooled samples and geographical distribution is given in Table 1 and Figure 1. No fish showed any visible signs of clinical disease during the post mortem examinations.

The Atlantic herring caught during the research cruises is part of the Norwegian Spring Spawning herring (NSS) stock. The average length and weight of the herring caught at trawling station Repparfjorden was 15.7 cm (±2.9 cm STDV) and 40 g (±20.1 STDV) and at trawling station Revsbotn 22.7 cm (±3.4 cm STDV) and 124.5 g (±47 STDV). The individual length and weight of the positive fish, included in Table 2, do not differ significantly from the rest of the fish in the same catch (data not shown). Based on analysis of length and age distribution of NSS it can be estimated that the VHSV positive Atlantic herring were 4 years or less [39].

Cell culture isolation

One VHS-virus isolate were obtained from the 453 organ pools tested using BF-2 cell culture inoculation (Tables 1, 2). The positive pool contained organs from five Atlantic herring *Clupea harengus* collected during cruise one at location Revsbotn in Finnmark county (Figure 1). Full CPE was observed within 2 weeks of incubation, and VHSV was confirmed in the culture supernatant by rRT-PCR (Tables 1, 2). No CPE was observed in the other virus cultures. Organ pools later found VHSV positive by rRT-PCR were additionally sub-cultivated with prolonged incubation time without any detection of CPE.

VHSV rRT-PCR detection

rRT-PCR screening of the 453 pooled samples revealed totally five VHSV positive pools (Table 2), four originating from Atlantic herring and one from silvery pout *Gadiculus argenteus*, all sampled on three trawling locations in Finnmark county within two days (Figure 1). Each pool consisted of organs from five fish. To follow up these positive findings all available organ samples from the Atlantic herrings and silvery pout represented in the pools (gills, heart, kidney, spleen and brain) were tested individually by rRT-PCR (Table 2). This revealed 1–2 positive fish per pool; respectively five Atlantic herring and one silvery pout.

Generally the Ct values obtained from the individual organs samples in positive fish show that the highest amount of VHSV RNA is present in the heart while gills showed the highest prevalence (Table 2). All individually sampled gills (n = 1369) and hearts (n = 1091) from all cruises were therefore tested by rRT-PCR. Due to low RNA yields on some of the gill samples (n = 183) results are only recorded from 1186 gills to avoid false-negative results (Table 2). Three additional Atlantic herring, two haddock *Melanogrammus aeglefinus* and one whiting *Merlangius merlangus* tested positive in gills, and they were all caught in the same trawl as the positive herring. The remaining sampled organs from these individual tested negative.

Sequencing

Unique G and N gene sequences were obtained from the PCR positive whiting, haddock, herring and silvery pout marked in bold in Table 2. This confirmed that they all belonged to genotype Ib and were closely related (99–100% identity, 6 nucleotide difference in the complete G-gene region). The partial sequences from whiting and haddock were too short to be included in the phylogenetic analysis. A ML phylogenetic tree based on complete G-gene sequences group the silvery pout and herring from this study together with other genotype Ib isolates reported from

Table 1. Species and number of fish and pooled organ samples from all five research cruises.

Family	Cruise 1		Cruise 2		Cruice 3		Cruice 4		Cruice 5		All cruises	
Species	**No.of sampled fish**	**No. of pools**	**No. of sampled fish**	**No. of pools**	**No. of sampled fish**	**No. of pools**	**No. of sampled fish**	**No. of pools**	**No. of sampled fish**	**No. of pools**	**Total no. of sampled fish**	**Total no. of pools**
Ammodytidae												
Small sandeel *Ammodytes tobianus*	10	2									10	2
Lesser sandeel *Ammodytes marinus*	6	2									6	2
Anarhichadidae												
Wolf-fish *Anarhichas lupus*	22	5							4	1	26	6
Argentinidae												
Argentine *Argentina silus*	176	37					9	2	5	1	190	40
Belonidae												
Garfish *Belone belone*					27	7					27	7
Carangidae												
Horse mackerel *Trachurus trachurus*	1	1					3	1			4	2
Clupeidae												
Atlantic herring *Clupea harengus*	170	37*¤					7	2	45	11	222	50*¤
Cyclopteridae												
Lumpsucker *Cyclopterus lumpus*			1	1	17	6					18	7
Gadidae												
Cod *Gadus morhua*	64	14	14	4					32	6	110	24
Blue whiting *Micromesistius poutassou*	104	24					5	1	5	1	114	26
Four-bearded rockling *Rhinonemus cimbrius*	1	1									1	1
Haddock *Melanogrammus aeglefinus*	121	26					5	1	15	3	141	30
Ling *Molva molva*	5	3	1	1			3	1			9	5
Norway pout *Trisopterus esmarkii*	336	68							5	1	341	69

Table 1. Cont.

Family	**Cruise 1**		**Cruise 2**		**Cruice 3**		**Cruice 4**		**Cruice 5**		**All cruises**	
Species	**No.of sampled fish**	**No. of pools**	**No. of sampled fish**	**No. of pools**	**No. of sampled fish**	**No. of pools**	**No. of sampled fish**	**No. of pools**	**No. of sampled fish**	**No. of pools**	**Total no. of sampled fish**	**Total no. of pools**
Pollack *Pollachius pollachius*			2	1			4	1			6	2
Poor cod *Trisopterus minutus*	12	4	15	3			1	1			28	8
Seith *Pollachius virens*	52	14	29	7			23	5	5	1	109	27
Silvery pout *Gadiculus argenteus*	60	12*									60	12*
Whiting *Merlangius merlangus*	54	12					4	1	13	3	71	16
Lophiidae												
Anglerfish *Lophius piscatorius*	10	6	4	2							14	8
Lotidae												
Tusk *Brosme brosme*	11	9	4	1			9	2			24	12
Blue ling *Molva dypterygia*			3	2							3	2
Merluccidae												
Hake *Merluccius merluccius*	2	1	23	6					4	1	29	8
Osmeridae												
Capline *Mallotus villosus*	79	16							10	2	89	18
Phycidae												
Greater forkbeard *Phycis blennoides*	3	1	3	1			4	1			10	3
Pleuronectidae												
Halibut *Hippoglossus hippoglossus*	1	1									1	1
Lemon sole *Microstomus kitt*	11	3									11	3
American plaice *Hippoglossoides platessoides*	10	2							5	1	15	3
Plaice *Pleuronectes platessa*	17	4	7	2			5	1	5	1	34	8
Witch *Glyptoceophalus cynoglossus*	10	4	5	1							15	5

Table 1. Cont.

Family	Cruise 1		Cruise 2		Cruice 3		Cruice 4		Cruice 5		All cruises	
Species	**No.of sampled fish**	**No. of pools**	**No. of sampled fish**	**No. of pools**	**No. of sampled fish**	**No. of pools**	**No. of sampled fish**	**No. of pools**	**No. of sampled fish**	**No. of pools**	**Total no. of sampled fish**	**Total no. of pools**
Scombridae												
Macerel *Scomber scombrus*					5	1					5	1
Sebastinae												
Small redfish *Sebastes viviparus*	73	16	10	2			5	1			88	19
Golden redfish *Sebastes marinus*	20	5					5	1			25	6
Sebastes sp.	52	11									52	11
Scophthalmidae												
Megrim *Lepidorhombus whiffiagonis*	3	2	1	1							4	3
Squalidae (dogfish sharks)												
Piked dogfish *Squalus acanthias*			5	1							5	1
Sternoptychidae												
Pearlsides *Maurolicus muelleri*			2	1							2	1
Triglidae												
Grey gurnard *Eutrigla gurnardus*	4	2							3	1	7	3
Zoarcidae												
Eelpout *Lycodes vahlii gracilis*	1	1									1	1
Total number	1501	346	129	3	49	14	92	22	156	34	1927	453

Samples were found VHSV positive by virus isolation (¤) and rRT-PCR analysis (*).

Table 2. VHSV rRT-PCR Ct values of pooled and individual organ samples from positive fish.

				Ct values individual organ samples						Ct values pooled organ samples	
Trawling location	**Species**	**Length cm**	**Weight gr**	**Gills**	**Heart**	**Kidney**	**Spleen**	**Brain**	**Gonads**	**Pool no.**	
Repparfjorden	Herring	16.0	44	28.48	24.75	33.61	31.23	33.03	N.A	63	37.17
Repparfjorden	Herring	17	42	37.44	32.27	-	-	-	♀	63	37.17
Revsbotn	Herring	21.0	87	37.75	-	-	-	-	N.A	66	-
Revsbotn	Herring	25.5	164	38.32	-	-	-	-	N.A	67	-
Revsbotn	Herring	26.0	171	38.76	-	-	-	-	♂	69	-
Revsbotn	Herring	27.0	200	37.68	-	-	-	-	N.A	70	33.17
Revsbotn	Herring	19.5	69	28.06	25.49	29.54	29.83	34.29	N.A	71	36.08
Revsbotn	Herring	17.0	44	28.21	28.96	34.06	31.8	36.53	N.A	72*	**26.74**
Revsbotn	Haddock	56.0	1152	**36.31**	-	-	-	-	N.A	73	-
Revsbotn	Haddock	44.5	849	**33.90**	-	-	-	-	N.A	74	-
Revsbotn	Whiting	40.0	586	**32.09**	-	-	-	-	♀	75	-
Altafjorden	Silvery pout	17.0	47	37.62	**23.41**	37.9	39.99	38.17	N. A	102	37.65

VHSV rRT-PCR Ct values of separate organ samples from positive individuals, in addition to the corresponding pooled organ sample in which the individual samples were included. Fish length, weight and trawling location, from which each individual were caught, are included. Pool no. = identity pool number Herring = Atlantic herring *Clupea harengus*, silvery pout *Gaduculus argenteus*, haddock *Melanogrammus aeglefinus*, whiting *Merlangius merlangus*.
- = No Ct. N.A = not available.
* = sample positive for VHSV by both virus isolation and rRT-PCR. VHSV sequences for verification were obtained from the samples indicated in bold.

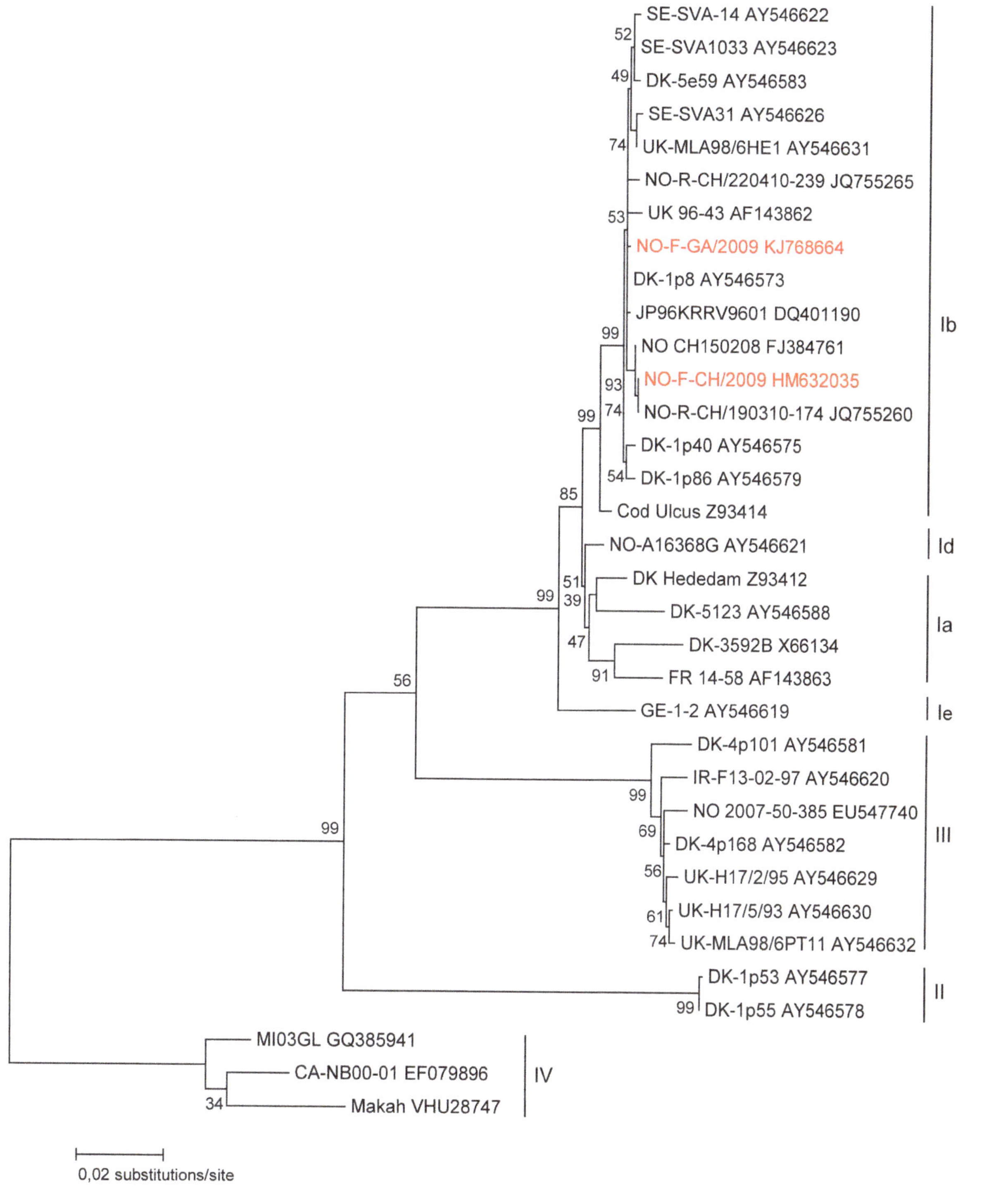

Figure 2. Maximum Likelihood (ML) phylogenetic tree showing the relationship of the new VHSV G-gene sequences (indicated in red) and other VHSV genotype representatives. Sequences are labelled by isolate name and GenBank accession number. The obtained tree was bootstrapped with 500 replicates and genotype IV was used as outgroup.

Atlantic herring and other fish species in the North Sea (Figure 2). The genotype Ib group also includes G-gene sequences detected in Atlantic herring of the Norwegian spring-spawning stock (Acc. no. JQ755260, JQ755265) [28].

Discussion

Sampling during five research cruises was conducted to investigate the presence of VHSV in wild fish along the Norwegian coastline. The present screening survey is the first to look for VHSV in coastal areas and fjord systems in Norway. In total, samples from 12 fish including Atlantic herring *Clupea harengus*, silvery pout *Gadiculus argenteus*, haddock *Melanogrammus aeglefinus* and whiting *Merlangius merlangus* were found positive for VHSV. The VHSV positive fish were sampled on geographically close trawling locations in the northern county of Finnmark. This is the first observation of VHSV this far north, which again points out the large natural marine reservoir for VHSV.

To our knowledge this is the first report of sampling and testing for VHSV in silvery pout. Silvery pout is a small deep water fish of maximum 15 cm length that lives in the Northeast Atlantic and is used for industrial purposes such as fish meal and fish oil. Both Atlantic herring and silvery pout occur in large shoals which can migrate long distances. It is therefore uncertain whether the VHSV-infected fish were infected in Finnmark or elsewhere, and then migrated into the sampling area. It is interesting to note that these two highly different fish species are found in the same area carrying closely related virus isolates (Figures 1, 2).

The haddock and whiting only tested positive for VHSV in gills and it has not been confirmed if these fish where truly infected or only passive carriers of the virus in the gill mucus. None of the fish testing positive for VHSV in this study showed any clinical signs of disease. This is in line with numerous other isolations of VHSV from asymptomatic wild marine fish; including Atlantic herring caught in the English Channel [40], North Sea [21], Baltic Sea [24,25] and Skagerrak and Kattegat [24]. The first marine isolate of VHSV was isolated in Danish coastal waters in 1979 from cod showing "ulcus syndrome", however no evidence of an association between ulcers and VHS have been demonstrated [3,4,21, 24,25,41].

During the five research cruises samples were taken from 39 different species. The number of fish of each species is highly variable due to the fact that both bottom and pelagic trawling was used, and the trawling was performed in various areas at different times of the year. In addition, fish were sampled at random from random hauls and the need for fresh samples had to be prioritised. As high prevalence of VHSV has been found in Atlantic herring on previous screenings, Atlantic herring were prioritised for sampling when available [21,24,25,28].

The 453 pooled organ samples (spleen, kidney and brain) were tested both with cell culturing and rRT-PCR, revealing a higher detection rate with rRT-PCR. Successful virus isolation in cell culture was only obtained from one pool of organs, including samples from five Atlantic herring. This sample was the strongest positive when tested using rRT-PCR (Ct 26), reflecting that overall the viral amount is relatively low. In addition, individual organs (spleen, kidney, brain, heart, gills and gonads) were sampled from most fish. rRT-PCR testing indicates that the highest amount of VHSV RNA was found in the heart and the highest prevalence were detected in gills. Heart samples are suitable for detecting VHSV in various fish species as it is an important target organ for VHSV [18,42–45]. According to OIE (Commissions decision 2001/183) heart and/or brain samples should be included in screenings surveys. Based on our results from this present screening survey and previous findings, it is suggested that heart samples should always be included when sampling marine fish species.

Seven fish, including Atlantic herring, haddock and whiting, tested positive for VHSV by rRT-PCR in gills only. Gills have in various species proven to be a good organ for detecting VHSV carrying fish [28,43,46]. The results obtained by Cornwell et al. [46] further indicate that the sensitivity of detecting VHSV in gill samples might vary between species. The ability of VHSV to infect the gills has also been correlated to virulence [47]. A study testing the viral load of VHSV in various tissues of rainbow trout at various stages during the course of infection showed that fish surviving a VHSV infection had the highest amount of virus in gill and brain tissue [48].

This study detected a low prevalence of VHSV in Atlantic herring (8 positive/222 sampled) and this is in accordance with several other surveys in the waters in the northern part of Europe (reviewed in [1,21,24,25]). High prevalence in Atlantic herring on the west coast of Norway were found during the spawning season in 2010 [28]. Whether this was caused by a generally higher prevalence during spawning seasons or an outbreak in this population is unknown and needs to be further investigated. Isolation of VHSV from Pacific herring *Clupea pallasii* populations is well known [14,26,27]. Studies have come to contradictory conclusions whether VHSV play a role in stock variations of Pacific herring or not ([13], reviwed in [14]).

There are indications that age plays an essential role in the susceptibility of VHSV. Higher prevalence of VHSV has been found in young wild caught Pacific herring compared to older [14,49]. Another explanation can be that herring are infected during the early life stages and that the virus amount decrease over time to a non-detectable level in surviving fish. In captivity asymptomatic Pacific herring larvae and juveniles has developed VHS within a week after confinement [26] indicating a latent virus infection in the fish triggered by and developed in captivity. These findings supports the suggestion that young herring is more susceptible to VHSV infection than older. The age of the Atlantic herring tested in this study was estimated based on length and age distribution of NSSH [39]. This indicates that the VHSV positive Atlantic herring were less than 4 years. It can also be noted that the positive Atlantic herring samples were among the smallest specimen tested from each catch, and these results may be consistent with young herring being more susceptible to VHSV. In the present study the low amount of virus found and lack of clinical signs on the VHSV positive fish indicates that these individuals are asymptomatic carriers of the virus.

Few detections of VHSV from previous screening surveys could be related to methodology. According to Dixon [50] some VHSV isolates prefer BF-2 cells, while others produce highest titers in EPC cells. Isolation of marine VHSV, both of genotype Ib and III, is more successful in BF-2 cells compared to EPC [5], which was therefore the choice of cells used in the present study. Earlier screening surveys in wild marine fish have mostly tested pooled organ samples by cell culture isolation [21,24,25]. One exception is the screening performed by Dixon et al. [23] in which cell culture supernatant or dilutions of homogenized tissue from pooled organ samples were tested by RT-PCR. As in the present study an increased number of positive pooled samples ($n = 4$) were found using RT-PCR compared to cell culture isolation. Experimental testing of the sensitivity of cell culturing versus PCR-methodology has demonstrated RT-PCR the most sensitive [51,52]. The development of PCR assays with higher sensitivity and a broader detection range to several genotypes of VHSV has made this the preferred method for VHSV detection in most laboratories [36].

Screening for VHSV based on testing of individual organs from marine fish is not common. Pooling of organ samples lower the sensitivity of the detection methods but at the same time allows testing of a larger amount of samples. Testing of all individual organs sampled during this study ($n = 6848$) was not feasible in our laboratory. The decreased sensitivity of the pooling strategy has been partly compensated by individual testing of all gill and heart samples. All individual organs from all available fish in the VHSV positive pools where tested by rRT-PCR revealing 1–2 positive fish per pool. This illustrates the difficulty of estimating prevalence based on pooled organ samples when virus yield and prevalence is low. The results further indicate that individual organ samples tested by rRT-PCR provide the best estimate for a true prevalence in the population. This difference in methodology may have led to an underestimation of the prevalence of VHSV in the European marine environment as methods with high sensitivity is required to detect carrier fish with low viral titers. Due to sampling and storage capacity during cruise one only three out of the five individuals included in each pool were sampled on RNAlater (n = 943). This could have affected the total number of positive results as additional positive individuals could have been included in the pool.

Sequencing based on the G and N gene revealed that the positive samples belonged to VHSV genotype Ib and were closely related. The ML phylogenetic tree group the strain from silvery pout and herring with isolates occurring in various fish species in the North Sea and Baltic Sea, including strains detected in Atlantic herring of the Norwegian spring-spawning stock [12,28]. No genotype III positive fish were found in the present sampling, but such positive haddock and whiting are known from previous screenings in the North Sea [37]. The positive fish originated from sampling at three close trawling locations. The transmission pattern of VHSV from fish to fish is by direct contact, in water or by ingestions of infected material [53,54]. In theory virus could transmit between fish while in the trawl. This is unlikely since internal organs also tested positive, indicating true carrier status. In addition, with one exception all individuals testing positive for VHSV came from different pool of fish, showing that the possibility of contamination between samples is limited. All positive pools also contain several negative fish showing that contamination was limited.

The risk of inter-species transmission of VHSV is always present in the marine environment. Evidence of VHSV transmission from wild to farmed fish has recently been studied by phylogenetic analysis that verified a closely genetic linkage between VHSV genotype IVa isolates from wild and farmed fish in the British Columbia area [15]. In the fjord systems were the positive marine samples were found in this study both wild and farmed Atlantic salmon are present (data from the Directorate of Fisheries). In addition Altafjorden and Repparfjorden are regulated fjord systems established for the protection of wild salmon. Although Atlantic salmon in general has shown low susceptibility to VHSV, relatively high virus titer (between 1×10^4 and 1×10^6 pfu/g) of VHSV was demonstrated in Atlantic salmon 10 weeks after i.p. injection, immersion and cohabitation challenge with VHSV genotype IV [33]. Three of 12 Atlantic salmon i.p. injected with the genotype III isolate from Storfjorden were still VHSV positive 29 days after infection [10]. This susceptibility and persistency of VHSV in Atlantic salmon demonstrated by Lovy et al. [33] together with the possibility of transmitting the virus back to Pacific herring indicates that salmon has the ability to serve as a vector and reservoir of VHSV. Although VHSV genotype III isolates does not normally cause disease in anadromous fish species, the outbreak in rainbow trout in Norway in 2007 shows the adaptation capacity of this virus for salmonids [10]. Therefore, it is important to keep farmed fish free of any VHSV by avoiding continuous production of the same fish species for several generations and keep generations at separate locations. The possibility of inter-species transmission of VHSV between fish species in close contact is highly relevant especially with the increased use of cleaner fish as biological control of sea lice in Atlantic salmon farms. Although all Atlantic salmon from a farm site that experienced an outbreak of VHSV on wrasse tested negative for VHSV, the potential of VHSV to adapt and cause disease outbreaks in Atlantic salmon farms, should be taken into account [55].

The fish testing positive for VHSV in the current study appeared asymptomatic carriers. The healthy carriers could represent survivors from an earlier disease outbreak with high mortality rates. Mortality rates in wild fish populations are often not detected due to removal of diseased fish by predators [56]. It is also possible that the virus carrier fish is weakened and thereby more susceptible to other disease problems or predators. Little is known about how VHSV affect the health situation of wild fish populations and further research is needed. Possible transfer of VHSV between wild and farmed fish will always be a potential risk and knowledge about marine reservoirs is therefore essential. It is therefore of major importance to conduct surveillance studies for VHSV to ensure early detection and eradication of the virus from farmed fish. History has taught us that such control is important to avoid adaptation of the virus into more virulent strains.

Acknowledgments

The authors would like to thank Helle Frank Skall and Mike Snow for helpful discussions during the designing of the sampling protocol. The authors are also grateful for the sampling assistance given by Ann Kristin Jøranlid, Rolf Hetlelid Olsen, Trygve Poppe, Arve Kristiansen and to Karen Gjertsen for creating Figure 1.

Author Contributions

Conceived and designed the experiments: NS ØB RJ NJO. Performed the experiments: NS ØB. Analyzed the data: NS RJ BG IM. Contributed reagents/materials/analysis tools: NS BG IM NJO ØB RJ. Contributed to the writing of the manuscript: NS BG IM NJO ØB RJ.

References

1. Skall HF, Olesen NJ, Mellergaard S (2005) Viral haemorrhagic septicaemia virus in marine fish and its implications for fish farming - a review. Journal of Fish Diseases 28: 509–529.
2. Olesen NJ, Skall HF (2013) Viral haemorrhagic septicaemia virus. In: Munir M, editor. Mononegaviruses of veterinary importance Vol I: Pathobiology and molecular diagnosis: CAB International. pp. 323–336.
3. Vestergard Jørgensen PE, Olesen NJ (1987) Cod ulcus syndrome rhabdovirus is indistinguishable from the Egtved (VHS) virus. Bulletin of the European Association of Fish Pathologists 7: 73–74.
4. Jensen NJ, Larsen JL (1979) Ulcus-syndrome in cod (*Gadus morhua*). I. A pathological and histopathological study. Nordisk Veterinaer Medicin 31: 222–228.
5. Anonymous (2014) Manual of Diagnostic Tests for Aquatic Animals 2014. Available: http://www.oie.int/fileadmin/Home/eng/Health_standards/aahm/current/2.3.09_VHS.pdf: World Organisation of Animal Health.
6. Walker PJ, Benmansour A, Calisher CH, Dietzgen R, X FR, et al. (2000) Family Rhabdoviridae. In: van Regenmortel MHV, Fauquet CM, Bishop DHL, Carstens EB, Estes MK et al., editors. Virus taxonomy; Seventh report of the international committee for taxonomy of viruses. San Diego: Academic Press. pp. 563–583.
7. Nordblom B (1998) Report on an outbreak of viral haemorrhagic septicaemia in Sweeden. Sweedish Board of Agriculture, Department of animal Production and Health.
8. Nordblom B, Norell AW (2000) Report on an outbreak of viral haemorrhagic septicaemia in farmed fish in Sweeden. Report for the standing veterinary committee. Sweedish Board of Agriculture, Department of animal Production and Health.
9. Gadd T, Jakava-Viljanen M, Tapiovaara H, Koski P, Sihvonen L (2011) Epidemiological aspects of viral haemorrhagic septicaemia virus genotype II isolated from Baltic herring, *Clupea harengus* membras L. Journal of Fish Diseases 34: 517–529.
10. Dale OB, Ørpetveit I, Lyngstad TM, Kahns S, Skall HF, et al. (2009) Outbreak of viral haemorrhagic septicaemia (VHS) in seawater-farmed rainbow trout in Norway caused by VHS virus Genotype III. Diseases of Aquatic Organisms 85: 93–103.

11. Hall LM, Smith RJ, Munro ES, Matejusova I, Allan CET, et al. (2013) Epidemiology and control of an outbreak of viral haemorrhagic septicaemia in wrasse around Shetland commencing 2012. The Scottish Government. 1–46 p.
12. Pierce LR, Stepien CA (2012) Evolution and biogeography of an emerging quasispecies: Diversity patterns of the fish Viral Hemorrhagic Septicemia virus (VHSv). Molecular Phylogenetics and Evolution 63: 327–341.
13. Elston RA, Meyers TR (2009) Effect of viral hemorrhagic septicemia virus on Pacific herring in Prince William Sound, Alaska, from 1989 to 2005. Diseases of Aquatic Organisms 83: 223–246.
14. Marty GD, Quinn TJ, Carpenter G, Meyers TR, Willits NH (2003) Role of disease in abundance of a Pacific herring (*Clupea pallasi*) population. Canadian Journal of Fisheries and Aquatic Sciences 60: 1258–1265.
15. Garver KA, Traxler GS, Hawley LM, Richard J, Ross JP, et al. (2013) Molecular epidemiology of viral haemorrhagic septicaemia virus (VHSV) in British Columbia, Canada, reveals transmission from wild to farmed fish. Dis Aquat Organ 104: 93–104.
16. Thompson TM, Batts WN, Faisal M, Bowser P, Casey JW, et al. (2011) Emergence of Viral hemorrhagic septicemia virus in the North American Great Lakes region is associated with low viral genetic diversity. Diseases of Aquatic Organisms 96: 29–43.
17. Bain MB, Cornwell ER, Hope KM, Eckerlin GE, Casey RN, et al. (2010) Distribution of an invasive aquatic pathogen (Viral Hemorrhagic Septicemia Virus) in the Great Lakes and its relationship to shipping. Plos One 5: 8.
18. Al-Hussinee L, Lord S, Stevenson RMW, Casey RN, Groocock GH, et al. (2011) Immunohistochemistry and pathology of multiple Great Lakes fish from mortality events associated with viral hemorrhagic septicemia virus type IVb. Diseases of Aquatic Organisms 93: 117–127.
19. Gagne N, MacKinnon AM, Boston L, Souter B, Cook-Versloot M, et al. (2007) Isolation of viral haemorrhagic septicaemia virus from mummichog, stickleback, striped bass and brown trout in eastern Canada. Journal of Fish Diseases 30: 213–223.
20. Mortensen HF, Heuer OE, Lorenzen N, Otte L, Olesen NJ (1999) Isolation of viral haemorrhagic septicaemia virus (VHSV) from wild marine fish species in the Baltic Sea, Kattegat, Skagerrak and the North Sea. Virus Research 63: 95–106.
21. King JA, Snow M, Smail DA, Raynard RS (2001) Distribution of viral haemorrhagic septicaemia virus in wild fish species of the North Sea, north east Atlantic Ocean and Irish Sea. Diseases of Aquatic Organisms 47: 81–86.
22. Brudeseth BE, Evensen Ø (2002) Occurrence of viral haemorrhagic septicaemia virus (VHSV) in wild marine fish species in the coastal regions of Norway. Diseases of Aquatic Organisms 52: 21–28.
23. Dixon PF, Avery S, Chambers E, Feist S, Mandhar H, et al. (2003) Four years of monitoring for viral haemorrhagic septicaemia virus in marine waters around the United Kingdom. Diseases of Aquatic Organisms 54: 175–186.
24. Skall HF, Olesen NJ, Mellergaard S (2005) Prevalence of viral haemorrhagic septicaemia virus in Danish marine fishes and its occurrence in new host species. Diseases of Aquatic Organisms 66: 145–151.
25. Mortensen HF, Heuer OE, Lorenzen N, Otte L, Olesen NJ (1999) Isolation of viral haemorrhagic septicaemia virus (VHSV) from wild marine fish species in the Baltic Sea, Kattegat, Skagerrak and the North Sea; pp. 95–106.
26. Kocan RM, Hershberger PK, Elder NE, Winton JR (2001) Epidemiology of viral hemorrhagic septicemia among juvenile pacific herring and Pacific sand lances in Puget Sound, Washington. Journal of Aquatic Animal Health 13: 77–85.
27. Meyers TR, Short S, Lipson K, Batts WN, Winton JR, et al. (1994) Association of viral hemorrhagic septicemia virus with epizootic hemorrhages of the skin in Pacific herring *Clupea harengus pallasi* from Prince William Sound and Kodiak Island, Alaska, USA. Diseases of Aquatic Organisms 19: 27–37.
28. Johansen R, Bergh Ø, Modahl I, Dahle G, Gjerset B, et al. (2013) High prevalence of viral haemorrhagic septicaemia virus (VHSV) in Norwegian spring-spawning herring. Marine Ecology Progress Series 478: 223–230.
29. Duesund H, Nylund S, Watanabe K, Ottem KF, Nylund A (2010) Characterization of a VHS virus genotype III isolated from rainbow trout (*Oncorhychus mykiss*) at a marine site on the west coast of Norway. Virology Journal 7: 1–15.
30. King JA, Snow M, Skall HF, Raynard RS (2001) Experimental susceptibility of Atlantic salmon *Salmo salar* and turbot *Scophthalmus maximus* to European freshwater and marine isolates of viral haemorrhagic septicaemia virus. Diseases of Aquatic Organisms 47: 25–31.
31. Snow M, Cunningham CO (2000) Virulence and nucleotide sequence analysis of marine viral haemorrhagic septicaemia virus following in vivo passage in rainbow trout *Onchorhynchus mykiss*. Diseases of Aquatic Organisms 42: 17–26.
32. De Kinkelin P, Castric J (1982) An experimental study of the susceptibility of atlantic salmon fry salmo-salar to viral hemorrhagic septicemia. Journal of Fish Diseases 5: 57–66.
33. Lovy J, Piesik P, Hershberger PK, Garver KA (2013) Experimental infection studies demonstrating Atlantic salmon as a host and reservoir of viral hemorrhagic septicemia virus type IVa with insights into pathology and host immunity. Veterinary Microbiology 166: 91–101.
34. Groocock GH, Frattini SA, Cornwell ER, Coffee LL, Wooster GA, et al. (2012) Experimental infection of four aquacultured species with viral hemorrhagic septicemia virus type IVb. Journal of the World Aquaculture Society 43: 459–476.
35. Schönherz AA, Lorenzen N, Einer-Jensen K (2013) Inter-species transmission of viral hemorrhagic septicemia virus (VHSV) from turbot (*Scophthalmus maximus*) to rainbow trout (*Onchorhynchus mykiss*). Journal of General Virology 94: 869–875.
36. Jonstrup SP, Kahns S, Skall HF, Boutrup TS, Olesen NJ (2013) Development and validation of a novel Taqman-based real-time RT-PCR assay suitable for demonstrating freedom from viral haemorrhagic septicaemia virus. Journal of Fish Diseases 36: 9–23.
37. Einer-Jensen K, Ahrens P, Forsberg R, Lorenzen N (2004) Evolution of the fish rhabdovirus viral haemorrhagic septicaemia virus. Journal of General Virology 85: 1167–1179.
38. Tamura K, Peterson D, Peterson N, Stecher G, Nei M, et al. (2011) MEGA5: Molecular evolutionary genetics analysis using maximum likelihood, evolutionary distance, and maximum parsimony methods. Molecular Biology and Evolution 28: 2731–2739.
39. Silva FFG, Slotte A, Johannessen A, Kennedy J, Kjesbu OS (2013) Strategies for partition between body growth and reproductive investment in migratory and stationary populations of spring-spawning Atlantic herring (*Clupea harengus* L.). Fisheries Research 138: 71–79.
40. Dixon PF, Feist S, Kehoe E, Parry L, Stone DM, et al. (1997) Isolation of viral haemorrhagic septicaemia virus from Atlantic herring *Clupea harengus* from the English Channel. Diseases of Aquatic Organisms 30: 81–89.
41. Smail DA (2000) Isolation and identification of Viral Haemorrhagic Septicaemia (VHS) viruses from cod *Gadus morhua* with the ulcus syndrome and from haddock *Melanogrammus aeglefinus* having skin haemorrhages in the North Sea. Diseases of Aquatic Organisms 41: 231–235.
42. Iida H, Mori K, Nishizawa T, Arimoto M, Muroga K (2003) Fate of viral hemorrhagic septicemia virus in Japanese flounder *Paralichthys olivaceus* challenged by immersion. Fish Pathology 38: 87–91.
43. Sandlund N, Johansen R, Fiksdal IU, Einen A-CB, Modahl I, et al. (Submitted) Susceptibility and pathology in juvenile Atlantic cod *Gadus morhua* to a marine viral haemorrhagic septicaemia virus isolated from diseased rainbow trout *Oncorhynchus mykiss*. Diseases of Aquatic Organisms.
44. Isshiki T, Nishizawa T, Kobayashi T, Nagano T, Miyazaki T (2001) An outbreak of VHSV (viral hemorrhagic septicemia virus) infection in farmed Japanese flounder *Paralichthys olivaceus* in Japan. Diseases of Aquatic Organisms 47: 87–99.
45. Nishizawa T, Savas H, Isidan H, Ustundag C, Iwamoto H, et al. (2006) Genotyping and pathogenicity of viral hemorrhagic septicemia virus from free-living turbot (*Psetta maxima*) in a Turkish coastal area of the Black Sea. Applied and Environmental Microbiology 72: 2373–2378.
46. Cornwell ER, Bellmund CA, Groocock GH, Wong PT, Hambury KL, et al. (2013) Fin and gill biopsies are effective nonlethal samples for detection of Viral hemorrhagic septicemia virus genotype IVb. Journal of Veterinary Diagnostic Investigation 25: 203–209.
47. Brudeseth BE, Skall HF, Evensen Ø (2008) Differences in virulence of marine and freshwater isolates of viral hemorrhagic septicemia virus in vivo correlate with in vitro ability to infect gill epithelial cells and macrophages of rainbow trout (*Oncorhynchus mykiss*). Journal of Virology 82: 10359–10365.
48. Oidtmann B, Joiner C, Stone D, Dodge M, Reese RA, et al. (2011) Viral load of various tissues of rainbow trout challenged with viral haemorrhagic septicaemia virus at various stages of disease. Diseases of Aquatic Organisms 93: 93–104.
49. Hershberger PK, Kocan RM, Elder NE, Meyers TR, Winton JR (1999) Epizootiology of viral hemorrhagic septicemia virus in Pacific herring from the spawn-on-kelp fishery in Prince William Sound, Alaska, USA. Diseases of Aquatic Organisms 37: 23–31.
50. Dixon PF (1999) VHSV came from the marine environment: Clues from the literature, or just red herrings? Bulletin of the European Association of Fish Pathologists 19: 60–65.
51. Knüsel R, Bergmann SM, Einer-Jensen K, Casey J, Segner H, et al. (2007) Virus isolation vs RT-PCR: which method is more successful in detecting VHSV and IHNV in fish tissue sampled under field conditions? Journal of Fish Diseases 30: 559–568.
52. Hope KM, Casey RN, Groocock GH, Getchell RG, Bowser PR, et al. (2010) Comparison of quantitative RT-PCR with cell culture to detect viral hemorrhagic septicemia virus (VHSV) IVb Infections in the Great Lakes. Journal of Aquatic Animal Health 22: 50–61.
53. Schönherz AA, Hansen MHH, Jorgensen HBH, Berg P, Lorenzen N, et al. (2012) Oral transmission as a route of infection for viral haemorrhagic septicaemia virus in rainbow trout, *Oncorhynchus mykiss* (Walbaum). Journal of Fish Diseases 35: 395–406.
54. Kurath G, Winton J (2011) Complex dynamics at the interface between wild and domestic viruses of finfish. Current Opinion in Virology 1: 73–80.
55. Munro ES, Allan CET, Matejusova I, Murray AG, Raynard RS (2013) An outbreak of viral haemorrhagic septicaemia (vhs) in wrasse cohabiting with Atlantic salmon in the Shetland Isles, Scotland. Report on the 17th Annual Meeting of the National Reference Laboratories for Fish Diseases Copenhagen, Denmark, May 29–30, 2013: 22–23.
56. Bergh Ø (2007) The dual myths of the healthy wild fish and the unhealthy farmed fish. Diseases of Aquatic Organisms 75: 159–164.

17

Quantifying Fish Assemblages in Large, Offshore Marine Protected Areas

Nicole A. Hill[1]*, Neville Barrett[1], Emma Lawrence[2], Justin Hulls[1], Jeffrey M. Dambacher[3], Scott Nichol[4], Alan Williams[5], Keith R. Hayes[3]

1 Institute for Marine and Antarctic Studies, University of Tasmania, Hobart, Tasmania, Australia, 2 Digital Productivity Flagship, Commonwealth Scientific and industrial Research Organisation (CSIRO), Brisbane, QLD, Australia, 3 Digital Productivity Flagship, Commonwealth Scientific and industrial Research Organisation (CSIRO), Hobart, Tasmania, Australia, 4 Geoscience Australia, Canberra, ACT, Australia, 5 Oceans and Atmosphere Flagship, Commonwealth Scientific and industrial Research Organisation (CSIRO), Hobart, Tasmania, Australia

Abstract

As the number of marine protected areas (MPAs) increases globally, so does the need to assess if MPAs are meeting their management goals. Integral to this assessment is usually a long-term biological monitoring program, which can be difficult to develop for large and remote areas that have little available fine-scale habitat and biological data. This is the situation for many MPAs within the newly declared Australian Commonwealth Marine Reserve (CMR) network which covers approximately 3.1 million km^2 of continental shelf, slope, and abyssal habitat, much of which is remote and difficult to access. A detailed inventory of the species, types of assemblages present and their spatial distribution within individual MPAs is required prior to developing monitoring programs to measure the impact of management strategies. Here we use a spatially-balanced survey design and non-extractive baited video observations to quantitatively document the fish assemblages within the continental shelf area (a multiple use zone, IUCN VI) of the Flinders Marine Reserve, within the Southeast marine region. We identified distinct demersal fish assemblages, quantified assemblage relationships with environmental gradients (primarily depth and habitat type), and described their spatial distribution across a variety of reef and sediment habitats. Baited videos recorded a range of species from multiple trophic levels, including species of commercial and recreational interest. The majority of species, whilst found commonly along the southern or south-eastern coasts of Australia, are endemic to Australia, highlighting the global significance of this region. Species richness was greater on habitats containing some reef and declined with increasing depth. The trophic breath of species in assemblages was also greater in shallow waters. We discuss the utility of our approach for establishing inventories when little prior knowledge is available and how such an approach may inform future monitoring efforts within the CMR network.

Editor: David Mark Bailey, University of Glasgow, United Kingdom

Funding: This work was undertaken for the Marine Biodiversity Hub, a collaborative partnership supported through funding from the Australian Government's National Environmental Research Program (NERP). NERP Marine Biodiversity Hub partners include the Institute for Marine and Antarctic Studies, University of Tasmania; CSIRO, Geoscience Australia, Australian Institute of Marine Science, Museum Victoria, Charles Darwin University and the University of Western Australia. The funders had no role in study design, data collection and analysis, decision to publish, or preparation of the manuscript.

Competing Interests: The authors have declared that no competing interests exist.

* Email: Nicole.Hill@utas.edu.au

Introduction

Marine ecosystems face pressures from a range of sources, including pollution, fishing, habitat destruction, invasive species and a changing climate [1]. These pressures threaten the essential processes and resources that marine ecosystems provide, such as climate regulation, nutrient cycling and the provision of food, as well as threatening their intrinsic natural and cultural value [2]. As a consequence, management strategies have been adopted to safeguard marine ecosystems and their associated biodiversity. These strategies include the establishment of Marine Protected Areas (MPAs) which offer varying levels of protection according to their zoning or IUCN designation [3]. Countries such as the United States, Australia, South Africa, New Zealand, Kenya and the Philippines have a long history of establishing MPAs [4,5]. However, the absence of a truly global system of MPAs prompted the Convention of Biological Diversity [6] and the Jakarta Mandate [7] to provide a framework and renewed international impetus for protecting marine biodiversity. This framework has resulted in a significant increase in recent years in the declaration of MPAs worldwide, with many forming part of large networks or systems of MPAs [8].

The increasing global commitment to the sustainability of healthy marine ecosystems is encouraging, but the declaration alone of MPAs does not ensure the conservation of biodiversity or ecosystems [4]. Management authorities must: clearly articulate the objectives of individual MPAs and MPA networks, both of which may contain multiple IUCN zones; formulate and enforce management plans [9] and periodically revisit and evaluate their management plans, which in turn requires that they establish a long-term biological monitoring program [4]. Developing monitoring programs for individual MPAs within networks of MPAs is often a difficult task. There is generally a mismatch between the spatial scales of information used to design MPA networks and those required to manage individual MPAs. Inventories of the types and distribution of habitats, communities and species within individual MPAs are often required before monitoring programs can be designed. In addition, for offshore or difficult to access MPAs, there are significant logistical and statistical challenges to working in remote environments.

Australia, a signatory to the CBD, provides a good example of the challenges involved in ultimately developing long-term monitoring programs in large MPA networks. In 2012, the Australian government announced an expansion of its MPA network to include 33 new MPAs. This makes a total of 60 MPAs covering approximately 3.1 million square kilometres, divided between six planning bioregions [10]. Although called the "Commonwealth Marine Reserve" or "CMR" network, areas with IUCN zones ranging from 1a (Sanctuary Zone) to VI (General Use Zone) are distributed throughout the network. From this point forward we adopt the above terminology and refer to MPAs within this network as reserves regardless of their IUCN zoning. All of the reserves in the CMR are however, located in offshore waters, most are large and encompass a broad depth range, and many are remote.

The Australian CMR network aims to meet the principles of being Comprehensive, Adequate and Representative (the CAR principles) [11] and utilised the national Integrated Marine and Coastal Regionalisation of Australia (IMCRA; [12]). The IMCRA bio-regionalisation is based on breaks in the distribution of demersal fish communities (together with some physical datasets) and is the best available continental-scale regionalisation of Australian marine fauna. Hence it was the most appropriate mechanism for delineating management regions and informing the placement of reserves. When designing biological monitoring programs for individual reserves however, much finer scale information is needed on the spatial distribution of important habitats (such as reefs) and the biological communities they contain. For many reserves in the new network this level of information is simply unavailable. It is therefore necessary to provide a robust inventory of the biological communities and key species within the reserves, delineate their spatial distribution and quantify the relationship between these distributions and key environmental drivers to enable the development of monitoring programs that will ultimately evaluate the effectiveness of each reserve.

Inventory, and ultimately monitoring, of large, deep reserves is challenging and relatively expensive. It requires vessels large enough to conduct surveys safely in offshore waters, together with a suite of non-extractive survey methods that can be deployed at depth, and sampling designs that are efficient and flexible enough to accommodate multiple (and potentially changing) objectives. A suite of non-extractive survey methods are currently available including multibeam sonar (MBS) for characterising the seafloor and associated habitat types [13], and image-based methods, such as towed camera systems, cameras attached to autonomous or remotely operated vehicles and baited cameras, for observing benthic fauna. Here we trialled MBS and Baited Remote Underwater Video (BRUVs), deployed within a spatially balanced design, for inventorying demersal fish in the multiple use zone (IUCN VI) of one reserve in the Australian network, the Flinders CMR. We use BRUVs because of their demonstrated ability to survey fish in deep waters [14] and because they have been extensively used in Australian inshore environments for monitoring and other ecological applications [15–17]. We trialled a spatially-balanced (i.e. evenly spread out in space) design known as the Generalised Random- Tessellation Stratified (GRTS) design [18] because it is flexible and provides a good way to obtain a representative sample that respects the spatial distribution of habitats and communities in the target population [19].

In this study we: identify patterns in demersal fish assemblages across the continental shelf (a multiple-use zone, IUCN VI) of the Flinders CMR; quantify these patterns in relation to environmental gradients; examine the spatial distribution of assemblages across the reserve; and describe how our approach and the information gained may be useful to the development of a long-term biological monitoring program. We focus on fish communities because: broad-scale distributional records of demersal fish were instrumental in the marine bioregionalisation that underpins the design of the reserve network [20]; many species are of recreational and commercial value and are therefore likely to respond to current and future management actions within the reserve; and demersal fish are key components of benthic ecosystems.

Methods

Survey Area

The Flinders Commonwealth Marine Reserve (CMR) was established in 2007 and lies about 25 km offshore of northern Tasmania (Fig. 1). The reserve is 26,975 km^2 in size and extends as a west-east corridor from 35 m to > 3,000 m water depths. It consists of two zones: a multiple use zone (IUCN Category VI) that covers the majority of the continental shelf and slope; and a marine national park zone (IUCN Category II) that extends from the continental slope to the edge of the reserve at the 200 nautical mile limit of Australia's Exclusive Economic Zone (Fig. 1 inset). Our study area is the continental shelf, part of the multiple use zone (IUCN Category VI), where activities that impact on benthic habitats are prohibited, including demersal trawling, Danish seining and scallop dredging, or subject to permit requirements, for example mining activities [21]. Benthic habitats on the Flinders CMR shelf consist of sediment plains with patches of low profile and sand-inundated reefs, and steep rocky outcrops where canyon heads incise the shelf break (Lawrence, unpublished data). Within this environment, reefs and rocky outcrops have been identified as features likely to contain enhanced benthic diversity [21].

The Flinders CMR shelf straddles two provincial biogeographical regions: the Southeast Shelf Transition in the north and the Tasmanian Shelf Province in the South [12]. This region is characterised by variable but generally high exposure and strong tidal currents, especially in shallow areas between Flinders Island and the Tasmanian mainland [22,23]. The region is also influenced by southwards incursions of the East Australian Current (EAC) which brings warmer waters on to the shelf in summer [24]. The flora of the region is moderately rich and contains species common in cold temperate waters as well as low abundances of species common to warmer temperate waters [20].

Sampling design and methods

In contrast to the Flinders CMR continental slope, which has been comprehensively mapped using multibeam sonar (from which seafloor habitats have been inferred), very little spatially explicit habitat or biological information is available for the shelf. This represents a challenge for designing targeted sampling programs. Interpolated bathymetry for the Flinders shelf exists, but is at a gridded resolution of 250 m [25] (Fig. 1) which does not allow the identification of fine scale features of interest on the shelf, such as reefs. As a consequence of our limited *a priori* knowledge of the area, the inventory and description of fish assemblages

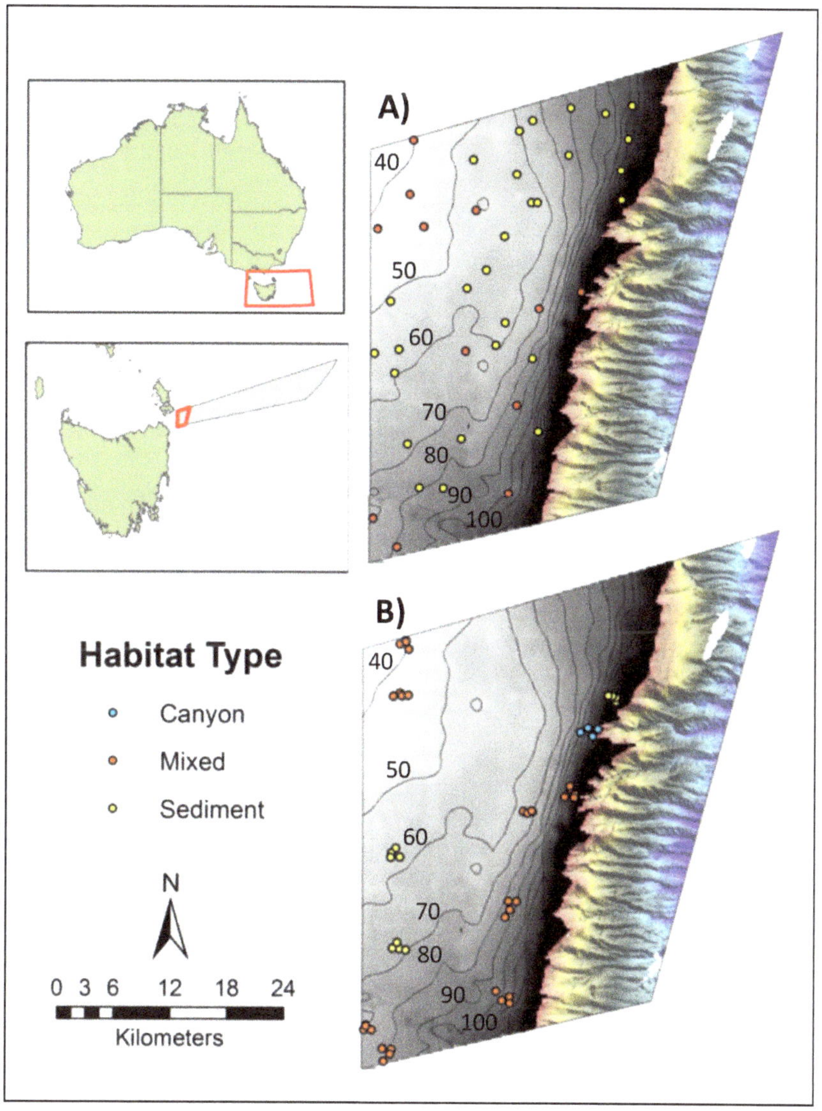

Figure 1. Location of the Flinders Commonwealth Marine Reserve (CMR) in Tasmania, Australia. Panels A and B show the CMR's Multiple Use Zone (IUCN VI) where survey work was conducted. The grey zone is the continental shelf (less than 200 m depth) where there was little pre-existing mapping data. Coarse bathymetry data (gridded at 250 m horizontal resolution) sourced from Geoscience Australia is shown with 10 m contour intervals overlain. The coloured area to the right shows the relatively steep and highly incised upper continental slope that had been mapped previously with multibeam sonar extending from 200 to 1500 m. A) Location of the 40 sites surveyed for habitat type in phase one of the sampling program. B) Location of the clustered sites where Baited Remote Underwater Videos (BRUVs) were deployed in phase two. Sites are coloured according to the broad habitat type: sediment (yellow); mixed, low-profile reef and sediments (red); canyon head (blue) recorded during phase one of sampling.

presented here forms one component of a multi-objective and multi-phase study within the Flinders CMR. Additional objectives of the broader study include examining the distribution and estimating the area of the different habitat types within the reserve, as well as quantifying benthic invertebrate communities using imagery captured by towed video and AUV, and these results will be reported elsewhere. To accommodate the objectives of the different components of the study, we used the probabilistic design, Generalised Random -Tessellation Stratified sampling technique (GRTS; [18]). GRTS ensures sampling sites are well spread out across the survey area (spatially-balanced), a desirable property for spatial sampling that enhances estimation efficiency and representativeness [18]. Other common sampling designs include simple or stratified random sampling or systematic sampling. However, simple random sampling often clumps sites [19] and stratified random sampling (based on habitat in our case) was not possible without more detailed knowledge of the study area. Systematic sampling prevents clumping but an accurate design-based variance estimator does not exist; a factor that was important for other aspects of the broader study [18,26]. provide a good discussion of the relative merits of various spatial sampling strategies. Another property of GRTS is that it produces an ordered list of sample sites and any set of consecutive sampling sites also maintain spatial balance, offering flexibility to adaptively change sample sizes in the field, if for example, sampling takes less time than expected. To date GRTS has primarily been used in natural resource assessment and monitoring in the United States e.g. [27].

Because of our lack of knowledge on the distribution of habitat types, we implemented a two-phase sampling program. In phase one, the distribution of habitat types across the reserve was quantified by characterising the first forty, 200 m square, sites from a master list of GRTS sites (essentially an ordered list of all possible GRTS sites in the study area. See [28] for a detailed discussion on the use of master lists) as either 'soft' sediments or 'mixed', low profile patchy reef using MBS and footage obtained from a drop camera (Fig. 1; Lawrence, unpublished data). In phase two of the sampling program, the subject of this paper, we sampled the fish assemblages using BRUVs. Ideally we would have revisited all forty of the phase one GRTS sites. The large size of the study area and the soak time required for each BRUV deployment however, meant that it was not possible to do so in the ship time available. Instead, we sampled clusters of sites surrounding a subset of the phase one GRTS- sites. In phase two, we sampled mixed reef habitats more intensively than sediment habitats because shelf reef systems within the reserve are recognised as an important biodiversity feature and a *priori* we expected that reefs would harbour a greater diversity of assemblage types than sediments. We utilised the fact that consecutive GRTS sites maintain spatial balance and selected (from the ordered list of phase one sites) the first eight mixed habitat sites and the first three sediment sites of the forty phase one sites as the basis of the phase two clusters (Fig. S1. shows the site numbers of the forty phase ones sites and lists those used as the basis of phase two clusters.) This amounts to selecting phase one sites for the BRUV sampling that are spatially balanced *within* each of the phase one habitat types, mixed reef and sediments, and presumably therefore also spread across the environmental gradients in the region. In addition, a site was added near the head of a canyon, to ensure representation of another identified feature of the reserve which occupies a relatively small proportion of space and is therefore less likely to be sampled in a probabilistic sampling design (Fig. 1B). Clusters of sites within 1 km of the subset of phase one sites were selected again using the GRTS master list, resulting in five sites per cluster (Fig. 1B).

At each site, one BRUV was deployed using systems and deployment conditions widely used in Australia [29]. Each stereo-BRUV consisted of a frame that houses a stereo camera pair, a bait bag attached to an arm within field of view of the cameras, a diode for synchronising imagery between camera pairs. In water deeper than approximately 70 m, a light with a blue filter was also attached [15]. Stereo- BRUVs were baited with 1 kg of crushed pilchards and deployed for 60 min (soak time). Appropriate ethics (University of Tasmania Animal Ethics Permit: A12514) and fieldwork (Australian Government Director of National Parks Approval of Research Activities in the Southeast Commonwealth Marine Reserve Network. Ref No 07/10622) approvals were obtained for this work.

Analysis

Imagery collected using stereo-BRUVs was scored using standard metrics including scoring the maximum number of fish occurring in any one frame for each species (MaxN) [15,30]. Scoring was completed with Event Measure software [31].

To determine the types of fish assemblages present at sites sampled on the Flinders CMR shelf, fuzzy clustering was performed on multivariate fish composition data using the 'cluster' package [32] in R [33]. Multivariate data were square root transformed to reduce the influence of the most abundant species and a Bray Curtis dissimilarity matrix generated. The number of fuzzy clusters was set to six, following preliminary analysis using several different clustering methods, and the membership exponent (which controls the 'fuzziness' of clusters) set to 1.3 after examining silhouette profiles of group membership.

A constrained ordination (Canonical Analysis of Principle Co-ordinates; CAP) [34] was used to test if the assemblage groups determined by fuzzy clustering were significantly different from each other (using a permutational test). In addition, jack-knife sampling was used to examine the overall and individual classification accuracy of the six groups. To enable the ecological interpretation of assemblages, the trophic level and habitat preference of species was defined. Species were assigned one of seven habitat preference categories: 1) pelagic; 2) wide-ranging demersal; 3) sediment associated; 4) reef and sediment associated; 5) reef and seagrass associated; 6) reef associated; and 7) demersal generalists (found in more than two habitats) based on species attributes described in [35,36]. The trophic level of species was extracted from FishBase [37]. Differences in the trophic level of species comprising the six groups was assessed with a one factor Analysis of Variance (ANOVA) after transforming data to satisfy the assumptions of ANOVA. To understand which species were the key contributors to each of the six assemblage groups, a SIMPER analysis was performed. SIMPER decomposes the Bray-Curtis dissimilarity matrix between all pairs within a group into the percentage contributions from each species [34].

Correlations between fish assemblages and their environment were examined against three easily quantifiable environmental variables that, from the literature, may influence benthic assemblages in our system; latitude, depth and the substratum type. At each cluster of five sites, only the central site had been surveyed with MBS and drop camera in phase one of the sampling program. Therefore, substratum type was derived from the BRUV field of view and characterised as either: 'sediment' if no hard substratum or no hard substratum-associated organisms (i.e. sponges etc.) were present; 'mixed' if some hard substratum was visible or if some hard substratum-associated organisms were present; and 'reef' if the majority of the field of view contained hard substratum or hard substratum-associated organisms. Latitude was taken from the recorded location of BRUV drops and depth was derived using the 250 m resolution bathymetry grid available from Geoscience Australia [25]. Relationships between assemblage groups and environmental variables were inferred by examining correlations with CAP axes and overlaying vectors of the environmental variables on CAP plots. Ordination and SIMPER analysis were performed in PRIMER + with PERMANOVA [34].

The influence of environmental variables in determining patterns in species richness (the number of species present) was examined using Generalised Linear Mixed Models (GLMMs). Observations were modelled using a Poisson distribution with a log link function and sampling cluster was included as a normally distributed random factor. Mixed models were fit in R [33] with penalised quasi-likelihood, using the glmmPQL function in the package MASS [38].

Results

Distribution of habitats across the CMR

The results of phase one sampling indicated that the majority (70%) of the 40 sites on the shelf were sedimentary habitats, and the remainder 'mixed' reef and sediment habitat. Mixed reef habitat occurred in the north-west, south-west, and near the edge of the shelf of the study area (Fig. 1A; Lawrence, unpublished data).

Table 1. List of species recorded in Baited Remote Underwater Video deployments across the continental shelf in the Flinders Commonwealth Reserve.

			Across CMR			Mean (SE) MaxN by group					
Family	Species name	Common name	Total MaxN	Max MaxN in any drop	Prevalence (% drops)	Group 1	Group 2	Group 3	Group 4	Group 5	Group 6
Blenniidae, Gobiidae, Tripterygiidae	*Blenniidae, Gobiidae, Tripterygiidae*	Blennies, Gobies, Triplefins	7	3	10	0.38 (0.18)	0.4 (0.31)	0	0	0	0
Callorhinchidae	*Callorhinchus milii*	Elephant shark	1	1	2	0	0.1 (0.1)	0	0	0	0
	Nemadactylus douglasii	Grey morwong	7	2	10	0	0.2 (0.13)	0.33 (0.24)	0.33 (0.33)	0	0
	Nemadactylus macropterus	Jackass morwong	379	43	59	0	10.7 (2.04)	23.22 (4.78)	0	5.25 (1.23)	0
Cyttidae	*Cyttus australis*	Silver dory	6	1	12	0	0.1 (0.1)	0	0	0.42 (0.15)	0
Diodontidae	*Diodon nicthemerus*	Globe fish	1	1	2	0	0	0	0	0.08 (0.08)	0
Gempylidae	*Thyrsites atun*	Barracouta	14	3	20	0.38 (0.37)	0.1 (0.1)	0.11 (0.11)	0	0.75 (0.22)	0
Gerreidae	*Parequula melbournensis*	Silverbelly	79	10	31	0	0.5 (0.27)	0	0	5.83 (0.6)	0.67 (0.67)
Heterodontidae	*Heterodontus portusjacksoni*	Port Jackson shark	4	1	8	0	0.3 (0.15)	0	0	0.08 (0.08)	0
Labridae	*Ophthalmolepis lineolatus*	Southern maori wrasse	1	1	2	0	0	0.11 (0.11)	0	0	0
	Pseudolabrus rubicundus	Rosy wrasse	72	13	25	0	1.4 (0.88)	0.67 (0.67)	0	4.33 (1.38)	0
Lamnidae	*Isurus oxyrinchus*	Shortfin Mako shark	1	1	2	0.12 (0.12)	0	0	0	0	0
Latridae	*Latris lineata*	Striped trumpeter	95	27	22	0	0.2 (0.13)	10.11 (3.47)	0.33 (0.33)	0	0
Loliginidae	*Sepioteuthis australis*	Southern calamari	21	3	29	0	0.2 (0.13)	0	0	1.5 (0.23)	0.17 (0.17)
Monacanthidae	*Acanthaluteres vittiger*	Toothbrush leatherjacket	15	3	18	0	0	0	0	1.25 (0.28)	0
	Meuschenia freycineti	Six-spine leatherjacket	3	1	6	0	0.2 (0.13)	0	0	0.08 (0.08)	0
	Meuschenia scaber	Velvet leatherjacket	97	11	53	0	3.3 (0.87)	0.78 (0.43)	0	4.25 (0.97)	1 (0.63)
	Meuschenia venusta	Stars and stripes leatherjacket	1	1	2	0	0.1 (0.1)	0	0	0	0
	Thamnaconus degeni	Degen's leatherjacket	471	89	41	0	1 (0.42)	0.11 (0.11)	0	36.92 (7.37)	2.83 (1.9)
Moridae	*Pseudophycis barbata*	Bearded cod	8	2	12	0	0.1 (0.1)	0.78 (0.28)	0	0	0
Mullidae	*Upeneichthys vlamingii*	Southern goatfish	16	8	16	0	0.3 (0.21)	0.11 (0.11)	0	1 (0.65)	0
Neosebastidae	*Neosebastes scorpaenoides*	Common gurnard perch	44	4	41	0	1.1 (0.31)	0	0	1.5 (0.42)	2.5 (0.5)
Ommastrephidae	*Ommastrephidae sp.*	Squid	9	2	12	0.5 (0.27)	0	0	0.83 (0.4)	0	0
Ostraciidae	*Aracana aurita*	Shaw's cowfish	7	2	12	0	0.1 (0.1)	0	0	0.5 (0.19)	0
Ostraciidae	*Aracana ornata*	Ornate cowfish	1	1	2	0	0.1 (0.1)	0	0	0	0
Palinuridae	*Jasus edwardsii*	Southern rocklobster	2	1	4	0	0	0.22 (0.15)	0	0	0
Paraulopidae	*Paraulopus nigripinnis*	Blacktip cucmberfish	47	12	22	2 (0.65)	0.1 (0.1)	0	5 (2.16)	0	0
Pinguipedidae	*Parapercis allporti*	Barred Grubfish	36	5	35	2.12 (0.52)	0.8 (0.25)	0.44 (0.44)	0.5 (0.34)	0	0.67 (0.42)
Platycephalidae	*Platycephalus bassensis*	Sand flathead	33	6	24	0.12 (0.12)	0	0	0.17 (0.17)	0.83 (0.42)	3.5 (0.96)
	Platycephalus richardsoni	Tiger flathead	21	4	25	0	0.2 (0.13)	0	1.67 (0.56)	0.17 (0.17)	1.17 (0.31)
Pristiophoridae	*Pristiophorus cirratus*	Longnose sawshark	2	1	4	0	0	0	0	0.08 (0.08)	0.17 (0.17)
Rajidae	*Dentiraja lemprieri*	Thornback skate	2	1	4	0	0	0	0	0.17 (0.11)	0

Table 1. Cont.

Family	Species name	Common name	Across CMR			Mean (SE) MaxN by group					
			Total MaxN	Max MaxN in any drop	Prevalence (% drops)	Group 1	Group 2	Group 3	Group 4	Group 5	Group 6
	Spiniraja whitleyi	Melbourne skate	9	1	18	0	0.3 (0.15)	0	0.33 (0.21)	0.33 (0.14)	0
Scyliorhinidae	*Cephaloscyllium laticeps*	Draughtboard shark	63	4	57	0.12 (0.12)	1.7 (0.5)	0.44 (0.34)	1.33 (0.49)	2.33 (0.31)	0.83 (0.48)
Sebastidae	*Helicolenus percoides*	Red gurnard perch	65	12	31	0.25 (0.25)	2.5 (0.81)	4.22 (1.28)	0	0	0
Serranidae	*Caesioperca lepidoptera/rasor*	Butterfly and Barber perch	140	60	14	0	6.1 (5.99)	2.11 (1.42)	0	5 (2.79)	0
Squalidae	*Squalus acanthias*	Spiny dogfish	1	1	2	0	0	0	0.17 (0.17)	0	0
	Squalus megalops	Shortnose spurdog	15	3	14	0.75 (0.49)	0	0	1.5 (0.43)	0	0
Triakidae	*Mustelus antarcticus*	Gummy shark	18	2	33	0.25 (0.16)	0	0	1.17 (0.17)	0.25 (0.13)	1
Triglidae	*Lepidoperca pulchella*	Tasmanian perch	11	5	10	0	0.1 (0.1)	1.11 (0.56)	0	0	0
	Lepidotrigla sp	Gurnards	3	1	6	0	0.1 (0.1)	0.11 (0.11)	0.17 (0.17)	0	0
Urolophidae	*Urolophus cruciatus*	Banded stingaree	2	1	4	0	0.1 (0.1)	0	0	0	0.17 (0.17)
	Urolophus paucimaculatus	Sparsely spotted stingaree	2	1	4	0	0.2 (0.13)	0	0	0	0
	Urolophidae sp.	Stingarees	4	1	8	0.12 (0.12)	0.2 (0.13)	0	0	0	0.17 (0.17)

The Total number of individuals recorded at Max N across all deployments, the greatest MaxN recorded in any single deployment and the prevalence (the percentage of deployments where a species was recorded) are presented as well as the mean (and standard error) MaxN for each assemblage group.

Table 2. Comparison between the number of sites assigned to each of the six groups using fuzzy clustering and Canonical Analysis of Principle coordinates (CAP).

		CAP Classified Group						Total	Correct
		1	2	3	4	5	6		(%)
Fuzzy Group	1	8	0	0	0	0	0	8	100
	2	0	9	1	0	0	0	10	90
	3	0	1	8	0	0	0	9	89
	4	0	0	0	5	0	1	6	83
	5	0	0	0	0	12	0	12	100
	6	0	0	0	0	0	6	6	100

Jack knife validation in the CAP procedure was used to assess overall and group-wise accuracy of the six groups determined by fuzzy clustering. Classification accuracy of all groups is good, indicating that the groups determined by the fuzzy clustering are robust.

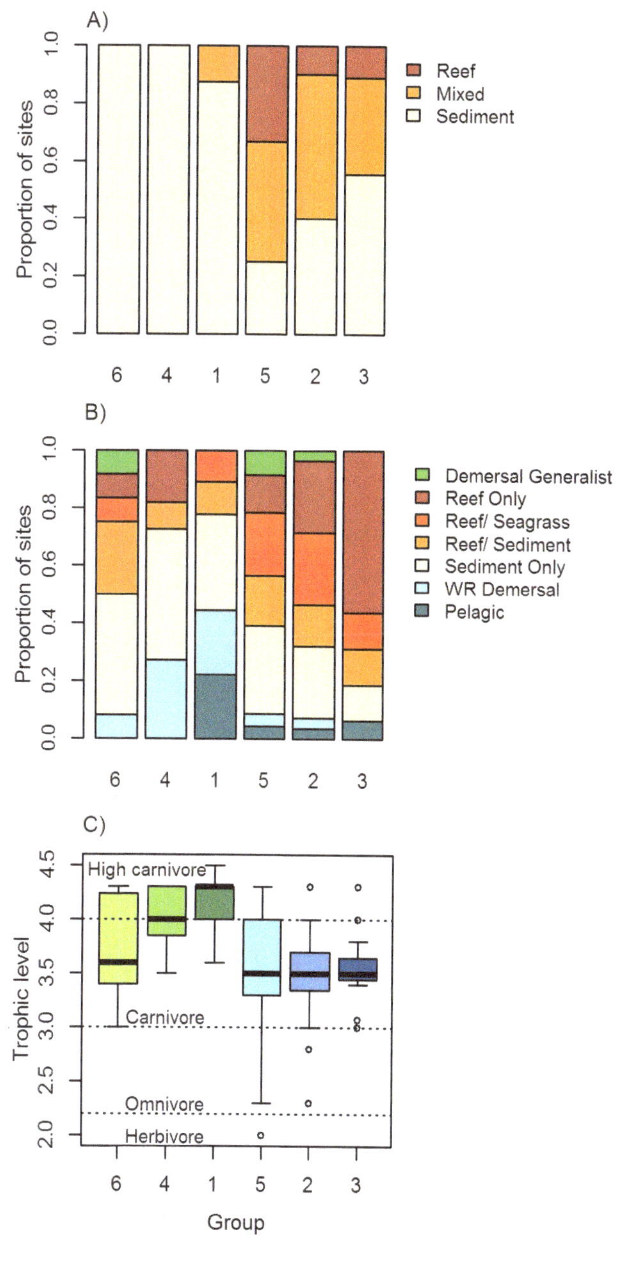

Figure 2. Characteristics of the six assemblage groups identified using fuzzy-clustering. A) Proportion of sites within each group classified as sediment, mixed or reef habitat based on the BRUVs footage; B) the habitat preference of species within each assemblage group; C) boxplot of the trophic level of species within each assemblage group; the box represents the 1st and 3rd quartiles, and circles denote potential outliers. In all panels, groups are ordered based on dominant habitat contained within groups (sediment versus mixed reef/reef) and from shallow to deep within each broad habitat type (based on spatial distribution of groups presented in Fig. 5) In plot C) sediment- associated groups are coloured green and reef-associated groups, blue.

Fish assemblages present in the CMR

Sixty BRUV deployments were completed in phase two, of which 51 yielded video footage of sufficient quality for scoring fish assemblage composition. Overall, 45 species were observed, with a total of 1,837 individuals recorded (at Max N) (Table 1). Of these species, 66% are endemic to Australia. Leatherjackets (Monacanthidae) were the most numerically abundant group, and

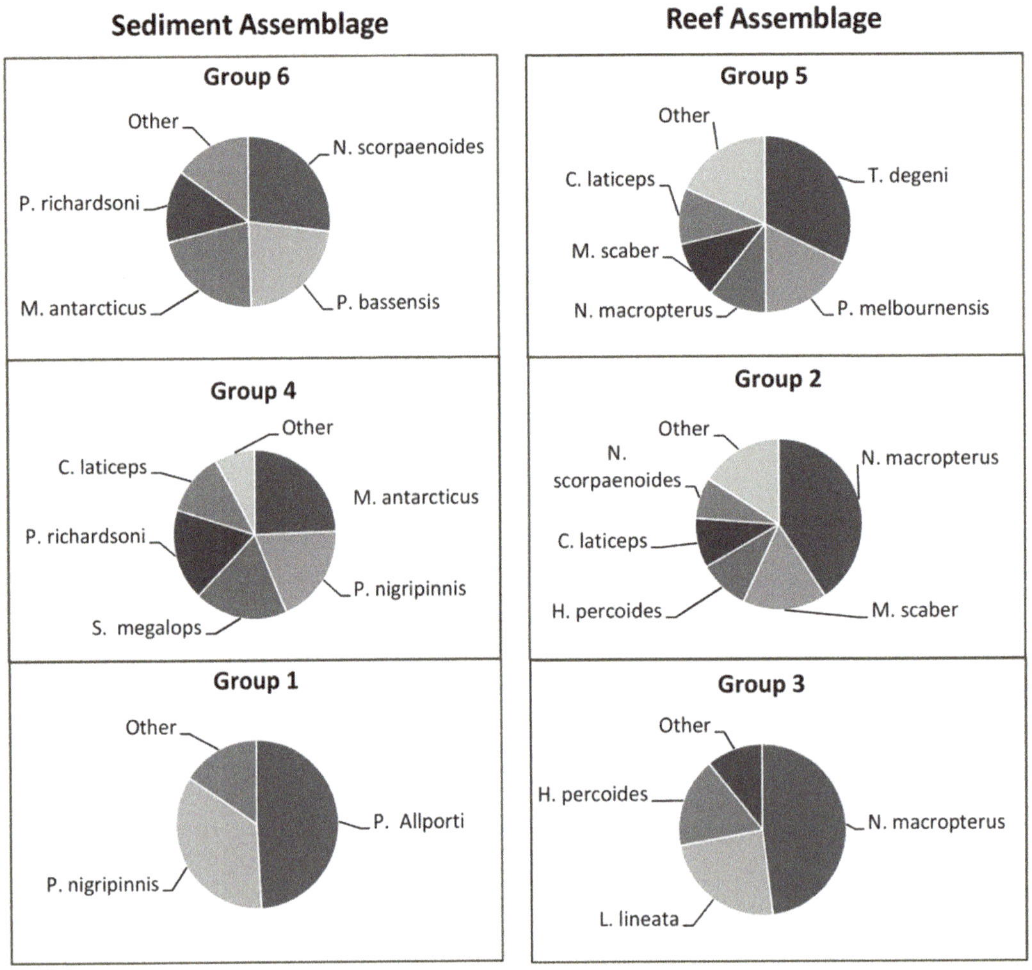

Figure 3. Species contributing up to 80% towards the similarity of the predominantly sediment- and predominantly reef-associated groups identified using fuzzy clustering.

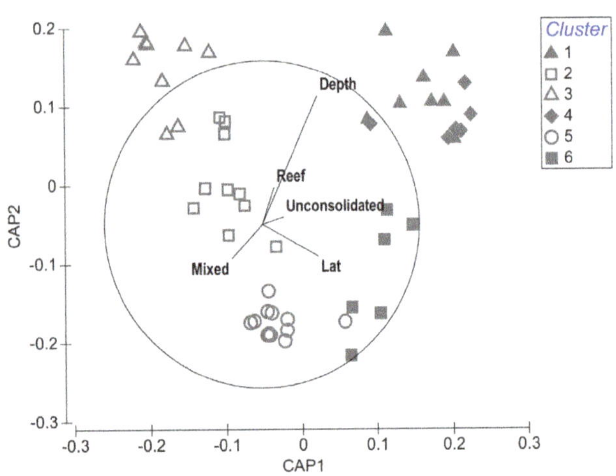

Figure 4. Canonical Analysis of Principle coordinates (CAP) plot discriminating sites based on fuzzy clustering. Each site was assigned to the cluster for which it had the highest probability of membership. Sites are shown as symbols: filled grey symbols represent sand-associated assemblages and open symbols represent reef-associated assemblages. Vectors of all environmental variables are overlaid and are proportional to their correlation with either CAP axis one or two.

comprised mostly *Thamnaconus degeni* (Degan's leatherjacket) and *Meuschenia scaber* (velvet leatherjacket) with, respectively, a total of 471 and 97 individuals recorded and MaxN of 89 and 11. *Nemadactylus macropterus* (jackass morwong), and *Caesioperca Lepidoptera and C. razor* (butterfly and barber perch) were the next most abundant species (379 and 140 individuals recorded and MaxN of 89 and 60). Other abundant species included *Latris lineata* (striped trumpeter), *Parequula melbournensis* (silverbelly), *Pseudolabrus rubicundus* (rosy wrasse), *Helicolenus percoides* (reef ocean perch), and *Cephaloscyllium laticeps* (draughtboard shark). The five most prevalent species were *N. macropertus* (59% of sites), *C. laticeps* (57% of sites), *M. scaber* (53% of sites), *Neosebastes scorpaenoides* (common gurnard perch; 41% of sites) and *T. degeni* (41% of sites).

The structure of fish assemblages surveyed was adequately represented by six groups. The overall classification accuracy of the six assemblage groups was high (94%) and a majority of sites (82%) were also assigned to groups with a high degree of confidence (greater than 0.85 probability of membership). Groups four and three were the least certain, but still had a high classification accuracy of 83% and 89%, respectively (Table 2). In addition, these six groups were statistically meaningful with significantly different multivariate community composition (P = 0.001).

Table 3. Results of Poisson GLMM for species richness recorded in BRUV deployments.

Factor	Co-efficient	Std Error	df	t-value	p-value
Depth	−0.277	0.067	36	−4.1489	0.0002
Mixed Habitat	0.330	0.092	36	3.575	0.0010
Reef Habitat	0.382	0.121	36	3.169	0.0031

Predictor variables have been scaled and centered and co-efficients are on the scale of the link function (log). Spatial clusters introduced as part of the sampling design were included as a random effect and the effects of habitat type (sediment, mixed or reef) are presented relative to sediment habitat.

The six assemblage groups were broadly aligned with habitats observed in the BRUV footage. Sites belonging to three groups (groups 1, 4, 6) contained exclusively or predominately sediment habitat (Fig. 2A) and contained greater numbers of species with a preference for sediment habitats. However, species with a strong affinity for reef were also found in the predominantly sediment-dominated habitat group, with groups 4 and 6 containing disproportionately more reef species than suggested by the available habitat (Fig. 2B). Sediment- associated assemblages (particularly groups 1 and 4) also contained a relatively high proportion of wide-ranging demersal species. Groups 2, 3, and 5 contained moderate to high proportions of mixed reef and reef habitat with the highest proportion of reef-associated species observed in group 3 (Fig. 2A, B).

Species representing a range of trophic levels were recorded in BRUV footage. Carnivores were the most numerous species, as expected given the use of baits. Even so, the average trophic level of species differed between the six assemblage groups (df = 5, F = 5.46, P<0.001; Fig. 2C). Groups 1 and 4 (sediment groups) contained species with significantly higher trophic level (higher carnivores) than groups 2 and 3 (mixed-reef) which contained primarily carnivores. The remaining two groups (5 and 6) contained species from a relatively wider range of trophic levels (Fig. 2C) with group 5 represented by the largest range in values and the only herbivore observed.

The characteristic species of each assemblage group (making the highest contributions to within-group similarity) re-enforced the broad habitat-related patterns described above (Fig. 3). Sediment-associated species such as flathead (*Platycephalus bassensis, Neoplatycephalus richardsoni*) and sharks (*Mustelus antarcticus, Squalus megalops*) as well as the common gurnard perch (*Neosebastes scorpaenoides*) and cucumber fish (*Paraulopus nigripinnis*) were characteristic of groups 4 and 6. Group 1 also consisted of sediment-associated fish (eastern- barred grubfish - *Parapercis allporti*- as well as *P. nigripinnis*), but was differentiated from groups 4 and 6 by relatively low abundances and very few species. Of the predominantly mixed-reef associated groups, groups 2 and 3 were both dominated by jackass morwong (*Nemadactylus macropterus*). Group 2 was also composed of a variety of others species including the velvet leatherjacket (*Meuschenia scaber*), reef ocean perch (*Helicolenus percoides*) and draughtboard sharks (*Cephaloscyllium laticeps*). Group 3 was further characterised by relatively high abundances of striped trumpeter (*Latris lineata*) as well as reef ocean perch. Group 5 was differentiated from other groups by large numbers of leatherjackets (primarily *Thamnaconus degeni*), but was also characterised by silverbelly (*Parequula melbournensis)* and jackass morwong.

Relationships with environmental variables

Environmental factors correlated well with the fish assemblage groupings (Fig. 4). Depth delineated groups most strongly (a correlation with CAP axis 2 of 0.78) and groups 1, 4 and 3 were generally found at greater depth. Substratum type also influenced assemblages, with sediment-associated groups (1, 4 and 6) falling on the right hand-side of the CAP plot, confirming the patterns seen in the composition of species above. Latitude was also moderately correlated with axis one (0.35) and appears to separate the mixed/reef associated groups 2, 3 and 5.

Species richness was also correlated with environmental factors. Richness decreased with depth and mixed and reef substratum also supported more species than sediments (Table 3). Latitude however, was not correlated with species richness (Table 3). This resulted in spatial patterns as depicted in Fig. 5, where assemblages are generally richer towards the shallow, western side of the reserve.

Spatial distribution of fish assemblages across CMR

Fish assemblages showed distinct spatial patterns across the Flinders CMR (Fig. 5). When combined with depth, the assemblages can be broadly categorised as: shallow, reef-associated (group 5); intermediate reef-associated (group 2); deeper, reef-associated (group 3); shallow sediment-associated (group 6); and deeper, reef-associated (groups 1 and 4). While sites within the same spatial cluster often belonged to the same fish assemblage, this was not always the case. Clusters of sites close to the shelf edge and in the south of the reserve were often more heterogeneous, composed of two to three different assemblages within a 1 km radius.

Discussion

This study trialled the use of BRUVs and a spatially- balanced survey design to provide the first quantitative description of demersal fish species and assemblages occurring the multiple use zone of the Flinders CMR shelf. Non-extractive sampling with BRUVs identified characteristic species (including jackass morwong, leatherjackets, draughtboard sharks, reef ocean perch and common gurnard perch), several species of commercial interest (including gummy sharks, reef ocean perch, striped trumpeter and jackass morwong) and species from trophic groups ranging from omnivores through to higher predators. While many of the species present in the study area are widely distributed across southern and/or south-eastern Australia, the majority are endemic to Australian waters. This highlights the unique biodiversity represented in temperate Australia, primarily as a result of its relatively isolated evolutionary history [39,40], and reinforces the global importance of conserving representative fauna.

Spatial and environmental patterns in fish assemblages

Six assemblages with a distinct spatial pattern were observed at sites across the reserve. Many species contributed to more than one assemblage and differences in the relative abundance of

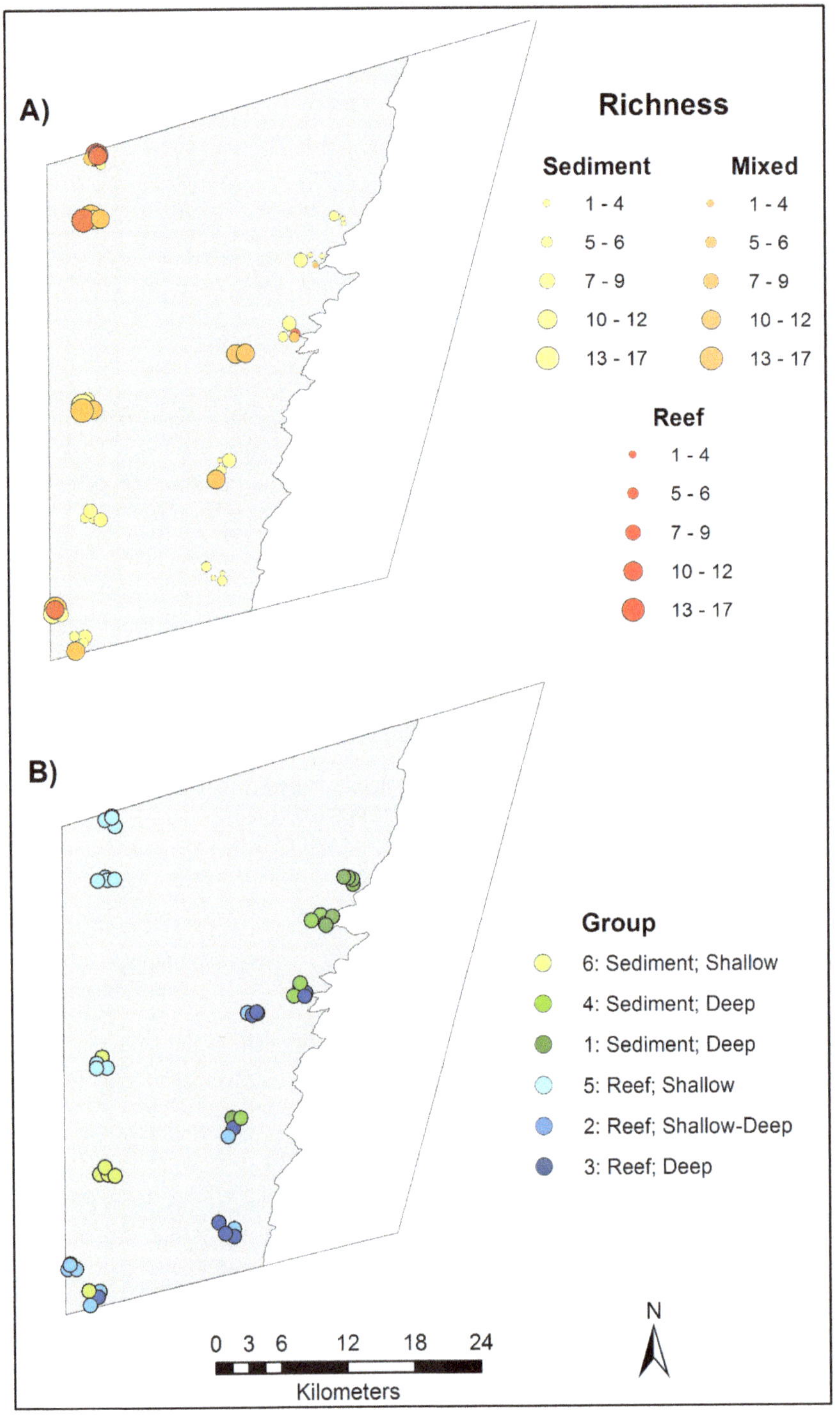

Figure 5. Spatial distribution of A) species richness B) fish assemblage groups on the Flinders CMR shelf (shaded grey). Increasing size of symbols in A) indicate increasing species richness. Symbols are colour coded according to the observed substratum type in BRUV footage: yellow = sediment; orange = mixed; red = reef. Assemblages in B) are coded by colour, with predominantly sediment-associated assemblages coloured green and predominantly reef-associated assemblages coloured blue.

species defined the differences between several assemblages. Assemblages were structured by depth and habitat type, and could be heterogeneous at relatively small scales (i.e. within a sampling cluster).

Depth-related assemblage transitions were most distinct and occurred in waters greater than approximately 80 m for sediment-associated assemblages and approximately 60 m for reef-associated assemblages, which is largely consistent with other studies on the southeast Australian shelf [41]. While some species were found across almost the full depth range of the study area (e.g. jackass morwong and the flathead, *Platycephalus richardsoni*) several species characterised particular substratum and depth zones. Leatherjackets (*T. degeni* in particular) for example, were characteristic of shallow reef sites; striped trumpeter were indicative of deeper, reef-associated assemblages; and cucumber fish were indicative of deeper, sediment-associated assemblages. These results accord with other studies where depth has a major influence on the distribution of fish assemblages [17,42].

As well as influencing assemblage composition, depth also influenced the diversity and ecology of assemblages. Species richness generally declined with depth consistent with observations elsewhere [39,43,44] and the trophic range of species occurring in shallow assemblages was generally greater than deeper assemblages. In reef-associated assemblages for example, larger trophic ranges of the shallow water assemblage were due to the co-occurrence of species such as leatherjackets (many of which are omnivorous) and draughtboard sharks (which are carnivores). Similarly, other studies have noted trophic-related changes with depth, such as increases in body size [42] and a transition towards higher –order predators [34].The combination of assemblage level patterns and changes in ecology and diversity with depth as seen here, have been used to suggest that the edge of the continental shelf forms an important faunal break. Consequently the shelf and slope environments should be considered separate management units [8,31] and not subsumed within the same management zone as currently occurs within the Flinders CMR.

Although a ubiquitous pattern, the processes involved in the depth structuring of benthic fish assemblages can be varied. In the shelf environments of the Flinders CMR, we suggest these are partly attributable to the strong tidal currents which affect the shallow, western section on the reserve [22]. In this region, low-profile reefs are dominated by tall biogenic structures with flexible forms such as sea whips and erect branching sponges (personal observation) that are indicative of high energy environments. This 3D biogenic structure is much reduced towards the deeper, eastern extent of the study area. These differences in reef structure [45], as well as physical limitations posed by increased current velocities [46], may have a large influence on the assemblages found in shallow regions of the reserve. Other physical and oceanographic conditions also co-vary with depth. For example, light availability decreases with depth, which affects primary productivity and therefore the range of food sources available to fish and appears to correspond with the observed changes in the trophic composition of assemblages. In addition, the oceanography in this region is complex, driven largely by inter-annual and inter-seasonal variability in the southward penetration of the warm-water, East Australian Current (EAC). The EAC can incur on the shelf or form fronts at the shelf break [24] and affects processes such as recruitment [47] and primary productivity [48] that are in turn likely influence the composition of fish assemblages, particularly on the outer shelf. Regardless of the proximal processes involved, the depth structuring of assemblages observed in the study area suggests that future sampling would benefit by breaking the region into shallow and deep strata around the 70 m depth mark.

Habitat or substratum type was also influential in discriminating between the assemblages observed in the Flinders CMR shelf. Despite preferentially sampling clusters of sites surrounding our initial mixed/reef sites, a large number of all sites surveyed contained only sediments, at least within the field of view of the BRUVs. This occurred because low-profile shelf reefs in the CMR are small scale features (tens of meters) and their distribution is highly patchy. Sediment sites were generally less species rich than those found on mixed or reef substrata. However, the composition of these sites varied enough to form three distinct assemblage groups, primarily on the basis of their depth distribution. The lower diversity of sediment assemblages has been attributed to lower habitat complexity with fewer niches to support co-existing species [49]. Still, sediment-associated assemblages contained some species with affinity for reef habitats. This may be because reef habitat was present outside the BRUVs field of view but within the swimming distance of fishes resulting in a 'halo' effect where species with an affinity for reef and sediments co-occur [49]. In subtropical shallow reefs this halo can extend 200 m from the reef edge [49]. The importance of local substratum in structuring fish assemblages highlights the need to develop comprehensive, high resolution habitat maps for the reserve. Assembling these habitat maps for marine reserves will require a commitment to strategic mapping of the seabed using multibeam sonar, so that over time complete coverage is achieved. For the continental shelf, this mapping could take years to decades [50]. In the meantime, management will need to adopt spatially representative sampling such as presented in this study to provide habitat and biodiversity inventories and to underpin monitoring programs.

Geomorphic features of the seafloor, such as submarine canyons, can substantially alter water flows, enhance local productivity [51] [52] and influence fish assemblage patterns [53]. Canyons are a conservation feature of the Flinders CMR and several canyons intersect the shelf edge. Canyons were represented in our sampling by a cluster of sites near a canyon head in the north of the survey area. Assemblages adjacent to canyon heads were not however, unique to canyon heads and were also found in other deeper sections of the shelf. Neither were these assemblages particularly species rich. Despite our preliminary findings, a better understanding of the ecological significance of canyon heads in this system will require a more targeted and intensive sampling strategy within and around the canyon head systems, e.g. [51].

Implications for Monitoring and Management

An objective of the Flinders CMR – and the larger Southeast Marine Reserve Network of which it is part - is to protect the biodiversity, natural and cultural values of the region. These include "representative examples of the ecosystems, communities and habitats" associated with the Tasmanian Shelf and Southeast Shelf Transition Provinces [21]. While these values are defined in broad terms from the national bioregionalisation, these have yet to be translated into inventories of conservation values, with specific metrics for monitoring their status. This study is a first step towards developing both an inventory of biological values, and elements of an effective biological monitoring program, including relevant environmental influences, for fish communities represented in the shelf component of the Flinders CMR. We suggest that representative demersal fish communities include those found on dynamic, low- profile reefs in both shallow (<70 m) and deep (70–160 m) areas of the shelf as well as those found on sediments in the same depth strata, and here we have characterised each of these assemblages. Several prevalent commercial and recreational species, including striped trumpeter and jackass morwong, which are likely to be responsive to management actions may be good candidates as indicator species and this work will provide the basis for further research into deriving specific monitoring targets.

We used a non-standard sampling design, GRTS, to survey fish assemblages and to satisfy the objectives of the broader survey of the Flinders CMR. Due to the cost and logistics associated with sampling in offshore and/or remote environments, the use of survey designs that are efficient, representative and flexible enough to accommodate multiple objectives is likely to become increasingly important. This is particularly true when simultaneously considering the inventory and monitoring needs of multiple reserves within networks. In the present study, the GRTS sampling ensured that clusters of reef and sediment sites sampled with BRUVs are well spread out across the reserve (within the distribution of each habitat type) and consequently encompass broad spatial and environmental gradients. This in turn provides confidence in the patterns observed and provides a better basis for additional or repeated sampling in a monitoring context. The site-ordered sampling of GRTS has the potential disadvantage of using ship time inefficiently if there is a large cumulative distance between deployments. This would have been the case in our study, but was circumvented by clustering samples around a subset of original sites. Conversely, the site ordering of GRTS allows dynamic altering of samples according to situations in the field, while maintaining spatial balance within strata.

The GRTS methodology will also be useful when choosing samples for future sampling, typically as part of an adaptive monitoring program. The master list of samples, that encompasses all possible sampling locations [28], can be subset in different ways by varying the inclusion probabilities for factors such as habitat types or depth, to accommodate new research questions or changing management objectives. While such changes lead to higher or lower sampling rates for different habitats or strata, the set of sites selected always maintains spatial balance within the strata. Inclusion probabilities are taken into account when deriving representative estimates of the occurrence, proportions or abundance of metrics or indicators across the entire sampling area [18]. Once suitable indicator species and metrics have been selected, the GRTS master sample can provide the basis for rotating panel designs [28] for repeat sampling that aims to detect changes in the abundance of these species, and possibly also major shifts in the distribution of assemblage types. Another advantage of GRTS in this respect, is its use of a local neighbourhood variance estimator, which is generally unbiased and more precise than variance estimators from other types of sampling designs [26]. Future sampling within the context of monitoring however, will need to incorporate comparable areas outside of the multiple use management zone of the Flinders CMR. In addition sufficient sample sizes will be needed to provide enough power to distinguish between natural variability and biological trends in order to be able to assess whether current management controls are effective in achieving their stated objectives.

The use of baited systems to survey fish communities has sometimes been criticised because of the bias towards sampling scavenging and predatory species [54]. However, all fish sampling methods have selectivity biases, and alternatives for sampling deep continental shelf fishes (e.g. trawls, meshnets and traps [41]) are extractive (and some destructive to the habitats the reserve is intended to protect). We recognise that by using BRUVs we have preferentially observed a subset of species that is attracted to, or undeterred by, baits. However, BRUVs data have been shown to clearly discriminate between fish assemblages in a variety of environmental settings [43] and have better statistical power to detect spatial and temporal changes in assemblage structure and abundances of individuals than unbaited methods [54]. In the future, we may complement BRUVs with other non-extractive video techniques (e.g. forward looking cameras on remotely or autonomously operated vehicles) to capture other components of the community. Never-the-less, in our present study BRUVs successfully observed a broad range of species, including key ecological species and potential indicator species. Standardised methodologies, in combination with robust sampling designs, will be essential for ongoing monitoring and assessment across multiple reserves in Australia's reserve network. Finally, the challenges we outline for the inventory and long-term monitoring of large and remote marine reserves are not unique to the Australian reserve network, nor exclusively to marine reserves. Therefore the approach that we have demonstrated may be usefully applied in other systems and under other management scenarios.

Supporting Information

Figure S1 The location of the forty GRTS sites sampled across the Flinders CMR shelf in phase one. Sites are coloured according to the broad habitat type: sediment (yellow); mixed, low-profile reef and sediments (red); canyon head (blue) recorded during phase one of sampling. Sites are labelled with their GRTS site number which, when sampled in order, represents a spatially balanced sample. The table shows the subset of sites sampled as the basis of clusters in the phase two sampling with BRUVs. These sites are the first three GRTS sites classified as sediment and the first eight classified as mixed reef.

Data S1 Matrix of the species observed at each BRUV deployment. Environmental variables used in analyses are also included.

Data S2 Attributes of species used in analysis, including trophic level, habitat preference and endemicity. Summary statistics on the prevalence and abundance of species across the survey area are also given.

Acknowledgments

The authors would like to thank the crew of *Challenger*, Matt Francis and Tim Green, for their expertise and dedication during fieldwork, Richard Zavalas for assistance with fieldwork and video scoring, Johnathan Kool for assistance with fieldwork, and Vanessa Lucieer for assistance in the planning phase. SN publishes with permission of the Chief Executive Officer of Geoscience Australia.

Author Contributions

Conceived and designed the experiments: NAH NB EL JMD SN KRH. Performed the experiments: NAH NB JH SN. Analyzed the data: NAH NB EL JH KRH. Contributed reagents/materials/analysis tools: NAH NB EL JH SN AW KRH. Wrote the paper: NAH NB EL JMD SN AW KRH.

References

1. Halpern BS, Walbridge S, Selkoe KA, Kappel CV, Micheli F, et al. (2008) A global map of human impact on marine ecosystems. Science 319: 948–952.
2. Halpern BS, Longo C, Hardy D, McLeod KL, Samhouri JF, et al. (2012) An index to assess the health and benefits of the global ocean. Nature: 615–620.
3. Day J, Dudley N, Hockings M, Holmes G, Laffoley D, et al. (2012) Guidelines for applying the IUCN Protected Area Management Categories to Marine Protected Areas. Gland, Switzerland: IUCN. 36 p.
4. Kemp J, Gibson R, Atkinson R, Gordon J, Hughes R, et al. (2012) Measuring the performance of spatial management in marine protected areas. Oceanography and Marine Biology: An Annual Review 50: 287–314.
5. Gubbay S, editor (1995) Marine Protected Areas: principles and techniques for management. London: Chapman and Hall.
6. CBD (1992) Convention on Biological Diversity. 5 June 1992 Rio de Janerio (Brazil).
7. CBD (1995) The Jakarta Mandate on Marine and Coastal Biological Diversity. Decisions of the Second Meeting of the Conference of the Parties to the Convention on Biological Diversity. Jakarta, Indonesia, 6–17 November 1995: UNEP.
8. Spalding MD, Meliane IN, Milam A, Fitzgerald C, Hale LZ (2013) Protecting marine spaces: Global targets and changing approaches. Ocean Yearbook Online 27: 213–248.
9. Edgar GJ, Stuart-Smith RD, Willis TJ, Kininmonth S, Baker SC, et al. (2014) Global conservation outcomes depend on marine protected areas with five key features. Nature advance online publication.

10. Barr LM, Possingham HP (2013) Are outcomes matching policy commitments in Australian marine conservation planning? Marine Policy 42: 39–48.
11. ANZECC (1998) Guidelines for establishing the National Representative System of Marine Protected Areas. Canberra: Australian and New Zealand Environment and Conservation Council (ANZECC) Task Force on Marine Protected Areas.
12. Commonwealth of Australia (2006) A Guide to the Integrated Marine and Coastal Regionalisation of Australia Version 4.0. In: Heritage DotEa, editor. Canberra, Australia.
13. Lucieer V, Lamarche G (2011) Unsupervised fuzzy classification and object-based image analysis of multibeam data to map deep water substrates, Cook Strait, New Zealand. Continental Shelf Research 31: 1236–1247.
14. Bailey DM, King NJ, Priede IM (2007) Cameras and carcasses: historical and current methods for using artificial food falls to study deep-water animals. Marine Ecology Progress Series 350: 179–191.
15. Willis TJ, Babcock RC (2000) A baited underwater video system for the determination of relative density of carnivorous reef fish. Marine and Freshwater Research 51: 755–763.
16. Langlois TJ, Harvey ES, Meeuwig JJ (2012) Strong direct and inconsistent indirect effects of fishing found using stereo-video: Testing indicators from fisheries closures. Ecological Indicators 23: 524–534.
17. Harvey ES, Cappo M, Kendrick GA, McLean DL (2013) Coastal fish assemblages reflect geological and oceanographic gradients within an Australian zootone. Plos One 8.
18. Stevens DL, Olsen AR (2004) Spatially balanced sampling of natural resources. Journal of the American Statistical Association 99: 262–278.
19. Olsen AR, Kincaid TM, Payton Q (2012) Spatially balanced survey designs for natural resources. In: Gitzen RA, Millspaugh JJ, Cooper AB, Licht DS, editors. Design and Analysis of Long-term Ecological Monitoring Studies. Cambridge, England: Cambridge University Press. pp. 126–150.
20. Commonwealth of Australia (2006) A Guide to the Integrated Marine and Coastal Regionalisation of Australia Version 4.0. Canberra, Australia: Department of the Environment and Heritage. 16 p.
21. Director of National Parks (2013) South-east Commonwealth Marine Reserves Network management plan 2013–23. Director of National Parks, Canberra.
22. Fandry C (1983) Model for the three-dimensional structure of wind-driven and tidal circulation in Bass Strait. Marine and Freshwater Research 34: 121–141.
23. National Tidal Centre (2014) Tide predictions for Australia, South Pacific and Antarctica.
24. Harris G, Nilsson C, Clementson L, Thomas D (1987) The water masses of the East Coast of Tasmania: Seasonal and interannual variability and the influence on phytoplankton biomass and productivity. Australian Journal of Marine and Freshwater Research 38: 569–590.
25. Whiteway TG (2009) Australian Bathymetry and Topography Grid. Canberra, Australia: Geoscience Australia. 46 p.
26. Stevens DL, Olsen A (2003) Variance estimation for spatially balanced samples of environmental resources. Environmetrics 14: 593–610.
27. Dambacher JM, Jones KK, Larsen DP (2009) Landscape-level sampling for status review of great basin redband trout. North American Journal of Fisheries Management 29: 1091–1105.
28. Larsen D, Olsen A, Stevens D (2008) Using a Master Sample to Integrate Stream Monitoring Programs. Journal of Agricultural, Biological, and Environmental Statistics 13: 243–254.
29. Harvey ES, Newman SJ, McLean DL, Cappo M, Meeuwig JJ, et al. (2012) Comparison of the relative efficiencies of stereo-BRUVs and traps for sampling tropical continental shelf demersal fishes. Fisheries Research 125–126: 108–120.
30. Cappo M, De'ath G, Speare P (2007) Inter-reef vertebrate communities of the Great Barrier Reef Marine Park determined by baited remote underwater video stations. Marine Ecology Progress Series 350: 209–221.
31. Seager J (2014) Transect Measure. 2.30 ed: SeaGIS.
32. Maechler M, Rousseeuw P, Struyf A, Hubert M, Hornik K (2013) cluster: Cluster Analysis Basics and Extensions. R package. 1.14.4 ed.
33. R Development Core Team (2008) R: A language and environment for statistical computing. Vienna, Austria: R Foundation for Statistical Computing.
34. Anderson MJ, Gorley RN, Clarke KR (2008) PERMANOVA+ for PRIMER: Guide to software and statistical methods. Plymouth, UK: PRIMER-E. 214 p.
35. Edgar GJ (2008) Australian Marine Life: the plants and animals of temperate waters. New Holland, Sydney. Sydney: Reed New Holland.
36. Gomon MF, Bray DJ, Kuiter RH (2008) Fishes of Australia's Southern Coast. Chatswood, Australia: Reed New Holland.
37. (2014) FishBase. In: Froese R, Pauly D, editors. Available: www.fishbase.org: World Wide Web electronic publication.
38. Venables WN, Ripley BD (2002) Modern Applied Statistics with S. New York: Springer.
39. Last PR, White WT, Gledhill DC, Pogonoski JJ, Lyne V, et al. (2011) Biogeographical structure and affinities of the marine demersal ichthyofauna of Australia. Journal of Biogeography 38: 1484–1496.
40. Phillips J (2001) Marine macroalgal biodiversity hotspots: why is there high species richness and endemism in southern Australian marine benthic flora? Biodiversity and Conservation 10: 1555–1577.
41. Williams A, Bax N (2001) Delineating fish-habitat associations for spatially based management: an example from the south-eastern Australian continental shelf. Marine and Freshwater Research 52: 513–536.
42. Fitzpatrick BM, Harvey ES, Heyward AJ, Twiggs EJ, Colquhoun J (2012) Habitat specialization in tropical continental shelf demersal fish assemblages. PLoS ONE 7: e39634.
43. Zintzen V, Anderson MJ, Roberts CD, Harvey ES, Stewart AL, et al. (2012) Diversity and composition of demersal fishes along a depth gradient assessed by Baited Remote Underwater Stereo-Video. PLoS ONE 7: e48522.
44. Chatfield BS, Van Niel KP, Kendrick GA, Harvey ES (2010) Combining environmental gradients to explain and predict the structure of demersal fish distributions. Journal of Biogeography 37: 593–605.
45. Gratwicke B, Speight MR (2005) The relationship between fish species richness, abundance and habitat complexity in a range of shallow tropical marine habitats. Journal of Fish Biology 66: 650–667.
46. Fulton CJ, Bellwood DR (2004) Wave exposure, swimming performance, and the structure of tropical and temperate reef fish assemblages. Marine Biology 144: 429–437.
47. Bruce BD, Evans K, Sutton CA, Young JW, Furlani DM (2001) Influence of mesoscale oceanographic processes on larval distribution and stock structure in jackass morwaong (*Nemodactlus macropertus:* Cheilodactylidae). ICES Journal of Marine Science 58: 1072–1080.
48. Bax NJ, Burford M, Clementson L, Davenport S (2001) Phytoplankton blooms and production sources on the south-east Australian continental shelf. Marine and Freshwater Research 52: 451–462.
49. Schultz AL, Malcolm HA, Bucher DJ, Smith SDA (2012) Effects of reef proximity on the structure of fish assemblages of unconsolidated substrata. Plos One 7.
50. Kloser RJ, Williams A, Butler A (2006) Exploring surveys of seabed habitats in Australia's deep ocean using remote sensing – needs and realities. In: Todd BJ, Greene HG, editors. Mapping the Seafloor for Habitat Characterization. pp. 93–109.
51. Currie DR, McClatchie S, Middleton JF, Nayar S (2012) Biophysical factors affecting the distribution of demersal fish around the head of a submarine canyon off the Bonney Coast, South Australia. Plos One 7.
52. Vetter EW, Smith CR, De Leo FC (2010) Hawaiian hotspots: enhanced megafaunal abundance and diversity in submarine canyons on the oceanic islands of Hawaii. Marine Ecology 31: 183–199.
53. Leathwick JR, Elith J, Francis MP, Hastie T, Taylor P (2006) Variation in demersal fish species richness in the oceans surrounding New Zealand: an analysis using boosted regression trees. Marine Ecology Progress Series 321: 267–281.
54. Harvey E, Cappo M, Butle RJ, Hall N, Kendrick G (2007) Bait attraction affects the performance of remote underwater video stations in assessment of demersal fish community structure. Marine Ecology Progress Series 350: 245–254.

18

Hydrostatic Pressure and Temperature Effects on the Membranes of a Seasonally Migrating Marine Copepod

David W. Pond[1]*, Geraint A. Tarling[2], Daniel J. Mayor[3]

1 Scottish Association for Marine Science, Oban, Argyll, United Kingdom, **2** British Antarctic Survey, Natural Environment Research Council, Cambridge, United Kingdom, **3** Institute of Biological and Environmental Sciences, Oceanlab, University of Aberdeen, Newburgh, Aberdeenshire, United Kingdom

Abstract

Marine planktonic copepods of the order Calanoida are central to the ecology and productivity of high latitude ecosystems, representing the interface between primary producers and fish. These animals typically undertake a seasonal vertical migration into the deep sea, where they remain dormant for periods of between three and nine months. Descending copepods are subject to low temperatures and increased hydrostatic pressures. Nothing is known about how these organisms adapt their membranes to these environmental stressors. We collected copepods (*Calanoides acutus*) from the Southern Ocean at depth horizons ranging from surface waters down to 1000 m. Temperature and/or pressure both had significant, additive effects on the overall composition of the membrane phospholipid fatty acids (PLFAs) in *C. acutus*. The most prominent constituent of the PLFAs, the polyunsaturated fatty acid docosahexanoic acid [DHA – 22:6(n-3)], was affected by a significant interaction between temperature and pressure. This moiety increased with pressure, with the rate of increase being greater at colder temperatures. We suggest that DHA is key to the physiological adaptations of vertically migrating zooplankton, most likely because the biophysical properties of this compound are suited to maintaining membrane order in the cold, high pressure conditions that persist in the deep sea. As copepods cannot synthesise DHA and do not feed during dormancy, sufficient DHA must be accumulated through ingestion before migration is initiated. Climate-driven changes in the timing and abundance of the flagellated microplankton that supply DHA to copepods have major implications for the capacity of these animals to undertake their seasonal life cycle successfully.

Editor: Adrianna Ianora, Stazione Zoologica Anton Dohrn, Naples, Italy

Funding: The authors thank the officers and crew of the RRS James Clark Ross for their assistance and professionalism whilst at sea. This research was supported by grants from the Marine Alliance for Science and Technology for Scotland (MASTS) and the NERC (NE/J007803/1). The research is also a contribution to the ECOSYSTEMS programme of the British Antarctic Survey. The funders had no role in study design, data collection and analysis, decision to publish, or preparation of the manuscript.

Competing Interests: The authors have declared that no competing interests exist.

* Email: David.Pond@sams.ac.uk

Introduction

Aquatic environments are characterised by gradients of temperature and pressure [depth], both of which affect the functioning of biological membranes. Phospholipids are major constituents of cellular membranes and their fatty acid composition is regulated to maintain membrane order, otherwise termed, membrane 'fluidity' [1]. The maintenance of membrane order and a liquid-crystalline state is of critical importance for the function and integrity of cell membranes, since it controls bilayer permeability and also the mobility and function of embedded enzymes [2]. Typical responses to both increasing pressure and decreasing temperature are to increase the proportions of unsaturated fatty acids in the membranes, a process termed homeoviscous adaptation (HVA; [1,3–4]).

Most work on adaptive changes in the lipid bilayers of poikilotherms has been conducted on strains of deep sea bacteria which are amenable to experimental manipulation. These organisms increase the proportions of mono- and polyunsaturated fatty acids in their membrane phospholipids in response to increased pressure and/or reduced temperature [5–7]. A comparison of mitochondrial membranes isolated from shallow and deep sea fish species has similarly identified changes in unsaturation that are consistent with pressure dependent HVA [8–9]. Pressure induced membrane adaptations have largely been considered independently to temperature effects [10]. However, vertical gradients in the oceans frequently encompass changes in both temperature and pressure. Calanoid copepods, which dominate marine zooplankton biomass at latitudes greater than 30° north and south [11], undergo a migration to depth during winter [12–13] in what is arguably the largest seasonal migration on Earth. These animals may encounter a 10°C and ~350 bar change in temperature and pressure respectively over the course of only a few days or weeks, providing a natural system for studying the functional responses of lipid bilayers to these environmental stressors.

This study examined the overarching hypothesis that temperature and pressure affect the composition of fatty acids in the membrane phospholipids of the dominant large calanoid copepod in the Southern Ocean, *Calanoides acutus* (Fig. 1). These animals descend to 500–3500 m each austral summer to overwinter winter in a state of dormancy [14]. During late winter, the copepods ascend to the upper ocean to mature into adults and commence feeding and reproductive activity.

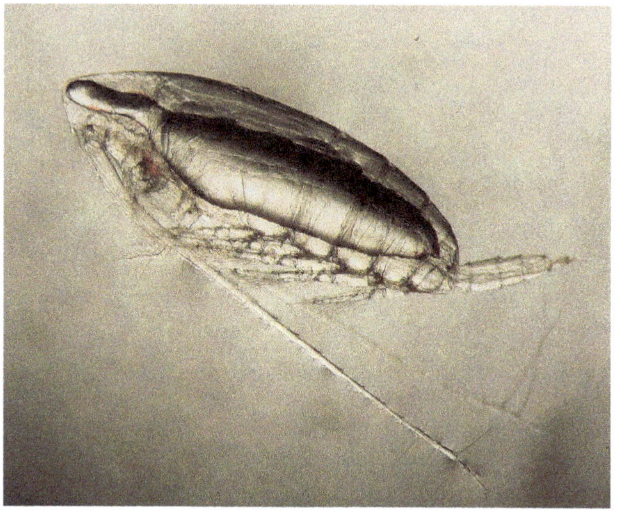

Figure 1. *Calanoides acutus*, a marine copepod that descends into the deep sea to overwinter in a state of dormancy. The large oil sac occupying the abdomen is thought to provide energy storage and aid buoyancy control. Body length 3 mm.

Methods

Specific permissions are not required for sampling of marine invertebrates

Sample collection. Copepods were collected from four stations during a cruise of the RRS James Clark Ross to the Scotia Sea in January and February 2008 (Fig. 2, Table S1, [15–16]). Vertical profiles of pressure are linear and effectively constant between stations whereas temperature profiles, determined at dawn using a SeaBird 911+CTD, differed between stations (Fig. 3). This sampling design enabled us to reliably distinguish between the effects of pressure and temperature; for any given pressure we had observations across a range of temperatures. Stratified samples were collected during the night from 375 m to 1000 m in 125 m depth horizons using a MOCNESS multi-net. Copepods from the photic zone water column were collected using Bongo nets deployed immediately after the CTD deployments and were hauled vertically at ~0.22 ms^{-1} from 400 m to the surface [16]. A total of 85 individual *C. acutus* (pre-adult 5th copepodite stage) were collected over four stations and six depth horizons. Copepods were individually sorted onboard and stored in 500 µl of chloroform:methanol (2:1 v/v) in 1.1 ml tapered vials (Chromacol) at −80°C until analysis. All samples were analysed in laboratories in Cambridge <1 year after sample collection.

Lipid analysis. Total lipid was extracted following Folch et al. [17]. Given the comparatively small amounts of lipid contained within an individual copepod, extraction procedures were simplified to reduce losses and minimise the introduction of contaminants into the samples [15]. After the addition of 125 mL of KCl (0.88% w/v), samples were initially whirlimixed before being centrifuged for 2 min at 1500 rpm to promote phase separation. The lower chloroform phase, containing the total lipid extract was removed using a 500 mL Hamilton glass syringe equipped with a teflon tipped plunger. The needle of the syringe was pushed through the upper methanol–water layer and the chloroform phase taken up into the syringe, leaving the aqueous phase as waste in the original vial. Tapered vials greatly facilitate this procedure. Phospholipids, which are major constituents of membrane lipids were purified from total lipid using high performance thin layer chromatography (HPTLC) in conjunction with a 18:2:0.2 (hexane/diethyl ether/acetic acid, v/v/v) solvent system. Phospholipids were then transesterified to generate fatty acid methyl esters and analysed on a TRACE 2000, Thermo Electron gas chromatograph (GC). The GC was equipped with oncolumn injection, a Stabilwax column (Restek 30 m×0.32 mm internal diameter) and hydrogen was used as the carrier gas. Data are expressed as percent composition of total fatty acids. See Purać et al. [18] for full details of the methods used.

Data analysis. Preliminary data exploration was undertaken to identify outliers and instances where explanatory variables were highly correlated [19,20]. The effects of depth (mid-sampling depth), temperature and station identity on the percentage composition of phospholipid fatty acids (PLFAs) in *C. acutus* were investigated using redundancy analysis (RDA).

The significance of each model term was assessed using a permuted (n = 9999) forwards selection procedure [19]. Linear regression techniques were used to examine how depth, temperature and a depth×temperature interaction affected the percentage abundance of the major (mean>5%) PLFAs in *C. acutus*. Where necessary, station identity was included as a random effect, thereby accounting for correlations between observations made within each sampling station. Variance covariates were also included in the random structure of the regression models where problems with unequal variance were identified. The procedures for model selection using generalised least squares- (GLS) and linear mixed-effects (LME) regression models are described elsewhere [20–22]. In brief, the random structures of the statistical models were established using a likelihood ratio (L. ratio) test using restricted maximum likelihood (REML) estimation. The fixed structure of each model was subsequently determined using a hierarchical backwards selection procedure based on the L. ratio test using maximum likelihood estimation. Model parameters were generated using REML estimation. The underlying statistical assumptions were verified as follows: theoretical quantiles were plotted against standardized residuals (Q-Q plot) to assess normality of residuals; residual values were plotted against fitted values and each covariate to verify homogeneity of variance and independence of observations respectively. Statistical modelling was conducted in the 'R v2.11.1' programming environment [23] using the 'vegan' [24] and 'nlme' [25] packages.

Results

Oceanography

Sampling stations covered a latitudinal range of around 5° and water depths of between 1400 m and 2700 m. Two of the stations (Su 2.1 and Su 4.0) lay south of Southern Boundary in the coldest Southern Ocean water mass. Station Su 5.0 lay north of this front but south of the Southern Antarctic Circumpolar Current Front (SACCF) while Su 6.0 was located just north SACCF. Near-surface temperatures were at least a whole degree colder in the two southerly stations compared to the two northerly stations (−0.5°C to −0.6°C compared to 0.5°C to 1.4°C, Fig. 3). Nevertheless, the vertical water mass structure below the surface was relatively similar across all sampling stations. Temperatures initially declined steeply until reaching a temperature minimum at around 100 m depth (−1.4°C in the south, 0°C to the north). This decline is the product of old winter water that lies beneath Antarctic surface waters. With greater depth, temperature increased through the mixing of winter water with warmer upper circumpolar deep water. No further temperature increase occurred below 500 m after which there was only a very gradual decline in temperature until the maximum sampling depth. Overall, after experiencing

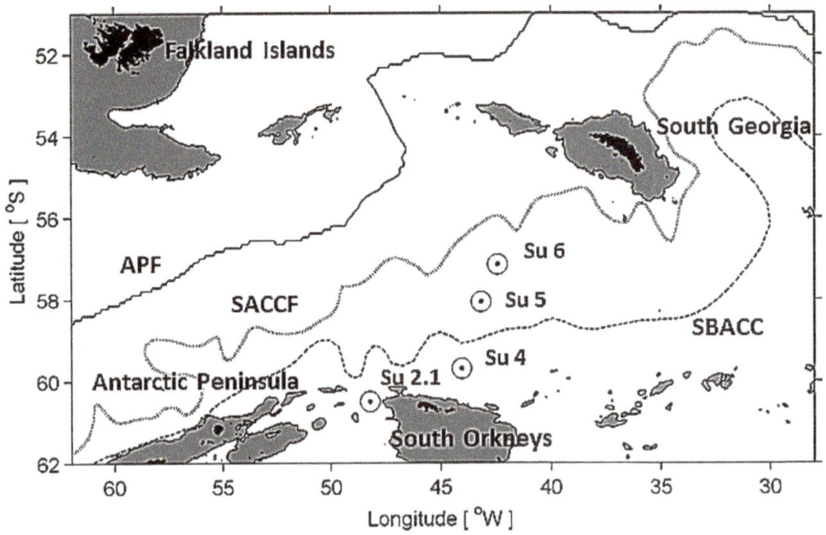

Figure 2. Map indicating locations of the four study sites in the Southern Ocean. (APF = Antarctic Polar Front, SACCF = Southern Antarctic Circumpolar Current Front, SBACC = Southern Boundary of the Antarctic Circumpolar Current).

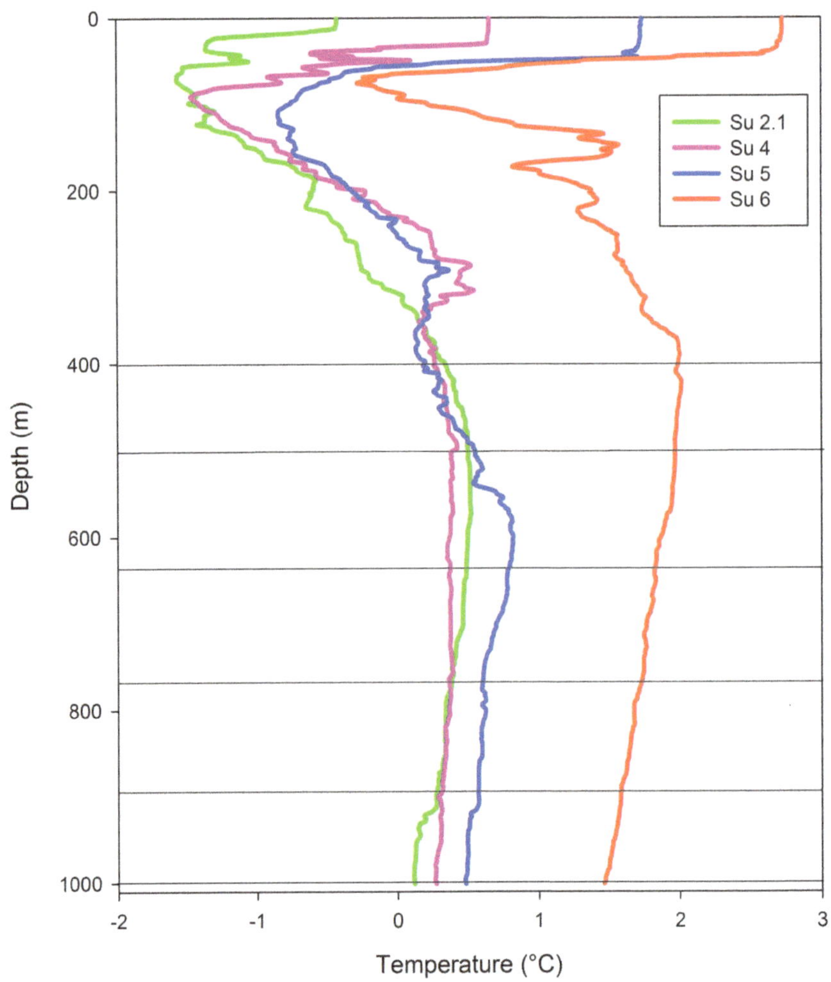

Figure 3. Profiles of temperature at the four biological sampling stations in the Scotia Sea (Jan–Feb 2008). Horizontal lines depict the sampling intervals of the Bongo net (0–400 m) and the MOCNESS multinet (375–1000 m).

Table 1. Average fatty acid composition of phospholipids in *C. acutus* from 4 stations (n = number of individual copepods sampled from each station, * position of double bonds uncertain).

Fatty acid	Station									
	2.1 (n = 24)	**(SE)**	**4 (n = 30)**	**(SE)**	**5 (n = 13)**	**(SE).**	**6 (n = 18)**	**(SE).**	**Overall mean**	**(SE)**
14:0	3.8	(0.3)	3.4	(0.3)	2.0	(0.2)	2.9	(0.2)	3.2	0.2
15:0	0.4	(0.1)	0.4	(0.1)	0.3	(0.1)	0. 7	(0.1)	0.4	0.1
16:0	17.2	(0.6)	17.7	(0.5)	20.8	(0.8)	20.5	(0.4)	18.6	0.3
16:1(n-7)	1.8	(0.2)	1.8	(0.1)	1.6	(0.1)	1.8	(0.1)	1.8	0.1
16:2(n-4)	5.9	(1.0)	3.8	(0.7)	2.0	(0.3)	8.0	(0.7)	5.0	0.4
18:0	5.2	(0.6)	3.8	(0.3)	4.3	(0.5)	15.0	(1.3)	6.6	0.6
18:1(n-9)	4.2	(0.2)	3.6	(0.4)	3.8	(0.3)	3.2	(0.2)	3.7	0.1
18:1(n-7)	2.8	(0.1)	2.7	(0.1)	2.4	(0.2)	2.2	(0.35)	2. 6	0.1
18:1(n-5)	5.1	(0.3)	4.2	(0.2)	4.2	(0.3)	5.6	(0.4)	4.7	0.3
18:2(n-6)	0.9	(0.1)	1.0	(0.1)	0.7	(0.04)	1.4	(0.4)	1.0	0.1
18:4(n-3)	0.3	(0.1)	0.2	(0.1)	0.6	(0.1)	0.3	(0.03)	0.3	0.03
20:1(n-9)	1.4	(0.1)	1.3	(0.1)	1.2	(0.1)	0.4	(0.1)	1.1	0.1
20:4(n-6)	0.1	(0.02)	0.4	(0.1)	0.3	(0.1)	0.1	(0.04)	0.2	0.03
20:4(n-3)	0.8	(0.1)	0.6	(0.1)	0.5	(0.1)	0.4	(0.1)	0.6	0.1
20:4*	1.3	(0.5)	1.8	(0.6)	1.2	(0.1)	0. 5	(0.1)	1.3	0.3
20:5(n-3)	17.3	(1.1)	20.8	(0.8)	24.8	(0.7)	13.8	(0.9)	18.9	0.6
22:1(n-11)	0.2	(0.03)	0.2	(0.03)	0.3	(0.1)	0.4	(0.1)	0.2	0.04
22:1(n-9)	1.0	(0.2)	0.6	(0.1)	0.6	(0.1)	0.2	(0.1)	0.6	0.1
22:5(n-3)	0.3	(0.1)	0.7	(0.1)	1.0	(0.1)	0.2	(0.03)	0.5	0.1
22:6(n-3)	30.2	(1.8)	31.4	(1.2)	27.7	(1.3)	23.1	(1.0)	28.7	0.7

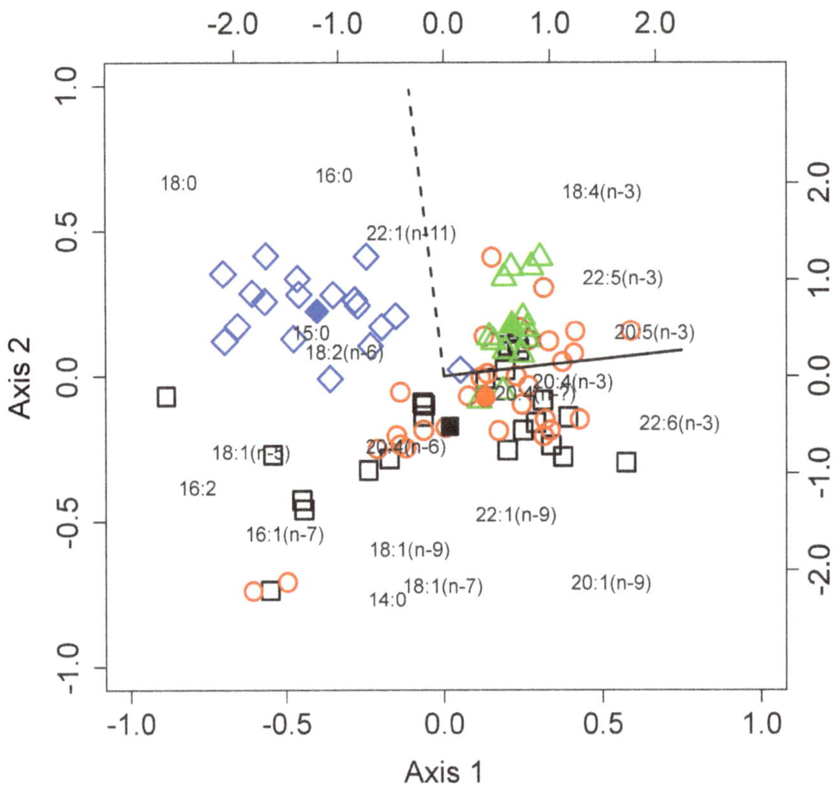

Figure 4. Redundancy analysis distance triplot of the phospholipid fatty acid data from 85 individual *C. acutus* ('sites') using 20 fatty acids ('species'). The effects of depth and temperature are plotted as vectors (solid- and dashed lines respectively). Symbols and colours denote Station identity, with the overall effect of each Station being represented by filled symbols. The primary and secondary sets of axes relate to the sites and species loadings respectively.

some variation in temperature while descending through the upper 400 m of the water column, a change of less than 0.4°C is experienced through the remainder of the water column in any one location. Between locations, there was a variation of around 1.5°C at comparative depths between the most southerly and northerly sampling stations (Fig. 3).

Membrane fatty acids

The PLFAs of 85 individual copepods were analysed in this study. These animals were sampled from 4 stations across temperature and depth gradients of 2.5°C and 1000 m respectively. A summary of the PLFA data is presented in Table 1. The polyunsaturated (PUFA) fatty acid docosahexaenoic [DHA; 22:6(n-3)] dominated the PLFAs of *C. acutus*, constituting up to 46.1% (28.7±0.8%; mean ± SE). Other fatty acids that contributed significantly (>5% of mean values) to the phospholipids were the PUFA eicosapentaenoic acid [EPA; 20:5(n-3)] (18.9±0.6%) and the saturated fatty acids (SFAs) palmitic acid [16:0] (18.6±0.3%) and stearic acid [18:0] (6.6±0.6%).

In order of importance, depth ($F = 11.34$, $df = 1$, $p < 0.001$), temperature ($F = 7.29$, $df = 1$, $p < 0.001$) and station identity ($F = 5.18$, $df = 3$, $p < 0.001$) all increased the amount of variance

Table 2. Summary of fixed-, random$^{(\dagger)}$ and variance covariate$^{(\ddagger)}$ terms in the optimal statistical models.

Response	Model term	df	L. ratio	p
22:6(n-3)	Depth×Temperature	1	5.590	0.018
	Temperature‡	1	5.561	0.018
16:0	Depth	1	24.145	<0.001
	Temperature	1	19.167	<0.001
18:0	Depth	1	39.619	<0.001
	Station†	1	56.278	<0.001
	Station‡	3	35.387	<0.001
20:5(n-3)	Temperature	1	8.479	0.004
	Station†	1	33.612	<0.001

Table 3. The effects of depth and temperature on the percentage composition of 22:6(n-3) in the phospholipids of *C. acutus*.

Model parameter	Coefficient (± SE)	t	p
Intercept	19.341±1.607	12.038	<0.001
Depth	0.024±0.003	7.821	<0.001
Temperature	0.608±1.388	0.438	0.663
Depth×Temperature	−0.008±0.003	−2.343	0.022

Table 4. The effects of depth and temperature on the percentage composition of 16:0 in the phospholipids of *C. acutus*.

Model parameter	Coefficient (± SE)	t	p
Intercept	20.313±0.612	33.186	<0.001
Depth	−0.006±0.001	−5.190	<0.001
Temperature	1.613±0.354	4.552	<0.001

explained in the PLFA compositional data. These explanatory variables accounted for 32% of the variation in the data and 26% was explained by the first two axes of the redundancy analysis. The RDA triplot visualises how the explanatory variables influence the composition of PLFAs (Fig. 4). The first axis, which was positively correlated with depth, had strong positive loadings of the PUFAs EPA and DHA and negative loadings of the SFAs 18:0 and 16:0. The second axis had strong positive loadings of the SFAs 16:0 and 18:0 and was positively correlated with temperature.

The PLFAs DHA, 16:0 and 18:0 were all significantly affected by depth (Tables 2–5). The PUFA DHA increased with depth, whereas the SFAs 16:0 and 18:0 both decreased (Figure 5 A–C; Tables 2–5). In the case of DHA, the effects of depth were dependent upon the effect of temperature; the effect of depth on this fatty acid increased as temperatures decreased (Fig. 5A; Tables 2 and 3). Temperature had a significant, additive effect on the SFA 16:0 (Fig. 5B; Tables 2 and 4); at any given depth, the percentage abundance of this compound increased as temperature increased (Fig. 5C). The PUFA EPA was not affected by depth (L. Ratio = 1.853, df = 1, $p = 0.174$) but increased significantly with temperature (Fig. 6; Tables 2 and 6).

Discussion

Our data demonstrate that the composition of fatty acids in the membranes of a seasonally migrating copepod, *C. acutus*, is influenced by a range of factors, including depth, temperature and geographic location. We found that the membrane phospholipids of these animals are dominated by DHA, with the proportions of DHA increasing with depth. This effect is exacerbated at colder temperatures. Such an interaction between temperature and pressure is entirely consistent with homeoviscous adaptation theory (HVA) theory, and is the first time that such a relationship has been demonstrated in a metazoan organism.

Knowledge of pressure and temperature adaptations in biological membranes has largely been derived from deep-sea bacteria [5–6,26–27]. Monounsaturated fatty acids (MFAs) are most commonly reported as the agents involved in maintaining membrane order, with the addition of further double bonds having little effect on the phase transition properties and hence fluidity of the lipids [28]. The predominance of DHA in the membranes of *C. acutus* indicates that this PUFA is a central part of their adaptive response to the effects of both temperature and pressure on membrane functionality. DHA is amongst the most unsaturated fatty acids found in nature. It contributes to the hyperfluidization of membranes and mediates processes catalysed by membrane proteins [7,29]. We therefore hypothesize that, in addition to maintaining membrane fluidity, DHA confers temperature and pressure tolerance to other essential properties of membrane function that are linked to the activity of membrane proteins and metabolism [29].

Calanoid copepods cannot synthesize DHA de novo [30] and this fatty acid is therefore considered to be essential, i.e. it must be derived from their diet. Flagellated microplankton are the predominant producers of DHA in the marine environment [31] and are likely to be the ultimate source of this compound in *C. acutus*. However, seasonally migrating copepods do not feed, but they sequester large lipid reserves within an internal sac during the spring and summer plankton blooms and later utilise this as an

Table 5. The effect of depth on the percentage composition of 18:0 in the phospholipids of *C. acutus*.

Model parameter	Coefficient (± SE)	t	p
Intercept	9.463±2.437	3.883	<0.001
Depth	−0.005±0.001	−7.215	<0.001

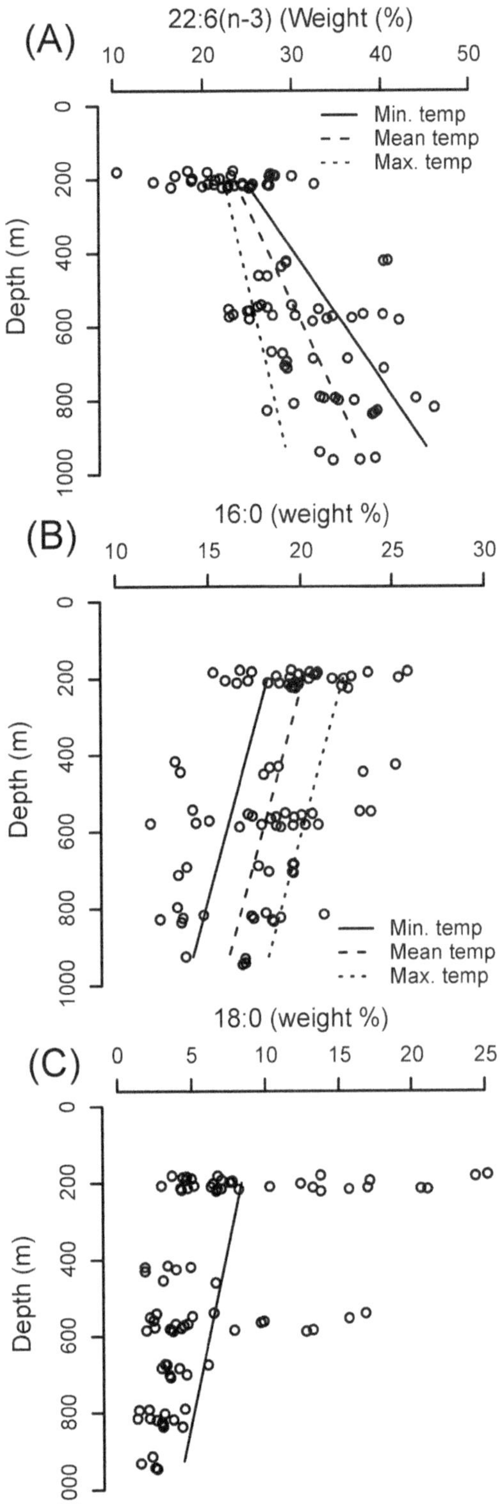

Figure 5. Model-predicted effects of depth and temperature on the percentage composition of (a) 22:6(n-3), (b) 16:0 and (c) 18:0 in the phospholipid fatty acids of *C. acutus*. Data points are presented for guidance only.

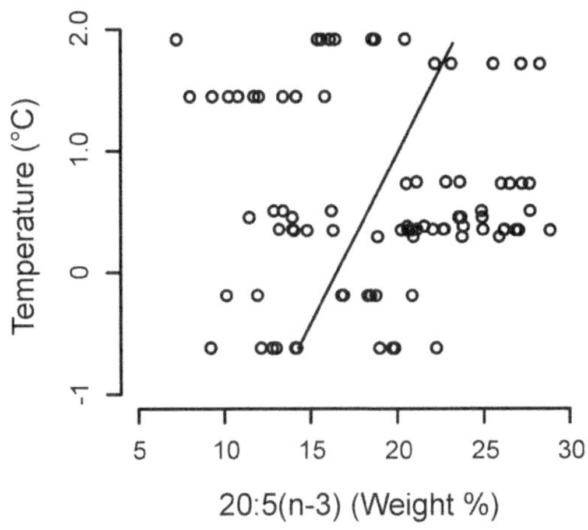

Figure 6. Model-predicted effects of temperature on the percentage composition of 20:5(n-3) in the phospholipid fatty acids of *C. acutus*. Data points are presented for guidance only.

energy reserve whilst undertaking their seasonal migration (Fig. 1). The lipid reserve of *C. acutus* contains appreciable quantities of DHA [16], and is most likely the source of fatty acids used for maintaining membrane order during migration. Despite the fatty acid composition of biological membranes being tightly regulated, we found geographical differences in the phospholipid fatty acid signatures of *C. acutus* (Fig. 4). This probably reflects differences in microplankton communities at the different locations [16].

The composition of lipids in the large oil sac was previously thought to reflect the accumulation of dietary lipids with little control on its composition by the copepod. More recent evidence suggests that the composition of the lipid sac is regulated since this determines its solid-liquid phase transition and enables the copepod to achieve neutral buoyancy in the deep sea [15,32]. Neutral buoyancy is essential if the copepods are to minimise swimming activity, which would attract predators and consume valuable energy reserves during the period of non-feeding. EPA is thought to play a key role in modulating the response of the oil sac to pressure and hence the extent to which this sac provides hydrostatic lift [15,32]). EPA is an essential dietary component [30] and ultimately derived from diatoms [15]. Collectively, these findings indicate two major functional roles for fatty acids derived from two distinct algal groups. EPA from diatoms is used to regulate buoyancy and the depth at which neutral buoyancy is achieved whilst DHA from flagellated microplankon is used to maintain biological membrane function. Climate-driven phenological change in the plankton, which is causing a temporal mismatch between zooplankton and the availability of their phytoplankton prey [33] therefore not only has implications for trophic transfer, but also for the successful overwintering of seasonally migrating copepods.

The additive effects of temperature and/or depth on the fatty acids 16:0 and 18:0 reported herein are consistent with the understanding that organisms manipulate the level of unsaturation of their membranes in order to maintain membrane functionality. Indeed, the inverse relationship between the PUFA DHA and SFAs (16:0 and 18:0) suggests that *C. acutus* respond to changes in their physical environment largely by changing the relative proportions of these compounds. It is possible that 16:0 and 18:0 are used interchangeably, as these compounds were similarly affected by pressure in the current study. However, the additional, additive effect of temperature on 16:0 suggests that the biophysical properties of this compound may differ to those of 18:0. The insignificant effect of depth on EPA is not entirely surprising given recent observations which suggest that this PUFA is not necessarily

Table 6. The effect of depth on the percentage composition of EPA (20:5(n-3)) in the phospholipids of *C. acutus*.

Model parameter	Coefficient (± SE)	t	p
Intercept	16.519±3.116	5.3018	<0.001
Temperature	3.567±1.114	3.138	0.0024

required for maintaining membrane fluidity, but rather plays a beneficial role in cell division and membrane organisation [34]. Indeed, the central role of EPA in buoyancy regulation may necessitate that this compound is only used for membrane maintenance when sufficient DHA is not immediately available. This interpretation is also consistent with the conformational properties and functions of EPA and DHA [35] and may explain why we did not find interactive effects between temperature and depth for EPA. Further investigation is required to address this more fully.

Membranes of mammals and birds also contain high proportions of DHA which co-vary with the MFA 18:1(n-9) [29]. In the copepods 18:1(n-9) was a minor component of PLFAs, and was substituted with the SFAs 16:0 and 18:0. These differences in the unsaturation of membrane bilayer composition between birds, mammals and invertebrates could plausibly be linked to metabolic activity [2]. Extrapolation of data in Fig. 5A suggests that *C. acutus* could potentially reach 3,000 m before its phospholipids would solely constitute DHA. This seems unlikely given that such high levels of DHA in biological membranes have not previously been found in nature. Teleosts have been reported at 8,000 m, and crustaceans are well known to thrive at full ocean depth, of around 11,000 m [36]. These observations apparently confirm that additional and/or alternative strategies for the maintenance of membrane functionality must be in existence and require further investigation. A study of cold tolerance and phospholipid saturation in the nematode, *Caenorhabditis elegans*, concluded that lipid saturation accounted from only 16% of observed differences in a variety of parameters for animals held at 10°C and 25°C. [37]. At the level of the whole organism, adaptation to pressure and thermal stressors is likely to be complex, with the unsaturation of membrane phospholipids being just one aspect, although it is clear that DHA plays an important role in the adaptation to pressure of *C. acutus* during its extensive vertical migrations through the ocean.

The ability to adapt to pressure suggests the existence of a mechanism by which copepods actually sense pressure. Some calcified crustaceans utilise statocysts to detect hydrostatic pressure [38]. Rudimentary statocysts are also present in calanoid copepods [39], although their potential role in the sensing of pressure is not known. Several pressure-regulated operons have been reported in deep-sea bacteria that regulate the synthesis of proteins that confer barotolerance, yet our understanding of the controlling mechanisms remains limited [40]. Clearly there is much to be learnt about the mechanism(s) through which aquatic organisms sense and respond to pressure.

We have demonstrated that DHA plays an important adaptive role in the life cycle of a species of marine copepod that undertakes extensive vertical migrations, and hence experiences changes in environmental temperature and pressure. The majority of DHA in the marine environment originates in flagellated microplankton, which must be accumulated in sufficient quantities via the diet before migration is initiated. Future, climate-driven changes in oceanic productivity and microplankton species composition could have major implications for the capacity of these animals to complete their seasonal life cycle.

Author Contributions

Conceived and designed the experiments: DWP GAT. Performed the experiments: DWP GAT. Analyzed the data: DJM DWP. Contributed reagents/materials/analysis tools: DWP DJM. Wrote the paper: DWP DJM GAT.

References

1. Guschina IA, Harwood JL (2006) Mechanisms of temperature adaptation in poikilotherms. FEBS Letters 580: 5477–5483.
2. Hulbert AJ (2007) Membrane fatty acids as pacemakers of animal metabolism. Lipids 42: 811–819.
3. Sinensky M (1974) Homeoviscous adaptation - a homeostatic process that regulates the viscosity of membrane lipids in *Escherichia coli*. Proc Nat Acad Sci 71: 522–525.
4. Cossins AR, Macdonald AG (1989) The adaptation of biological membranes to temperature and pressure: fish from the deep and cold. J Bioenerg Biomembr 21:115–35.
5. DeLong EF, Yayanos AA (1985) Adaptation of the membrane lipids of a deep-sea bacterium to changes in hydrostatic pressure. Science 228: 1101–1103.
6. Yano Y, Nakayama A, Ishihara K, Saito H (1998) Adaptive changes in membrane lipids of barophilic bacteria in response to changes in growth pressure. Appl Environ Microbiol 64: 479–485.
7. Valentine RC, Valentine DL (2004) Omega-3 fatty acids in cellular membranes: a unified concept. Prog Lipid Res 43: 383–402.
8. Cossins AR, Macdonald AG (1986) Homeoviscous Adaptation under Pressure. The Fatty-Acid Composition of Liver Mitochondrial Phospholipids of Deep-Sea Fish. Biochimica et Biophysica Acta 860: 325–335.
9. Vettier A, Labbe C, Amérand GA, Da Costa G, Le Rumeur E, et al. (2006) Hydrostatic pressure effects on eel mitochondrial functioning and membrane fluidity. Undersea Hyperbaric Medicine 33: 149–156.
10. Hazel JR, Williams EE (1990) The role of alterations in membrane lipid composition in enabling physiological adaptation of organisms to their physical environment. Prog Lipid Res 29:167–227.
11. Longhurst A (1998) Ecological geography of the sea. Academic Press, San Diego 398 p.
12. Dahms UC (1995). Dormancy in the copepod - an overview. Hydobiologia 306: 199–211.
13. Hirche HJ (1996) Diapause in the marine copepod *Calanus finmarchicus*—a review. Ophelia 44: 129–143.
14. Tarling GA, Shreeve RS, Ward P, Atkinson A, Hirst A (2004) Life-cycle phenotypic composition and mortality of *Calanoides acutus* (Copepoda: Calanoida) in the Scotia Sea: a modelling approach. Mar Ecol Prog Ser 272: 165–181.
15. Pond DW, Tarling GA (2011) Phase transitions of wax esters adjust buoyancy in diapausing *Calanoides acutus*. Limnol Oceanogr 56: 1310–1318.
16. Pond DW, Tarling GA, Ward P, Mayor DJ (2012) Wax ester composition influences the diapause patterns in the copepod, *Calanoides acutus*. Deep Sea Res II 59–60: 93–104.

17. Folch J., Lees M, Sloane-Stanley GHA (1957) A simple method for the isolation and purification of total lipid from animal tissues. J Biol Chem 226: 497–509.
18. Purać J, Pond DW, Grubor-Lajšić G, Kojic D, Blagojević DP, et al. (2011) Cold hardening induces transfer of fatty acids between polar and non-polar lipid pools in the Arctic collembolan *Megaphorura arctica*. Physiol. Entomol 36: 135–140.
19. Zuur AF, Ieno EN, Smith GM (2007) Analysing Ecological Data. Springer, New York.
20. Zuur AF, Ieno EN, Elphick CS (2010) A protocol for data exploration to avoid common statistical problems. Method in Ecology and Evolution 1. 3–14.
21. Mayor DJ, Thornton B, Hay S, Zuur AF, Nicol GW, et al. (2012) Resource quality affects carbon cycling in deep-sea sediments. The ISME Journal 6: 1740–1748.
22. Mayor DJ, Gray NB, Elver-Evans J, Midwood AJ, Thornton B (2013) Metal-macrofauna interactions determine microbial community structure and function in copper contaminated sediments. PLOS ONE 8: e64940.
23. R Development Core Team (2010) R: A Language and Environment for Statistical Computing. R Foundation for Statistical Computing. Vienna. Austria. ISBN 3-900051-07-0.
24. Oksanen J, Blanchet FG, Kindt R, Legendre P, O'Hara RB, et al. (2010) Vegan: Community Ecology Package. R package version 1.17–4.
25. Pinheiro J, Bates D, Debroy S, Sarkar D, R Development Core Team (2010) nlme: Linear and Nonlinear Mixed Effects Models. R package version 3.1–97.
26. Casadei MA, Manas P, Niven G, Needs E, Mackey BM (2002) Role of membrane fluidity in pressure resistance of *Escherichia coli* NCTC 8164. Appl Environ Microbiol 68: 5965–5972.
27. Kaye JZ Baross JA (2004) Synchronous effects of temperature and hydrostatic pressure, and salinity on growth, phospholipid profiles, and protein patterns of four *Halomonas* species isolated from the deep-sea hydrothermal vent and sea surface environments. Appl Environ Microbiol 70: 6220–6229.
28. Hazel JR (1995) Thermal adaptation in biological membranes: is homeoviscous adaptation the explanation? Annual Rev Physiol 57: 19–42.
29. Hulbert AJ (2003) Life, death and membrane bilayers. J Exp Biol 206: 2303–2311.
30. Bell MV, Dick JR, Anderson TR, Pond DW (2007) Application of liposome and stable isotope tracer techniques to study polyunsaturated fatty acid biosynthesis in marine zooplankton. J Plankton Res 29: 417–422.
31. Sargent JR, Bell MV, Bell JG, Henderson RJ, Tocher DR (1995) Origins and functions of n-3 polyunsaturated fatty acids in marine organisms. In: Ceve G, Paltauf F (eds) Phospholipid characterisation, metabolism and novel biological applications. American Oil Society Press. Champaign, IL, p 248–259.
32. Pond DW (2012) The physical properties of lipids and their role in controlling the distribution of zooplankton in the oceans. Journal of Plankton Research. Horizons 34: 443–453.
33. Edwards M, Richardson AJ (2004) Impact of climate change on marine pelagic phenology and trophic mismatch. Nature 430: 881–884.
34. Kawamoto J, Kurihara T, Yamomoto K, Nagayasu M, Tani Y, et al. (2009) Eicosapentaenoic acid plays a beneficial role in membrane organisation and cell division of a cold-adapted bacterium, *Shewanella livingstonensis* Ac10. Journal of Bacteriology 191: 632–640.
35. Rabinovich AL, Ripatti PO (1990) On the conformation properties and functions of docosahexanoic acid. Dokl RAN 314: 752–756.
36. Jamieson AJ, Fujii T, Mayor DJ, Solan M, Priede IG (2010) Hadal Trenches: the ecology of the deepest places on Earth. Trends in Ecology and Evolution 25: 190–197.
37. Murray P, Hayward ALH, Govan GG, Gracey AY, Cossins AR (2007) An explicit test of the phospholipid saturation hypothesis of acquired cold tolerance in *Caenorhabditis elegans*. Proc Nat Acad Sci 104: 5489–5494.
38. Fraser PJ, Macdonald AG (1994) Crab hydrostatic pressure sensors. Nature 371: 383–384.
39. Lowe E (1935) On the anatomy of a marine copepod, *Calanus finmarchicus* (Gunnerus). Trans Royal Soc Edinburgh 58: 561–603.
40. Li L, Kato C, Nogi Y, Horikoshi K (1998) Distribution of the pressure-regulated operons in deep-sea bacteria. FEMS Microbiol Letters 159: 159–166.

19

Adaptive Radiation along a Thermal Gradient: Preliminary Results of Habitat Use and Respiration Rate Divergence among Whitefish Morphs

Kimmo Kalevi Kahilainen[1,2]*, William Paul Patterson[3], Eloni Sonninen[4], Chris Harrod[5,6], Mikko Kiljunen[7]

1 Kilpisjärvi Biological Station, University of Helsinki, Kilpisjärvi, Finland, **2** Department of Environmental Sciences, University of Helsinki, Helsinki, Finland, **3** Department of Geological Sciences, Saskatchewan, Isotope Laboratory, University of Saskatchewan, Saskatoon, Saskatchewan, Canada, **4** Finnish Museum of Natural History, University of Helsinki, Helsinki, Finland, **5** School of Biological and Chemical Sciences, Queen Mary University of London, London, United Kingdom, **6** Instituto de Ciencias Naturales Alexander Von Humboldt, Universidad de Antofagasta, Antofagasta, Chile, **7** Department of Biological and Environmental Sciences, University of Jyväskylä, Jyväskylä, Finland

Abstract

Adaptive radiation is considered an important mechanism for the development of new species, but very little is known about the role of thermal adaptation during this process. Such adaptation should be especially important in poikilothermic animals that are often subjected to pronounced seasonal temperature variation that directly affects metabolic function. We conducted a preliminary study of individual lifetime thermal habitat use and respiration rates of four whitefish (*Coregonus lavaretus* (L.)) morphs (two pelagic, one littoral and one profundal) using stable carbon and oxygen isotope values of otolith carbonate. These morphs, two of which utilized pelagic habitats, one littoral and one profundal recently diverged via adaptive radiation to exploit different major niches in a deep and thermally stratified subarctic lake. We found evidence that the morphs used different thermal niches. The profundal morph had the most distinct thermal niche and consistently occupied the coldest thermal habitat of the lake, whereas differences were less pronounced among the shallow water pelagic and littoral morphs. Our results indicated ontogenetic shifts in thermal niches: juveniles of all whitefish morphs inhabited warmer ambient temperatures than adults. According to sampling of the otolith nucleus, hatching temperatures were higher for benthic compared to pelagic morphs. Estimated respiration rate was the lowest for benthivorous profundal morph, contrasting with the higher values estimated for the other morphs that inhabited shallower and warmer water. These preliminary results suggest that physiological adaptation to different thermal habitats shown by the sympatric morphs may play a significant role in maintaining or strengthening niche segregation and divergence in life-history traits, potentially contributing to reproductive isolation and incipient speciation.

Editor: Syuhei Ban, University of Shiga Prefecture, Japan

Funding: Funding of this study was obtained from Ministry of Agriculture and Forestry, Academy of Finland (projects #140903, #1268566, and #134139) and Emil Aaltonen Foundation. The funders had no role in study design, data collection and analysis, decision to publish, or preparation of the manuscript.

Competing Interests: The authors have declared that no competing interests exist.

* Email: kimmo.kahilainen@helsinki.fi

Introduction

A major task in evolutionary biology is to understand the formation of new species. Adaptive radiation is recognised as a key mechanism in speciation [1]. A central tenet of adaptive radiation is phenotypic adaptation to a specific environment [1]. Here, disruptive selection linked to adaptation to different environments is considered an important mechanism, both with regard to the initial divergence, as well as the maintenance of differences [1]. The final outcome of disruptive selection may lead to the formation of new species via ecological speciation [2]. Speciation research conducted over recent decades has demonstrated that many avian, terrestrial and aquatic species are now considered to have arisen via adaptive radiation and ecological speciation [1–3].

Most of this speciation research has focused on ecomorphological divergence, whereas the contribution of thermal adaptations in ecological speciation is generally poorly understood [4]. Many species distributions of closely related species exhibit distinct correlations with ambient temperature rather than ecological niche segregation [5]. Many famous adaptive radiations including Darwin's finches and *Anolis* lizards occur on volcanic island environments displaying pronounced altitudinal and vegetation gradients, resulting in different temperature and precipitation conditions [6,7]. However, even in these model systems most of the research has focused on ecomorphological divergence, largely neglecting the potential selective importance of temperature and related thermal adaptations. Such adaptations are likely most important for poikilotherms, such as amphibians, reptiles and fish.

Polymorphic fishes inhabiting post-glacial lakes have become important models for understanding how disruptive selection acts on phenotypic divergence and how it is related to accumulation of reproductive isolation between diverging species [8–10]. Across different fish lineages, a common pattern displayed is morphological divergence correlated to specific resource utilization in pelagic-littoral resource axis in lakes [e.g. 11]. In more complex cases, morphological divergence can be extended to a third principal lake habitat, the profundal [12–14]. In contrast to morphological divergence, which is well studied, far less is known about physiological adaptations to specific niches. Because freshwater fishes can be divided into cold, cool and warm water adapted species according to species-specific thermal optima, tolerance ranges and habitat selection [15], it is also likely that physiological adaptions should arise early in the divergence process. One of the best experimental studies comes from the salmonid fishes, where the closely related and morphologically similar cisco (*Coregonus fontanea*) and vendace (*Coregonus albula*) from Lake Stechlin in Germany differ in their optimum temperature preferences and standard metabolic adaptations [16]. Beyond this, little is generally known about temperature preferences and metabolic adaptions of polymorphic fish.

Adaptive radiation and ecological speciation is common throughout the natural range of European whitefish (*Coregonus lavaretus* (L.)) [17–19]. Whitefish are known for their ability to diverge, allowing the utilisation of pelagic and benthic habitats in lakes, with up to five sympatric morphs in the same lake [13,17,20]. Ecomorphological divergence of whitefish into morphs using the three major habitats (i.e. pelagic, littoral and profundal zones) in large, deep lakes is particularly pronounced in northern Fennoscandian lakes [12,13,21]. Apart from variation in food resources [22] and basal sources of energy [21], these habitats also differ with regard to their thermal conditions. Thermal stratification of the water column in these lakes means that littoral and pelagic habitats undergo large seasonal changes in water temperature between warm summer months (water temperature range 10–20°C) and cold winter months (0–2°C), when lakes are ice-covered [22]. Profundal fish occupy a cool/cold habitat (2–8°C) throughout the year. Reproductive isolation among whitefish morphs is generally based on spatial and temporal differences in spawning place and time [14,17,19].

Whitefish morphs dwelling in these different habitats differ in the number of gill rakers, which is commonly correlated to trophic ecology and habitat use in each of sympatric whitefish morphs and is also a heritable trait [9,17,23,24]. Gill raker number is closely correlated with overall head and body morphology: pelagic morphs are typically fusiform with a higher number of gill rakers, whereas benthic morphs have more robust body shape and a lower number of gill rakers [21,23]. Much of our understanding of polymorphic whitefish ecology relates to studies conducted during summer [21,23,25], and equivalent studies of habitat selection during winter months (i.e. when lakes in these regions are ice covered) are lacking. Furthermore, data regarding lifetime patterns of thermal niche selection from the point of hatching to maturation are not currently available from polymorphic fish.

Such studies of long-term thermal habitat selection could include the use of electronic data storage tags. However, these approaches are not suitable for delicate and small bodied fishes such as pelagic and profundal whitefish morphs. Here, it is more suitable to use natural anatomical recorders from individual fish such as balancing bones (e.g. otoliths), that document information such as secular variation in seasonal metabolism [26], thermal preferences [27], and growth of fish [28]. Furthermore, otolith microchemistry provides a chemical record of the annual changes in temperature related to metabolism that can be used to quantify variation in the life-time thermal niche via stable isotope analysis [29]. Oxygen stable isotope values ($\delta^{18}O$) have also been used to calculate temperatures encountered during the lifetime of fishes living in the distant past [e.g. 30]. Furthermore, the inorganic carbon isotope value ($\delta^{13}C$) of otoliths is correlated with the specific respiration rate of individual fish [e.g. 26,31,32], and analysed in parallel with $\delta^{18}O$ values. The use of micromilling techniques facilitates sampling of otoliths at a subannual scale throughout the lifetime of the fish by discretely isolating aliquots of carbonate that may represent days to weeks, depending on the growth rate and size of the otolith [29,33]. While otolith $\delta^{18}O$ and $\delta^{13}C$ values have been successfully used to track the lifetime thermal niche and respiration rates for many marine and freshwater species [32,34–36], we are unaware of any comparative studies of polymorphic species.

In this preliminary study, we were interested in putative differences in lifetime thermal niche selection among individuals of different sympatric whitefish morphs inhabiting a subarctic lake and whether such whitefish morphs displayed ontogenetic thermal shifts. If differences in thermal niche selection are apparent, we were interested in determining any correlative differences in respiration rates among whitefish morphs. Finally, we evaluated the likelihood of different hatching times among whitefish morphs using micromilling of otolith nuclei. According to previous data collected during summer sampling campaigns relating to temperature and habitat selection, the profundal whitefish morph was hypothesized to dwell in the coldest and the most annually thermally stable environment relative to the littoral and pelagic morphs [21,25]. If the profundal morph is adapted to a cold and thermally stable environment, we predicted that its respiration rates should be both lower and less variable than morphs that inhabit shallower, warmer habitats. Following the results of a previous otolith isotope study conducted on an alpine monomorphic whitefish population in Lake Annecy, France [32], we hypothesized that all whitefish morphs likely undertake an ontogenetic niche shift from a warmer juvenile thermal niche to a cooler adult thermal niche and that in all morphs respiration rates should decrease with age. Because all sympatric whitefish morphs investigated in this study lake display genetic differences and partial reproductive isolation according to neutral nuclear markers [14,37], we expected to find measurable differences in estimated hatching temperatures among the morphs, reflecting differences in spawning time and inter-morph reproductive isolation.

Materials and Methods

Study area and fish fauna

Samples were collected from the subarctic oligotrophic (total phosphorus 6 μg/L and total nitrogen 160 μg/L) Lake Paadar (68°52′N, 26°35′E), located in the Paatsjoki watercourse in Finnish Lapland (Fig. 1). The drainage area comprises a mixture of peatland, pine (*Pinus silvestris*) and birch (*Betula* spp.) forest. The surface area of the lake is circa 21 km^2, the maximum depth is 56 m, with a mean depth of 11.7 m [40]. Proportions of principal habitat types in this lake are littoral 38% and pelagic/profundal 62% [21]. The fish fauna consists of nine species, whitefish, brown trout (*Salmo trutta* L.), grayling (*Thymallus thymallus* (L.)), pike (*Esox lucius* L.), burbot (*Lota lota* (L.)), perch (*Perca fluviatilis* L.), three-spined stickleback (*Gasterosteus aculeatus* L.), nine-spined stickleback (*Pungitius pungitius* (L.)) and minnow (*Phoxinus phoxinus* (L.)) [40]. Whitefish dominate the fish community, contributing ~90% of the total catch by number [23]. Lake Paadar whitefish stock has evolved into two pelagic and two benthic morphs, each are genetically differentiated from each other [37], according to neutral nuclear DNA (range of F_{st} = 0.02–0.08).

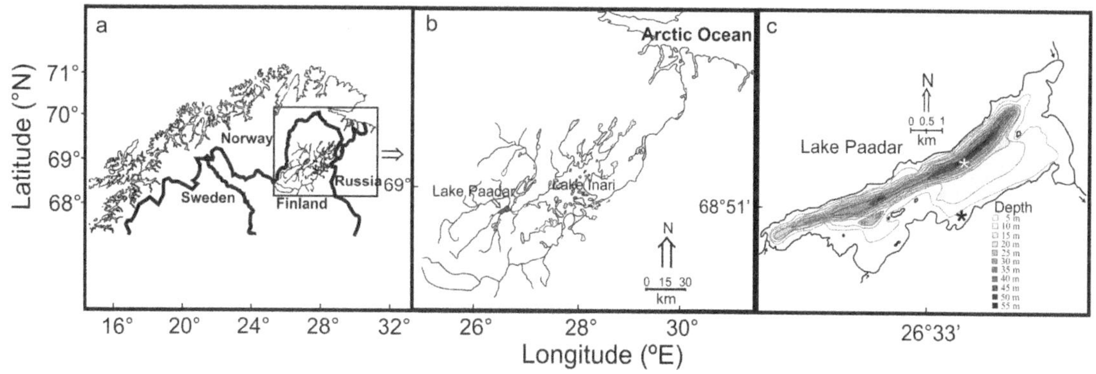

Figure 1. Study area in northern Fennoscandia (a), highlighting the Paatsjoki watercourse (b), and study lake (c). The asterisk in Fig. b indicates the position of water column temperature measurements in 1999–2004 Lake Inari (see Figure 2) and in Fig. c, the position of water column isotope sampling in the pelagic (white asterisk) and littoral (black asterisk) habitats in Lake Paadar (data presented in Figure 3).

Sampling

Whitefish were sampled for research purposes only in September 2004 using a gill net series combined of eight gill nets, each 30 m long and 1.8 m high, with different mesh sizes (12, 15, 20, 25, 30, 35, 45 and 60 mm). Gill net series were set in littoral, pelagic and profundal habitats in the evening and collected the following morning. Fish were immediately removed from gill nets, euthanized by cerebral concussion according to the Finnish Animal Conservation Law (32§ 9.8.2013/584) conducted by lead author (KKK) and subsequently placed on ice for measurements in the laboratory. Fishing rights in Finland belong to the land owner according to the Finnish Fishing Law (5§ 27.5.2011/600) and lead author (KKK) get a permit for research fishing from the land owner (Forest and Park Service, permit number 14.5.2004, 1386/5713-2004). No ethical permission is required for described scientific sampling with gill nets according to the Finnish Animal Conservation Law (7§ 28.6.2013/498).

Fish were divided by species (whitefish by morphs) and total length (±1 mm) and wet mass (±0.1 g) were measured from every captured fish. Whitefish were classified into morphs in the field reflecting differences in gill raker, body and head shape [12]. The profundal small sparsely rakered (SSR) benthivorous whitefish morph are small bodied with the lowest relative number (circa 15–20) of short, bent and very widely spaced gill rakers. They also have a characteristic sub-terminal mouth and large eyes positioned high on the head, large fins and a robust body shape [12,21]. The littoral benthivorous morph, that we refer to as the large sparsely rakered (LSR) morph has an intermediate number (c. 20–30) of short and widely space gill rakers, characteristic whitefish coloration and a terminal mouth. A similar body and head shape is typical for the large densely rakered (LDR) morph, but compared to the LSR morph, gill rakers are more numerous (c. 30–36), longer and more densely spaced. The fourth morph is the densely rakered (DR) whitefish, with a slender fusiform body, superior mouth, and numerous (c. 30–40), fine, long and very densely spaced gill rakers [12,21]. A size-based subsample (c. 40–50 individuals) reflecting the length distribution of the catch was frozen (−20°C) for more detailed analysis. The main prey items of the different whitefish morphs were collected for carbon and nitrogen stable isotope analyses [21]. Pelagic zooplankton was collected with vertical hauls using 25 cm diameter (50 μm mesh size) plankton net. Littoral (sampling depths 1–3 m) and profundal (10–40 m) invertebrate samples were collected using Ekman grab. Water samples for oxygen and inorganic carbon stable isotopes were collected at the end of July 2012 from pelagic and littoral habitats (Fig. 1). The oxygen isotope values of water in the pelagic zone is unlikely to change considerably despite the time delay between otolith (2004) and water sampling (2012) due to the long residence time of the large study lake and the sampling time in the middle of summer [38,39]. For inorganic carbon isotopes, there are obviously some seasonal differences between thermally stratified and non-stratified seasons [32] and thus our results from summer thermal stratification should be considered as a snap shot. Temperature and oxygen profiles were recorded at the same time.

Laboratory work

Whitefish samples were thawed in the laboratory and positioned for photographs taken of the left flank of each fish [21]. The first left gill arch was subsequently dissected and gill rakers were counted under a microscope. The middle gill raker length was also measured for comparison of pelagic and benthic morphs [12]. Scales and sagittal otoliths were removed for the determination of age and growth. A section of dorsal muscle was dissected for carbon ($\delta^{13}C$) and nitrogen ($\delta^{15}N$) stable isotope analyses. Muscle and invertebrate samples were dried (60°C) for 48 hours and later ground into a fine powder. A small amount (0.5–0.6 mg) of powder was encapsulated in tin cups. Water samples were collected from littoral (depths 0–3 m) and pelagic/profundal (0–33 m) habitats (Fig. 1) and directly stored in pre-combusted scintillation vials. Water samples for dissolved inorganic carbon were collected and directly injected into pre-phosphoric acid loaded and helium flushed Exetainer vials. All tissue and water samples were analyzed using an isotope mass spectrometer.

Whitefish were aged using both scales and otoliths [41]. According to these estimates of age, we selected a single individual of similar age (5+) from each morph to assess the lifetime thermal habitat use and respiration rates through otolith isotope analysis. Initially, we aimed to run several individuals per morph, but due to time and financial constraints this was not possible. However, of the majority of similar studies typically rely on small sample sizes and an individual-based approach without population-level comparisons [27,32,34]. Due to the lack of this type of data from other polymorphic poikilothermic populations, we hope that our preliminary results would promote more research on thermal adaptation.

One otolith of each morph was glued (Crystalbond©) to a microscope slide, sanded and polished flat on one side until the otolith nucleus was reached. Otoliths were photographed under a stereomicroscope, and the location and spacing of summer and winter circuli determined in order to characterize subsequent sampling paths via computer-controlled micromilling. This was conducted using a custom three-dimensional micromilling system in the Saskatchewan Isotope Laboratory facilities at the University of Saskatchewan [33]. This system permitted the sampling of 33–40 paths concordant with the growth banding in each otolith. The required sample size for isotope analysis (c. 7–25 μg) was obtained from almost all sampling paths (total 140 paths from 4 otoliths), with only four samples yielding insufficient material for analysis. Precision of carbon and oxygen analyses was 0.04 and 0.07‰, respectively.

Temperature Calculations

Isotopic values for fish, otolith, invertebrate and water samples were calculated using:

$$\delta X = \frac{(R_{sample} - R_{standard})}{R_{standard}} \times 1000$$

where δX is the stable isotope ratio of dorsal muscle or otolith carbonate and R is the isotope ratio of carbon, nitrogen or oxygen in the *sample* and the *standard*. As muscle isotope values show variations in C:N ratios, suggesting differences in lipid concentrations, we arithmetically corrected the values [42].

We used the average habitat-specific water and micro-milled otolith $\delta^{18}O$ values to calculate whitefish lifetime thermal niche. Species-specific models are not available for whitefish: hence we estimated values using a general model [29] and models developed for closely related salmonids, brook charr [*Salvelinus fontinalis*; 43] and Arctic charr [*Salvelinus alpinus*; 44]. Here, the latter non-linear model by Godiksen et al. [44] for Arctic charr provided the most similar values compared to annual water column temperature datasets during the lifetime of sample fish from the same watercourse (Fig. 2). The other models gave on the average 0.5–5.0°C higher temperature values (Table S1) than the non-linear model for Arctic charr [44]. Otolith $\delta^{18}O$ values reported relative to Vienna PeeDee Beleminite (VPDB) were transformed to the Vienna standard mean ocean water (VSMOW) scale:

$$\delta^{18}O_{otolith(VSMOW)} = 1.03092 * \delta^{18}O_{otolith(VPDB)} + 30.92$$

We used $\delta^{18}O_{water}$ values from the water samples collected from each habitat type and whitefish otoliths to calculate the fractionation factors (α) for each sample:

$$\alpha = \frac{\delta^{18}O_{otolith(VSMOW)} + 1000}{\delta^{18}O_{water(VSMOW)} + 1000}$$

where $\delta^{18}O_{otolith(VSMOW)}$ is the isotope value of otolith samples and $\delta^{18}O_{water(VSMOW)}$ is the isotope value of water samples (both presented relative to the VSMOW scale). These values were then utilized in the polynomial model [44]:

$$1000\ln\alpha = a + \frac{b}{T} + \frac{c}{T^2}$$

where T is temperature (10^3/K), a, b and c are empirical constants. The empirical constants a, b and c were 1089.24, 89.9 and −617.19 [44].

Respiration calculations

We used photographs to measure and calculate caudal aspect ratio, a morphometric index of fish swimming activity [45]:

$$K_{caud} = \frac{h^2}{s}$$

where h is the height of the opened caudal fin and s is caudal fin surface area. Here, higher values indicate streamlined and active fish such as pelagic cruisers, and low values represent less active, benthivores [45]. In general, K_{caud} can be used successfully to

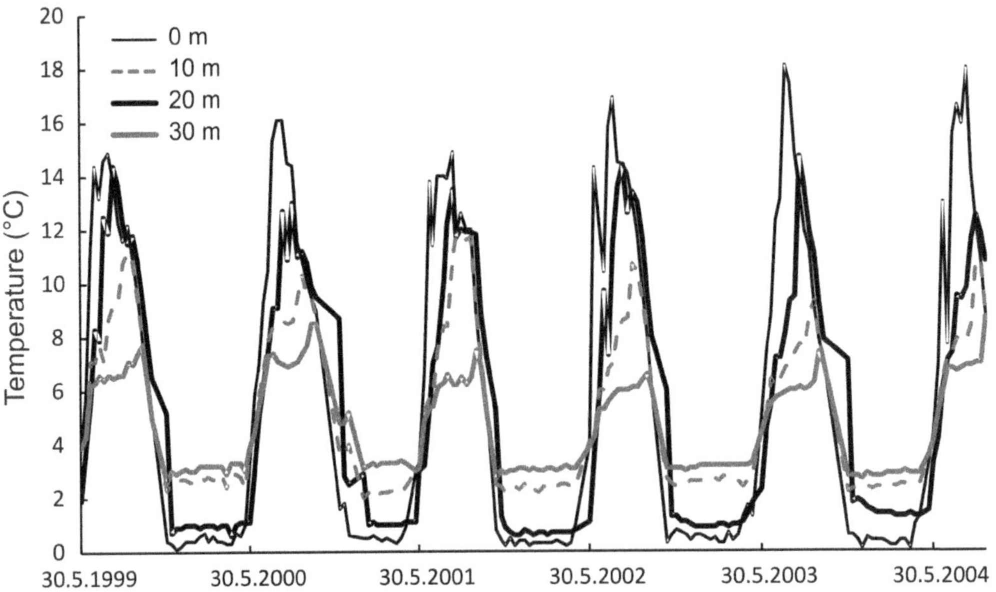

Figure 2. Annual temperature profiles (0–30 m) from Lake Inari, used as a proxy for Lake Paadar annual temperatures, during the life-time (1999–2004) of sampled whitefish morphs.

predict otolith $\delta^{13}C$ values [46]. Solomon *et al.* [31] further developed this relationship to directly estimate proportion of otolith carbon that is metabolically derived (*M*):

$$M = 0.025 + 0.066K_{caud}$$

In addition, estimated of lifetime metabolism of the different whitefish morphs were calculated using the Wisconsin bioenergetics model with parameters for the generalized coregonid model [47]. To model parameters of consumption and to empirically estimate respiration, the model requires data that reflect lifetime temperature, growth in terms of wet mass, proportion of prey items consumed, as well as the energy density of whitefish morphs and invertebrate prey. For the input data, we used temperature estimates from the $\delta^{18}O_{otolith}$ values, i.e. starting the modelling at the estimated time of hatching (1th of June 1999, first modelling day = 150) and continued until the time of death (15th of September 2004 = modelling day 2085). The age and length data from the total sample of each whitefish morph was used to calculate von Bertalanffy non-linear growth curves [48]:

$$L_t = L_\infty \left[1 - e^{-K(t - t_0)}\right]$$

where L_t is the expected total length at time t, L_∞ is the asymptotic total length (i.e. theoretical total length to which a fish would grow if time permitted), K is the rate at which the growth curve approaches the asymptote and t_0 is the hypothetical time at which L_t was zero (i.e. hatching time).

The estimated lengths were transformed into wet mass with corresponding whitefish morph specific mass-length relationships [49]:

$$W = aL^b$$

where W = wet mass (g), L = total length (cm), a = a constant, b = an exponent. Due to a lack of seasonal diet data, we assumed that all whitefish morphs consumed zooplankton prey during the first two months of life and thereafter ate prey associated with their respective habitat, i.e. DR whitefish consumed pelagic zooplankton, SSR whitefish fed on profundal benthic macroinvertebrates and LSR whitefish ate littoral benthic macroinvertebrates [21,25]. As LDR whitefish diet consists of both pelagic zooplankton and insect pupae or adults, we used 50% pelagic and 50% littoral diet proportions. We calculated the energy density (*Y*) of whitefish morphs and invertebrates using the proportion of elemental carbon (%) derived from the isotope analysis of dorsal muscle and invertebrates [50]:

$$Y = -8.415 + 0.62X$$

where X is the proportion of carbon (%). Energy densities of whitefish morphs and their invertebrate prey were kept as constant. The calculated energy contents for the SSR, LSR, LDR and DR whitefish were 5227, 5134, 5088 and 5150 J/g, respectively. Energy contents for littoral, profundal and pelagic invertebrates were 4694, 4341 and 2747 J/g.

The average gill raker count, gill raker length and isotope values of dorsal muscle ($\delta^{13}C$, $\delta^{15}N$) were statistically compared using analysis of variance and pairwise comparisons with Tukey's test.

Results

The total survey catch was 939 whitefish, of which 179 (40–50 individuals per morph) were subjected to more detailed analysis (Table 1). Whitefish morphs showed differences in gill raker number (ANOVA, $F_{3,175} = 850$, $p<0.001$) and the length of the longest raker (ANOVA, $F_{3,175} = 208$, $p<0.001$): the morphs differed from each other with regard to both measures (Tukey's pairwise comparisons $p<0.05$), except SSR and LSR whitefish which overlapped in gill raker length. There were also significant differences in muscle tissue $\delta^{13}C$ values (ANOVA, $F_{3,175} = 167$, $p<0.001$) and $\delta^{15}N$ values (ANOVA, $F_{3,175} = 200$, $p<0.001$). Here, all morphs differed (Tukey's pairwise comparisons $p<0.05$), except SSR and DR whitefish in $\delta^{13}C$ values. The ecomorphological traits of the individuals selected for the otolith micromilling from each morph were in the range of values for the sample of the larger population (Table 1).

There was a clear variation in $\delta^{18}O_{H2O}$ and $\delta^{13}C_{DIC}$ values in the water column (Fig. 3). As expected, the highest $\delta^{13}C_{DIC}$ values were observed near the surface (−6.4‰ $_{VPDB}$) and the lowest at the deepest sampling depth of 33 m (−12.4‰$_{VPDB}$). $\delta^{18}O_{H2O}$ values were less variable, ranging between −13.7 and −12.6‰$_{VSMOW}$ (Fig. 3). Mean $\delta^{18}O_{H2O}$ values were calculated for each whitefish morph habitat i.e. littoral (mean for depths 0, 1, 2, 3 m, −13.1‰$_{VSMOW}$), pelagic (mean for depths 0, 1, 3, 6, 9 m, −13.3‰ $_{VSMOW}$) and profundal (mean for depths 21, 24, 27, 30, 33 m, −13.6‰ $_{VSMOW}$). At the time of sampling, there was evidence for water column thermal stratification, with water temperatures ranging from 15°C at the surface to 8°C at 30 m (Fig. 3). Oxygen concentrations at this depth interval were *ca.* 9 mg l^{-1}.

$\delta^{18}O_{otolith}$ and $\delta^{13}C_{DIC}$ values of otoliths indicated differences among the whitefish morphs (Fig. 4). The large bodied littoral LSR whitefish had consistently higher otolith $\delta^{13}C$ values than the other morphs and also displayed the most pronounced seasonal variations. The individual of the pelagic LDR whitefish morph also exhibited strong seasonal variation in otolith $\delta^{13}C$ values, as well as some indication of higher otolith $\delta^{13}C$ values with age. The DR whitefish showed seasonal variation in otolith $\delta^{13}C$ and $\delta^{18}O$ values within the two first years of its life, but not in subsequent years. In contrast to the other morphs, the profundal SSR whitefish had the most stable otolith $\delta^{13}C$ and $\delta^{18}O$ values, exhibiting variation in only the first year of life (Fig. 4).

$\delta^{18}O_{otolith}$ values indicated seasonal water temperature variability for the 5+ individuals examined of each of the four of the whitefish morphs, which taken together, represents a time period that extended across six summer and five winter seasons (Fig. 5). Differences between summer and winter habitat temperatures were the most pronounced for the littoral LSR and the pelagic LDR whitefish, suggesting that both inhabit near-surface habitats. The pelagic DR whitefish display pronounced seasonal differences in the two first years of life, with apparently reduced seasonality in subsequent years. The profundal SSR whitefish individual displayed the most distinct thermal history, living at temperatures <11°C throughout their entire lifetime and with a pattern of decreasing temperatures after the first two years of life (Fig. 2 and 5). Although ontogenetic temperature niche shifts were most pronounced for the profundal SSR whitefish and the pelagic DR whitefish, there was a sign of decreased temperature variation for both the LSR and LDR whitefish too. $\delta^{18}O$ values of the kernel (nucleus of otoliths), indicated that the two benthic morphs (profundal SSR whitefish and littoral LSR whitefish) hatched at higher temperatures (11–12°C) than the pelagic (DR and LDR whitefish) morphs (8–9°C).

Table 1. Basic data from whitefish morphs (SSR = small sparsely rakered, LSR = large sparsely rakered, LDR = large densely rakered, DR = densely rakered whitefish).

Variable	SSR whitefish	LSR whitefish	LDR whitefish	DR whitefish
N	40	48	41	50
Total length (cm)	17.5±3.6 11.1-27.4 (16.7)	18.5±4.1 12.2-27.0 (25.9)	29.3±1.9 21.6-32.1 (25.0)	14.0±1.5 11.3-17.8 (15.4)
Total mass (g)	48.1±39.6 10.8-218 (33.6)	52.9±36.4 13.0-160.0 (141.1)	202.9±32.8 85.3-252.6 (124.2)	20.2±6.2 10-37.7 (25.0)
Gill raker number	18.3±1.3 16-21 (19)	24.4±2.2 20-31 (26)	33.7±1.9 30-38 (33)	36.3±2.0 32-40 (38)
Gill raker length (mm)	2.4±0.5 1.5-4.0 (2.6)	2.7±0.6 1.6-4.3 (3.4)	6.3±0.8 4.3-7.6 (5.3)	4.1±0.6 2.8-5.4 (4.6)
$\delta^{13}C$ (‰$_{VPDB}$) dorsal muscle	−28.4±1.1-31.9-26.8 (−28.0)	−24.5±1.5-27.5-21.9, (−22.6)	−26.8±0.7-28.1-25.5 (−26.5)	−28.4±0.2-28.7-27.5 (−28.6)
$\delta^{15}N$ (‰$_{AIR}$) dorsal muscle	9.9±0.4 9.0-10.8 (10.2)	7.1±0.6 6.0-8.1 (6.6)	7.6±0.8 5.5-9.8 (7.4)	8.6±0.5 7.8-9.9 (8.7)
Mean $\delta^{18}O$ (‰$_{VPDB}$) otolith	(−12.4)	(−12.6)	(−12.8)	(−12.8)
Mean $\delta^{13}C_{DIC}$ (‰$_{VPDB}$) otolith	(−15.0)	(−11.6)	(−15.1)	(−16.1)
Invertebrate prey $\delta^{13}C$ (‰$_{VPDB}$)	−29.4±1.1	−23.4±1.9	−30.2±0.6, −23.4±1.9	−30.2±0.6
Invertebrate prey $\delta^{15}N$ (‰$_{AIR}$)	6.6±0.5	2.8±1.2	5.3±0.5, 2.8±1.2	5.3±0.5
M, estimated	0.260	0.286	0.322	0.349

The average (±SD) and range (min–max) for each morph are indicated with the individual used for the otolith micromilling in parenthesis. $\delta^{18}O$ and $\delta^{13}C_{DIC}$ values indicate the average value from all drilled transects. Isotopic values of invertebrate prey of LDR whitefish indicate both pelagic zooplankton and littoral macroinvertebrate values. *M* is the proportion of metabolically derived otolith carbon estimated from the caudal aspect ratio (K_{caud}) [31].

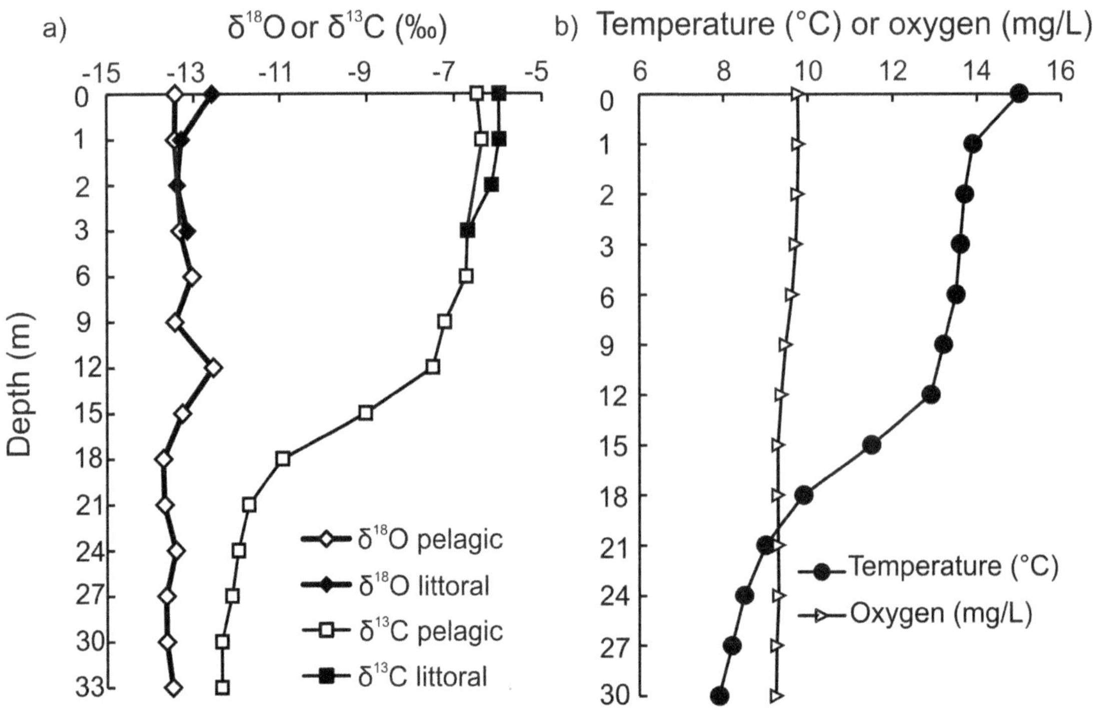

Figure 3. Oxygen ($\delta^{18}O$) and dissolved inorganic carbon ($\delta^{13}C_{DIC}$) stable isotope values and profiles of temperature and oxygen (b) measured from the littoral (0–3 m) and pelagic (0–33 m) water column of Lake Paadar.

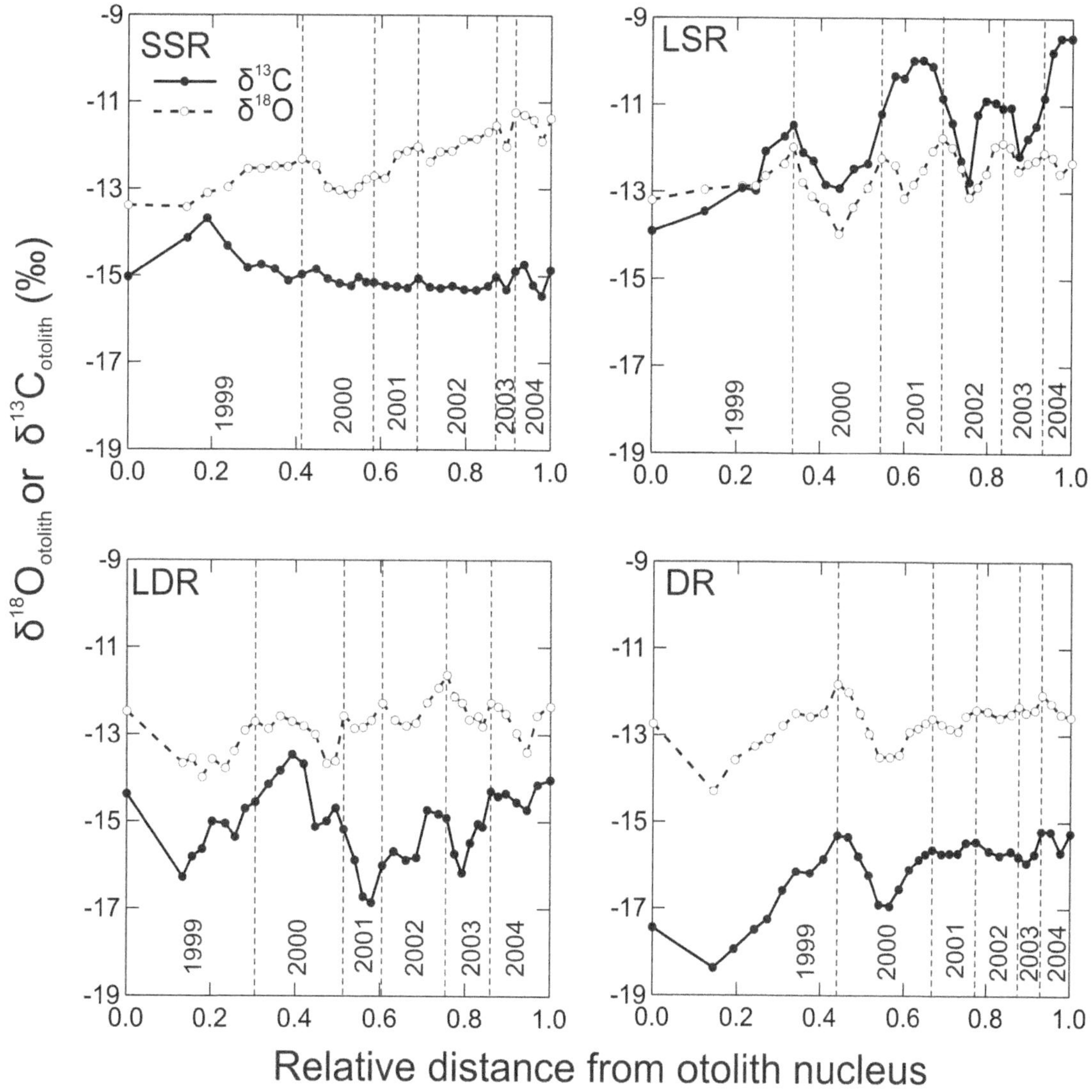

Figure 4. δ^{18}O and δ^{13}C values estimated from individual whitefish morph otoliths (SSR = small sparsely rakered, LSR = large sparsely rakered, LDR = large densely rakered and DR = densely rakered whitefish). The first x-axis value indicates the nucleus.

The estimated proportion of metabolically derived otolith carbon (*M*) calculated from caudal fin morphology was the lowest for profundal SSR whitefish and the second lowest for littoral LSR whitefish (Table 1). The pelagic morphs had markedly higher *M*, with peak values observed for DR whitefish. Bioenergetic modelling revealed clear growth differences among the morphs. Somatic growth (assessed with wet mass in the model) was the fastest for the large bodied morphs (pelagic LDR whitefish and littoral LSR whitefish), both of which continue growing after six growing seasons (Fig. 6). Somatic growth was much lower for SSR whitefish, which displayed steadily increasing weight. The lowest growth was observed in the pelagic DR whitefish, which attained their estimated maximum size after three years of life. The corresponding estimates of specific respiration rates indicated that seasonal fluctuations were the highest for littoral LSR whitefish and the lowest for profundal SSR whitefish (Fig. 6). The specific respiration rates of pelagic LDR whitefish and DR whitefish were similar, and seasonal fluctuations were generally lower than those estimated from the littoral LSR whitefish.

Discussion

We found support for our predictions that the different whitefish morphs occupied distinct thermal niches associated with differences in estimated respiration rates. The profundal SSR whitefish occupied the coldest, most stable thermal niche and had the lowest respiration estimates, whereas annual fluctuations in the thermal niche were the most pronounced in the littoral LSR whitefish. The thermal niche of the pelagic LDR and DR whitefish also showed clear seasonal fluctuations: respiration rates in individuals from these morphs were generally higher than the SSR whitefish and lower than the LSR whitefish. All morphs used a warmer thermal niche earlier in their ontogeny relative to later life, suggesting ontogenetic movement into colder water. Finally, we demonstrated that pelagic morphs hatched at lower temperatures than benthic morphs, according to the first otolith measurement in the nucleus, which likely relate to temporal and spatial differences in spawning.

In general, our evidence for thermal niche segregation among whitefish morphs provides a further dimension to previously

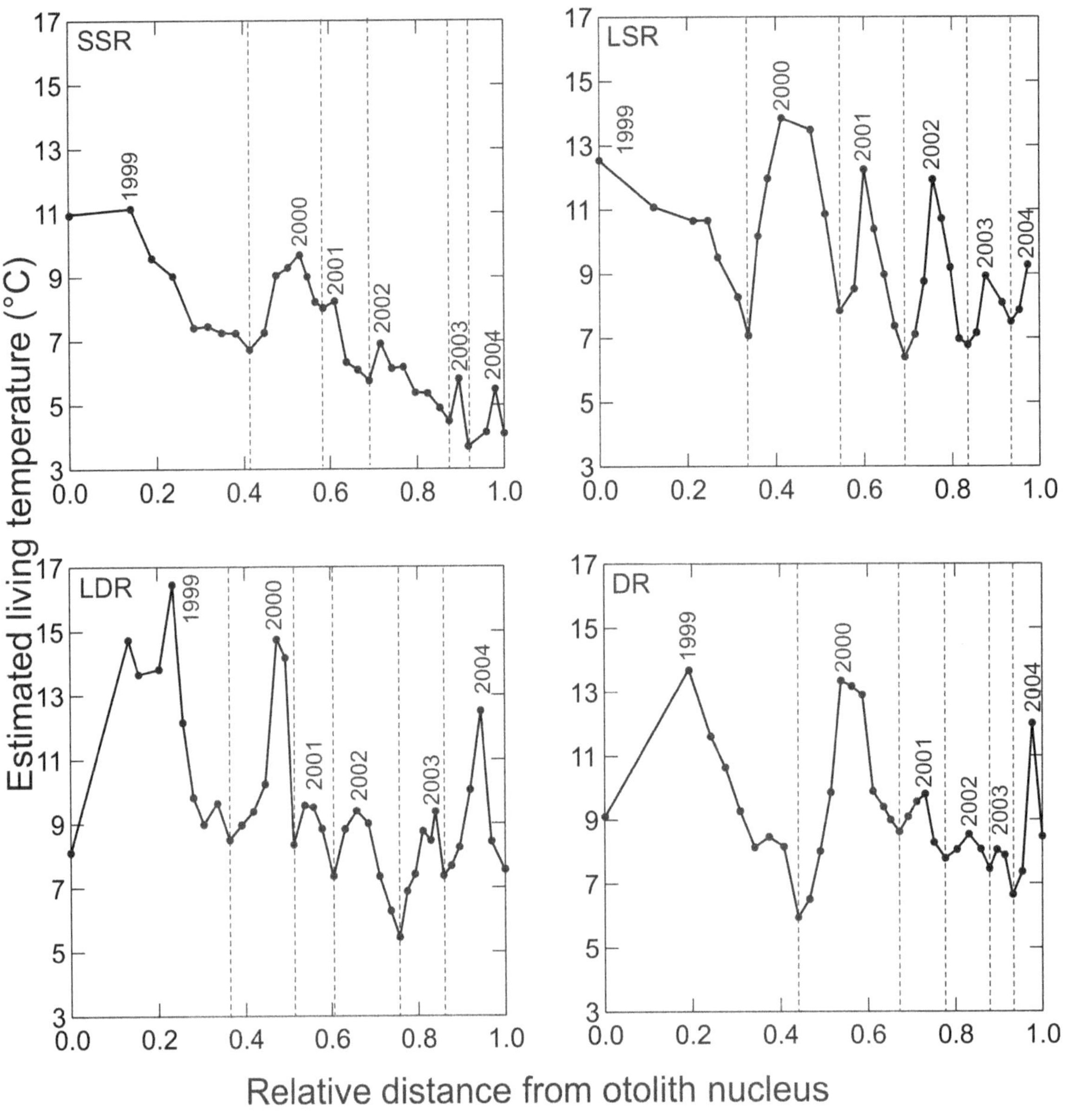

Figure 5. Lifetime temperature estimates for the sympatric whitefish morphs (SSR = small sparsely rakered, LSR = large sparsely rakered, LDR = large densely rakered and DR = densely rakered whitefish) according to stable isotopes of oxygen in otoliths. The first x-axis value indicates the nucleus i.e. estimated temperature at hatching.

documented ecomorphological and life-history divergence of these whitefish morphs [12,21,25]. Sympatric whitefish morphs have diverged and now exploit the three principal habitat types found in lakes (littoral, pelagic and profundal). LSR whitefish are ubiquitous to lakes in the study area, and when found in sympatry with other morphs is a littoral benthivore [21,41]. Otolith $\delta^{18}O$ values indicated that LSR whitefish tolerated pronounced yearly temperature variability, typical for shallow water habitats, suggesting a life-time occupation of littoral habitats. The results from the other benthivorous morph, the SSR whitefish [41] indicate consistent occupation of cold-water profundal niches. Both pelagic morphs displayed evidence for annual variability in their thermal niche, typical for thermally labile pelagic habitats. Results from the LDR whitefish showed more pronounced fluctuations than those shown by the DR whitefish, most likely related to near surface niche occupation in the LDR rather than the DR whitefish [25,51]. In case of whitefish morphs, thermal adaptations are correlated to other ecomorphological divergence, which may be the case with other taxa as well.

There is some recent evidence that adaptive benefit of larger bill size of birds in warm environments is a metabolic benefit via heat loss [52]. Traditionally, bill size is very well known to correlate with foraging, song and mate selection for example in Darwin's finches [6,53]. Different *Anolis* lizard species in the Caribbean Islands utilize distinct parts of trees, reflecting different microclimates [7]. Thermal preferences may have at least partly originated via use of these microclimates and furthermore, may influence relative ecomorphological performance in their respective niche. It is apparent that ecomorphological and thermal adaptations can

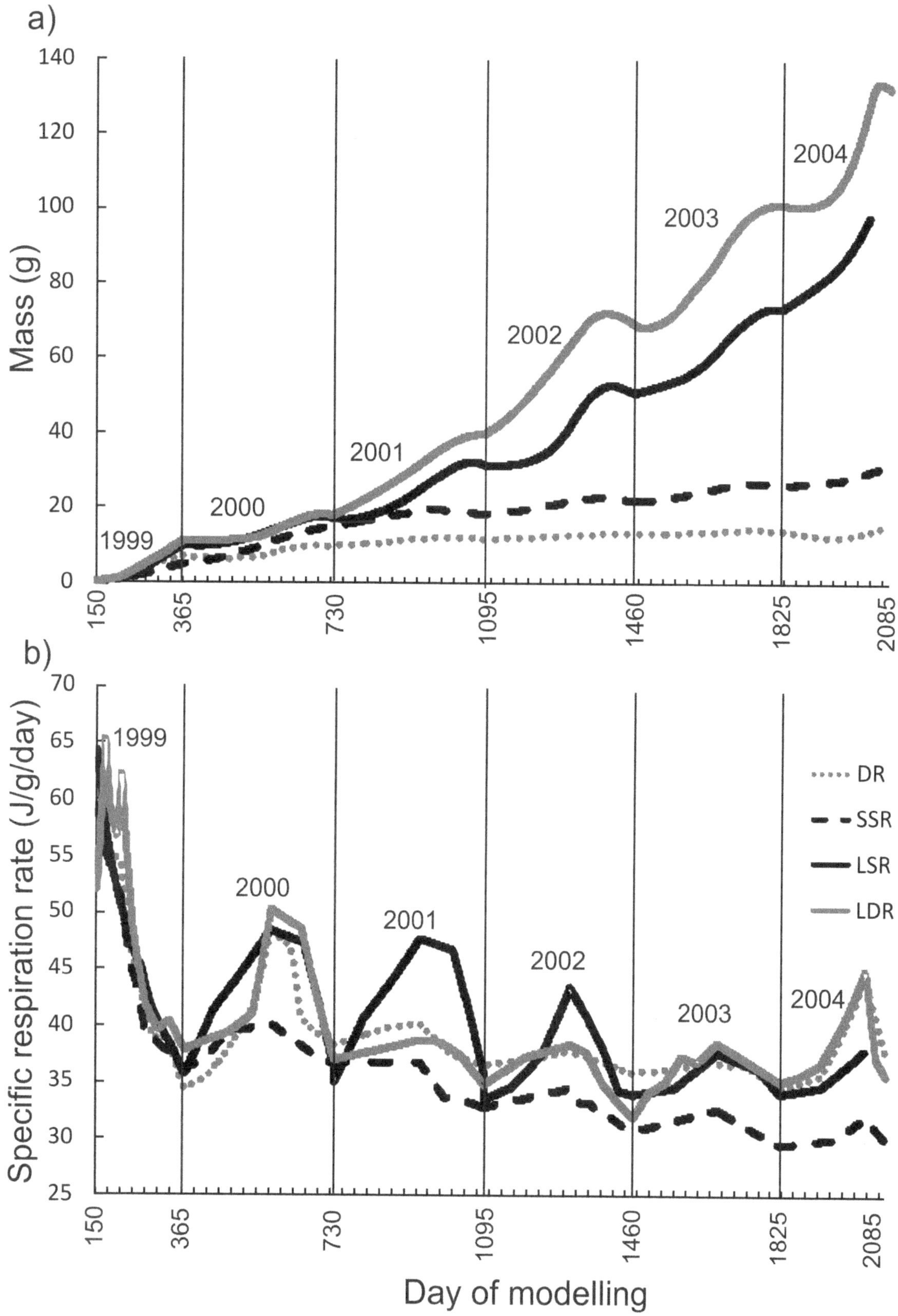

Figure 6. Bioenergetic modelling results of the weight gain (a) and the specific respiration rate (b) of whitefish morphs (SSR = small sparsely rakered, LSR = large sparsely rakered, LDR = large densely rakered and DR = densely rakered whitefish) from hatching (1.6.1999 i.e. day 150) to death (15.9.2004 i.e. day 2085). Years are separated with vertical lines.

both work in the same direction and it is therefore difficult to judge which is more important with regard to the origin of divergence.

Resource competition is considered as an obvious driver of initial divergence of morphs in general [1,54,55] and could be followed by physiological divergence at a later stage. The different habitat and thermal niche selection may result from resource competition among whitefish morphs, in which pelagic zooplankton prey are more efficiently consumed by those pelagic morphs with more gill rakers compared to the sparsely rakered benthic morphs [23]. Previous evidence from allopatric populations has indicated that in the absence of DR whitefish, the pelagic habitat is frequently occupied by LSR whitefish [21,56]. The strong ecomorphological divergence [21 and this study] among the morphs may also suggest physiological adaptations to different thermal niches. For example, in vendace morphs, the deeper water morph is clearly adapted to colder water, with a lower respiration rate relative to that of the shallow water morph [16]. An alternative scenario, especially in allopatric speciation and in the case of secondary contact [57], suggests that physiologically divergent species/morphs colonize and segregate the niches in the same lake or island according to their thermal preferences.

The SSR whitefish inhabiting cold profundal lake habitats exhibited the lowest estimated respiration rates according to calculations based on both morphology and bioenergetics. This is supported by differences in lifespan, as the profundal morph commonly lives for more than 20 years [58]. More variable metabolic rates were calculated for littoral LSR whitefish, which according to morphology (*M*) had the second lowest respiration rate. Both LSR and SSR whitefish forage on benthic macroinvertebrate prey [41], which requires the detection and capture of prey hiding in sediment. Benthic foraging behaviour is less energetically expensive than pelagic foraging, which relies on constant swimming to find zooplankton [45]. However, bioenergetic modelling indicated that the specific respiration rate of LSR whitefish, living in the habitat with the greatest temperature variability, was the highest. The difference between methods likely reflects the strong influence of living temperature on the results of bioenergetic modelling results. Similar differences between approaches were seen in both pelagic morphs: these had higher respiration rates than benthic morphs according to morphology, but were intermediate according to bioenergetics.

Thus, the most consistent finding is that profundal SSR whitefish have the lowest respiration rate, likely related to cold thermal habitats and the lowest risk of predation in profundal zones [59]. In contrast, pelagic morphs are subject to continuous risk of predation by piscivorous brown trout [40,60]. The risk of predation by brown trout is reduced by diel vertical migration by DR whitefish [51], which may drive the highest estimated respiration rate for any of the four morphs. The pelagic DR morph also has the shortest average lifespan, rarely living more than six years [61]. Higher metabolic rates have been estimated for pelagic than benthic morphs through bioenergetic modelling in the closely related lake whitefish (*Coregonus clupeaformis*) [62]. Also, recent genetic results of lake whitefish indicate that metabolic traits are inherited and the pelagic morph consistently exhibits over-expression of genes related to metabolism and oxygen transport compared to benthic morphs, related to higher swimming activity required in pelagic habitats [63,64]. There is an urgent need to evaluate the metabolic differences in other diverging taxa. Reptile and amphibian diversity is correlated with temperature, but very little is known about actual niche-specific metabolic adaptations [7,65].

Individuals of each of the whitefish morphs examined here used warmer water layers during the summer in the first two years of life. This has been observed for many fish species in the first year of life, in both field and experimental studies [35,66]. Dufour *et al.* [32] studied peri-alpine whitefish otoliths using the same techniques as used here and found that whitefish occupied warmer habitats during the first year of life. Whitefish larvae are known to concentrate in near shore areas after hatching and soon start to feed on small zooplankton [67]. Their diet may subsequently shift towards benthic macroinvertebrates or they may continue to consume zooplankton, depending on the individual lake and whitefish morph [68]. In nearby Lake Muddus, surface trawl catches in early autumn frequently contain 0+ DR whitefish in the pelagic habitat, whereas in sheltered shallow littoral bays 0+ LSR whitefish are caught (K. Kahilainen pers. obs.). To date, no 0+ SSR whitefish have been captured during these surveys, suggesting that a shift to morph-specific niches may occur during the first year of life. After the first two summer seasons, all morphs apparently choose cooler waters, shifting to their adult thermal niche. All the specimens studied here had reached or were very close to reaching sexual maturity, which occurs at different ages and sizes at the population level. Here, DR whitefish reach sexual maturity at the lowest age (2–3 years old), SSR whitefish at the highest age (6–7 years old) and the two other morphs at an intermediate 5–6 years old. Little is known about the exact spawning times and places for the different morphs, but field observations of spawning shoals in shallow water close to the main basin during field sampling in September 2004 indicated that DR whitefish were the first morphs to spawn in Lake Paadar. Observations in maturation and gonad state from the other morphs suggest that LDR and LSR whitefish spawned after the DR whitefish, with the SSR whitefish spawning last (K. Kahilainen pers. obs.). In coregonids, such inter-morph differences in spawning times are important factors driving reproductive isolation [17,19,63].

The general differences in spawning time shown by pelagic and benthic morphs may be indirectly related to the role of temperature governing the availability of invertebrate prey resources: pelagic zooplankton are only available in large amounts for a short period each summer, whereas benthic invertebrates are more consistently available year round [22,41,61]. The late spawning time shown by SSR whitefish may reflect temperatures in profundal habitats, which are the highest during the autumnal turnover of the water column. Physiological adaptation to a particular thermal regime may also be an important component of reproductive isolation in many species, but our current understanding is very limited [4]. Recently, niche segregation along thermal gradients associated with variation in both altitude and latitude has been shown to be efficient reproductive isolation barriers even in face of gene-flow in birds, amphibians and fish [69–71].

A recent modelling study also suggests that such physiological adaptations to different thermal habitat may be important in the formation of reproductive isolation and subsequent speciation [72]. Here, adaption to divergent thermal niches is likely to be of crucial importance to the relative capacity to assimilate the energy required to reach sexual maturity and develop gonads. If fish morphs are physiologically adapted to their individual thermal niche, this may cascade affecting spawning times and furthermore hatching times of eggs. The onset of spawning in whitefish is influenced by both temperature and light regimes [73]. An adaptation to lower temperatures may delay spawning times, as observed here for SSR whitefish, the most morphologically and genetically distinct morph [21,37]. The isotope values from the otolith nucleus suggest potentially different hatching temperatures for benthic and pelagic morphs, perhaps related to either temporal or/and spatial differences in spawning. In experiments on lake whitefish morphs, hatching time was demonstrated to be genetically determined, with very small differences between pelagic and benthic morphs [63]. However, little is known about the early development of polymorphic fish eggs in nature, especially with regard to potential differences of degree-day counts among the morphs in relation to their physiological adaptations. These are potentially important thermal adaptations, as in amphibian thermal adaptations in early larval development that apparently separate different populations [71].

Our study indicates that otolith $\delta^{18}O$ and $\delta^{13}C$ analysis provide an excellent tool for understanding lifetime thermal niche and respiration values of whitefish morphs, when $\delta^{18}O_{H2O}$ and $\delta^{13}C_{DIC}$ values are obtained from lake water. Our preliminary study evaluates patterns based on snap shot sampling of water column during the summer stratification, whereas seasonal sampling should be done to track annual differences in $\delta^{13}C_{DIC}$ values [32]. The results also verified the age determination estimates made previously from the same otoliths. Although, larger sample sizes are needed to reveal within-morph variation in thermal niche and respiration between multiple individuals, we hope that our preliminary results based on individual fish promote more studies on thermal adaptation from other polymorphic populations. Even though fine micromilling techniques were utilized to get 33–40 samples from each otolith, the resolution decreased due to declining fish growth rates after the first two years of life. At the very beginning of an individual fish's life, it is unclear at which point during the egg development and what rate otolith growth starts. Reflecting this, do the first measurements taken from the otolith in nucleus indicate hatching temperatures or combine temperatures from the late egg development and hatching periods? Despite the lack of a clear answer, we have observed differences among the morphs, indicating distinct thermal habitats at the very beginning of their lifecycles.

All whitefish morphs can be considered as slow growing fish when compared to the fast growing marine fishes, such as cod (*Gadus morhua*) and halibut (*Hippoglosus stenolepis*) previously examined in similar studies [34,36]. The estimated winter temperatures from otoliths are clearly too high as water temperatures in winter are usually <4°C even in the deepest parts of subarctic lakes [22]. This error is most likely due to the extremely small winter increments in the otoliths and subsequent integration of autumn and spring otolith growth into the same sample. Alternatively, these fish may have a shutdown temperature <4°C that precludes any otolith accretion during the cooler season. Otolith carbonate also records metabolism with winter rings clearly visible in whitefish, including the individuals analysed here. Based on difficulties in estimating timing and duration of a potential winter break in otolith growth and thus thermal estimation, we decided to present the full data in our results and discuss here the uncertainties involved. In addition, it should be noted that we used a fractionation equation developed for Arctic charr, which is adapted to colder temperatures than whitefish [44,74]. An experimental approach using whitefish in general or even whitefish morphs is urgently needed to build species or whitefish morph specific equations between otolith $\delta^{18}O$ values and temperature.

To conclude, we found a clear segregation of thermal niches between profundal and non-profundal whitefish morphs, but our capacity to differentiate between pelagic and littoral morphs was less successful. However, all individuals of the four whitefish morphs displayed evidence for ontogenetic thermal niche shifts, from warm juvenile summer habitats to cooler temperatures. Interestingly, estimated respiration rates were clearly correlated to foraging ecology and life-history, where the profundal SSR whitefish has lower rates than other morphs. These results also provide a useful starting point to conduct further studies of thermal niche life-histories with other polymorphic populations.

Acknowledgments

Authors would like to thank O. Aikio, D. Besic, K. Johansson, P. Jääskeläinen, J. Marttila, K. Mäenpää, J. Niemistö, M. Salonen, S. Timsic and A. Tuomaala for their help during the field and laboratory work. Muddusjärvi Research Station kindly provided needed facilities during the field work. Funding of this study was obtained from Ministry of Agriculture and Forestry, Academy of Finland (projects #140903, #1268566, #134139) and Emil Aaltonen Foundation.

Author Contributions

Conceived and designed the experiments: KKK WPP MK. Performed the experiments: KKK WPP ES CH MK. Analyzed the data: KKK WPP ES CH MK. Contributed reagents/materials/analysis tools: KKK WPP ES CH MK. Contributed to the writing of the manuscript: KKK WPP ES CH MK.

References

1. Schluter D (2000) The ecology of adaptive radiation. Oxford University Press, Oxford.
2. Nosil P (2012) Ecological speciation. Oxford University Press. Oxford.
3. Price T (2008) Speciation in birds. Roberts and Company Publishers, Colorado.
4. Keller I, Seehausen O (2012) Thermal adaptation and ecological speciation. Mol Ecol 21: 782–799.
5. Svensson EI (2012) Non-ecological speciation, niche conservatism and thermal adaptation: how are they connected? Org Div Evol 12: 229–240.
6. Grant PR, Grant BR (2008) How and why species multiply? The radiation of Darwin's Finches. Princeton University Press, New Jersey.
7. Losos JB (2009) Lizards in an evolutionary tree: ecology and adaptive radiation of anoles. University of California Press, Berkeley and Los Angeles.
8. Østbye K, Amundsen PA, Bernatchez L, Klemetsen A, Knudsen R, et al. (2006) Parallel evolution of ecomorphological traits in the European whitefish *Coregonus lavaretus* (L.) species complex during postglacial times. Mol Ecol 15: 3983–4001.
9. Rogers SM, Bernatchez L (2007) The genetic architecture of ecological speciation and the association with signatures of selection in natural lake whitefish (*Coregonus* sp. Salmonidae) species pairs. Mol Biol Evol 24: 1423–1438.
10. Ravinet M, Prodöhl PA, Harrod C (2013) Parallel and nonparallel ecological, morphological and genetic divergence in lake-stream stickleback from a single catchment. J Evol Biol 26: 186–204.
11. Schluter D, McPhail JD (1993) Character displacement and replicate adaptive radiation. Trends Ecol Evol 8: 197–200.
12. Kahilainen K, Østbye K (2006) Morphological differentiation and resource polymorphism in three sympatric whitefish *Coregonus lavaretus* (L.) forms in a subarctic lake. J Fish Biol 68: 63–79.
13. Siwertsson A, Knudsen R, Kahilainen KK, Præbel K, Primicerio R, et al. (2010) Sympatric diversification influenced by ecological opportunity and historical contingency in a young species lineage of whitefish. Evol Ecol Res 12: 929–947.
14. Præbel K, Knudsen R, Siwertsson A, Kahilainen KK, Østbye K, et al. (2013) Ecological speciation in postglacial European whitefish: rapid adaptive radiations into the littoral, pelagic and profundal lake habitats. Ecol Evol 3: 4970–4986.
15. Magnuson JJ, Crowder LB, Medvick PA (1979) Temperature as an ecological resource. Am Zool 19: 331–343.
16. Ohlberger J, Mehner T, Staaks G, Hölker F (2008) Temperature-related physiological adaptations promote ecological divergence in a sympatric species pair of temperate freshwater fish, *Coregonus* spp. Funct Ecol 22: 501–508.
17. Svärdson G (1979) Speciation of Scandinavian *Coregonus*. Rep Inst Freshw Res Drott 57: 1–95.

18. Østbye K, Bernatchez L, Næsje TF, Himberg KJM, Hindar K (2005) The evolutionary history of European whitefish (*Coregonus lavaretus* L.) as inferred from mtDNA phylogeography and gillraker numbers. Mol Ecol 14: 4371–4387.
19. Hudson AG, Vonlanthen P, Müller R, Seehausen O (2007) Review: The geography of speciation and adaptive radiation of coregonines. Arch Hydrobiol Spec Iss Adv Limnol 60: 111–146.
20. Vonlanthen P, Bittner D, Hudson AG, Young KA, Müller R, et al. (2012) Eutrophication causes speciation reversal in whitefish adaptive radiations. Nature 482: 357–363.
21. Harrod C, Mallela J, Kahilainen KK (2010) Phenotype-environment correlations and character displacement in a putative fish radiation. J Anim Ecol 79: 1057–1068.
22. Hayden B, Harrod C, Kahilainen KK (2013) The effects of winter ice cover on the trophic ecology of whitefish (*Coregonus lavaretus* L.) in subarctic lakes. Ecol Freshw Fish 22: 192–201.
23. Kahilainen KK, Siwertsson A, Gjelland KØ, Knudsen R, Bøhn T, et al. (2011) The role of gill raker number variability in adaptive radiation of coregonid fish. Evolutionary Ecology 25: 573–588.
24. Kahilainen KK, Østbye K, Harrod C, Shikano T, Malinen T, et al. (2011) Species introduction promotes hybridization and introgression in *Coregonus*: is there sign of selection against hybrids? Mol Ecol 20: 3838–3855.
25. Kahilainen K, Malinen T, Tuomaala A, Lehtonen H (2004) Diel and seasonal habitat and food segregation of three sympatric whitefish (*Coregonus lavaretus* (L.)) forms in a subarctic lake. J Fish Biol 64: 418–434.
26. Wurster CM, Patterson WP (2003) Metabolic rate of late Holocene freshwater fish: evidence from $\delta^{13}C$ values from otoliths. Paleobiology 29: 492–505.
27. Wurster CM, Patterson WP, Stewart DJ, Bowlby JN, Stewart TJ (2005) Thermal histories, stress, and metabolic rates of chinook salmon (*Oncorhynchus tshawytscha*) in Lake Ontario: evidence from intra-otolith stable isotope analyses. Can J Fish Aquat Sci 62: 700–713.
28. Campana SE (1999) Chemistry and composition of fish otoliths: pathways, mechanisms and applications. Mar Ecol Progr Ser 188: 263–297.
29. Patterson WP, Smith GR, Lohmann KC (1993) Continental paleothermometry and seasonality using the isotopic composition of aragonitic otoliths of freshwater fishes. In: P. Swart P, Lohmann K, McKenzie J, Savin S, editors. Climate change in Continental Isotopic Records. American Geophysical Union,Washington. 191–202.
30. Patterson WP (1998) North American continental seasonality during the last millennium: high-resolution analysis of sagittal otoliths. Palaeogeogr Palaeoclim Palaeoecol 138: 271–303.
31. Solomon CT, Weber PK, Cech JJ, Ingram BL, Conrad ME, et al. (2006) Experimental determination of the sources of otolith carbon and associated fractionation. Can J Fish Aquat Sci 63: 79–89.
32. Dufour E, Gerdeaux D, Wurster CM (2007) Whitefish (*Coregonus lavaretus*) respiration rate governs intra-otolith variation of $\delta^{13}C_{oto}$ values in Lake Annecy. Can J Fish Aquat Sci 64: 1736–1746.
33. Wurster CM, Patterson WP, Cheatham MM (1999) Advances in micromilling techniques: a new apparatus for acquiring high-resolution oxygen and carbon stable isotope values and major/minor elemental ratios from accretionary carbonate. Comput Geosci 25: 1155–1162.
34. Weidman CR, Millner R (2000) High resolution stable isotope records from North Atlantic cod. Fish Res 46: 327–342.
35. Joukhadar Z, Patterson WP, Todd TN, Smith GR (2002) Temperature history of *Coregonus artedi* in the St. Marys River, Laurentian Great Lakes, inferred from oxygen isotopes of otoliths. Arch Hydrobiol Spec Iss Adv Limnol 57: 453–461.
36. Gao Y, Beamish RJ (2003) Stable isotope variation in otoliths of Pacific halibut (*Hippoglossus stenolepis*) and indications of the possible 1990 regime shift. Fish Res 60: 393–404.
37. Couton M (2012) Adaptive radiation of the European whitefish, *Coregonus lavaretus* (L.), in the Pasvik watercourse: the genetic description of a new morph. MSc thesis, University of Tromsø, Tromsø.
38. Forsström L, Sorvari S, Rautio M, Sonninen E, Korhola A (2007) Changes in physical and chemical limnology and plankton during the spring melt period in a subarctic lake. Internat Rev Hydrobiol 92: 301–325.
39. Jonsson CE, Leng MJ, Rosqvist GC, Seibert J, Arrowsmith C (2009) Stable oxygen and hydrogen isotopes in sub-Arctic lake waters from northern Sweden. J Hydrol 376: 143–151.
40. Jensen H, Kahilainen KK, Amundsen PA, Gjelland KØ, Tuomaala A, et al. (2008) Predation by brown trout (*Salmo trutta*) along a diversifying prey community gradient. Can J Fish Aquat Sci 65: 1831–1841.
41. Kahilainen K, Lehtonen H, Könönen K (2003) Consequence of habitat segregation to growth rate of two sparsely rakered whitefish forms (*Coregonus lavaretus* (L.)) in a subarctic lake. Ecol Freshw Fish 12: 275–285.
42. Kiljunen M, Grey J, Sinisalo T, Harrod C, Immonen H, et al. (2006) A revised model for lipid-normalising $\delta^{13}C$ values from aquatic organisms, with implications for isotope mixing models. J Appl Ecol 43: 1213–1222.
43. Storm-Suke A, Dempson JB, Reist JD, Power M (2007) A field-derived oxygen isotope fractionation equation for *Salvelinus* species. Rapid Comm Mass Spec 21: 4109–4116.
44. Godiksen J, Svenning MA, Dempson BJ, Marttila M, Storm-Suke A, et al. (2010) Development of a species-specific fractionation equation for Arctic charr (*Salvelinus alpinus* (L.)): an experimental approach. Hydrobiologia 650: 67–77.
45. Webb PW (1984) Form and function in fish swimming. Sci Am 251: 58–68.
46. Sherwood GD, Rose GA (2003) Influence of swimming form on otolith $\delta^{13}C$ in marine fish. Mar Ecol Prog Ser 258: 283–289.
47. Hanson PC, Johnson TB, Schindler DE, Kitchell JF (1997) Fish Bioenergetics 3.0. University of Wisconsin, Wisconsin.
48. von Bertalanffy L (1938) A quantitative theory of organic growth. Human Biol 10: 181–243.
49. LeCren ED (1951) The length-weigth relationship and seasonal cycle in gonad weight and condition in the perch (*Perca fluviatilis*). J Anim Ecol 20: 201–219.
50. Salonen K, Sarvala J, Hakala I, Viljanen ML (1976) The relation of energy and organic carbon in aquatic invertebrates. Limnol Oceanog 21: 724–730.
51. Kahilainen KK, Malinen T, Lehtonen H (2009) Polar light regime and piscivory govern diel vertical migrations of planktivorous fish and zooplankton in a subarctic lake. Ecol Freshw Fish 18: 481–490.
52. Greenberg R, Cadena V, Danner RM, Tattersall G (2012) Heat loss may explain bill size differences between birds occypying different habitats. PloS One 7: e40933.
53. Podos J (2001) Correlated evolution of morphology and vocal signal structure in Darwin's finches. Nature 409: 185–187.
54. Amundsen PA, Knudsen R, Klemetsen A, Kristoffersen R (2004) Resource competition and interactive segregation between sympatric whitefish morphs. Ann Zool Fenn 41: 301–307.
55. Bolnick DI, Lau OL (2008) Predictable patterns of disruptive selection in stickleback in postglacial lakes. Am Nat 172: 1–11.
56. Kahilainen KK, Malinen T, Tuomaala A, Alajärvi E, Tolonen A, et al. (2007) Empirical evaluation of phenotype-environment correlation and trait utility with allopatric and sympatric whitefish (*Coregonus lavaretus* (L.)) populations in subarctic lakes. Biol J Linn Soc 92: 561–572.
57. Coyne JA, Orr HA (2004) Speciation. Sinauer Associates, Sunderland, Massachusetts.
58. Kahilainen K, Lehtonen H (2002) Habitat use and growth of three sympatric forms of European whitefish, *Coregonus lavaretus* (L.), in subarctic Lake Muddusjärvi. Arch Hydrobiol Spec Iss Adv Limnol 57: 277–290.
59. Kahilainen K, Lehtonen H (2003) Piscivory and prey selection of four predator species in a whitefish dominated subarctic lake. J Fish Biol 63: 659–672.
60. Kahilainen K, Lehtonen H (2002) Brown trout (*Salmo trutta* L.) and Arctic charr (*Salvelinus alpinus* (L.)) as predators on three sympatric whitefish (*Coregonus lavaretus* (L.)) forms in the subarctic Lake Muddusjärvi. Ecol Freshw Fish 11: 158–167.
61. Kahilainen K, Alajärvi E, Lehtonen H (2005) Planktivory and diet-overlap of densely rakered whitefish (*Coregonus lavaretus* (L.)) in a subarctic lake. Ecol Freshw Fish 14: 50–58.
62. Trudel M, Tremblay A, Schetagne R, Rasmussen JB (2001) Why are dwarf fish so small? An energetic analysis of polymorphism in lake whitefish (*Coregonus clupeaformis*). Can J Fish Aquat Sci 58: 394–405.
63. Bernatchez L, Renaut S, Whiteley AR, Derome N, Jeukens J, et al. (2010) On the origin of species: insights from the ecological genomics of lake whitefish. Phil Trans R Soc B 365: 1783–1800.
64. Evans ML, Præbel K, Peruzzi S, Bernatchez L (2012) Parallelism in the oxygen transport system of the lake whitefish: the role of physiological divergence in ecological speciation. Mol Ecol 21: 4038–4050.
65. Cadena CD, Kozak KH, Gómez JP, Parra JL, McCain CM, et al. (2012) Latitude, elevational climatic zonation and speciation in New World vertebrates. Proc R Soc London B 279: 194–201.
66. Hokanson KEF (1977) Temperature requirements of some percids and adaptations to the seasonal temperature cycle. J Fish Res Bd Can 34: 1524–1550.
67. Næsje TF, Sandlund OT, Jonsson B (1986) Habitat use and growth of age-0 whitefish, *Coregonus lavaretus*, and cisco, *C. albula*. Env Biol Fish 15: 309–314.
68. Lindström T (1988) Diet changes during metamorphosis in whitefish. Finn Fish Res 9: 89–105.
69. Culumber ZW, Shepard DB, Coleman SW, Rosenthal GG, Tobler M (2012) Physiological adaptation along environmental gradients and replicated hybrid zone structure in swordtails (Teleostei: *Xiphophorus*). J Evol Biol 25: 1800–1814.
70. Caro LM, Caycedo-Rosales PC, Bowie RCK, Slabbekoorn H, Cadena CD (2013) Ecological speciation along and elevational gradient in a tropical passerine bird? J Evol Biol 26: 357–374.
71. Muir AP, Biek R, Thomas R Mable BK (2014) Local adaptation with high gene flow: temperature parameters drive adaptation to altitude in the common frog (*Rana temporaria*). Mol Ecol 23: 561–574.
72. Ohlberger J, Brännström Å, Dieckmann U (2013) Adaptive phenotypic diversification along a temperature-depth gradient. Am Nat 182: 359–373.
73. Wanzenböck J, Pamminger-Lahnsteiner B, Winkler K, Weiss SJ (2012) Experimental evaluation of the spawning periods of whitefish (*Coregonus lavaretus* complex) in Lake Mondsee, Austria. Fund Appl Limnol Adv Limnol 63: 89–97.
74. Siikavuopio S, Knudsen R, Amundsen PA (2010) Growth and mortality of Arctic charr and European whitefish reared at low temperatures. Hydrobiologia 650: 255–263.

Biochemical Indices and Life Traits of Loggerhead Turtles (*Caretta caretta*) from Cape Verde Islands

Sara Vieira[1], Samir Martins[1], Lucy A. Hawkes[2], Adolfo Marco[3], M. Alexandra Teodósio[1]*

1 Centro de Ciências do Mar, Universidade do Algarve, Campus de Gambelas, Faro, Portugal, **2** University of Exeter, College of Life and Environmental Sciences, Penryn Campus, Cornwall, United Kingdom, **3** Estación Biológica de Doñana (CSIC), Américo Vespucio, Sevilla, Spain

Abstract

The loggerhead turtle (*Caretta caretta*) is an endangered marine reptile for whom assessing population health requires knowledge of demographic parameters such as individual growth rate. In Cape Verde, as within several populations, adult female loggerhead sea turtles show a size-related behavioral and trophic dichotomy. While smaller females are associated with oceanic habitats, larger females tend to feed in neritic habitats, which is reflected in their physiological condition and in their offspring. The ratio of RNA/DNA provides a measure of cellular protein synthesis capacity, which varies depending on changes in environmental conditions such as temperature and food availability. The purpose of this study was to evaluate the combined use of morphometric data and biochemical indices as predictors of the physiological condition of the females of distinct sizes and hatchlings during their nesting season and how temperature may influence the physiological condition on the offspring. Here we employed biochemical indices based on nucleic acid derived indices (standardized RNA/DNA ratio-sRD, RNA concentration and DNA concentration) in skin tissue as a potential predictor of recent growth rate in nesting females and hatchling loggerhead turtles. Our major findings were that the physiological condition of all nesting females (sRD) decreased during the nesting season, but that females associated with neritic habitats had a higher physiological condition than females associated with oceanic habitats. In addition, the amount of time required for a hatchling to right itself was negatively correlated with its physiological condition (sRD) and shaded nests produced hatchlings with lower sRD. Overall, our results showed that nucleic acid concentrations and ratios of RNA to DNA are an important tool as potential biomarkers of recent growth in marine turtles. Hence, as biochemical indices of instantaneous growth are likely temperature-, size- and age-dependent, the utility and validation of these indices on marine turtles stocks deserves further study.

Editor: Jacob Lawrence Kerby, University of South Dakota, United States of America

Funding: The authors thank the Cape Verde Ministry of Environment (General Direction for the Environment), INDP (National Fisheries Institution), the Canary Islands Government (D.G. Africa and D.G. Research and Universities), ICCM (Canarian Institution for Marine Sciences), the Andalusian Government (Andalusian Environmental Office) and AEGINA PROJECT (INTERREG IIIB) for funding and hosting them during this study. The authors also thank the European Regional Development Fund (ERDF) through the COMPETE - Operational Competitiveness Programme, and national funds through FCT - PEst-C/MAR/LA0015/2011 for supporting the biochemical analysis. The funders had no role in study design, data collection and analysis, decision to publish, or preparation of the manuscript.

Competing Interests: The authors have declared that no competing interests exist.

* Email: mchichar@ualg.pt

Introduction

The impact of anthropogenic activity on biodiversity has now been demonstrated throughout terrestrial and aquatic ecosystems [1,2]. Impacts include, but are not limited to, historic and ongoing harvest (legal and illegal) [3], accidental take (e.g. fisheries by-catch) [4], habitat destruction and degradation [5] and climate change [6]. In assessing recovery from such impacts to inform conservation management, it is imperative to gain information on population growth rate and health status. Life history traits determine the schedule and duration of key events in an organism's lifetime that are shaped by natural selection to produce the largest possible number of surviving offspring. The ratio of RNA to DNA (known as 'standardized RNA/DNA ratio'- sRD) has been widely used as a biochemical index to assess potential survival and growth and in determining the ecophysiological condition of marine organisms [7]. This index has quickly become a particularly promising biochemical tool that not only reflects physiological condition, but also allows for the estimation of instantaneous growth rates in a great variety of organisms [8,9,10,11].

The usefulness of sRD as a measure of physiological condition is related to the fact that the concentration of cellular DNA is relatively constant in the somatic cells regardless of any changes in the organism's environment, while the RNA content of a cell increases as the cellular demand for protein synthesis and growth increases [7]. The ratio of RNA/DNA thus provides a measure of cellular protein synthesis capacity, which varies depending on changes in environmental conditions such as temperature and food availability. Poor nutritional condition contributes to low protein synthesis and slow growth, resulting in a low RNA/DNA ratio [9,10]. Consequently, individuals in good nutritional condition generally have high levels of RNA/DNA, whereas individuals undergoing dietary restriction have a lower amount of RNA in their cells and hence a lower rate of RNA/DNA [12].

Despite its widespread use in marine fishes and invertebrates, sRD has mainly been validated for application to studies with small marine organisms. The application of this method in studies with large migratory organisms, such as marine vertebrates, should offer important insights, particularly in view of the conservation concern for most marine vertebrate taxa [7]. Moreover, sRD yields specific information about the physiological condition and growth rate of animals in the different areas they occupy throughout their life cycle, with a minimal amount of distress to the animal.

Of the marine megavertebrates, marine turtles are one of the best studied taxa, with all seven species having now been tracked worldwide [13,14]. However, management of sea turtles is still hindered by a lack of key data on turtle biology, population status and environmental threats [15]. The loggerhead sea turtle, *Caretta caretta*, is an endangered species with a circumglobal distribution. Human exploitation of this species over the last few centuries has led to dramatic population declines [16]. Assessing the current status and predicting the viability of wild populations of marine turtles requires the quantification of demographic parameters such as individual growth rate [17,18]. However, gathering such growth rates for marine turtles typically involves mark and recapture programs in which recapture probabilities can be quite low [19] or repeated sampling of a population for use in length-frequency analyses [20].

Satellite tracking has now demonstrated that many populations of marine turtles exhibit plasticity with multiple foraging strategies seen in populations of adult turtles. For example, in loggerhead turtles in some areas, larger adult turtles may be more likely to forage in shallow neritic habitats, whereas smaller adult turtles appear to be more likely to forage in oceanic habitats [21,22,23,24,25,26]. It is not yet understood whether these differences in size are related to differences in age, as oceanic and neritic foragers have been suggested to reach sexual maturity at similar ages [23]. As differing strategies may be expected to have different growth rates and nutritional status, loggerhead sea turtles provide an excellent model system with which to test this methodological approach on a marine vertebrate species.

Establishing techniques with which to supplement mark-recapture and length-frequency analyses for estimating recent growth rates of turtles upon first capture would also substantially improve our ability to evaluate the status of loggerhead turtle populations on a shorter timescale. In addition, loggerhead turtles expend considerable energy migrating from foraging areas to breeding and nesting areas. Loggerhead turtles nest on average every 12 to 17 days during the breeding season and return to breed every 2 to 4 years, presumably because the energetic cost of doing so prohibits breeding more frequently [27]. It seems that food availability and regional or temporal variations may limit growth rate and physiological condition, which may be reflected in breeding frequency and on the fitness of offspring [28]. The purpose of this study was to evaluate the combined use of morphometric, behavioral data and biochemical indices based on nucleic acid derived indices (*e.g.*, standardized RNA/DNA ratio-sRD, RNA concentration and DNA concentration) as predictors of the physiological condition of the females and hatchlings during their nesting season, and to assess how temperature may influence the physiological condition of offspring. We suggest this may help towards development of minimally invasive growth and viability measurement strategies, which are essential in planning future conservation strategies.

In this study, we aimed to investigate the following hypotheses: regarding adult turtles, (i) Does the physiological condition of females decrease during the nesting season (A1)?; (ii) Is the physiological condition of "large" adult sea turtles associated with neritic habitats different to the physiological condition of "small" adult sea turtles associated with oceanic habitats (A2)? Regarding hatchling sea turtles, (i) Are the hatching and emergence successes of "small" adult sea turtles lower than in "large" adult sea turtles (O1)?; (ii) Are hatchlings produced by "large" females larger than the hatchlings produced by "small" females (O2)?; (iii) Is hatchling vigour correlated with physiological condition (O3)?; and (iv) Does incubation temperature influence the physiological condition of the hatchlings (O4)?

Methods

Study area

The Cape Verde Islands are located in the eastern North Atlantic between 14° 45′ and 17° 18′ N and 22° 38′ and 25° 22′ W, 500 km off the coast of Senegal (Fig. 1). The local weather conditions give this tropical region a moderate subtropical climate. The loggerhead nesting season in Cape Verde is prolonged, extending from June through October. This volcanic archipelago hosts the third largest rookery of loggerhead turtles in the world [29] and the population is reproductively isolated from the other Atlantic loggerhead rookeries [30]. Furthermore, the Cape Verde population constitutes the most endangered regional management unit for this species in the Atlantic [31].

Ethics Statement

The present study was carried out on Ervatão beach, within the *Reserva Natural da Tartaruga*, which is included in the national protected area network of Cabo Verde (Decreto-Regulamentar n° 14/2013 de 9 de Maio), operated by the "Direcção Geral do Ambiente", an agency of the Cape Verdian Ministry of the Environment. The work was ethically approved by the "Direcção Geral do Ambiente" and loggerhead sea turtle tissues were imported to Portugal (University of Algarve) under a CITES permit.

Sampling and data collection

We collected a total of 57 tissue samples from the right front flipper from randomly selected nesting females during the beginning (7th to 12th of July) and end (19th of September to 10th of October) of the 2012 nesting season from one of the main nesting beaches, Praia Ervatão. The skin tissue samples (<2 mg) were taken using a 5-mm biopsy punch, while the female was covering her completed nest with sand. We stored samples in a RNAlater solution at 4°C. Nesting females were measured (curved carapace length: CCL_{min}) using a fiberglass tape measure (±0.1 cm) and turtles classified as "small" (≤85 cm CCL_{min}; 55% of turtles; n = 31) and "large" (>85 cm CCL_{min}; 45% of the total; n = 26; based on neritic/oceanic separation by Eder *et al.* 2012 [32]). Nests laid by females at the beginning of the season were immediately translocated to an artificial hatchery, taking care to maintain the original vertical orientation of the eggs during transport, and reburied at 40 cm depth (mean nest depth for Boavista [25]). The artificial hatchery (50×15 m^2) was established on Benguinho Beach, a natural nesting beach area adjacent to Praia Ervatão.

In order to investigate if the temperature might influence hatchling physiological condition (as measured using sRD) we manipulated nest temperature as follows: We marked and randomly assigned the 20 nests to one of two experimental treatments: either shading them under a shade cloth made of plastic mesh over the sand surface above the nests ('shaded'; n = 10); or leaving them uncovered ('unshaded'; n = 10) (Fig. 2).

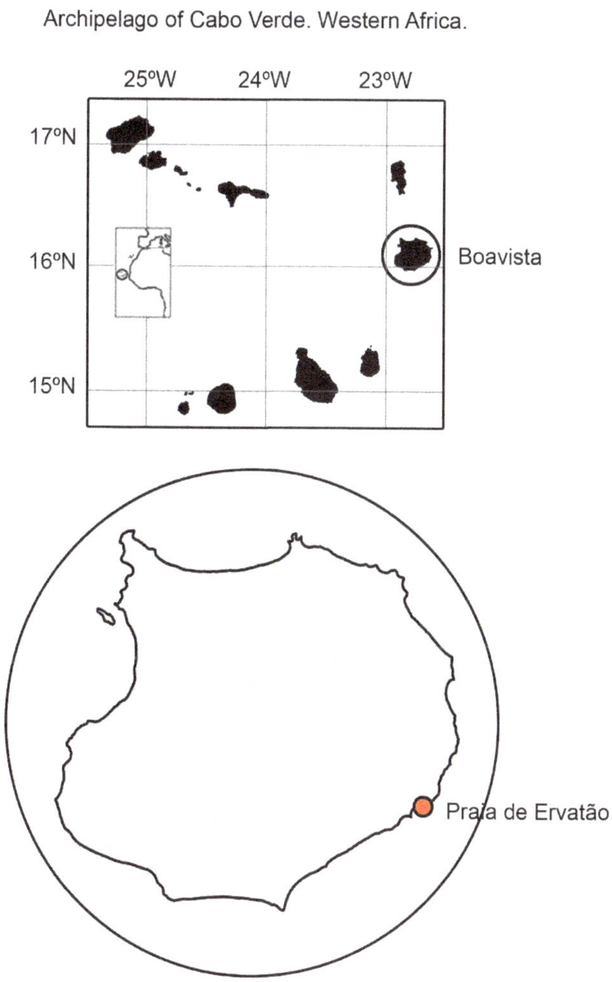

Figure 1. Map of the Cape Verde Archipelago, Boavista Island (Republic of Cape Verde, western Africa) and the study site (framed area).

We assumed that shading significantly lowered nest temperatures [32], by reducing solar infra-red radiation, and arranged nests in parcels of 1 m^2 for every nest, distributing nests of different treatments in a block design of five nests per treatment block. We placed a round plastic net (45 cm in diameter, 50 cm in height) over all nests 45 days after egg-laying to collect hatchlings after emergence. Nests were checked throughout the night and again at daybreak to check for hatchling emergence. Within two hours of emergence, we measured hatchling straight carapace length (SCL) with a digital caliper (±0.1 mm), assessed hatchling vigour and collected tissue samples from 3 randomly selected hatchlings from each nest. To assess hatchling vigour we placed hatchlings (n = 10 per nest) on their dorsal side and the time taken for hatchlings to 'right' themselves was recorded [33]. The skin samples (<1 mg) were taken on a sanitized polyethylene board (10×10 cm^2) with one front flipper held flat against the board and isopropyl alcohol applied to disinfect the biopsy site. A single sample was then taken using a 2-mm biopsy punch (with plunger) from the trailing edge of the front flipper proximal to the body out to midway along the flipper [35]. Handling time did not exceed 15 minutes. Hatchlings were released immediately after the sampling event.

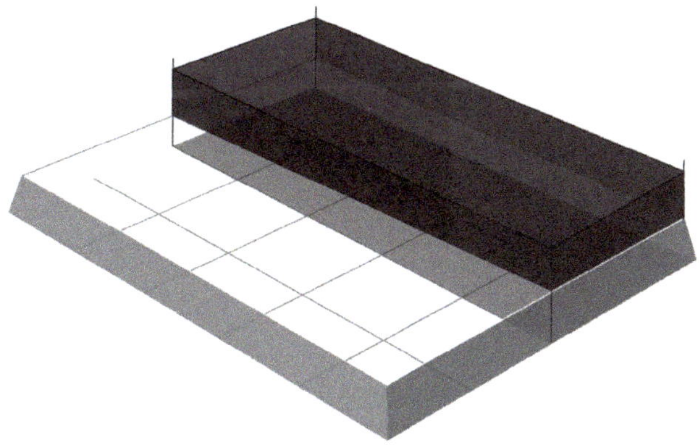

Figure 2. Diagram of experimental design illustrating the two experimental treatments: shaded (under a shade cloth made of plastic mesh) and unshaded (uncovered nests).

Determination of nucleic acids, concentrations and ratios

Nutritional condition was assessed using the following nucleic acid acid-derived indices: sRD (standardized RNA/DNA ratios), DNA/mg and RNA/mg per tissue. Nucleic acid analysis was carried out for ecophysiological condition of breeding females in the beginning and at the end of the breeding season and for hatchling condition at the shaded/unshaded treatments. The procedures used to quantify nucleic acids in individual females and hatchlings are outlined in Caldarone *et al.* (2001) [36] and Esteves *et al.* (2000) [37] using muscle tissue [38]. Briefly, tissues were mechanically and chemically homogenized and subsequent fluorescence-photometric measurements were taken using ethidium bromide (EB) as a specific nucleic acid fluorochrome dye. Fluorescence was measured on a microplate reader (Biotek synergy HT model SIAFRTD) using an excitation wavelength of 365 nm and an emission wavelength of 590 nm. We measured endogenous fluorescence (before EB addition) from the first set of samples from each tissue, but this was found to be negligible, so it was disregarded in the calculations of nucleic acid concentrations. Total fluorescence was first read, and then samples were incubated with ribonuclease A (Type-II A) at 37°C for 30 minutes, and then cooled to room temperature before reading. The fluorescence due to total RNA, mainly ribosomal, was calculated as the difference between total fluorescence (RNA and DNA) and the fluorescence measured after ribonuclease treatment, which is assumed to be due to DNA. Concentrations were determined by running standard curves of DNA–EB (Ethidium bromide) and RNA–EB with known concentrations of u-phagus DNA (0.25 ug uLl–1) and 16S–23S s *E. coli* RNA (4 ug uLl–1) (Roche), in the appropriate range of values. The average ratio of DNA and RNA slopes (mean ± SE) was 2.84±0.20 SE. RNA/DNA ratios were standardized (sRD) using this information and the reference slope ratio of 2.4, according to Caldarone *et al.* (2006) [39].

Data Analyses

All graphics and statistical analyses were performed using the open source software R version 3.0.0 (R Development Core Team; www.r-project.org).

Data were analysed using linear models, after removal of statistical outliers and testing for normality, and the influence of independent factors (environmental/behaviour; beginning/end of breeding season, large/small female, shaded/unshaded) and independent variables (time to righting) on dependent variables (ecophysiological condition of hatchlings and adult females based on sRD ratio, size, and hatching success) were analyzed. We also performed two regression analyses to investigate the relationship among the ecophysiological condition of hatchlings and the hatching success of their nests and their righting time. Significance was assigned at $p<0.05$.

Results

Adult females

Ecophysiological condition of nesting females was higher at the beginning of the season (sRD = 2.9±0.95) compared to the end of the season (sRD = 1.9±0.98) ($F_{1,45} = 23.4$, $p<0.001$; Fig. 3A). Nesting females from the beginning of the season also exhibited higher RNA and DNA concentrations than nesting females sampled at the end of the season ($F_{1,55} = 25.95$, $p<0.001$; $F_{1,55} = 7.70$, $p = 0.008$, respectively) such that females sampled at the beginning of the season ([RNA] = 17.11±9.52, [DNA] = 7.10±3.94) had more than the double the RNA concentration than the females sampled at the end ([RNA] = 6.51±9.21, [DNA] = 4.23±4.06).

Although RNA concentration was not significantly different for small and large females ($CCL_{min} \leq 85$ cm and >85 cm respectively) ($F_{1,43} = 0.81$, $p = 0.37$), the analysis of sRD for the two distinct size groups reveal significant differences for the biochemical index (Table 1, Fig. 3B).

Hatchling turtles

Hatching success of nests laid by "small" turtles was not significantly different to that of "large" turtles (Table 1). However, hatchlings produced by "large" females were significantly larger than those produced by "small" females ($F_{1,35} = 23$, $p<0.001$; Table 1, Fig. 3C). Although hatching success was not correlated with the size of the nesting females, hatching success was positively correlated with the ecophysiological condition of the hatchlings (sRD) ($F_{1,52} = 4.282$, $p<0.05$) (Fig. 4).

The ecophysiological condition of the hatchlings (sRD) was negatively correlated with righting time ($F_{1,52} = 3.932$,$p<0,0001$; Fig. 5) such that hatchlings with a lower average condition (sRD) took significantly longer to right (Table 1).

Unshaded versus Shaded nests

Although hatchlings from shaded nests exhibited higher DNA concentration (mean ± s.d. = 3.27±2.85) than those from unshaded nests (2.76±2.75) and hatchlings from unshaded nests had a higher average RNA concentration (3.19±2.77) than hatchlings from shaded nests (3.03±2.86), they were not statistically significantly different ($F_{1,56} = 0.50$, $p = 0.48$; $F_{1,56} = 0.05$, $p = 0.8$). On average, hatchlings from unshaded nests had a higher sRD (1.39±0.75) than the hatchlings from shaded nests (0.9±0.5) ($F_{8,105} = 11.12$, $p<0.001$; Table 1, Fig. 6).

Discussion

The major finding of our research was to demonstrate that physiological condition of females decrease during the nesting season (Hypothesis A1). In fact, the purpose of this study was to evaluate the ecophysiological condition of loggerhead turtles using novel and cost effective biochemical behavioral indices to infer life history traits. We were able to support the utility of this approach using a population for which a strong dichotomy in foraging ecology is known to exist (Hypothesis A2). Validation of assays for this method with substantial predictive power for estimating growth rates would provide a less invasive alternative to mark and recapture programs for marine vertebrates and would feed into

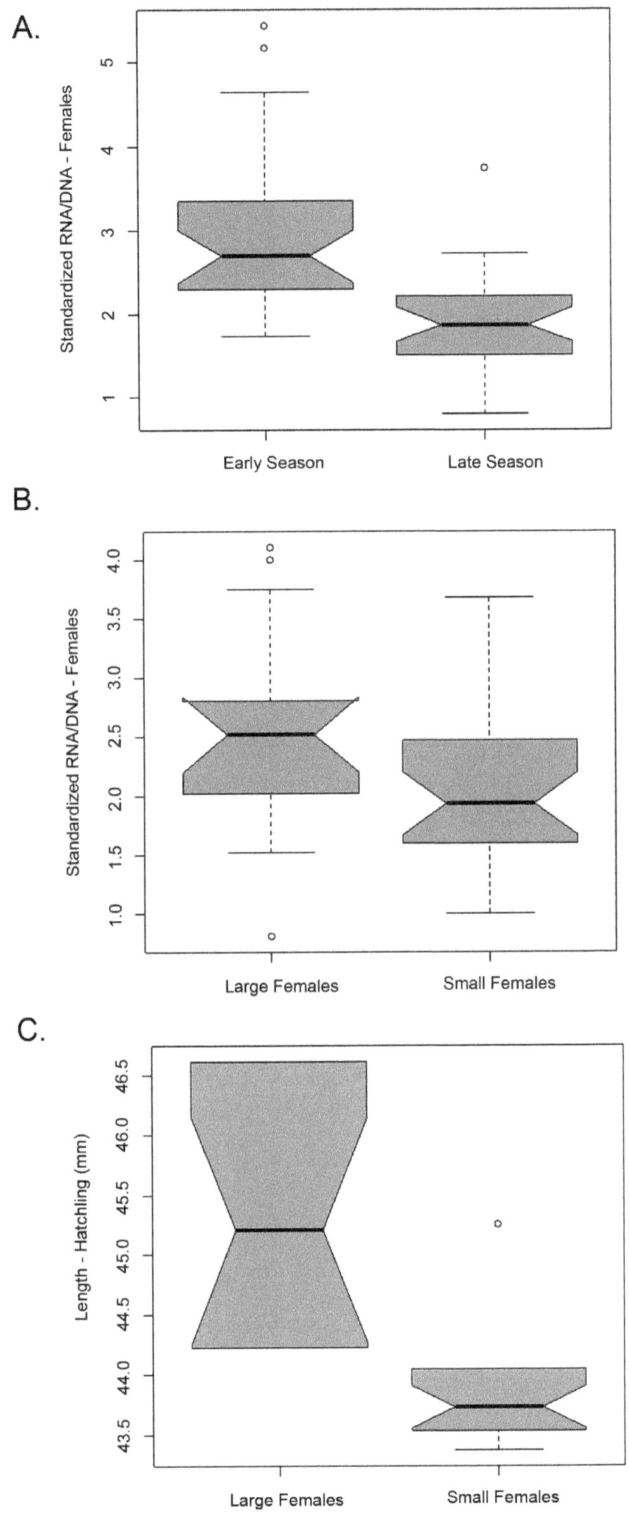

Figure 3. Biochemical condition of sea turtles and Hatchling length. (**A**). Standardized RNA/DNA ratio (mean ± standard error) for sea turtles in the beginning of the season (Early Season, n = 28) (7th–12th of July) and at the end of the season (Late Season, n = 29) (19th of September–10th of October); (**B**). Standardized RNA/DNA ratio (mean ± standard error) for "large" sea turtles ($CCL_{min}>85$ cm, n = 26) and for "small" sea turtles ($CCL_{min} \leq 85$ cm, n = 31); (**C**). Hatchling carapace length (mean ± standard error) for "large" sea turtles ($CCL_{min}>85$ cm) and for "small" sea turtles ($CCL_{min} \leq 85$ cm).

Table 1. Results of one way-ANOVA for hypotheses in the present study: (A1) Does the physiological condition of females decrease during the nesting season?; (A2) Is the physiological condition of "large" adult sea turtles associated with neritic habitats different to the physiological condition of "small" adult sea turtles associated with oceanic habitats?

Hypothesis	ANOVA	SS	DF	F-stat	p
A1	Time of season	14.020	1	23.4	1.59e-5***
	Residuals	26.979	45		
A2	Length Female	2.20	1	3.919	0.05.
	Residuals	23.019	41		
O1	Hatching Sucess	0.064	1	1.020	0.32
	Residuals	3.281	52		
O2	Length Hatchling	17.093	1	23	$2.964e^{-5}$***
	Residuals	26.006	35		
O3	Time Righting	321.589	1	3.923	$5e^{-45}$***
	Residuals	55.628	52		
O4	Shading	16.8	8	11.12	$2.813e^{-11}$***
	Residuals	19.824	105		

Regarding hatchling sea turtles, (O1) Are the hatching and emergence successes of "small" adult sea turtles lower than in "large" adult sea turtles?; (O2) Are hatchlings produced by "large" females larger than the hatchlings produced by "small" females?; (O3) Is hatchling vigour correlated with physiological condition?; and (O4) Does incubation temperature influence the physiological condition of the hatchlings? DF- degrees of freedom; SS- sum of squares; F-stat - F-statistic; p- significance value (Significant codes: 0 '***' 0.001 '**' 0.01 '*' 0.05 '.').

monitoring programs for populations of endangered species. This study represents the first application of tissue nucleic acid content analysis to evaluate the physiological condition or potential growth of loggerhead sea turtles.

Loggerhead turtles gather the energy necessary for reproduction over several years while in their foraging areas, before they migrate to mate and nest. Loggerhead turtles return to beaches in the region in which they hatched (termed 'philopatry'), typically after intervals of two to four years (although much longer intervals have been recorded), thought to depend in part on the quality and quantity of food available in foraging areas [27,28]. Marine turtles likely consume little food during the migration and nesting period and they breed, on average every 12 to 17 days during the breeding season [28]. A substantial decrease in food intake during the nesting season is expected to influence biochemical parameters, including RNA and DNA concentrations. In the present study, nesting females from the beginning of the nesting season exhibited higher nucleic acid concentrations and better physiological condition (sRD) than the females sampled at the end of the season. Similar findings have also been reported in leatherback turtles [40], hawksbill turtles [41] and green turtles [42]. In fact,

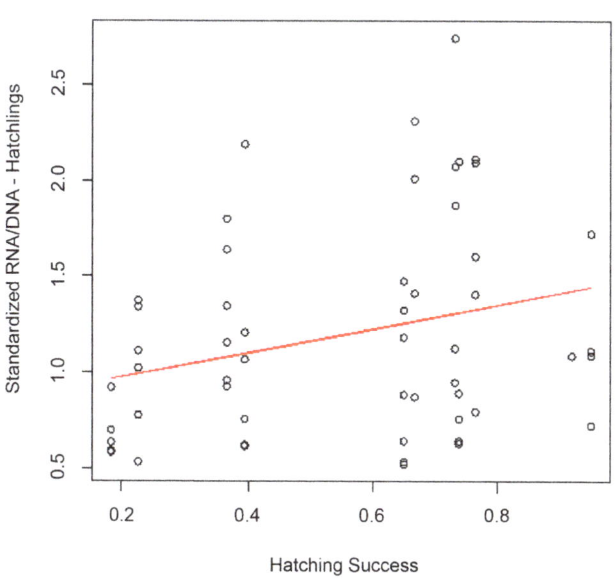

Figure 4. Relationship between the average condition of hatchlings (standardized RNA/DNA ratio) and the hatching success of the nest they came from ($R^2 = 0.06$, $p<0.001$, $F_{1,52} = 4.28$, $p = 0.04$).

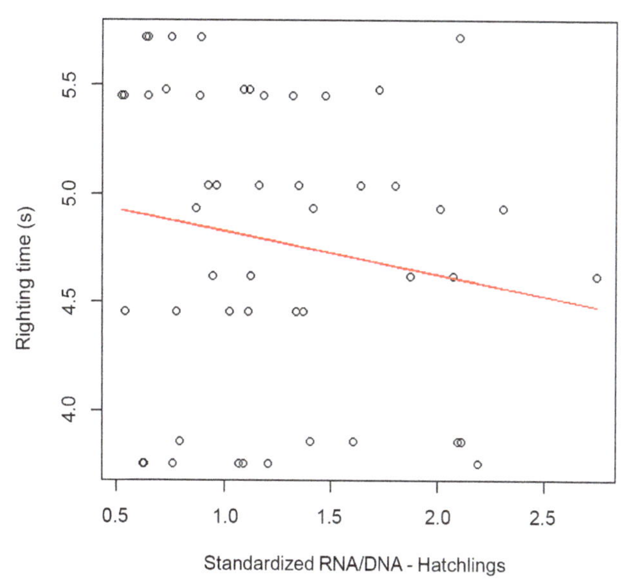

Figure 5. Relationship between the average condition (standardized RNA/DNA ratio) of the hatchlings and their righting time ($R^2 = 0.077$, $p<0.0001$, $F_{1,52} = 3.923$, $p<0.0001$).

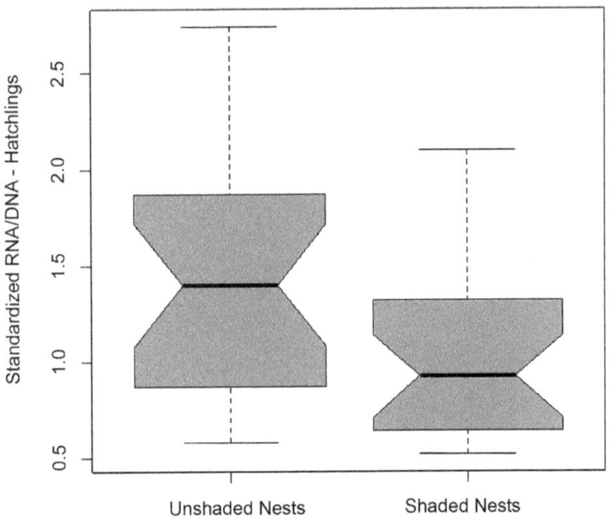

Figure 6. Standardized RNA/DNA ratio (mean ± standard error) for hatchlings from shaded and unshaded nests.

marine turtles rely on the mobilization of fat stores during the nesting season, which is supported by a large decline in plasma triglycerides [42]. Perraut *et al.* (2014) [40] found that gravid leatherbacks may experience protein loss during the nesting season. This trend is also common in birds, where total protein decreases during egg production [44].

Meyer *et al.* (2012) [45] analysed the starvation-induced changes in biochemical condition in early life stages on 9 species of marine fishes. In all cases, the mean biochemical condition (sRD) decreased exponentially with starvation time, regardless of their initial condition. In addition, they demonstrated that muscle growth during feeding was highly correlated with sRD because of its high protein synthesis rate. Therefore, a starvation signal can be strongly expressed in muscle tissue even though it only accounts for a part of the physiological response to resource limitation.

Larger nesting females exhibited a higher biochemical index (sRD) than "small" nesting females, which may mean that large females have a higher growth potential, due to differences in their diets. Adult female loggerhead sea turtles show a size-related behavioural and trophic dichotomy within several populations: oceanic planktivory by small females and neritic benthivory by large females [21,22,23,24,25]. Although oceanic adult females are more prevalent in Cape Verde [24], adult neritic foragers apparently have better fitness, as evidenced by larger carapace length, larger clutch size and higher reproductive success [21,32,46].

Although "large" nesting females exhibited a higher growth potential than smaller nesting females, there were no significant differences in hatching or emergence success (Hypothesis O1). This may suggest that egg quality was similar between the two foraging groups [47] but the findings of the present study differ from Eder *et al.* 2012 [32], who showed that clutch volume was higher in neritic turtles than oceanic type turtles and was significantly correlated with trophic foraging level, as indicated using stable isotope ratios. This latter result may have been due to the fact that we used relocated nests, which may have a drastic effect on hatching success [48,49].

Buskirk & Crowder (1994) [50] hypothesized that loggerhead sea turtles that experience a nutritional advantage during development may grow large and realize their increased reproductive potential by laying larger eggs and producing larger hatchlings. Although we did not compare egg size among large neritic foraging and small oceanic foraging females, our results showed that hatchlings produced by "large" females were significantly larger than those produced by "small" females (Hypothesis O2). Eder *et al.* (2012) [32] noted that body size of nesting loggerhead turtles at Cape Verde was not correlated with the sizes of their eggs, but was with the body size of their hatchlings, contrary to some previous studies [47,51,52]. Theoretically, offspring size is considered to be an important determinant of fitness in many reptiles, and larger size may have several advantages. The larger size may allow hatchlings to escape gape-limited predators, swim faster, and to successfully handle larger prey items [53,54], which may lead to enhanced survival during development [55]. Conditions during incubation and emergence are thought to have a significant effect on the size and locomotor performance of turtle hatchlings [56]. In the present study, hatchlings exhibiting a lower ecophysiological condition (sRD) took longer to right themselves than hatchlings with a higher sRD (Hypothesis O3). On land, sea turtle locomotion appears to be inefficient and, as a result hatchlings may overturn or become trapped by marine debris [57]. If they remain on beaches for long periods, particularly during the day, hatchlings may experience an increased risk of desiccation, overheating from sun exposure and predation [58]. Therefore, the ability of hatchling sea turtles to right themselves is critical. The present study shows that hatchlings with a better physiological condition have a better chance of surviving the journey from nest to sea. Additionally, the righting response has been considered as an indicator of fitness in freshwater turtles [34,58].

The results of the present study also indicate that shaded nests produce hatchlings of lower ecophysiological condition (Hypothesis O4). Other studies have suggested that temperature is one of the main factors with the potential to influence reproductive fitness in marine turtles, as well as a number of phenotypic traits, including hatchling sex [28]. For example, Booth *et al.* (2004) [59] showed incubation temperature influenced hatchling fitness in green sea turtles such that cooler nests produced males with inferior swimming ability compared with warmer nests that produced females, which may be critical during the 'hatchling frenzy' (the first 30–60 min of swimming in the shallow water surrounding natal beaches, where predation is highest) [54]. Therefore, it may not be surprising that nesting beach sand temperatures from the study population and from other Atlantic populations suggest a female-biased hatchling sex ratio [59,60]. Understanding the micro-environmental factors that influence hatching success and hatchling survival can further our knowledge of the habitat requirements for successful sea turtle reproduction.

Nucleic acid concentrations and ratios have become an important tool as biomarkers of recent growth in fish [9], crustaceans [8], bivalves [11] and plankton [10]. Combining morphometric measurements with biochemical analyses of skin and other easily obtainable tissues (including blood) could provide a minimally invasive technique for estimating recent growth rates in marine vertebrates. Hence, as biochemical indices of instantaneous growth are likely temperature and size-dependent, the utility and validation of these indices on marine turtles stocks deserves further study.

Author Contributions

Conceived and designed the experiments: SV SM AM MAT. Performed the experiments: SV SM LH. Analyzed the data: MAT SV. Contributed reagents/materials/analysis tools: MAT SV AM. Wrote the paper: SV LH MAT.

References

1. Halpern BS, Walbridge S, Selkoe KA, Kappel CV, Micheli F, et al. (2008) A Global Map of Human Impact on Marine Ecosystems. Science 319: 948–952.
2. Haberl H, Krausmann F, Gaube V, Bondeau A, Plutzar C, et al. (2007) Quantifying and mapping the human appropriation of net primary production in earth's terrestrial ecosystems. PNAS 104: 12942–12947.
3. Broad S, Mulliken T, Roe D (2003) Nature and Extent of Legal and Illegal Trade in Wildlife. In: Oldfield S, editor. The Trade in Wildlife: Regulation for Conservation. Earthscan Publications Ltd, London, UK, pp. 3–22.
4. Lewison RL, Crowder LB, Read AJ, Freeman SA (2004) Understanding impacts of fisheries bycatch on marine megafauna. Trends Ecol Evol 19, 598–604.
5. Lotze HK, Lenihan HS, Bourque BJ, Bradbury RH, Cooke RG, et al. (2006) Depletion, Degradation and Recovery Potencial of Estuaries and Coastal Seas. Science 312: 1806–1809.
6. IPCC (2014) Climate Change 2014: Impacts, Adaptation, and Vulnerability. Part A: Global and Sectoral Aspects. Contribution of Working Group II to the Fifth Assessment Report of the Intergovernmental Panel on Climate Change. Field CB, VR . Barros, DJ . Dokken , KJ . Mach, MD . Mastrandrea, TE . Bilir, M . Chatterjee, KL . Ebi, YO . Estrada, RC . Genova, B . Girma, ES . Kissel, AN . Levy, S . MacCracken, PR . Mastrandrea, LL . White, eds. Cambridge University Press, Cambridge, United Kingdom and New York, NY, USA, pp. 1–32
7. Chícharo MA, Chícharo L (2008) RNA/DNA Ratio and Other Nucleic Acid Derived Indices in Marine Ecology. Int J Mol Sci 9: 1453–1471.
8. Amaral V, Cabral HN, Paula J (2008) Implications of habitat specific growth and physiological condition for juvenile crab population structure. Mar Freshwater Res 59: 726–734.
9. Buckley LJ, Caldarone EM, Clemmesen C (2008) Multi-species larval fish growth model based on temperature and fluorometrically derived RNA/DNA ratios: results from a meta-analysis. Mar Ecol-Prog Ser 371: 221–232.
10. Ikeda T, Sano F, Yamaguchi A, Matsuishi T (2007) RNA/DNA ratios of calanoid copepods from the epipelagic through abyssopelagic zones of the North Pacific Ocean. Aquatic Biol 1: 99–108.
11. Chícharo MA, Chícharo L, Amaral A, Condinho S, Gaspar M (2003) Chronic effects of dredging-induced stress on the clam (*Spisula solida*): nucleic acid and lipid composition. Fish Res 1525: 1–6.
12. Frommel A, Clemmesen C (2009) Use of biochemical indices for analysis of growth in juvenile two-spotted gobies (*Gobiusculus flavescens*) of the Baltic Sea. Sci Mar 73: 59–170.
13. Pendoley KL, Schofield G, Whittock PA, Ierodiaconou D, Hays GC (2014) Protected species use of a coastal marine migratory corridor connecting marine protected areas. Mar Biol 161: 1455–1466.
14. Godley BJ, Blumenthal JM, Broderick AC, Coyne MS, Godfrey MH, et al. (2008) Satellite tracking of sea turtles: Where have we been and where do we go next? Endanger Species Res 4: 3–22.
15. Hamann M, Godfrey MH, Seminoff JA, Arthur K, Barata PCR, et al. (2010) Global research priorities for sea turtles: informing management and conservation in the 21st century. Endang Species Res 11: 245–269.
16. Seminoff JA, Shanker K (2008) Marine turtles and IUCN Red Listing: A review of the process, the pitfalls, and novel assessment approaches. J Exp Mar Biol Ecol 356: 52–68.
17. Hawkes LA, McGowan A, Godley BJ, Gore S, Lange A, et al. (2013) Estimating sex ratios in Caribean hawksbill turtles: testosterone levels and climate effects. Aquatic Biol 18: 9–19.
18. Roark AM, Bjorndal KA, Bolten BA, Leeuwenburgh C (2009) Biochemical indices as correlates of recent growth in juveniles green turtles (*Chelonia mydas*). J Exp Mar Biol Ecol 37: 59–67.
19. Limpus CJ (1992) Estimation of tag loss in marine turtle research. Wildlife Res 19: 457–469.
20. Bjorndal KA, Bolten AB, Coan AL, Kleiber P (1995) Estimation of green turtle (*Chelonia mydas*) growth rates from length-frequency analysis. Copeia 1: 71–77.
21. Hatase H, Omuta K, Tsukamoto K (2010) Oceanic residents, neritic migrants: a possible mechanism underlying foraging dichotomy in adult female loggerhead turtles (*Caretta caretta*). Mar Biol 157: 1337–1342.
22. Hatase H, Omuta K, Tsukamoto K (2007) Bottom or mid- water: alternative foraging behaviours in adult female loggerhead sea turtles. J Zool 273: 46–55.
23. Hatase H, Takai N, Matsuzawa Y, Sakamoto W, Omuta K, et al. (2002) Size-related differences in feeding habitat use of adult female loggerhead turtles *Caretta caretta* around Japan determined by stable isotope analyses and satellite telemetry. Mar Ecol-Prog Ser 233: 273–281.
24. Hawkes LA, Broderick AC, Coyne MS, Godfrey MH, Lopez-Jurado LF, et al. (2006) Phenotypically linked dichotomy in sea turtle foraging requires multiple conservation approaches. Curr Biol 16: 990–995.
25. Varo-Cruz N, Cejudo D, López-Jurado LF (2007) Reproductive biology of the loggerhead turtle (*Caretta caretta* L. 1758) on the Island of Boavista (Cape Verde, West Africa) In: López-Jurado LF, Liria-Loza A, editors. Marine Turtles, Recovery of Extinct Populations. Instituto Canário de Ciencias Marinas, Las Palmas, Spain. pp. 125–144.
26. Zbinden JA, Bearhop S, Bradshaw P, Gill B, Margaritoulis D, et al. (2011) Migratory dichotomy and associated phenotypic variation in marine turtles revealed by satellite tracking and stable isotope analysis. Mar Ecol-Prog Ser 421: 291–302.
27. Limpus CJ, Miller JD (1993) Family *Chelonidae*. In: Glasby CJ, Ross GJB, Beesley PL, editors. Fauna of Australia. Vol. 2A - Amphibia and Reptilia. Australian Government Publishing Service, Canberra, Australia, pp. 113–138.
28. Miller JD (1997) Reproduction in sea turtles. In: Lutz PL, Musick JA, editors. The biology of sea turtles. CRC Press, Boca Raton, FL, USA, pp. 51–82.
29. Marco A, Abella E, Liria-Loza A, Martins S, López O, et al. (2012) Abundance and exploitation of loggerhead turtles nesting in Boa Vista island, Cape Verde: The only substantial rookery in the eastern Atlantic. Anim Conserv 15: 351–360.
30. Monzón-Argüello C, Rico C, Naro-Maciel E, Varo-Cruz N, López P, et al. (2010) Population structure and conservation implications for the loggerhead sea turtle of the Cape Verde Islands. Conserv Genet 11: 1871–1884.
31. Wallace BP, DiMatteo AD, Bolten AB, Chaloupka MY, Hutchinson BJ, et al. (2011) Global Conservation Priorities for Marine Turtles. PLoS ONE 6(9): e24510. doi:10.1371/journal.pone.0024510
32. Eder E, Ceballos A, Martins S, Pérez-García H, Marín I, et al. (2012) Foraging dichotomy in loggerhead sea turtles *Caretta caretta* off northwestern Africa. Mar Ecol-Prog Ser 470: 113–122.
33. Wood A, Booth DT, Limpus CJ (2014) Sun exposure, nest temperature and loggerhead turtle hatchlings: Implications for beach shading management strategies at sea turtle rookeries. J Exp Mar Biol Ecol 451: 105–114.
34. Delmas V, Baudry E, Girondot M, Prevot-Julliard AC (2007) The righting response as a fitness index in freshwater turtles. Biol J Linn Soc 91: 99–109.
35. Dutton PH (1996) Methods for collection and preservation of samples for sea turtle genetic studies. In: Bowen BW, Witzell WN, compilers. Proceedings of the International Symposium on Sea Turtle Conservation Genetics. NOAA Tech Memo NMFS- SEFSC-396, pp. 17–24.
36. Caldarone EM, Wagner M, St. Onge-Burns J, Buckley LJ (2001) Protocol and Guide for Estimating Nucleic Acids in Larval Fish Using a Fluorescence Microplate Reader. National Marine Fisheries Service, Woods Hole, MA. Available: http://www.nefsc.noaa.gov/nefsc/publications/crd/crd0111/crd0111.pdf. Accessed 14 April 2013
37. Esteves E, Chícharo MA, Pina T, Coelho ML, Andrade JP (2000) Comparison of RNA/DNA ratios obtained with two methods for nucleic acid quantification in gobiid larvae. J Exp Mar Biol Ecol 245: 43–55.
38. Olivar MP, Diaz M, Chícharo MA (2009) Tissue effect on RNA/DNA ratios of marine fish larvae. Sci Mar 73: 171–182.
39. Caldarone EM, Clemmesen CM, Berdalet E, Miller TJ, Folkvord A, et al. (2006) Intercalibration of four spectrofluorometric protocols for measuring RNA/DNA ratios in larval and juvenile fish. Limnol Oceanogr-Meth 4: 153–163.
40. Perrault JR, Wyneken J, Page-Karjian A, Merrill A, Miller DL (2014) Seasonal trends in nesting leatherback turtle (*Dermochelys coriacea*) serum proteins further verify capital breeding hypothesis. Conserv Physiol 2: doi:10.1093/conphys/cou002
41. Goldberg DW, Leitão SAT, Godfrey MH, Lopez GG, Santos AJB et al. (2013) Ghrelin and leptin modulate the feeding behaviour of the hawksbill turtle *Eretmochelys imbricata* during nesting season. Conserv Physiol 1: doi: 10.1093/conphys/cot016
42. Hamann M, Limpus CJ, Whittier JM (2002) Patterns of lipid storage and mobilisation in the female green sea turtle (*Chelonia mydas*). J Comp Physiol B (2002) 172: 485–493. DOI 10.1007/s00360-002-0271-2
43. Plot V, Jenkins T, Robin J-P, Fossette S, Georges J-Y (2013) Leatherback Turtles Are Capital Breeders: Morphometric and Physiological Evidence from Longitudinal Monitoring. Physiological and Biochemical Zoology 86 (4): 385–397.
44. Cherel Y, Robin JP, Walch O, Karmann H, Netchitailo P, Le Maho Y (1988) Fasting in king penguin. I. Hormonal and metabolic changes during breeding. Am J Physiol 254: R170–R177.
45. Meyer S, Caldarone EM, Chícharo MA, Clemmesen C, Faria AM, et al. (2012) On the edge of death: Rates of decline and lower thresholds of biochemical condition in food-deprived fish larvae and juveniles. J Marine Syst 93: 11–24.
46. Rafferty AR, Santidria'n Tomillo P, Spotila JR, Paladino FV, Reina RD (2011) Embryonic Death Is Linked to Maternal Identity in the Leatherback Turtle (*Dermochelys coriacea*). PLoS ONE 6(6): e21038. doi:10.1371/journal.pone.0021038
47. Hatase H, Omuta K, Tsukamoto K (2013) Do loggerhead turtle (*Caretta caretta*) eggs vary with alternative foranging tactics? J Exp Mar Biol Ecol 455: 56–61.
48. Limpus CJ, Baker V, Paukstis GL (1979) Movement induced mortality of loggerhead eggs. Herpetologica 35: 335–338.
49. Mortimer JA (1999) Reducing threats to eggs and hatchlings: Hatcheries. In: Eckert KL, KA . Bjorndal, FA. Abreu-Grobois, M. Donnelly, editors. Research and Management Techniques fot the Conservation of Sea Turtles. IUCN/MTSG 4, pp.175–178.
50. Buskirk J, Crowder LB (1994) Life-History Variation in Marine Turtles. Copeia 1: 66–81.
51. Özdemir A, Ilgaz Ç, Kumlutaş Y, Durmuş SH, Kaska Y, et al. (2007) An assessment of initial body size in loggerhead sea turtle (*Caretta caretta*) hatchlings in Turkey. Zool Sci 24: 376–380.
52. Pinckney J (1990) Correlation analysis of adult female, egg and hatchling sizes in the loggerhead turtle, *Caretta caretta* (L.), nesting at Kiawah Island, South Carolina, USA. B Mar Sci 47: 670–679.

53. Ji X, Qiu QB, Diong CH (2002) Influence of incubation temperature on hatching success, energy expenditure for embryonic development, and size and morphology of hatchlings in the oriental garden lizards, *Calotes versicolor* (*Agamidae*). J Exp Zool 292: 649–659.
54. Gyuris E (2000) The relationship between body size and predation rates on hatchlings of the green turtle (*Chelonia mydas*): is bigger better? In: Pilcher N, Ismail M, editors. Sea Turtles of the Indo-Pacific: Research, Management and Conservation. Academic Press, New York, USA, pp. 143–147.
55. Janzen FJ (1993) An experimental analysis of natural selection on body size of hatchling turtles. Ecology 74: 332–341.
56. Packard GC, Packard MJ (1988) The physiological ecology of reptilian eggs and embryos. In: Gans C, editors. Biology of the reptilia, vol 16. Academic, New York, USA, pp. 523–605.
57. Davenport J (1997) Temperature and the life-history strategies of sea turtles. J Therm Biol 22: 479–488.
58. Steyermark AC, Spotila JR (2001) Effects of maternal identity and incubation temperature on hatching and hatchling morphology in snapping turtles, Chelydra serpentina. Copeia: 129–135.
59. Booth DT, Burgess E, McCosker J, Lanyon JM (2004) The influence of incubation temperature on post-hatching fitness characteristics of turtles. Int Congr Ser 1275: 226–233.
60. Hawkes LA, Broderick AC, Godfrey MH, Godley BJ (2009) Climate change and marine turtles. Endanger Species Res 7: 137–154.

Deep Vision: An In-Trawl Stereo Camera Makes a Step Forward in Monitoring the Pelagic Community

Melanie J. Underwood[1,2]*, Shale Rosen[1], Arill Engås[1,2], Elena Eriksen[1]

1 Institute of Marine Research, Bergen, Norway, **2** Department of Biology, University of Bergen, Bergen, Norway

Abstract

Ecosystem surveys are carried out annually in the Barents Sea by Russia and Norway to monitor the spatial distribution of ecosystem components and to study population dynamics. One component of the survey is mapping the upper pelagic zone using a trawl towed at several depths. However, the current technique with a single codend does not provide fine-scale spatial data needed to directly study species overlaps. An in-trawl camera system, Deep Vision, was mounted in front of the codend in order to acquire continuous images of all organisms passing. It was possible to identify and quantify of most young-of-the-year fish (e.g. *Gadus morhua, Boreogadus saida* and *Reinhardtius hippoglossoides*) and zooplankton, including Ctenophora, which are usually damaged in the codend. The system showed potential for measuring the length of small organisms and also recorded the vertical and horizontal positions where individuals were imaged. Young-of-the-year fish were difficult to identify when passing the camera at maximum range and to quantify during high densities. In addition, a large number of fish with damaged opercula were observed passing the Deep Vision camera during heaving; suggesting individuals had become entangled in meshes farther forward in the trawl. This indicates that unknown numbers of fish are probably lost in forward sections of the trawl and that the heaving procedure may influence the number of fish entering the codend, with implications for abundance indices and understanding population dynamics. This study suggests modifications to the Deep Vision and the trawl to increase our understanding of the population dynamics.

Editor: Athanassios C. Tsikliras, Aristotle University of Thessaloniki, Greece

Funding: The authors would like to acknowledge support provided by the Barents Sea Research programme at the Institute of Marine Research and Scantrol AS. The funders had no role in the study design, data collection and analysis, decision to publish, or preparation of the manuscript.

Competing Interests: The authors have read the journal's policy and have the following conflicts: One of the authors was employed by Scantrol AS during data collection and analysis. Scantrol AS had no role in the study design, data collection and analysis, decision to publish, or preparation of the manuscript.

* Email: melanie.underwood@imr.no

Introduction

Fishery management has shifted focus from a single-species approach towards an ecosystem approach that takes how human interventions and food web linkages affect ecosystems into account [1,2]. The monitoring programmes used as a basis for fisheries management advice have been forced to adapt to meet the data needs for ecosystem-based management by measuring a wide range of ecosystem components [1,3]. A number of methods and gears have been employed, ranging from water sampling to plankton nets, pelagic and demersal trawls, grabs and sledges, echo sounders and direct visual observations [4]. Even with modern research vessels, equipment and methods, limitations remain related to gear efficiency and documenting vertical distribution and overlap of organisms. There is thus a need to continue to develop tools and methods in order to overcome these limitations and increase our understanding of population dynamics.

The Norwegian-Russian Barents Sea Ecosystem Survey (BESS) is a comprehensive survey that gathers a wide range of measurements from the physical and biological components of the ecosystem [5]. One task of the BESS is to map the upper pelagic community (including young-of-the-year fish, large krill (Euphausiidae) and jellyfish) with a pelagic trawl to measure abundances and provide biomass indices [6]. However, current survey methods have their limitations, including a lack of spatial distribution data due to all species being collected in a single codend, an inability to identify and quantify less robust species that are destroyed by the codend (e.g. comb jellyfish, Ctenophora) [7] and the difference in the size-catchability performance of the various trawls used [8,9]. Therefore, it is necessary to develop new tools that can identify and quantify species that are easily damaged and that can sample all species and sizes at the same time.

An in-trawl camera system, Deep Vision (Scantrol AS, Bergen, Norway), has been developed to identify and measure species continuously as they pass inside the trawl [10,11]. The Deep Vision system has been used for several surveys to identify, quantify and measure the length of large fish such as adult Northeast Arctic cod (*Gadus morhua*), haddock (*Melanogrammus aeglefinus*) and Atlantic mackerel (*Scomber scombrus*) along the trawl track. Time-references for each image can be matched up with data such as geographic position and acoustic backscatter information collected by the vessel's echo sounder [12]. The stereo images can be processed to calculate the size of passing objects with high accuracy (less than 5% error; [10]) as well as the spatial distribution along the trawl path. Our goal was to evaluate if the current system can be used to identify, quantify and measure small and fragile organisms, and how it can be improved to meet the goal of collecting data to increase our understanding of population dynamics.

Materials and Methods

Ethics Statement

BESS is conducted as part of the Norwegian and Russian obligations under the international law for the monitoring of environmental changes and the management of living marine resources, approved by the Joint Norwegian-Russian Fisheries Commission and the Norwegian Ministry of Fisheries and Coastal Affairs. The collection of image data was carried out as part of the BESS. No endangered or protected species were encountered during the field studies, and fish acquired by trawling were immediately killed when they came onboard. Collection of the image data had no additional impact on the welfare of the organisms. The standard biological sampling procedures are routine work at sea, approved by the Institute of Marine Research Institutional Animal Care and Use Committee.

Study area and trawling procedure

Deep Vision observations were carried out during standard sampling hauls on the BESS inside the Isfjord and Billefjord areas at Svalbard in August 2012 with RV "Johan Hjort". The trawl was towed at three depths (with the headline at 0, 20 and 40 m) for 0.5 nautical miles each, at a speed over ground of 3 knots [6]. The survey uses a pelagic standard sampling trawl for young-of-the-year fish [Harstad sampling trawl; 8]. The four panel trawl consists of seven sections with mesh sizes ranging from 200 mm in the front of the trawl to 8 mm in the codend [13]. The Deep Vision section was attached to the trawl between the extension and the codend (Fig. 1). The trawl dimensions (i.e. vertical opening of the net mouth, wing spread and depth of the headline) were measured with acoustic trawl instrumentations (SCANMAR AS, Åsgårdstrand, Norway). The catch was measured using the BESS standard biological sampling procedure [14]. Images from Deep Vision were analysed post-cruise.

Deep Vision camera system

The Deep Vision camera system consists of two 1.4 megapixel digital colour cameras fitted with 4.8 mm focal length lenses and arranged in a parallel stereo orientation. The cameras are connected to a PC for control and data storage and placed inside a subsea housing rated to 200 bar pressure. Illumination is provided by two external light emitting diode (LED) strobes which generate 38 400 lumens. With such a high level of artificial illumination, there is very little difference between images collected at the surface and at depth. Batteries provide power for up to eight hours of operation. A pressure sensor continuously collects system depth data (5 m resolution) and images are time-stamped to match them with external sensor data.

Five images per second were collected, starting before the trawl was set out and ending when it came back onboard the vessel, saving a continuous time-referenced record of all objects that passed through the trawl during the shooting, trawling, and heaving phases. The system and its measurement accuracy are described in detail by Rosen *et al* [10] and Rosen and Holst [12]. In this study, the system was operated in autonomous mode without cable connection to the vessel.

The camera housing, lights, and battery were mounted inside a 90 cm high ×90 cm wide ×150 cm long neutrally buoyant frame made of 10 mm-thick high-density polyethylene (HDPE). The frame was placed in a 12.5 m-long four-panel net section between the extension and the codend (Fig. 1). The whole Deep Vision section was lined with 8 mm mesh similar to the codend in order to guide all the catch in front of the camera. The leading end of the net section was 190 cm in height and width, corresponding to a minimum sampling cross-section of 3.6 m^2 for individual organisms that were not herded by the larger meshes in the trawl. Tapering and lead nets immediately in front of the HDPE frame guided the catch through a passage within the field of view of both cameras at a range of 27 to 73 cm (field of view = 41 and 102 cm width, respectively). The camera side of the passage was made of 10 mm-thick transparent polycarbonate while the back wall, roof, and floor were opaque to retain light and provide contrast with the edges of the passing fish. The area surrounding the camera was also constructed of opaque white material in order to provide diffused, even illumination.

Image Analysis

Both images and biological samples were collected on eight hauls. Two hauls were selected for post-cruise image analysis to evaluate if the current system can be used to identify, quantify and measure small and fragile organisms. In order to provide the most diverse samples of fish and zooplankton, one haul with the greatest number of commercially important finfish (haul 04; 11:56 UTC) and one with the greatest number of zooplankton with finfish present (haul 06; 3:29 UTC) were selected. All images collected in the course of these hauls were reviewed in order to identify and

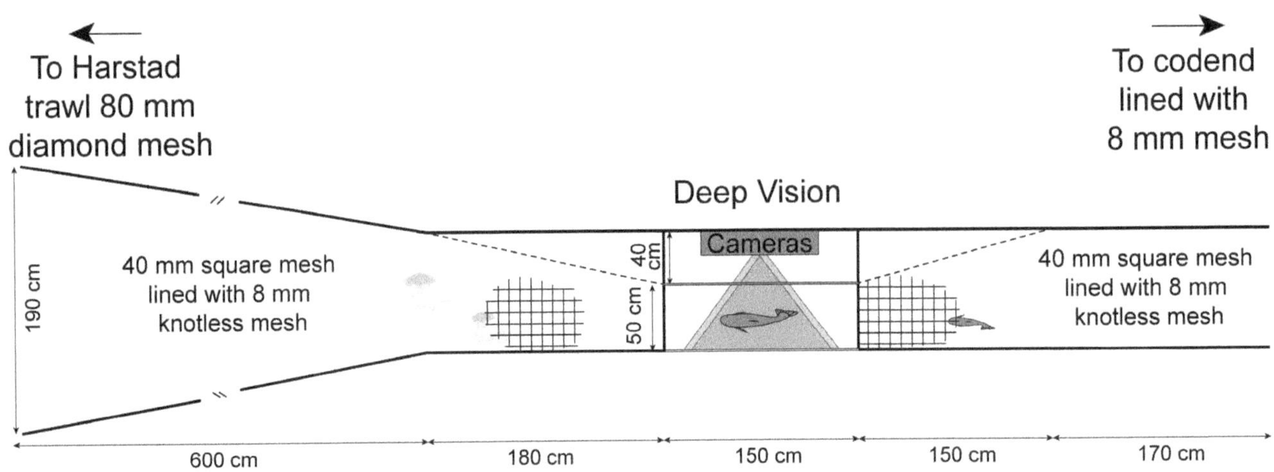

Figure 1. Schematic representations of the Deep Vision frame and trawl section.

quantify the finfish and zooplankton that passed the Deep Vision camera. The depth at which each individual was imaged was determined by matching the image timestamp with the time-referenced depth recorded from the pressure sensor in the camera housing. Data from haul 04 were analysed to compare spatial distributions between species, while data from haul 06 were analysed to compare length measurements made by the Deep Vision system and those of the actual catch.

A total of 21 030 images from haul 04 and 19 714 images from haul 06 were analysed in two ways. First, all finfish and lion's mane jellyfish (*Cyanea capillata*) were counted, in accordance with the BESS catch sampling protocol. Most individuals were imaged several times as they passed through the 41–102 cm field of view (generally taking 3 images or 600 ms to pass). In order to prevent double-counting, individuals were tracked across images (Fig. 2) and counted in the image when they first entered the field of view. Individuals that exited the field of view from the direction of the trawl entrance were subtracted from counts and added again when they re-entered the field of view. Individuals that could not be identified were recorded as 'unidentified'. Second, all other zooplankton were quantified by sampling the first image of each 30-second interval and counting the number of individuals present (a total of 140 images from haul 04 and 124 images from haul 06). This method was used because it was impossible to track small zooplankton between images when densities were high (hundreds of individuals in a single image). Since there was a 30-second interval between images, it is unlikely that any individuals were double-counted. To estimate the total count of zooplankton in each haul, each sub-sample representing the number of passages in 600 ms was multiplied by 50 to estimate total number of individuals passing in each 30-second interval.

Lengths of polar cod (*Boreogadus saida*), shorthorn sculpin (*Myoxocephalus scorpius*) and Northeast Arctic cod were measured using the Deep Vision software [10], and compared with the codend catch data to test the possibility of using the Deep Vision for measuring the lengths of small organisms. Measurements were limited to species with visible caudal fins (i.e. Greenland halibut were not used), since the catch data were recorded as total length (from the snout to the end of the caudal fin) and individuals presenting both snout and caudal fin to the camera. Since the catches of polar cod were large (4588 individuals), lengths were measured from a random sub-sample, consistent with the BESS protocol. The first Deep Vision image of every 15-second interval was analysed and lengths were calculated for all polar cod orientated such that they could be measured. The differences between the length measurements from the catch data and the image-estimated lengths of polar cod were analysed by one-way ANOVA.

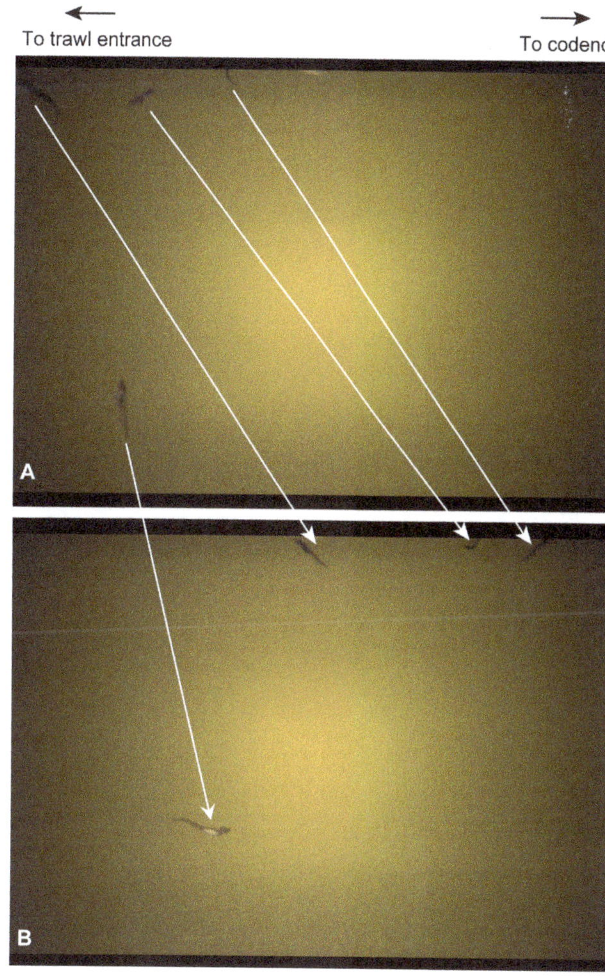

Figure 2. Example of tracking fish through sequential images. (A) Four polar cod (*Boreogadus saida*) enter the Deep Vision chamber and (B) move towards to the codend in the next image, taken 200 ms later. The white arrows show the movement by each individual.

Results

Trawl geometry and performance

The depth sensor showed that the path of the trawl in the water column deviated from the survey protocol (Fig. 3). Measurements of the trawl geometry also showed that the vertical opening and wing spread changed with depth. The vertical opening of the trawl diminished from approximately 16 to 10 m with the headline at 0 m and 40 m respectively, while the corresponding wingspread measurements increased from approximately 25 to 29 m.

Deep Vision images

We were able to identify and quantify from the Deep Vision images eleven species of juvenile finfish, including six species used for species index analysis, and five zooplankton species (including comb jellyfish; Table 1). However, some organisms could not be identified to species level and some were present in the catch but not observed in the images. We were unable to consistently discriminate between redfish (*Sebastes* spp.) and polar cod, and counts of polar cod may therefore include redfish. This was also the case with capelin (*Mallotus villosus*) and shannies (Stichaeidae), and counts of shannies may therefore include capelin. The problem was greatest when individuals passed the camera at the maximum range of 73 cm. Four haddock were found in the catch but not identified in the images. Total counts of small zooplankton per haul were generally overestimated compared to the catch data when extrapolated from the 30-second sub-sample.

Individuals were observed at all the depth layers during the standard 30-minute towing time, with most species increasing in number down to 30 m and then decreasing at greater depths (Fig. 4). However, numbers of Northeast Arctic cod and krill continued to increase as the depth increased. Species were observed to enter the codend in patches and with other species during haul 04 (Fig. 5). Polar cod and shannies were observed together at all depths throughout haul 04. A large number of polar cod and Greenland halibut were observed to pass the camera when the trawl was at the surface during heaving with up to 35%

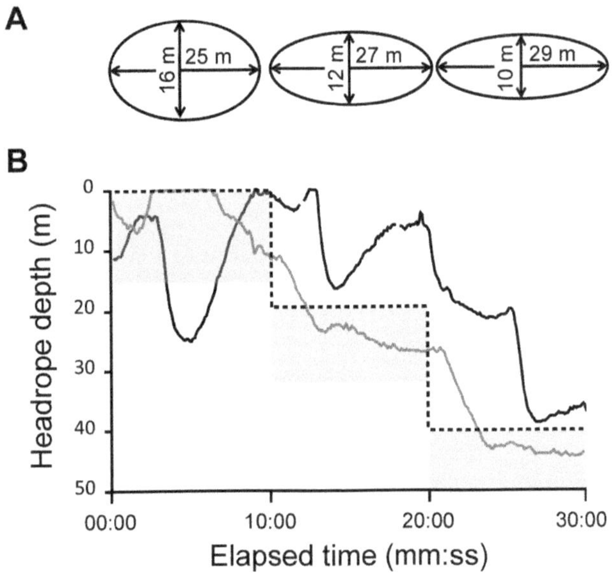

Figure 3. Schematic representation of the (A) net mouth geometry and (B) headline depth. The black line indicates the headline depth for haul 04, while the grey line indicates the headline depth for haul 06. The dashed line shows the stepwise protocol for BESS. The shaded areas under the dashed line indicate the height of the trawl mouth opening and the depths surveyed.

of polar cod and 80% of Greenland halibut passing outside of the designated 30-minute trawling time (Fig. 6). During heaving, when the trawl was already at the surface, more than 280 polar cod per second passed the Deep Vision system, with individuals moving ahead (towards the trawl entrance) in patches for short periods of time before re-entering the field of view. This made it difficult to quantify young-of-the-year fish during heaving when high densities and turbulent flow were observed. Counts from the images may therefore be underestimates. Individuals of all species were observed entering the codend in groups during heaving (at the surface) and some were seen to have damaged opercula.

Fifty polar cod were measured from the catch and 77 out of 115 individuals were measured from the images for length comparisons (the remaining 38 were not imaged in orientations where lengths could be estimated). Average lengths of polar cod were not significantly different in the images and catch data (mean length 34 mm for each method; $F = 0.034$, $p = 0.85$; Fig. 7). The single Northeast Arctic cod was measured at 53 mm in both the Deep Vision image and the catch data. Only the larger of the two shorthorn sculpin was passed in an orientation where it could be measured. Its length was calculated to be 43 mm in the Deep Vision image compared to 45 mm in the catch data (4.4% difference).

Discussion

The pelagic community is a vital component of the Barents Sea ecosystem, providing important links between lower and higher trophic levels [15,16]. Pelagic fish species such as capelin, herring

Table 1. List of species and families identified and quantified in the Deep Vision images and in the catch data.

	Counts from the Deep Vision and Catch data for each haul			
	Haul 04		**Haul 06**	
Species/family	**Images**	**Catch data**	**Images**	**Catch data**
Greenland halibut *Reinhardtius hippoglossoides*	50	51	6	5
Capelin *Mallotus villosus*[a]	NA	1184	NA	1
Norwegian spring spawning herring *Clupea harengus*	1	4	0	1
Northeast Arctic cod *Gadus morhua*	18	22	1	1
Redfishes *Sebastes* spp.[a]	NA	551	NA	0
Polar cod *Boreogadus saida*	19431	29268	2114	4588
Shorthorn sculpin *Myoxocephalus scorpius*	274	248	2	2
Shanny family Stichaeidae	1734	1680	3332	5797
Lumpfish Cyclopteridae	4	4	0	0
Snailfish family Liparidae	119	193	32	12
Atlantic poacher *Leptagonus decagonus*	1	1	12	4
Lion's mane jellyfish *Cyanea capillata*	44	38	144	81
Haddock *Melanogrammus aeglefinus*	0	4	0	0
Unidentified	25		3	
Krill *Thysanoessa* spp.[b]	2450	330	19000	9020
Comb jellyfish Ctenophora[b]	30000	NA	38250	NA
Amphipods *Themisto* spp.[b]	2200	4067	8700	7447
Sea butterfly Thecosomata[b]	600	413	950	111

Species and families unable to be quantified are marked as NA.
[a]Redfishes were difficult to distinguish from polar cod, and capelin were difficult to distinguish from the shanny family. Both were not counted.
[b]Small individual zooplankton were unable to be tracked between images and therefore were sub-sampled every 30 s. Counts from Deep Vision are estimated from the sub-sample.

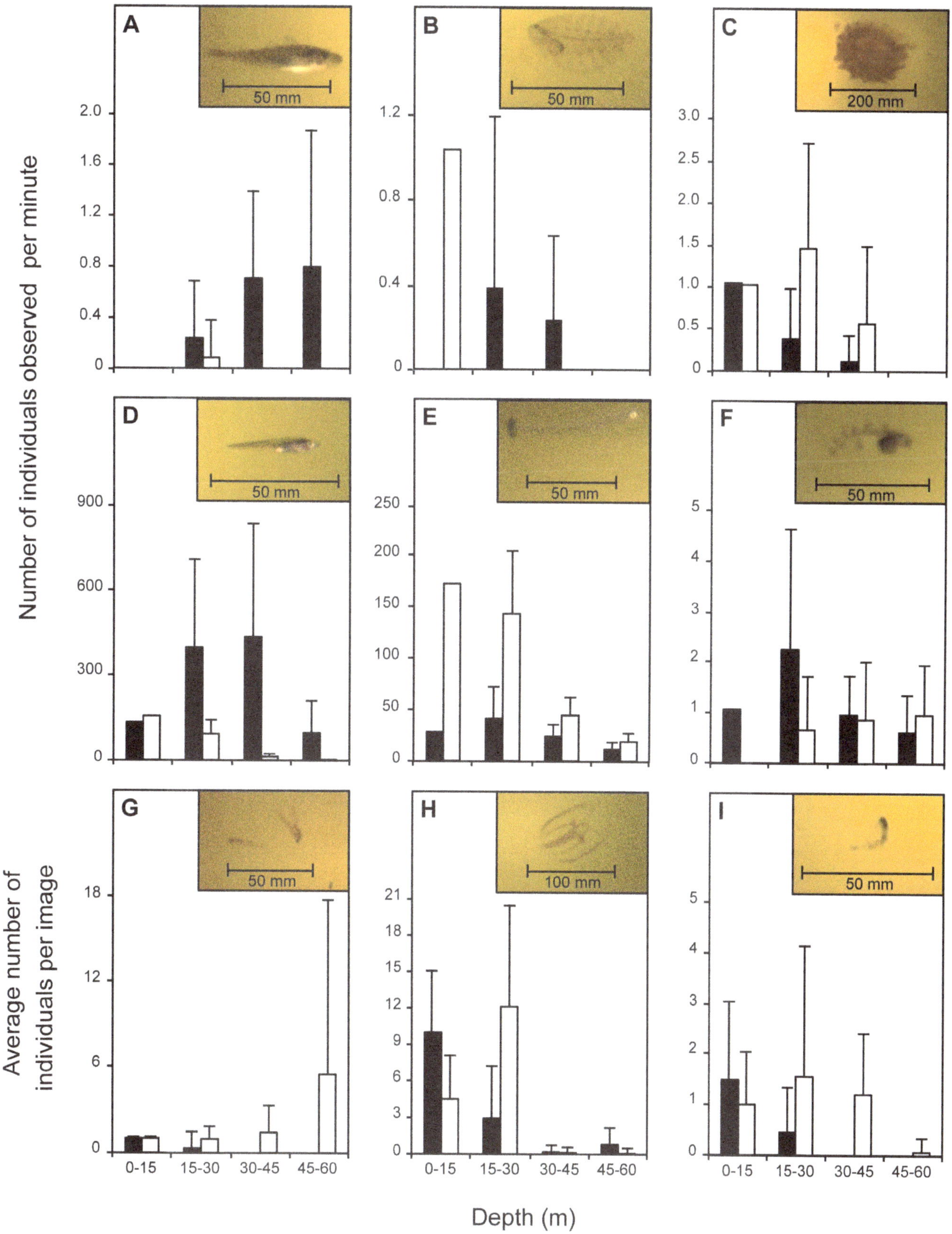

Figure 4. Species caught at various depths by two image-sampling methods during the haul. (A–F) sample all images and show the number of individuals observed per minute for each species/family and (G–I) are sub-sampled (first image of each 30-second interval) and indicate the average number of individuals per image, at four different depths. The dark column bars indicate haul 04 while haul 06 is shown by white column

bars. The species/families are (A) Northeast Arctic cod (*Gadus morhua*; Haul 04, $n=15$; Haul 06, $n=1$), (B) Greenland halibut (*Reinhardtius hippoglossoides*; Haul 04, $n=10$; Haul 06, $n=1$), (C) lion's mane jellyfish (*Cyanea capillata*; Haul 04, $n=10$; Haul 06, $n=25$), (D) polar cod (*Boreogadus saida*; Haul 04, $n=12633$; Haul 06, $n=1443$), (E) shanny family (Stichaeidae; Haul 04, $n=1162$; Haul 06, $n=2531$), (F) snailfish family (Liparidae; Haul 04, $n=58$; Haul 06, $n=23$), (G) krill (*Thysanoessa* spp.; Haul 04, $n=15$; Haul 06, $n=130$), (H) comb jellyfish (Ctenophora; Haul 04, $n=151$; Haul 06, $n=344$), and (I) amphipods (*Themisto* spp.; Haul 04, $n=25$; Haul 06, $n=66$). The error bars indicate the upper standard deviation. The trawl spent just one minute in the 0–15 m depth, therefore no standard deviation was calculated for (A–F).

(*Clupea harengus*) and polar cod make up the bulk of the total biomass and are mainly plankton-feeders (consuming primarily Euphausiacea and Amphipoda). The Barents Sea is also a nursery area for several commercially and ecologically important fish stocks, with young-of-the-year distributed in the upper water column during the summer and autumn. The pelagic community is surveyed by acoustics, plankton nets and trawls during BESS. The use of several sampling methods with different efficiencies, sampling volumes and deployment times makes it difficult to observe species overlaps, which are essential for studies of small scale processes, such as interspecies interactions (competition, mutualism, protocooperation, predation etc.), which are important to understand the population dynamics [17]. There is thus a need for new tools that can measure several components of the ecosystem simultaneously and can provide greater spatial resolution than single-codend trawling. The Deep Vision system has shown promising results for larger species [11] and this study is the first step towards verifying whether the current Deep Vision can be used to identify, quantify and measure small organisms continuously as they pass inside the trawl.

Benefits and limitations of the Deep Vision system

The current Deep Vision was able to identify, and quantify most passing small organisms, as well as species damaged in the codend (e.g. Ctenophora). The system showed potential for measuring the length of small organisms and also recorded the vertical and horizontal positions where individuals were imaged. Limitations became apparent with the current system and modifications should be made and tested.

The Deep Vision system was designed with a camera resolution and field of view suitable for observations of large opaque fish (40–70 cm), and some limitations were seen with observing smaller and transparent individuals within the Deep Vision chamber. For instance, it was difficult to identify species with similar body shape, particularly when they passed the camera at maximum range (73 cm). This was particularly noticeable with shannies as they could not be identified below family level and were difficult to distinguish from capelin. Similarly, redfish and polar cod were also difficult to tell apart. This problem may be mitigated by increasing the image resolution either by reducing the maximum range at which objects pass the camera or by using higher resolution cameras. Furthermore, adjustments to the lighting setup to make transparent organisms more visible in the images should be tested.

High densities of fish and turbulence inside the Deep Vision chamber made it difficult to track individuals during heaving. This was especially evident for polar cod and shannies, which showed the greatest differences between catch and image counts. Reducing the sampling period to during the designated haul (i.e. not to include setting out and heaving) would make tracking individuals more precise and increase the accuracy of quantifying fish using the Deep Vision system. Mismatch between the counts of small zooplankton from the images and codend catch may be the result of the image sub-sampling interval coinciding with the passage of high density patches or incomplete emptying of the codend when it was brought back onboard.

This study indicated the Deep Vision holds promise for measuring the length of small organisms. For polar cod which were oriented such that they could be measured with the Deep Vision, the mean length was the same as the catch data. For the Northeast Arctic cod and sculpin, the individual lengths differed by less than 5%, a similar result as for larger individuals by Rosen *et al* [10]. Due to the large number of individuals orientated poorly during high densities (i.e. parts of individuals hidden by other organisms), it became apparent that the current method would not be efficient for length studies of small organisms. Therefore, a more in-depth study on the length verification of small organisms using a modified trawl with the Deep Vision is recommended. It would be beneficial to establish a partial body to total length ratio

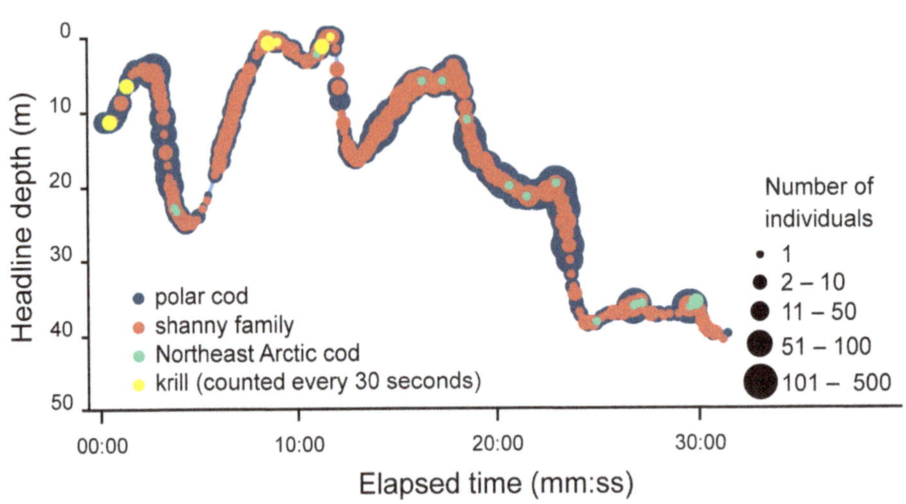

Figure 5. Species distribution and abundance throughout haul 04. Species include polar cod (*Boreogadus saida*), shanny family (Stichaeidae), Northeast Arctic cod (*Gadus morhua*) and krill (*Thysanoessa* spp.).

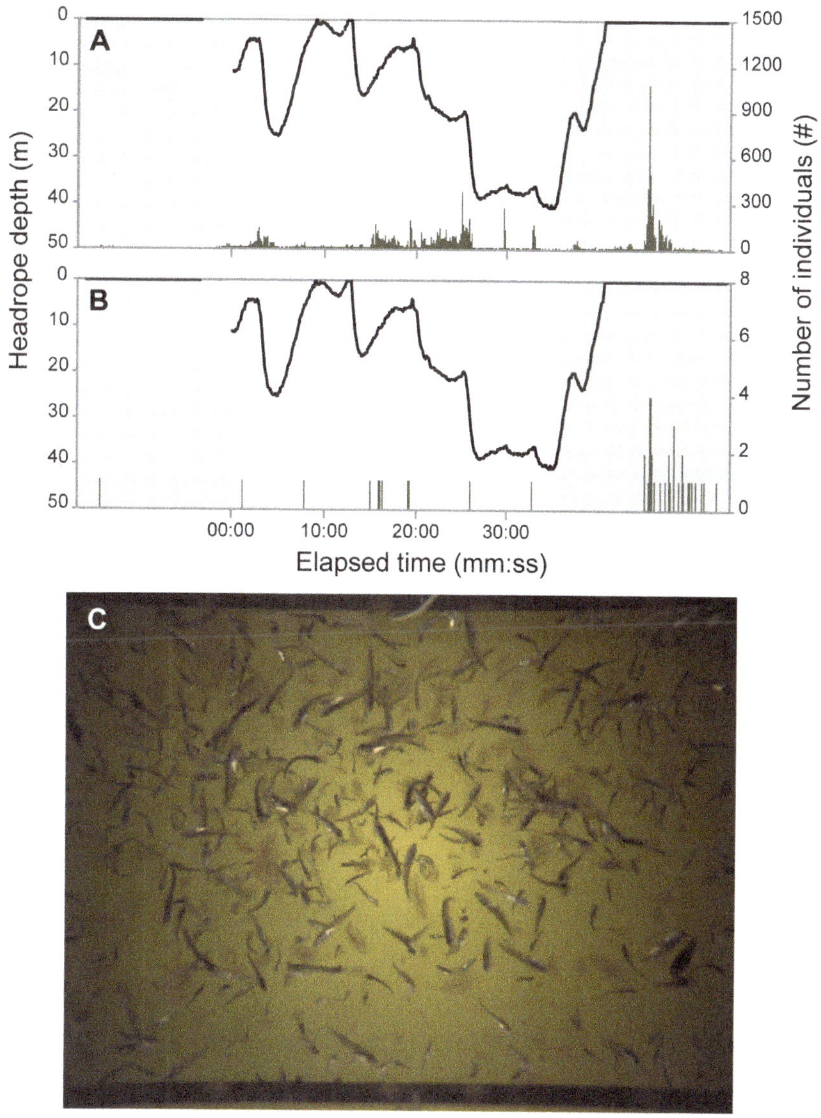

Figure 6. Density of two species during all phases of trawling, haul 04. Species include (A) Polar cod (*Boreogadus saida*) and (B) Greenland halibut (*Reinhardtius hippoglossoides*). The grey area on the left represents shooting and the grey area on the right signifies heaving. Lower image (C) is an example of an image during heaving.

for each species, as seen for larger fish in Rosen *et al* [10] to reduce the number of fish unable to be measured due to body position. In addition, a multi-sampling codend [18,19] can be used to collect sub-samples over shorter time periods for verifying the accuracy of the Deep Vision results. Finally, manually analysing 20,000 images per haul required a substantial amount of time and an automated system would be a major advance. Software for automating tasks such as eliminating empty images, object counting, measurement and species identification is currently under development.

Survey and ecological implications

Trawls are known to be species- and size- selective [20]. In order to use the Deep Vision with trawls for ecological studies, an understanding of the trawl efficiency and the rate at which organisms pass through the trawl is needed.

High numbers of fish were observed to pass the Deep Vision showing signs of damaged opercula. The damaged opercula indicate that the fish had been caught in the meshes before passing the camera, and it is assumed that a portion of the meshed fish were flushed out of the trawl rather than moving back past the camera and into the codend. This was supported by the observation of fish caught in the middle-sized meshes ahead of the Deep Vision section when the trawl was brought on deck. However, it was not possible to estimate the meshing rates for different species or sizes. This may have implications for ecological studies due to the delay of when meshed individuals entered the trawl and when they were imaged in the Deep Vision, as well as the loss of individuals during towing. For the small individuals observed in this study that did not contact the mesh, it is assumed that due to their poor swimming capacity, the passage rate back into the trawl is similar to the speed of the trawl through water [21].

Previously meshed fish were primary observed passing through the Deep Vision during heaving which may have been due to the way that the trawl was handled and/or sea state (i.e. netting alternately slack and taut from stopping/starting heaving and

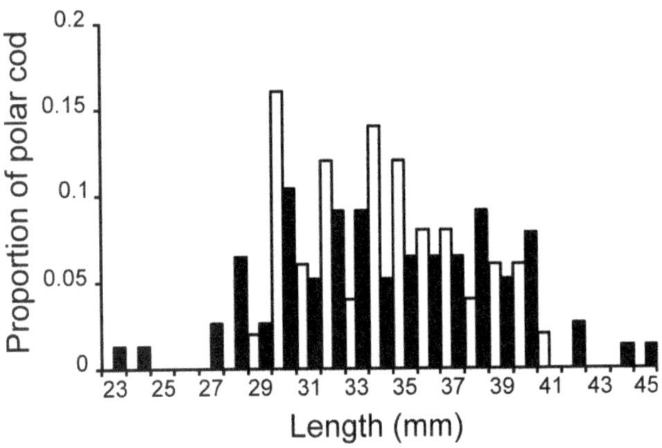

Figure 7. Length comparison of polar cod (*Boreogadus saida*) measured by two different methods. The dark column bars indicate measurements from the Deep Vision data ($n = 77$, mean = 34 mm), while the catch data measurements are indicated by white column bars ($n = 50$, mean = 34 mm).

wave activity). The proportion of fish entering the codend verses escaping may therefore have differed from haul to haul, which may have influenced species composition, number of individuals and length distribution of the catch, which in turn may have influenced the abundance index. The proportion of fish entering the codend rather than escaping during heaving could be reduced by using the information from the Deep Vision system to quantify species only during the designated trawling time and modifying the trawl construction to prevent individuals from becoming meshed.

The two hauls we analysed did not follow the standard protocol and the time spent at each depth range was unevenly distributed. The change in trawl geometry with depth meant that the full upper 60 m of the water column was not sampled, which may have produced a change in the catch efficiency of the trawl with depth. However, since each image captured by the Deep Vision has a depth associated with it, the catch data could be weighted by the amount of time spent at each depth range, reducing the importance of equal depth sampling.

The Deep Vision has the potential to be a valuable addition to the tools available for monitoring the upper pelagic community. Further development of this system to improve the analysis of images of small organisms and current survey methods could provide an increased understanding of population dynamics. In addition to further development of Deep Vision system itself, we need to design and test a trawl that performs consistently at all depths and that prevents organisms from becoming meshed before they enter the codend. One way forward might be to construct the front of the trawl with large square meshes to prevent herding and meshing of small organisms, combined with trawl doors that fully spread the trawl at the surface. Farther back in the trawl, the large square meshes would be lined with overlapping sections of small mesh netting to prevent meshing and escapes. Similar techniques are used in the commercial krill fishery in the Antarctic sea and were successfully tested during the 2013 BESS.

Acknowledgments

We thank the research vessel crew and the scientific staff of the Norwegian-Russian Barents Sea Ecosystem Survey for sharing valuable vessel time and helping to rig the Deep Vision system.

Author Contributions

Conceived and designed the experiments: SR AA EE. Performed the experiments: SR AA EE. Analyzed the data: MU SR. Contributed reagents/materials/analysis tools: MU SR. Wrote the paper: MU SR AA EE.

References

1. Pitcher TJ, Kalikoski D, Short K, Varkey D, Pramod G (2009) An evaluation of progress in implementing ecosystem-based management of fisheries in 33 countries. Marine Policy 33(2): 223–232. doi: 10.1016/j.marpol.2008.06.002.
2. Pikitch EK, Santora C, Babcock EA, Bakun A, Bonfil R, et al. (2004) Ecosystem-Based Fishery Management. Science 305: 346–347. doi: 10.1126/science.1098222.
3. Link JS, Bundy A, Overholtz WJ, Shackell N, Manderson J, et al. (2011) Ecosystem-based fisheries management in the Northwest Atlantic. Fish and Fisheries 12(2): 152–170. doi: 10.1111/j.1467-2979.2011.00411.x.
4. Eriksen E, Gjøsæter H (2013) A monitoring strategy for the Barents Sea. Report from project nr.14256 Survey strategy for the Barents Sea. Bergen: Institute of Marine Research. 9 p.
5. Michalsen K, Dalpadado P, Eriksen E, Gjosaeter H, Ingvaldsen RB, et al. (2013) Marine living resources of the Barents Sea - Ecosystem understanding and monitoring in a climate change perspective. Marine Biology Research 9(9): 932–947. doi: 10.1080/17451000.2013.775459.
6. Eriksen E, Prozorkevich D (2012) 0-group surveys. In: Jakobsen T, Ozhigin V, editors. The Barents Sea Ecosystem, Resources, Management Half a Century of Russian–Norwegian Cooperation. Trondheim: Tapir Academic Press. pp. 557–569.
7. Eriksen E, Prozorkevich D, Trofimov A, Howell D (2012) Biomass of Scyphozoan Jellyfish, and Its Spatial Association with 0-Group Fish in the Barents Sea. PLoS ONE 7(3): e33050. doi: 10.1371/journal.pone.0033050.
8. Godø OR, Valdemarsen JW, Engås A (1993) Comparison of efficiency of standard and experimental juvenile gadoid sampling trawls. ICES Marine Science Symposium 196: 196–201.
9. Hylen A, Korsbrekke K, Nakken O, Ona E (1995) Comparison of the capture efficiency of 0-group fish in pelagic trawls. In: Hylen A, editor. Precision and Relevance of Pre-recruit Studies for Fishery Management Related to Fish Stocks in th Barents Sea and Adjacent Waters, Proceedings of the Sixth IMR-PINRO Symposium. Bergen: Institute of Marine Research. pp. 145–156.
10. Rosen S, Jörgensen T, Hammersland-White D, Holst JC (2013) DeepVision: a stereo camera system provides highly accurate counts and lengths of fish passing inside a trawl. Canadian Journal of Fisheries and Aquatic Sciences 70(10): 1456–1467. doi: 10.1139/cjfas-2013-0124.
11. Williams K, Rooper CN, Towler R (2010) Use of stereo camera systems for assessment of rockfish abundance in untrawlable areas and for recording pollock behavior during midwater trawls. Fishery Bulletin 108(3): 352–362.
12. Rosen S, Holst JC (2013) DeepVision in-trawl imaging: Sampling the water column in four dimensions. Fisheries Research 148: 64–73. doi: 10.1016/j.fishres.2013.08.002.
13. Anonymous (2004) Proceedings of the international 0-group fish survey in the Barents Sea and adjacent waters in August-September 1965–1997. Murmansk: IMR/PINRO Joint Report Series, 2/2004. PINRO Press. 551 p.
14. Mjanger H, Alvsvåg J, Hestenes K, Svendsen BV, de Lange Wenneck TDL (2003) Håndbok for prøvetaking av fisk og krepsdyr. Version 3.14. Bergen: Institute of Marine Research. 185 p.
15. Skjoldal H, Rey F (1989) Pelagic production and variability of the Barents Sea ecosystem. In: Sherman K, Alexander L, editors. Biomass yields and geography of large marine ecosystems. Colorado: Westview Press. pp. 241–286.
16. Dolgov A, Orlova E, Johannesen E, Bogstad B, Rudneva G, et al. (2011) Planktivorous fishes. In: Jakobsen T, Ozhigin, VK, editors. The Barents Sea Ecosystem, Resources, Management Half a Century of Russian–Norwegian Cooperation. Trondheim: Tapir Academic Press. pp. 438–454.
17. Odum E, Barrett G (2005) Fundamentals of Ecology Thomson Brooks. California: Cole. 598 p.
18. Engas A, Skeide R, West CW (1997) The 'MultiSampler': a system for remotely opening and closing multiple codends on a sampling trawl. Fisheries Research 29(3): 295–298. doi: 10.1016/S0165-7836(96)00545-0.
19. Madsen N, Hansen KE, Frandsen RP, Krag LA (2012) Development and test of a remotely operated Minisampler for discrete trawl sampling. Fisheries Research 123: 16–20. doi: 10.1016/j.fishres.2011.11.016.
20. Winger PD, Eayrs S, Glass CW (2010) Fish behaviour near bottom trawls. In: He P, editor. Behavior of Marine Fishes: Capture Processes and Conservation Challenges. Oxford: Wiley-Blackwell. pp. 67–103.
21. Blaxter JHS, Dickson W (1959) Observations on the Swimming Speeds of Fish. ICES J Mar Sci 24(3): 472–479. doi: 10.1093/icesjms/24.3.472.

22

Seasonal Distributions and Migrations of Northwest Atlantic Swordfish: Inferences from Integration of Pop-Up Satellite Archival Tagging Studies

John D. Neilson[1]*, Josh Loefer[2], Eric D. Prince[3], François Royer[4], Beatriz Calmettes[5], Philippe Gaspar[5], Rémy Lopez[5], Irene Andrushchenko[1]

1 Fisheries and Oceans Canada, 531 Brandy Cove Road, St. Andrews, New Brunswick E5B 2L9 Canada, **2** Marine Resources Research Institute, South Carolina Department of Natural Resources, PO Box 12559, Charleston, South Carolina, United States of America, **3** National Marine Fisheries Service, Southeast Fisheries Science Center, 75 Virginia Beach Drive, Miami, Florida 33149, United States of America, **4** Datasio, 22 rue Mage, 31000 Toulouse, France, **5** Collecte Localisation Satellites, Parc Technologique du Canal, 8–10 rue Hèrmes, 31526 Ramonville, France

Abstract

Data sets from three laboratories conducting studies of movements and migrations of Atlantic swordfish (*Xiphias gladius*) using pop-up satellite archival tags were pooled, and processed using a common methodology. From 78 available deployments, 38 were selected for detailed examination based on deployment duration. The points of deployment ranged from southern Newfoundland to the Straits of Florida. The aggregate data comprise the most comprehensive information describing migrations of swordfish in the Atlantic. Challenges in using data from different tag manufacturers are discussed. The relative utility of geolocations obtained with light is compared with results derived from temperature information for this deep-diving species. The results show that fish tagged off North America remain in the western Atlantic throughout their deployments. This is inconsistent with the model of stock structure used in assessments conducted by the International Commission for the Conservation of Atlantic Tunas, which assumes that fish mix freely throughout the North Atlantic.

Editor: Moncho Gomez-Gesteira, University of Vigo, Spain

Funding: The work was jointly supported by the Canadian Department of Fisheries and Oceans, the United States National Marine Fisheries Service and the South Carolina Department of Natural Resources. The funders had no role in study design, data collection and analysis, decision to publish, or preparation of the manuscript.

Competing Interests: While two authors have affiliations with the commercial companies Datasio and Collecte Localisation Satellites, this does not alter their adherence to PLOS ONE policies on sharing data and materials.

* Email: John.Neilson@ dfo-mpo.gc.ca

Introduction

Compared with other large pelagic species such as bluefin tuna (*Thunnus thynnus*, Scombridae), the horizontal movement and migrations of Atlantic swordfish (*Xiphias gladius*, Xiphiidae) are poorly described. A recent (2012) search of the literature revealed only one paper whose title included "Atlantic swordfish movements" [1], and we are aware of two further relevant contributions [2,3], comprising a total of three papers in the primary literature as of that year. One of the three papers, [1], focussed largely upon vertical movements rather than horizontal migrations and distribution. This contrasts with the state of knowledge for bluefin tuna, where stock structure investigations using satellite tagging approaches is a particularly active field of investigation, with an analogous literature search revealing 24 published works.

Early studies using conventional "spaghetti" type tags indicated that while swordfish are capable of long distance migration, there were relatively few instances of trans-Atlantic movements (see summary in [4]). The results from those studies appear inconsistent with the stock model used by the organization responsible for management of the species (International Commission for the Conservation of Atlantic Tunas, ICCAT), who defined a management unit for stock assessment purposes of the entire North Atlantic, north of 5° N latitude. However, the absence of such trans-Atlantic movements could be related to limitations associated with conventional tagging that include low and variable rates of detection and return [5], and the often non-uniform distribution of effort that can complicate inferences of migration [6]. The development of pop-up satellite archival tags (PSATs) for the study of fish migrations has provided investigators with a tool that does not rely on the recapture and return of the tag by fishermen, thereby addressing some of the concerns surrounding the use of conventional tagging information for stock structure investigations.

A recent study [3] of swordfish migrations using the new technology found that swordfish tagged off Georges Bank in the northwest Atlantic occupy temperate waters north of 40° N latitude from early June through October, before commencing southward migration to the Caribbean Sea. The fish reach the Caribbean region by early January, where they reside until April, when the northward migration of the fish begins. While the tagged fish remained within the southern boundary of the management unit, no movement into the northeast Atlantic was found. The authors suggested this was inconsistent with the current management unit assumption.

The authors, however, acknowledged that the scope of their investigation reported in [3] was limited in several regards. It included a relatively small sample size ($n = 10$), and only one area of deployment (Georges Bank in the northwest Atlantic). Furthermore, there is the possibility that swordfish aggregations elsewhere in the western North Atlantic show greater affinities with those in the eastern North Atlantic. Another limitation of the conclusions reported in [3] was that the fish tagged were predominately large, mature individuals. This could be problematic, if mixing of the population occurs at younger stages of the life history. For example, long distance movement and migrations of Pacific halibut (*Hippoglossus stenolepis*) are thought to occur at relatively young ages [7].

In this contribution, we address the concerns of the earlier analyses by pooling results from three laboratories conducting PSAT studies of Atlantic swordfish, who have made deployments ranging from the Grand Banks of Newfoundland to waters off Florida. While taking a pooled approach provided a large number of tagged individuals for the analyses compared with other studies involving pop-up archival tags, it also provided a greater range of sizes of individuals, as well as diverse deployment locations.

Combining information from three agencies conducting PSAT-based investigations of swordfish movement has allowed us to present the most comprehensive synthesis of swordfish movements and seasonal distributions in the Northwest Atlantic produced to date. In addition, we access independent data sources such as fishery logbook information and conventional mark-recapture data to corroborate the inferences we draw from the PSAT data.

While integrating results from different investigations offers significant advantages, it also provides certain analytical challenges. For example, datasets collected from PSATs provided by different manufacturers follow different formats. Ambient variables such as pressure, temperature and light are also collected at different time resolutions, and further aggregated or binned following different strategies. High-level records include binned histograms, time series, temperature or light profiles that can be difficult to combine. The geolocation process can also differ depending on the manufacturer, based on light level fitting (e.g. Wildlife Computers), or solar event detection (e.g. Microwave Telemetry). This has implications on the error structure of light-based geolocations, and further warrants the need for filtering and smoothing to limit observation error. Several filtering methods have been published in the literature, allowing investigators to extract meaningful patterns from tagging datasets and limit the influence of the geolocation process (see, for examples [8–11]). These rely on filtering and smoothing observations using linear and non-linear Kalman Filter variants. Unknown parameters can be estimated via likelihood maximization, with results taking the form of a most probable track and its posterior variance estimates. For sparse datasets with complex, arbitrary constraints such as bathymetry and coastlines, an improved method was proposed in [12], using Hidden Markov Models (HMM) applied to a discretized spatial representation of the space. This method proved particularly efficient and robust when applied to tide-based geolocation of bottom fish. In this paper, we introduce an extension of this HMM-based method, using a more accurate movement model explicitly accounting for constraints and boundaries. In consequence, a secondary objective of this contribution is to describe how we integrated these diverse data.

Methods

PSAT Deployments

Field permits for the PSAT tagging operations were issued by Canada Department of Fisheries and Oceans and the US National Marine Fisheries Service for their respective national waters.

Three laboratories (Southeast Fisheries Science Center (Miami), South Carolina Department of Natural Resources (Charleston), and Department of Fisheries and Oceans Canada (St. Andrews)) pooled their PSAT data for the purposes of describing the movement of swordfish off the east coast of North America. All data were obtained between 2002 and 2009.

The Canadian PSATs were Wildlife Computers Mark 10 tags, deployed during between 2005–2009 inclusive. Satellite tags were attached to a modified harpoon pole following methods described in [13]. The harpoon pole was approximately 5 m long and constructed of aluminum. The distal end was modified to accept a stainless dart needle and a cushioned cradle to hold the tag. The tag was then attached to the pole using elastic bands with "breakaway" knots so that it would release when the fish was tagged. An experienced harpoon fishermen was employed to conduct the tagging, using a vessel specialized for this type of fishery [3].

The releases off Florida and Northeastern Bahamas were made by the National Marine Fisheries Service, Southeast Fisheries Science Center of the United States. Most of the PSATs were Wildlife Computers PAT4 or Mark 10 units. The PSATs were deployed off commercial pelagic longline vessels, as well as recreational vessels. Commercial buoy gear fishing vessels were also occasionally used as PSAT deployment platforms (hand-lines attached to free-floating buoys, which contain no more than two hooks per handline). Bait was squid *Illex* sp., small "tinker" mackerel *Scomber* sp., and little tunny *Euthynnus alletteratus*. Hooks used during recreational rod-and-reel fishing were size 11/0 J-style hooks, while sizes 14/0 and 16/0 non-offset circle hooks were used during the commercial longline and buoy gear operations. All PSAT tags were attached externally using a 181 kg monofilament tether and a double barb medical grade nylon anchor placed in the dorsal musculature as previously described in [14] and [15]. Some of the recent deployments used the same anchor equipped with elongated nylon toggles [1]. A 2.44 m long tagging pole was used to place the tag in the dorsal musculature, about 4–5 cm below the dorsal midline to a depth just short of exiting the opposite side of the fish, near the first dorsal fin [15], [16].

The PSATs deployed off of the Carolinas were manufactured by Microwave Telemetry Inc. (Columbia, MD, USA) and were attached to swordfish captured on pelagic longline gear deployed from the R/V Palmetto. Tags (Model PTT-100) were attached by harpooning a 6.25 cm titanium M-type dart anchor into the dorsal musculature approximately 5 cm below the midline of the base of the dorsal fin. The anchors were inserted using a 20 cm titanium tagging needle protected by a 6 cm diameter rubber stopper limiting penetration depth to 12 cm. Tags were tethered to the anchors with 30 cm of 1.66 mm diameter fluorocarbon monofilament (100 kg tensile strength). Monofilament was attached at either end with stainless steel crimps covered with adhesive-lined polyolefin heat shrink tubing. Fish were not removed from the water during tagging, and were released by removing the hook

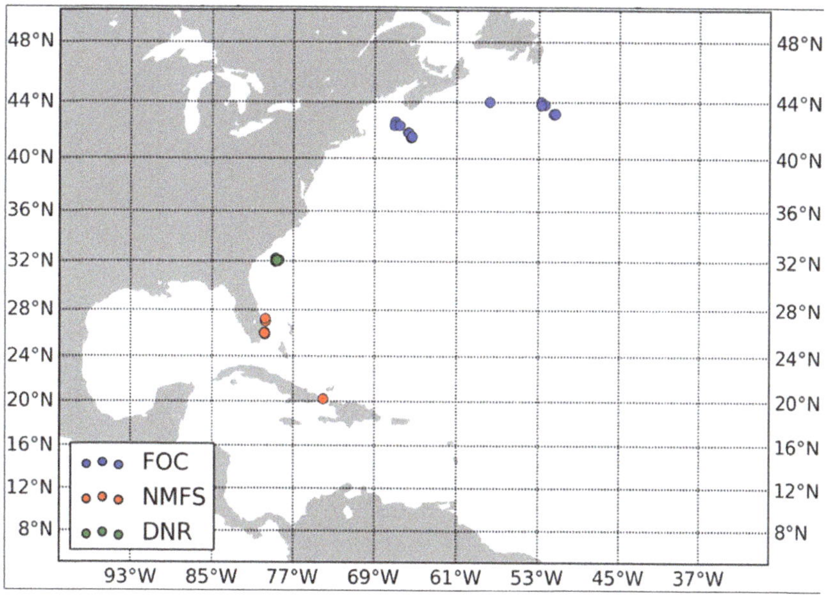

Figure 1. Location of releases of Atlantic swordfish marked with pop-up satellite archival tags by the National Marine Fisheries Service (NMFS), South Carolina Department of Natural Resources (DNR) and Fisheries and Oceans Canada (DFO).

(when possible) or cutting the hook leader within a few cm of the hook.

The areas where deployments were made by the three laboratories are shown on Fig. 1.

From an initial set of 78 deployed PSATs, we selected 54 individuals based on a minimum deployment length of 30 days. Unfortunately, sixteen of the selected tag returns contained corrupted data due to either Argos-transmission problems, faulty sensors (light, pressure or depth), or erroneous recordings of the sensor measurements. This further reduced the number of releases finally used in the analyses to 38. A summary of the temporal distribution of PSAT deployments by the three laboratories is provided in Fig. 2. The intended deployment lengths varied, according to differing study objectives of the three agencies. Realized deployment durations were variable in all cases, and averaged 99, 83 and 249 d for the National Marine Fisheries Service (NMFS), South Carolina Department of Natural Resources (SCDNR) and Fisheries and Oceans Canada (DFO) respectively (see summary statistics concerning deployment durations, Table 1).

Lengths or weights were estimated at the time of release, and where weights were supplied, they were converted to lower jaw fork length using conversion factors published by ICCAT. The length of fish released varied among the three labs. The SCDNR releases were, on average, the smallest swordfish tagged: 118 cm lower jaw-fork length (LJFL) (range 81–152 cm, standard deviation 19.6 cm). The NMFS releases were on average, 155 cm LJFL (range 109–280 cm, standard deviation 53.0 cm). The DFO releases were largest, on average (mean 204 cm, LJFL range 124–263 cm, standard deviation 40.5 cm).

Data Selection and Preprocessing

Light-based geolocation exploits sunrise and sunset times. They are linked to the animal's longitude and latitude by the following set of equations:

$$\cos(UT_{sr} + GMST - RA_{sr}) = \frac{\cos(\varepsilon) - \sin(\delta_{sr})\sin(\varphi)}{\cos(\delta_{sr})\cos(\varphi)}$$

$$\cos(UT_{ss} + GMST - RA_{ss}) = \frac{\cos(\varepsilon) - \sin(\delta_{ss})\sin(\varphi)}{\cos(\delta_{ss})\cos(\varphi)}$$

where φ is the animal's latitude, and δ_{sr}, δ_{ss} and RA_{sr}, RA_{ss} are the solar declination and right angle at sunrise and sunset. ε is the (predefined) sun's altitude at twilight. One can then solve for the universal time UT of sunrise and sunset, given the mean Greenwich sidereal time (GMST) which is computed from the animal's longitude λ (see [17] for a complete description of the procedure).

Methods for estimating location differ between Microwave Telemetry's PTT100 and Wildlife Computers' MK10-PAT. To derive location estimates, Microwave Telemetry's PTT100 relies on a proprietary fuzzy logic algorithm to extract timings of sunrise and sunset. Wildlife Computers' MK10-PAT transmits a subset of light levels near sunrise and sunset, to which a sun elevation model is later fitted. Subject to different noise processes and uncertainty sources, both approaches yield location estimates of similar accuracy, and are sensitive to the vertical/depth behavior of the animal. In particular, deep or repeated dives at critical times (near sunrise and sunset) yield noisy location estimates, particularly in latitude. Longitude estimates are less affected [11].

To reduce noise, sunrise and sunset times were cleaned using a Set Theoretic or Set membership approach [18–20] prior to applying the location processing. This method builds a daily set containing only feasible values of longitude and latitude. Starting from the deployment and pop-up locations of the tag as initial feasible values, the set is extended using forward propagation techniques, considering the maximum sustainable animal's speed, the land masks, the measured SST and the maximum measured depth. The resulting feasible set is then used to compute the likelihood of the sunrise/sunset times and to discard erroneous

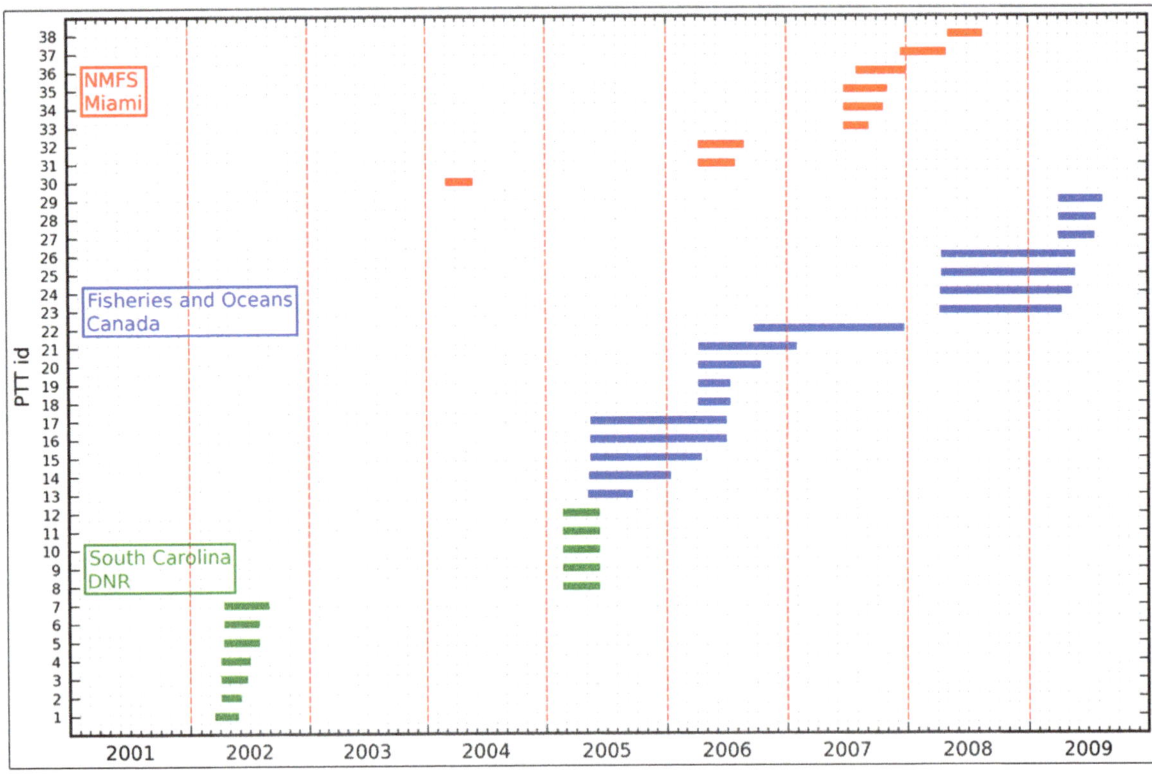

Figure 2. The temporal distribution of the releases of pop-up satellite archival tags on North Atlantic swordfish made by the National Marine Fisheries Service (NMFS), South Carolina Department of Natural Resources (DNR) and Fisheries and Oceans Canada (DFO). The length of the horizontal bar reflects the duration of the deployment.

data. The timings of sunrise and sunset retained for the processing are shown in Fig. 3. Daily SST estimates are computed as the median temperature measured between the surface and a depth of 15 m, when the fish was present in that water layer. The median is computed from temperature-at-depth profiles for MK10 tags, and from the temperature-depth time series for PTT-100 tags. The estimated SSTs are also shown in Fig. 3 (mean SST = 22.6°C±5.8°C). A time series of daily (SST, sunrise time, sunset time) measurements is then formed for each tag.

Unfortunately, most of these daily records were empty. Besides occasional data acquisition problems, the lack of data was essentially due to the fact that the tagged individuals spend most of the time at considerable depths (typically between 100 and 200 m), where light was so dim that measurements do not allow detection of sunrise/sunset events. The surface, or near surface, temperature measurements are also sparse as the tagged fish rarely visit the upper meters of the water column. Of a total of 6228 daily records (for 38 individuals) only 436 (7%) contain sunrise and/or sunset time measurements and 2553 (41%) contain SST data. Thus, Atlantic swordfish present a situation where positioning is essentially constrained by sparse SST measurements, only occasionally complemented by sunrise/sunset times. Under such circumstances, the determination of fish trajectories (i.e. the estimation of series of pointwise positions) cannot be reliably estimated. Instead, we implemented a grid filter approach providing estimates of the probability distribution of the successive fish positions instead of pointwise locations.

Location Inference via Grid Filtering

Hidden Markov Models (HMM), also known as Grid Filters in the geolocation literature, rely on discretization of the state-space to infer the probability density (or the Utilization Distribution, UD) of the (hidden) location, given a sequence of observations, in this case SST and sunrise/sunset times measurements. The

Table 1. Summary of deployments (days) by the three agencies contributing data.

Deployment Statistics by Laboratory

	N	Minimum	Maximum	Mean	Std. Deviation
SCDNR	12	31	123	82.6	27.6
DFO	17	77	411	248.8	135.4
NMFS	9	57	135	99.2	26.7

The agencies are National Marine Fisheries Service (NMFS), South Carolina Department of Natural Resources (SCDNR) and Fisheries and Oceans Canada (DFO).

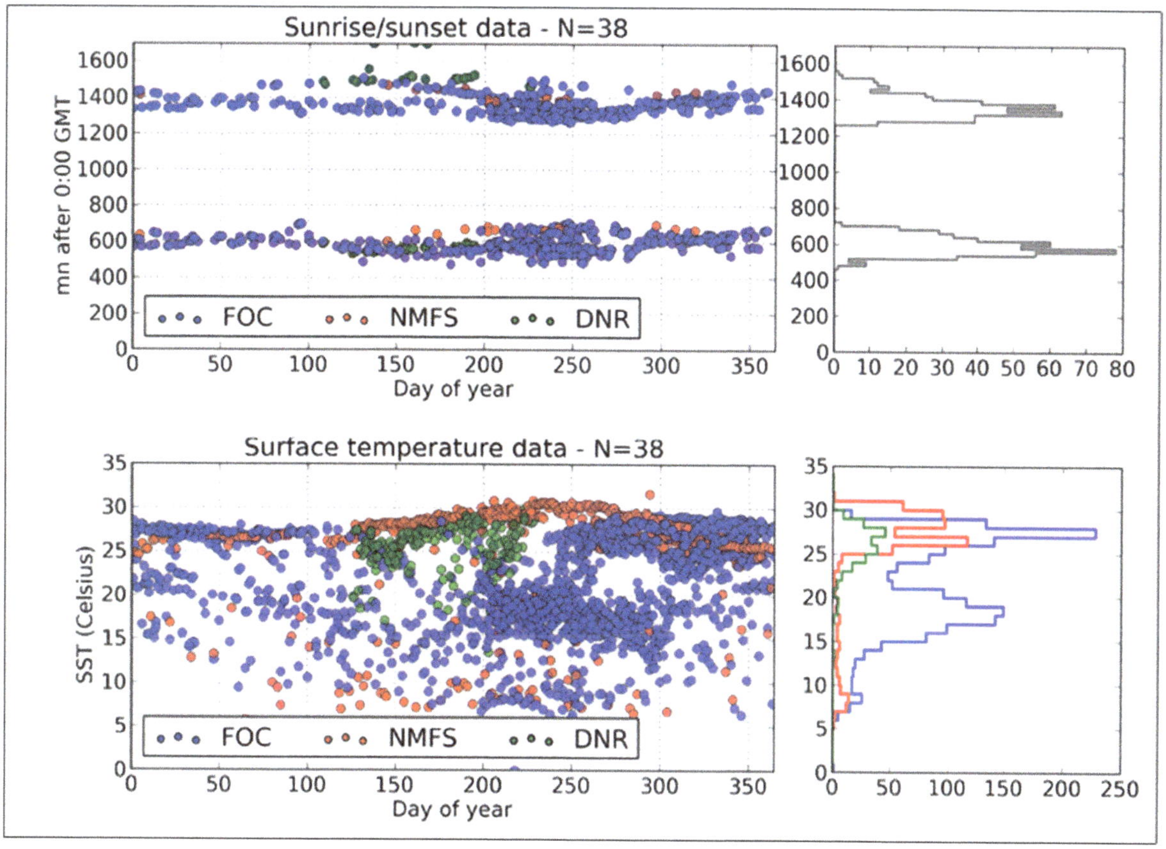

Figure 3. Edited sets of sunrise and sunset times from pop-up satellite archival tags deployed on North Atlantic swordfish deployed by the National Marine Fisheries Service (NMFS), South Carolina Department of Natural Resources (DNR) and Fisheries and Oceans Canada (DFO) (top panel) and SST recorded by the same tags (bottom panel). The corresponding frequency distributions of sunrise/sunset times and SST are shown on the right.

approach was first introduced in the field of fish tracking using depth measurements as the sole positioning information [12]. At each sampling time, the HMM method performs a position prediction step by numerically solving the advection-diffusion equation for the two-dimensional probability of animal's presence. A position update step is then performed to combine the predicted probability density with fish depth data to produce the posterior distribution of the animal. At the end of this forward filtering pass, the HMM method finally refines the filtered density estimates with a backward smoothing pass. The reader is referred to [12] for a more detailed presentation of the method.

In this study, we implement a somewhat modified version of the positioning method described in [12]. In the prediction step of the filter, the advection-diffusion equation for the probability of animal's presence is numerically solved imposing a zero-flux boundary condition both at the limits of the domain and along the coastlines. This naturally prohibits movement onto land, without leaking any probability mass. The complex boundaries of the solution domain are given by a detailed land mask at the model resolution (grid size = 9 km). This was not achieved in [12] where the authors assumed that the domain in which the equation was solved was large enough to neglect the probability that the fish crosses the boundary. The position update step of the filter is then performed as in [12], but using SST, sunrise and sunset time instead of fish depth to produce the posterior distribution of the individual. SST and sunrise/sunset time observations are modeled as Gaussian random variables where the mean of the associated distribution is the observation value and the standard deviation is set to 1°C for SST and 20 min for sunrise and sunset time. The assumed error on SST takes into account the usual measurement error of the temperature sensors plus a sampling error due to the scarcity of temperature measurement in the upper 15 meters of water. The relatively large (20 min) error used for sunrise/sunset times is justified by the fact that their estimation suffers from the low signal to noise ratio affecting the light measurements obtained at large average depths. Note, however, that the UDs presented here are relatively insensitive to the assumed sunrise/sunset time estimation error, largely because sunrise/sunset time measurements are few and can thus have only a limited impact on the utilization distributions' estimation.

Finally, we performed a smoothing pass that slightly differs from [12] as we use the two-filter smoother formula described in [21]. It combines the previously computed filtered densities together with predicted densities given by the same filter propagating backward in time. While the forward filter is initialized with the deployment position, the backward filter is initialised with the pop-up location. This allows us to take full advantage of the accurate information provided by the pop-up position.

Presentation of Results and Comparison with Independent Data Sets

On the assumption that the pattern of movement and migrations might differ between immature and mature fish, as observed for other teleosts [7], [22], we identified immature individuals as those less than 179 cm lower jaw fork length, and mature individuals were greater than or equal to 179 cm, following the ICCAT convention for this stock [23]. Results are presented separately for the two size-classes of swordfish.

Finally, we compared swordfish habitat use as inferred from the PSAT data to the distribution of conventionally-marked swordfish released and recaptured throughout the Atlantic since 1940. The conventional tagging database is maintained on behalf of participating countries by the International Commission for the Conservation of Atlantic Tunas (http://www.iccat.int/en/accessingdb.htm). We considered that the locations of tagging deployments and recaptures in the database provided an independent indication of the occurrence of swordfish that could be compared with the PSAT inferences.

Results

PSAT-tagged swordfish provided relatively few estimates of sunrise and sunset (Fig. 3) that were suitable for inferences of geolocation. The data supplied by DFO was the most complete in terms of providing year-round information, consistent with the longer periods of deployment associated with that program. The combined dataset had considerably more information on SST compared with sunrise and sunset (Fig. 3). Swordfish in the northwest Atlantic experience considerable variation in sea surface temperatures concurrently, ranging from about 5 to 30°C. The data associated with the DFO program also illustrate a bimodality of temperature frequency distributions (Fig. 3), which can be seen to be linked to seasonal migrations which will be described subsequently.

The overall distribution of immature swordfish is shown on Fig. 4. The centre of the distribution of such swordfish occurs relatively close to the eastern USA, south of Cape Cod to Florida and the Bahamas. The edges of the distribution include the Gulf of Mexico, the Caribbean Sea, and the northwest Atlantic east until about 50° W longitude.

The information in Fig. 4 is disaggregated into quarterly components in Fig. 5. The distribution of immature swordfish in Quarter One is centered in the Caribbean Sea region. In Quarter Two, the distribution has shifted northward, and is closely associated with coastal region of the eastern US, from southern Florida to Cape Cod. The northward movement continues in Quarter 3, and part of the distribution is now found further north and east, along the continent shelf edge off Nova Scotia and Newfoundland. The distribution of immature swordfish begins to retreat south in Quarter Four.

Fig. 6 shows the overall distribution of mature swordfish as inferred from the PSAT information. The centre of the distribution extends through higher latitudes than was the case for the immature fish, with evidence for swordfish occurring in latitudes as high as 50° N. latitude. There are also indications that swordfish occur in the Gulf of Mexico, and the Caribbean Sea. Mature swordfish did not appear to move further south than about 10° N. latitude.

The quarterly disaggregation of the information in Fig. 6 is presented in Fig. 7. In general, the larger mature swordfish show a more northerly distribution by quarter compared with the immature swordfish in our dataset. For example, in Quarter One, the centre of distribution extends as far north as about 40° N. latitude, compared with about 25° N latitude for the immature swordfish. By Quarter Two, the swordfish had moved north to 45° N latitude, and spread in an easterly direction around the Grand Banks of Newfoundland. By Quarter Three, the centre of distribution of the mature swordfish was in the Gulf of Maine, off Georges Bank, the Scotian Shelf and the Grand Banks of Newfoundland. By Quarter Four, the larger swordfish had begun their southward seasonal movement, with the southern margin of the distribution approaching the Caribbean Sea.

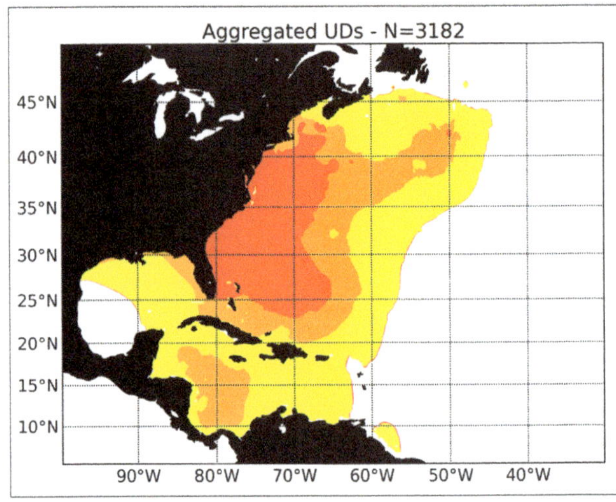

Figure 4. The overall distribution of immature swordfish (<179 cm) based on information from pop-up archival satellite results obtained by the National Marine Fisheries Service (NMFS), South Carolina Department of Natural Resources (DNR) and Fisheries and Oceans Canada (DFO). The contoured Utilization Distributions (UD) are normalized to one (100%), and the red areas represent normalized UD> 95%, the orange areas correspond with 95%> =UD>75%, the yellow areas are 75> =UD>50%, and the uncoloured areas are UD< = 50%. The same colour coding is employed in subsequent figures.

An independent verification of our results is provided in Fig. 8, which compares the seasonal PSAT information on swordfish distributions for both immature and mature fish with the locations of swordfish conventional tagging events, including cases when a recapture was made. In general, the two indicators of swordfish distribution show good agreement. There were 2473, 2499, 4545 and 3583 records of tag recaptures and deployments in the conventional tagging database, for the first to fourth quarter respectively. Of those records, the proportion that fell within the swordfish distributions inferred from the PSAT information was 93.7, 95.2. 85.6 and 96.5%, for the first to fourth quarters respectively. In addition, areas of predicted high usage based on the PSAT data agreed well with aggregations of swordfish apparent from the conventional tagging studies (see, for examples Fig. 8 first quarter, eastern Gulf of Mexico and Atlantic coastal waters and third quarter, Gulf of Maine). However, there were also some instances where there were mismatches between the distributions inferred from PSAT data, and the observed distributions of conventional tagging deployments and recaptures (see for examples, Fig. 8 third quarter, eastern Gulf of Mexico and fourth quarter, coastal Venezuela.

Discussion

Previous work on Atlantic swordfish estimated fish trajectories within the framework of Dynamic Generalized Linear Models [3]. The animal's locations were computed by maximizing a likelihood function taking into account SST constraints, the observations of sunrise and sunset times as well as the prior dynamics of the animal. The previous method is simpler, as the underlying

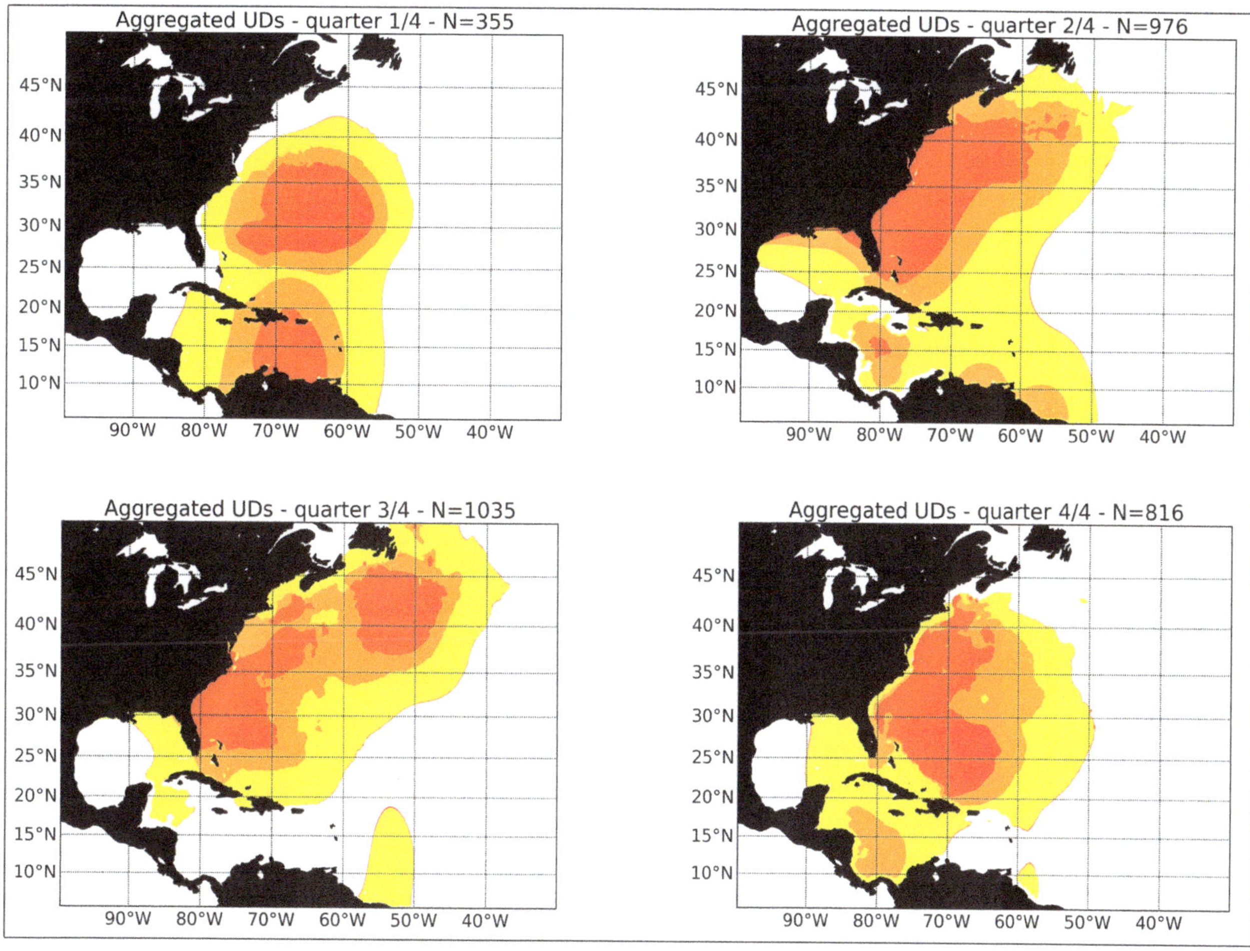

Figure 5. The seasonal distribution of immature swordfish (<179 cm) based on information from pop-up archival satellite results obtained by the National Marine Fisheries Service (NMFS), South Carolina Department of Natural Resources (DNR) and Fisheries and Oceans Canada (DFO).

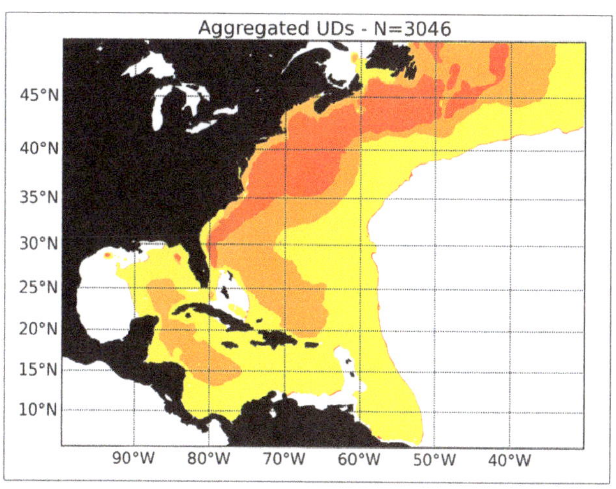

Figure 6. The overall distribution of mature swordfish (=>179 cm) based on information from pop-up archival satellite results obtained by the National Marine Fisheries Service (NMFS), South Carolina Department of Natural Resources (DNR) and Fisheries and Oceans Canada (DFO).

distributions are assumed Gaussian (i.e. parameterized by a mean and a covariance) and are not constrained by coastline data.

In contrast, the HMM method implemented here provides a non-parametric probability density of the animal's position. HMM can take into account arbitrary observations models, non-Gaussian error sources and highly non-linear constraints such as shorelines or islands. This scheme is adapted to manipulate multi-modal distributions and should be favoured in data-poor scenarios, as was the case here. Although this method does not readily provide individual trajectories, the analyses of the generated UDs has provided important new information on the seasonal distribution of Atlantic swordfish in the northwest Atlantic.

By combining disparate data from three laboratories, this study has provided the most comprehensive examination of swordfish migration in the North Atlantic to date. There are several conclusions that have implications for the current view of swordfish stock structure and fisheries management. An earlier study has concluded that the seasonal movements exhibited by swordfish in the Northwest Atlantic appeared to be inconsistent with the management unit employed by ICCAT, which assumes a single stock unit throughout the North Atlantic [3]. The results showed no movements across the North Atlantic, and fish remained in the Northwest Atlantic, west of 50° W longitude. Our new findings, which include a considerably larger data set

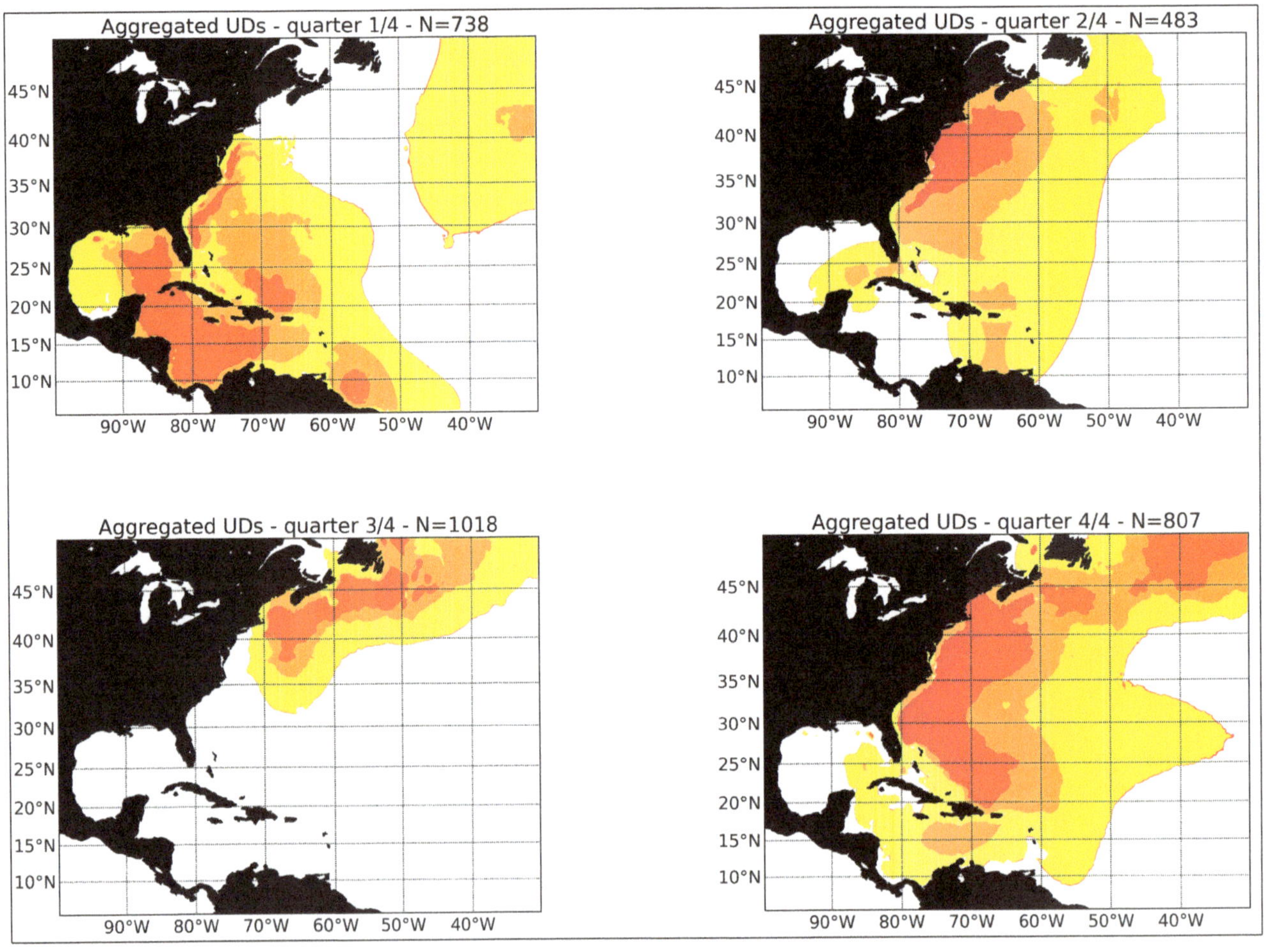

Figure 7. The seasonal distribution of mature swordfish (=>179 cm) based on information from pop-up archival satellite results obtained by the National Marine Fisheries Service (NMFS), South Carolina Department of Natural Resources (DNR) and Fisheries and Oceans Canada (DFO).

with a greater range of sizes and deployment locations (a recommendation made in [3]), generally support the conclusions of the earlier paper. We saw little evidence for movement of fish east of 40° W longitude (the greatest eastward movement shown by mature fish during Quarter Two (Fig. 7, see also the Supporting Information (Video S1, Video S2) illustrating the movement of two mature fish at liberty for more than 400 d). The Supporting Information also illustrates the previously noted tendency for swordfish to return to the same foraging areas on an annual basis [3].

While our data do not appear consistent with a population that is freely mixing in an east-west direction across the Atlantic, they are consistent with the convention for the boundary between north and south Atlantic stocks employed by ICCAT, which considers the division to be at 5° N. latitude. The exact location of the boundary is somewhat contentious, with some workers claiming that the division should be further north. For example, a genetic investigation of swordfish stock structure suggested that the boundary between the northern and southern stock may be located in the range of 10 to 20° N, and that individuals from the north and south populations may intermingle in the boundary zone [24]. Our results indicate that PSAT-tagged swordfish deployed in this study did not occur south of 5° N, thus supporting the current convention.

In spite of the absence of movements of PSAT-tagged swordfish south of 5° N, there are, however, some data from other investigations that suggest that the management unit boundary could be reviewed. For instance, changes in the spatial distribution of the industrial longline fleets [25] and new genetic information [26] have prompted some to suggest that the current boundary could be revised. Swordfish are caught by longline fleets near the management boundary in substantial quantities (Figure 9), to the extent that there is no spatial discontinuity in the distribution of the catch, as would be expected if the management boundary delineated different stocks. However, this area is associated with the Oxygen Minimum Zone (OMZ), thus there is potential for hypoxia-based habitat compression resulting in a condensed distribution of predators, preferred prey, and fishing effort in these areas [27]. Under such circumstances, spatial discontinuities in catch distributions usually indicative of stock boundaries might not be expected. PSAT deployments on swordfish near the existing boundary could allow further evaluation of the suitability of the management boundary and also allow a better understanding of swordfish habitat use in the equatorial Atlantic and how it differs from swordfish in higher latitudes.

The greater scope of this study compared with [3] has allowed us to better describe seasonal habitat use of swordfish. For example, the swordfish released on Georges Bank described in [3]

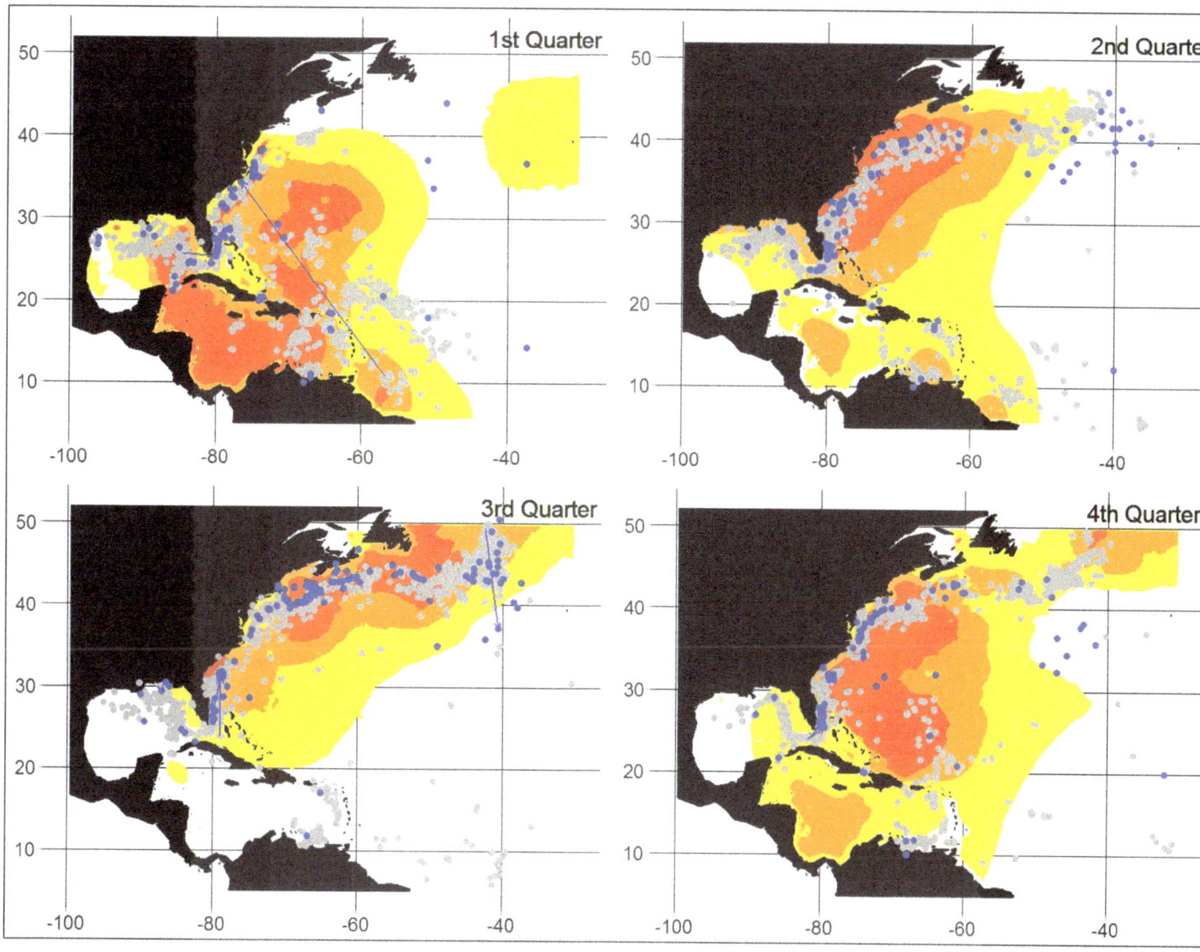

Figure 8. The seasonal distribution of both mature and immature swordfish obtained from the combined satellite archival tag data presented here, compared with conventional tagging data available from the International Commission for the Conservation of Atlantic Tunas, also shown by quarter. The release and recapture points for conventional tagging data are shown as grey and blue points, respectively. In the few cases where recaptures were made within the same quarter and year, an arrow is drawn between the release and recapture points.

showed no affinity with the Gulf of Mexico, a known spawning area [28]. However, with the larger number of deployments in this study including a broader range of release sites, the distribution of mature fish during Quarter One (which includes the peak period of spawning, Fig. 7) is shown to include the Gulf of Mexico, the Caribbean Sea, and waters off eastern North America from 40° N. latitude south to about 20° N latitude. The areas of fish that were assumed to be mature in the first quarter (Fig. 7) generally concurs with the distribution of known spawning areas [28], but the PSAT results presented here show a broader distribution of potential spawners in the Caribbean Sea than was previously described [28].

We have used an independent data set (a database that contains the aggregated conventional tagging data provided by Contracting Parties to ICCAT) to validate the inferences made from the PSAT data, the first time such a comparative approach has been used in descriptions of fish distribution from electronic tagging studies, to our knowledge. The distribution of tagging releases and recaptures compared well with the information from PSATs, as >85% of locations of conventionally-tagged swordfish releases or recaptures were found within the Utilization Distribution corresponding with 75>=UD>50% (Fig. 8). However, there were some mismatches between the two datasets, noted earlier (for examples, see Fig. 8 third quarter, eastern Gulf of Mexico and fourth quarter, coastal

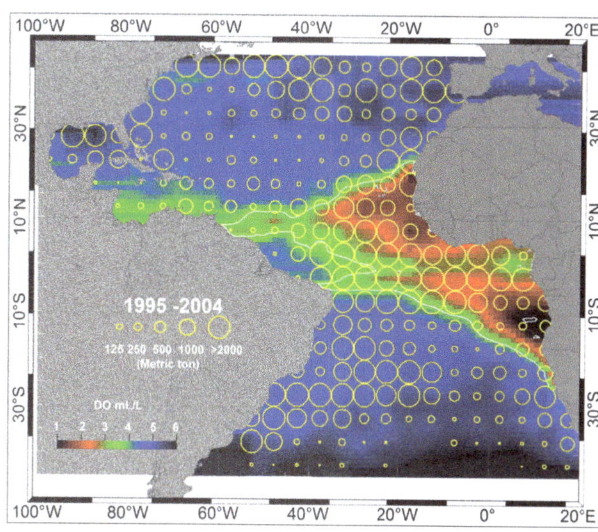

Figure 9. Atlantic Ocean-wide swordfish catches from the international longline fleets for the decade 1995–2004, shown in relation to dissolved oxygen concentration. Largest yellow circles are≥2000 mt. The catch data were provided by the International Commission for the Conservation of Atlantic Tunas.

Venezuela). Such apparent discrepancies, in part, could be due to the conventional tagging data being reliant on the distribution of fishing effort for recapture information, whereas the PSAT method is independent of the fishery. It could also reflect the fact that the data from the ICCAT conventional tagging database covered a considerably longer period (70 years, 1940 to 2009) than did the PSAT studies reported here (8 years, 2002 to 2009). There is evidence that swordfish occurred coastally during early periods of exploitation (see examples off eastern North America and South Africa, as shown in [28]), and in some cases, the coastal aggregations decreased markedly. Thus, during the 70-year period for the conventional tagging studies, the relative abundance of population components may have changed in comparison to their abundance during the period of PSAT studies.

Acknowledgments

Sean C. Smith of the Department of Fisheries and Oceans conducted the Canadian field component onboard the *Oran II* (Captain Gus Reyno) and the *Four Ladies* (Captain Larry Sears). Figure 9 was provided courtesy of John P. Hoolihan (Cooperative Institute for Marine and Atmospheric Studies, University of Miami, Miami, Florida), Jiangang Luo (Rosenstiel School of Marine and Atmospheric Science, University of Miami, Miami Florida), Carlos Palma (International Commission for the Conservation of Atlantic Tunas, Madrid, Spain), Lothar Stramma and Sunke Schmitko (Helmholtz Centre for Ocean Research, Kiel, Germany). This is Contribution 716 of the South Carolina Marine Resources Division.

Author Contributions

Conceived and designed the experiments: JN JL EP. Performed the experiments: JN JL EP. Analyzed the data: JN FR BC PG RL IA. Contributed reagents/materials/analysis tools: JN JL EP FR BC PG RL IA. Wrote the paper: JN JL EP FR BC PG RL IA.

References

1. Dewar H, Prince ED, Musyl MK, Brill RW, Sepulveda C, et al. (2011) Movements and behaviours of swordfish in the Atlantic and Pacific Oceans examined using pop-up satellite archival tags. Fish. Oceanogr. 20: 219–241.
2. Sedberry G, Loefer J (2001) Satellite tracking of swordfish, *Xiphias gladius*, off the eastern United States. Marine Biol. **139**, 355–360.
3. Neilson JD, Smith S, Royer R, Paul SD, Porter JM, et al. (2009) Investigations of horizontal movements of Atlantic swordfish using pop-up satellite archival tags. Rev.: Methods Technol. Fish Biol. Fish. 9: 145–159.
4. Sperling AT, Neilson JD, Carruthers EH, Stone HH (2005) Compilation and analyses of Canadian conventional tagging data for swordfish (*Xiphias gladius*), 1961–2004. Col. Vol. Sci. Pap. ICCAT **58**, 1483–1494.
5. Guy CS, Blakenship HL, Nielsen LA (1996) Fish Tagging. *In* Murphy, B.R. and Willis, D.W. (eds) Fisheries Techniques (Second Edition), pp. 353–383.
6. Armannsson H, Jonsson STh, Neilson JD, Marteinsdottir G (2007) Distribution and migration of saithe (*Pollachius virens*) around Iceland inferred from mark-recapture studies. ICES J. Mar. Sci. **64**, 1006–1016.
7. Skud BE (1977) Drift, migration and intermingling of Pacific halibut stocks. Int. Pac. Halibut Comm. Sci. Rep. Seattle. WA. **63**, 42.
8. Sibert JR, Musyl MK, Brill R (2003) Horizontal movements of bigeye tuna (*Thunnus obesus*) near Hawaii determined by Kalman filter analysis from archival tagging data. Fish. Oceanogr. 12: 141–151.
9. Nielsen A, Bigelow KA, Musyl MK, Sibert JR (2006) Improving light-based geolocation by including sea surface temperature. Fish Oceanogr. 15: 314–325
10. Nielsen A, Sibert JR (2007) State-space model for light-based tracking of marine animals. Can J Fish Aquat Sci 64: 1055–1068.
11. Royer F, Lutcavage M (2009) Positioning Pelagic Fish from Sunrise and Sunset Times: Complex Observation Errors Call for Constrained, Robust Modeling. Tagging and Tracking of Marine Animals with Electronic Devices.
12. Thygesen UH, Pedersen MW, Madsen H (2009) Geolocating fish using hidden Markov models and data storage tags. In Tagging and tracking of marine animals with electronic devices II. Volume 8 reviews: methods and technologies in fish biology and fisheries. Edited by JL Nielsen, H Arrizabalaga, N Fragoso, A Hobday, M Lutcavage, and J Sibert. Springer, The Netherlands.
13. Chaprales W, Lutcavage M, Brill R, Chase B, Skomal G (1998) Harpoon method for attaching ultrasonic and "popup" satellite tags to giant bluefin tuna and large pelagic fishes. Mar. Tech. Soc. J. **32**, 104–105.
14. Prince ED, Ortiz M, Venizelos A, Rosenthal DS (2002). In-water conventional tagging techniques developed by the Cooperative Tagging Center for large, highly migratory species. In American Fisheries Society Symposium (pp. 155–171).
15. Prince ED, Goodyear CP (2006). Hypoxia-based habitat compression of tropical pelagic fishes. Fisheries Oceanography, 15(6), 451–464.
16. Prince ED, Luo J, Goodyear P, Hoolihan JP, Snodgrass D, et al. (2010). Ocean scale hypoxia-based habitat compression of Atlantic istiophorid billfishes. Fisheries Oceanography, 19(6), 448–462.
17. Vallado DA (2001) Fundamentals of astrodynamics and applications (Vol. 12). Springer.
18. Milanese M, Vicino A (1991) Optimal estimation theory for dynamic systems with set membership uncertainty: an overview. Automatica, 27(6), 997–1009.
19. Walter E, Norton J, Piet-Lahanier H, Milanese M (1996) Bounding approaches to system identification. Perseus Publishing.
20. Garulli A, Vicino A (2001) Set membership localization of mobile robots via angle measurements. Robotics and Automation, IEEE Transactions on, 17(4), 450–463.
21. Briers M, Doucet A, Maskell S (2010) Smoothing algorithms for state-space models. Volume 62, Number 1, 61-89. Annals of the Institute of Statistical Mathematics.
22. Hein AM, Hou C, Gillooly JF (2012) Energetic and biomechanical constraints on animal migration distance. Ecology Letters 15: 104–110.
23. Abid N, Idrissi M (2006) Swordfish, Chapter 2.1.9 of the ICCAT Manual. http://www.iccat.int/Documents/SCRS/Manual/CH2/2_1_9_SWO_ENG.pdf. Accessed February 3 2012.
24. Chow S, Clarke S, Nakadate N, Okazaki M (2007) Boundary between the north and south Atlantic populations of the swordfish (*Xiphias gladius*) inferred by a single nucleotide polymorphism at calcodulin gene intron. Mar. Biol. 152: 87–93.
25. Goodyear CP, Bigelow KA (2011) Preliminary analysis of simulated longline Atlantic blue marlin CPUE with HBS and Generalized linear models. Collect. Vol. Sci. Pap. ICCAT, 68(4): 1510–1523.
26. B.L. Smith. Submitted. Genetic Multi-Locus Bayesian Assessment of Stock Admixture and Management Boundaries of Atlantic Swordfish (Xiphias gladius L.) Fish. Res.
27. Stramma L, Prince ED, Schmitko S, Juo J, Hoolihan JP, et al. (2012) Expansion of oxygen minimum zones may reduce valuable habitat for tropical pelagic fishes. Nature Climate Change, 2: 1, 33–37.
28. Neilson JD, Arocha F, Cass-Calay S, Mejuto J, Ortiz M, et al. (2013) The Recovery of Atlantic Swordfish: The Comparative Roles of the Regional Fisheries Management Organization and Species Biology. Rev. Fish. Sci. 21(2): 1–39.

23

Trends in Extinction Risk for Imperiled Species in Canada

Brett Favaro[1,2,3]*, Danielle C. Claar[1], Caroline H. Fox[4,5], Cameron Freshwater[1], Jessica J. Holden[1], Allan Roberts[6], UVic Research Derby[1,4,7]¶

1 Department of Biology, University of Victoria, Victoria, Canada, **2** Centre for Sustainable Aquatic Resources, Fisheries and Marine Institute of Memorial University of Newfoundland, St John's, Canada, **3** Department of Ocean Sciences, Memorial University of Newfoundland, St. John's, Canada, **4** Department of Geography, University of Victoria, Victoria, Canada, **5** Raincoast Conservation Foundation, Sidney, Canada, **6** Bamfield Marine Sciences Centre, Bamfield East, Canada, **7** School of Environmental Studies, University of Victoria, Victoria, Canada

Abstract

Protecting and promoting recovery of species at risk of extinction is a critical component of biodiversity conservation. In Canada, the Committee on the Status of Endangered Wildlife in Canada (COSEWIC) determines whether species are at risk of extinction or extirpation, and has conducted these assessments since 1977. We examined trends in COSEWIC assessments to identify whether at-risk species that have been assessed more than once tended to improve, remain constant, or deteriorate in status, as a way of assessing the effectiveness of biodiversity conservation in Canada. Of 369 species that met our criteria for examination, 115 deteriorated, 202 remained unchanged, and 52 improved in status. Only 20 species (5.4%) improved to the point where they were 'not at risk', and five of those were due to increased sampling efforts rather than an increase in population size. Species outcomes were also dependent on the severity of their initial assessment; for example, 47% of species that were initially listed as special concern deteriorated between assessments. After receiving an at-risk assessment by COSEWIC, a species is considered for listing under the federal Species at Risk Act (SARA), which is the primary national tool that mandates protection for at-risk species. We examined whether SARA-listing was associated with improved COSEWIC assessment outcomes relative to unlisted species. Of 305 species that had multiple assessments and were SARA-listed, 221 were listed at a level that required identification and protection of critical habitat; however, critical habitat was fully identified for only 56 of these species. We suggest that the Canadian government should formally identify and protect critical habitat, as is required by existing legislation. In addition, our finding that at-risk species in Canada rarely recover leads us to recommend that every effort be made to actively prevent species from becoming at-risk in the first place.

Editor: Clinton N. Jenkins, Instituto de Pesquisas Ecológicas, Brazil

Funding: BF was supported by a Liber Ero postdoctoral fellowship (http://liberero.ca/). CHF was supported by the Raincoast Conservation Foundation (http://www.raincoast.org/) and the National Sciences and Engineering Research Council of Canada (http://www.nserc-crsng.gc.ca/). DCC and CF were supported by University of Victoria Graduate Fellowships. Catering for the UVIC Research Derby event was provided by the University of Victoria Environmental Science and Biology departments. The funders had no role in study design, data collection and analysis, decision to publish, or preparation of the manuscript.

Competing Interests: The authors have declared that no competing interests exist.

* Email: brett.favaro@mi.mun.ca

¶ Membership of the UVic Research Derby is provided in the Acknowledgments.

Introduction

Unsustainable exploitation, climate change, ocean acidification and other anthropogenic impacts have resulted in a global extinction rate that is as much as 1000 times the historic background rate [1–3]. Given the irreversibility of extinctions, preventing or reversing the continuing decline of at-risk species is a major focus of conservation [4]. Preserving global biodiversity is also considered essential for human well-being and the maintenance of ecosystem processes [3,5–7].

Many countries have legislation that explicitly protects species at risk of extinction. In general, such legislation is designed to identify vulnerable taxa, establish recovery plans, prevent further declines, and promote recovery [8]. Recognizing that habitat loss is the leading cause of extinction [9–12], the identification and preservation of habitat is often required, contributing to the stabilization and recovery of threatened species [13]. Despite the implementation of laws and conservation programs, global biodiversity continues to decline [3,5,14].

In Canada, species at risk are identified and protected in a multi-step process. The process begins with the Committee on the Status of Endangered Wildlife in Canada (COSEWIC), an independent scientific body formed in 1977. COSEWIC assesses the status of candidate species that are potentially at risk of extinction or extirpation [15]. This body only considers scientific evidence relevant to a species' recovery potential, and ignores socioeconomic costs or benefits of protection [15]. Their assessments are based on the Canadian extent of the species' range, even if the species occurs in the United States or elsewhere [15]. Species assessed as at-risk by COSEWIC do not automatically secure legal protection from the committee's decision. Rather, protection is issued under the Species at Risk Act (SARA), which was passed in 2003, and which formalized the use of COSEWIC assessments as the scientific basis for listing decisions [16]. Upon receipt of a COSEWIC assessment, the Minister of the

Environment must issue a response statement indicating whether the species will be listed at the status recommended by the assessment, or whether more information is required [17]. The ultimate decision whether to list may also incorporate the socio-economic impacts of listing. If listed, it becomes illegal to kill or harm individuals of that species and the species' critical habitat must be identified and protected to the extent possible [18]. Protections derived from SARA automatically apply to federal lands (including oceans), but for these measures to apply outside of these areas (e.g. in provincial or private land) complementary protections must be implemented that applies to the species and its habitat [19]. If the Minister of the Environment decides that provincial or territorial law is insufficient to effectively protect a listed species, the federal government has the power to apply SARA's protections to provincial or territorial lands [20]. Recovery strategies and action plans (required by SARA) are mandated to be released within a specified timeframe, the length of which depends on the species' designation and when it was listed [21,22]. Furthermore, SARA requires that management plans be produced for species that are not at immediate risk of extinction or extirpation, but are nevertheless of 'special concern' since they may become at-risk in the future [23]. Species of special concern do not receive the full protections offered to species listed under SARA as threatened, endangered, or extirpated.

Previous studies have identified biases, limitations, and a general lack of implementation associated with at-risk species legislation in Canada (Table 1). For instance, harvested species and those found in northern territories are less likely to be listed under SARA [24,26], as are species threatened by biological resource use, including those that are unintentionally harvested [27]. In addition, recovery strategies and action plans are often not completed within deadlines established by SARA [22], leading to concerns that species are not receiving timely protections, despite being listed. This finding was confirmed in a recent federal court decision [28]. Moreover, COSEWIC itself has been criticized for its assessment criteria and apparent biases [31–32]. For example, species with low information availability tend to receive less severe assessments than would be suggested under the precautionary principle [30].

In this study, we examine trends in the status of species repeatedly assessed by COSEWIC, used here as a proxy for Canada's effectiveness in species conservation. Using records obtained from COSEWIC [33] and the SARA Public Registry [34], we analyze trends in the designations of species with two or more assessments to determine whether species are, on average, improving, deteriorating, or remaining stable in status. In addition, we examine outcome differences across taxonomic groups, and whether the basic obligation to identify critical habitat for listed species has been met. Finally, we assess whether species listed for longer have better outcomes. While this study builds on previous work reviewing the process of listing species under SARA (Table 1), it is the first to assess the overall trends in status of at-risk species that have been assessed more than once by COSEWIC.

Methods

COSEWIC Wildlife Species Database

Species, subspecies, and populations (hereafter 'species') that are assessed by COSEWIC are placed into one of five categories on a scale of increasing risk of extinction or extirpation [32]: not at risk, special concern, threatened, endangered, or extinct or extirpated (hereafter extirpated). A sixth category, data deficient, is used when there are insufficient data to classify a species. The criteria for categorization depends on a variety of factors including changes in total numbers of mature individuals, whether a species has a small and declining population size, and the changes in size of its range [35]. COSEWIC's assessment criteria were updated in 2001 [35]. Although the same assessment categories were employed after the revision in 2001, species required a more severe decline in total population to qualify as either endangered or threatened. For example, prior to 2001 the criteria for an Endangered assessment (Criteria A1: [35]) required a reduction of ≥50% in the total number of mature individuals over the last 10 years or 3 generations. As of 2001, this same category required a ≥70% decline [36].

The definition of full recovery that we employ here is a change in COSEWIC assessment status to 'not at risk.' Although this may not equate to a full ecological recovery (e.g. [37,38]) it allows us to broadly assess trends with a consistent metric.

Data collection

We identified all species that COSEWIC has assessed more than once using the Wildlife Species Search Engine on the COSEWIC website [33]. Assessments occurred between 1977 (COSEWIC's establishment) and December 2013 (when we conducted our search). We recognize that our analysis may be limited by biases associated with which species have been assessed multiple times and that COSEWIC listing criteria have changed since 2001. However, because the 2001 revision tightened criteria for more-severe listings, the expected bias introduced by the criteria change would be a greater proportion of less severe species listings. We used the 'change in status' tab on the website to include any species 'in a higher risk category' (N = 81); 'in a lower risk category' (N = 36); 'no longer at risk' (N = 21); 'changed' (N = 20); 'reassigned' (N = 69); and 'no change' (N = 272); populating our database with a total of 499 species. In our analysis, we excluded any species that had ever been assessed as 'data deficient' by COSEWIC (N = 30 species), or that had experienced some form of reassignment (e.g. a species split into multiple designatable units, N = 106 species), because changes in status for these species cannot be interpreted as a true change in extinction risk. We excluded six species that were both data deficient and reclassified, leaving a total of 369 species in our analysis.

We collected the history of COSEWIC assessments for each species on its respective species summary page. We recorded the date and status designation of each COSEWIC assessment for each species, and the taxonomic group to which that species belongs (due to small sample sizes, we combined molluscs and arthropods into a single 'invertebrates' category, and combined lichens and mosses). We then used the SARA public registry [34] to record whether species were SARA listed, and to access COSEWIC status reports for each species. To account for differences between species' life histories, COSEWIC and the IUCN place rates of decline into a biological context by scaling assessments by the generation time (GT) of each species [35,39]. Therefore we also recorded each species' generation time, if available. If the species was SARA-listed as extirpated, endangered, or threatened, we recorded whether a recovery strategy had been completed. If the recovery strategy was completed, we recorded whether the strategy indicated that critical habitat had been fully, partially, or not identified. If a recovery strategy was not complete for the species, we recorded that critical habitat had not been identified. For SARA-listed species, Environment Canada provided us with the dates when species were listed.

For some species, the COSEWIC summary explained that an apparent improvement in status was due to the discovery of new populations through increased sampling, and not because of an

Table 1. Summary of review papers related to endangered species assessment and legislation in Canada.

Publication	Reference #	Primary findings
Vanderzwaag and Hutchings (2005)	[51]	Review of SARA implementation related to marine fish. Paper advocates for biodiversity preservation by implementing marine protected areas and modernizing Fisheries Act.
Mooers et al. (2007)	[52]	First identification of taxonomic and regional biases in SARA listing. Northern species and marine fish and terrestrial mammals unlikely to receive SARA-listing.
Findlay et al. (2009)	[24]	Commercially harvested species, species managed by DFO, and species that occur entirely within Canada are less likely to receive listing.
Lukey et al. (2009)	[29]	Changes in assessment status from 'endangered' to 'threatened' often occur without sufficient justification for the change. Assessment criteria are not always applied consistently.
Lukey et al. (2010)	[30]	COSEWIC assessments do not follow the precautionary principle – lack of information is associated with assessments of species to lower risk categories.
Mooers et al. (2010)	[25]	Most SARA-listed species lack recovery plans. Scientific advice is insufficiently reflected in conservation policy for at-risk species. First review of changes in COSEWIC status across assessments.
Powles (2011)	[31]	General overview of marine fish assessed by COSEWIC.
Taylor and Pinkus (2013)	[53]	Only 17% of recovery strategies led by DFO included critical habitat, as opposed to 63% for those led by Environment Canada. Recovery strategies written after court judgments related to SARA were more likely to identify critical habitat.
Waples et al. (2013)	[8]	Comparison of United States' Endangered Species Act (ESA) with SARA. ESA should adopt a single national scientific body, while SARA should adopt strict deadlines for listing action. The emphasis on socioeconomic factors should also be reduced.
McCune et al. (2013)	[27]	Human disturbance, invasive species, residential development, and ultimately loss of habitat are major threats to the majority of SARA-listed species. Threats differ by taxonomic grouping.
Schultz et al. (2013)	[26]	Imperiled marine fish are unlikely to receive SARA listing if the forecasted cost of listing exceeds $90,000 per decade. The threshold for freshwater fish is $5,000,000. Rationale used in the decision to list was inconsistent between freshwater and marine fish.
Auditor General of Canada (2013)	[22]	All three government agencies involved with species at risk (Environment Canada, Fisheries and Oceans Canada, and Parks Canada) are not meeting obligations to complete recovery strategies, action plans, or management plans. Only seven of 97 required action plans were complete.

actual recovery in population. We made note of these cases (N = 20) so as to distinguish them from improvements due to conservation action.

Data analysis

We assigned each COSEWIC status a numerical value in descending order of severity (5 = not at risk, 4 = special concern, 3 = threatened, 2 = endangered, 1 = extirpated). We then calculated the overall change for each species, between the first and most recent COSEWIC assessment. For example, a species that was initially classified as special concern, and then deteriorated to endangered in its most recent assessment would receive a score of −2. For species requiring critical habitat designation (extirpated, endangered and threatened), we performed a Pearson's chi-squared test to determine whether critical habitat identifications (full, partial, or not identified) across the ten taxonomic groups differed from the proportions across all species. We used the statistical software R for computations and data plots [40].

We identified species whose most recent COSEWIC assessment occurred after the species had been listed under SARA for at least three of that species' GT, to identify species for which some recovery may be biologically possible. We then employed a cumulative link mixed model using the "ordinal" package in R [41] to test whether the number of GTs since SARA-listing was associated with a change in final COSEWIC status across species. Cumulative link mixed models test the influence of one or more independent variables (herein: number of generation times since listing) on an ordinal dependent variable (change in COSEWIC status), and are analogous to generalized linear mixed effects models in that they allow for the incorporation of a random effect, or 'grouping' factor (taxonomic order) [41–43].

Results

There were a total of 369 species across ten taxonomic groups that met our criteria for inclusion (Table S1, Fig. 1). For all taxonomic groups (except marine fish), the majority of species declined or remained the same in status (Table S1, Fig. 2). However, species trajectories varied substantially based on their initial assessment (Fig. 3). For species initially classified as not at risk, endangered, or extirpated, the most common outcome was to remain at their initially assessed status. However, for species initially classified as special concern or threatened, deterioration in status was the most common outcome. For example, 47% of species listed as special concern deteriorated in status (Fig. 3). Only 20 species (5.4%) received a 'not at risk' assessment after previously being listed in an at-risk category (three terrestrial mammals, nine birds, four freshwater fish, one marine fish, two vascular plants, and one lichen). Five of these cases were not due to conservation action, but were instead due to increased sampling effort. A total of 221 species are listed under SARA as threatened, endangered, or extirpated (Fig. 4), and therefore their critical habitat should be identified. Overall, and for all taxonomic groups, critical habitat has not been fully identified for more than half of SARA-listed species (Fig. 4). Further, critical habitat identifications (full, partial, or not identified) significantly varied across the ten taxonomic groups ($\div^2 = 51.90$, df = 18, $p<0.001$).

There were 163 species that met our criteria for inclusion in the Generation Time analysis (listed as extirpated, endangered, or

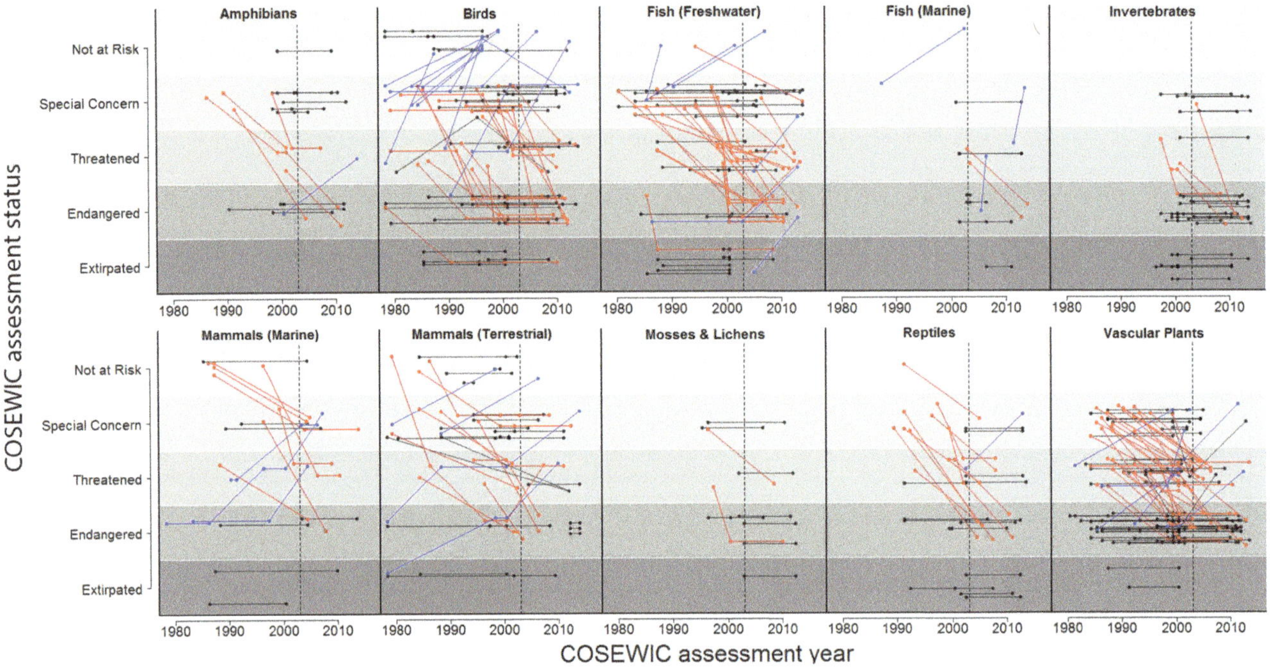

Figure 1. Overview of COSEWIC assessment statuses for all species that have been assessed more than once, and that have never been data deficient or taxonomically reassigned. Each species is represented by points (assessment date and outcome) connected by lines. Species that deteriorated in status from their first to final COSEWIC assessments are red, species that have improved are blue, and species that have remained constant are black. The vertical dotted line indicates 2003, or the passing of SARA. Species whose apparent recovery was due to increased sampling effort, and not biological recovery (N = 20) are not shown.

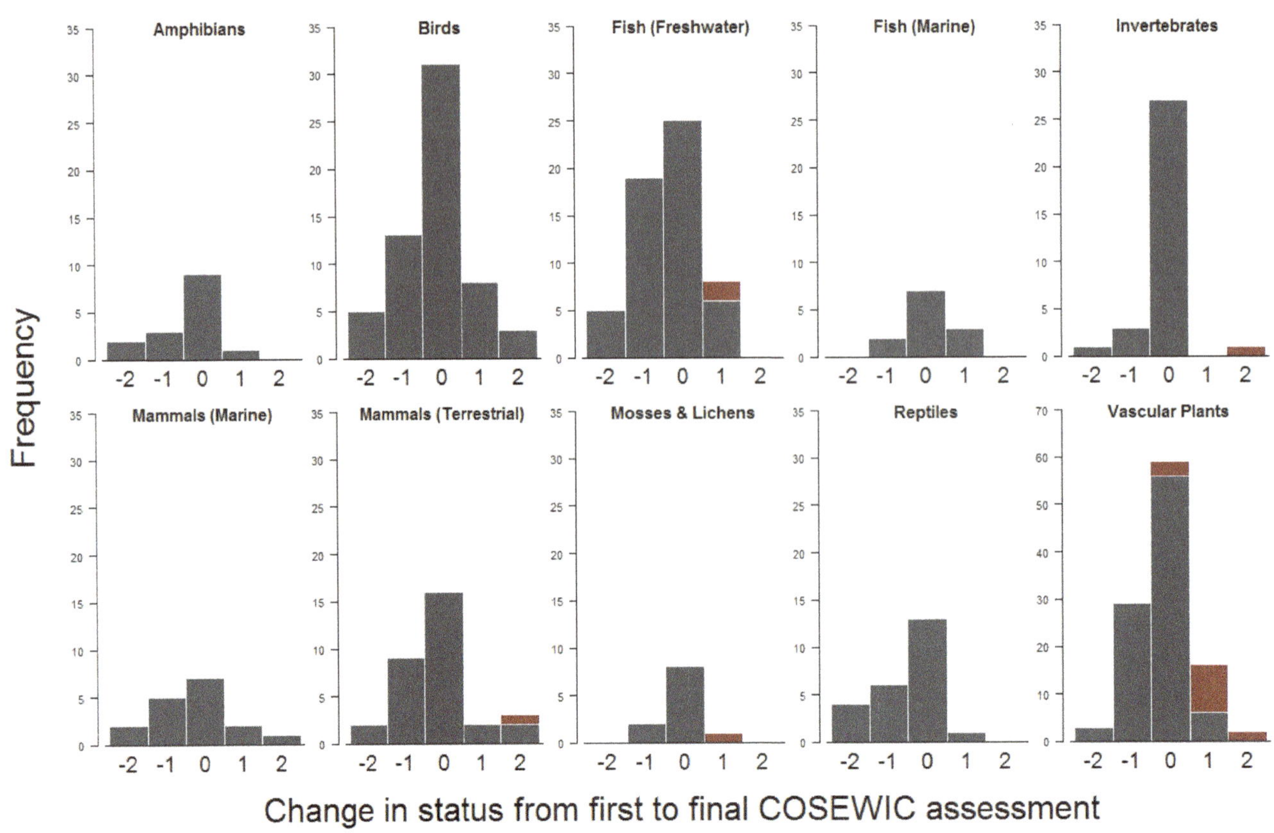

Figure 2. Frequencies of occurrence of change in COSEWIC assessed status for each taxonomic group. Positive numbers indicate improvement (e.g. a transition from endangered to threatened would be +1, while endangered to special concern would be +2), negative numbers indicate deterioration, and zero indicates no change across assessments. Red bars indicate apparent recoveries due to increased sampling efforts. Note that the y-axis for vascular plants is scaled differently from other taxonomic groups.

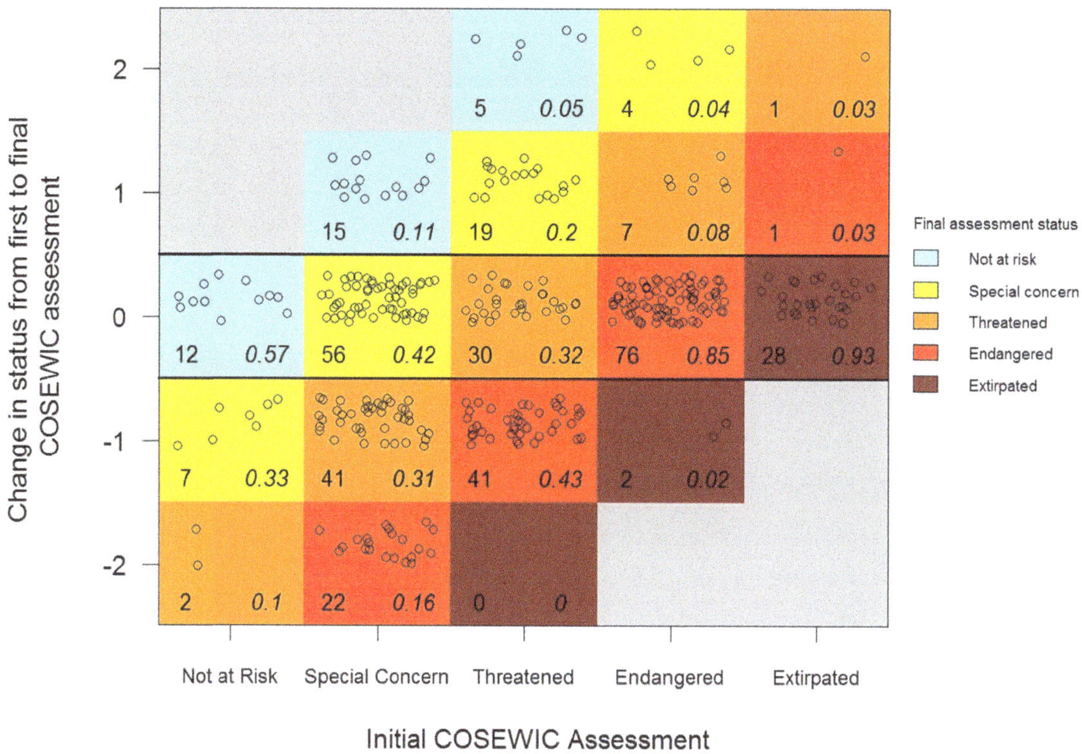

Figure 3. Trends in COSEWIC statuses for each species, grouped by initial assessments. Proportions are grouped by initial assessment status (column). Change from first to final assessment is indicated on the Y axis, and colours indicate the final assessment status. For example, of species that were initially assessed as threatened (third column), 19 of them (or a proportion of 0.2) improved by one status level (+1 on Y-axis), ultimately placing them into the special concern category (yellow).

threatened; GT reported; schedule 1), and 69 that have been listed for three or more GT. The number of GT since listing ranged from 0.11 to 35.8 (mean ± 1 S.D., 3.7 ± 4.3, Figure 5). If a species has been adequately protected, the probability of a species improving in status should increase with the number of generations since listing. However, there was an estimated decline in the probability of a species improving as the number of generations since listing increased ($\hat{\beta} = -0.05$, $\{\hat{\theta}_j\} = \{-2.40, -0.18, 2.44, 4.31\}$, 95% C.I. $= -0.124$ to 0.008), but it was not significant (Wald $p = 0.084$). This model had a small absolute gradient ($7.4 * 10^{-6}$) and a condition number of the Hessian of 720, indicating that the model was able to converge and was well-defined, respectively [42]. The cumulative odds ratio was between 0.88 and 1.01 (95% C.I.), meaning that for each unit increase in GT since listing, the odds of a species improving in status ranged from increasing by 1% to decreasing by 12%.

Discussion

For species that have been assessed more than once by COSEWIC, improvement was rare, and recovery to a 'not at risk' status occurred in only 5.7% of cases. Moreover, one quarter of those were apparent recoveries due to increased sampling, rather than increases in population size driven by conservation action. Contrary to the intent of endangered species legislation in Canada, the probability of a species improving in status did not increase with the number of generations since initial listing under SARA. Moreover, species had a greater probability of deteriorating in status with the number of generations since listing, although this relationship was marginally statistically non-significant (Wald $p = 0.084$). In contrast, species recovery in the United States was strongly correlated with the number of years of protection under the Endangered Species Act [44], indicating that endangered species legislation can be effective. These results suggest a potential failure of Canadian legislation, its subsequent implementation, or both. While COSEWIC's assessment criteria were modified in 2001 [35], the revisions made it harder to qualify for more-severe categories. Therefore we have no evidence to suggest that observed declines in status were an artifact of changes in criteria.

The lack of observed recovery for SARA-listed species may be due to the lack of implementation of the law. For example, for those species without identified critical habitat, the habitat protection provisions of SARA cannot be fully implemented. Overall, the proportion of species with critical habitat identified was higher than what has been reported in a previous study [25], although our study focused on a subset of species that have been assessed multiple times, and that have never been data deficient. These are the species for which it should have been most feasible to identify critical habitat, as they have been scrutinized for a relatively long time and have sufficient data to complete an accurate assessment. The fact remains that despite the importance of critical habitat identification, the proportion of species whose critical habitat was fully identified remained low, and varied significantly across the ten taxonomic groups. This pattern suggests that there are considerable differences between the protections that species should receive under SARA, and what is actually achieved.

Since our data demonstrate that it is rare for at-risk species to recover in Canada, it is essential that substantial efforts be made to prevent species from becoming at-risk in the first place. Given that

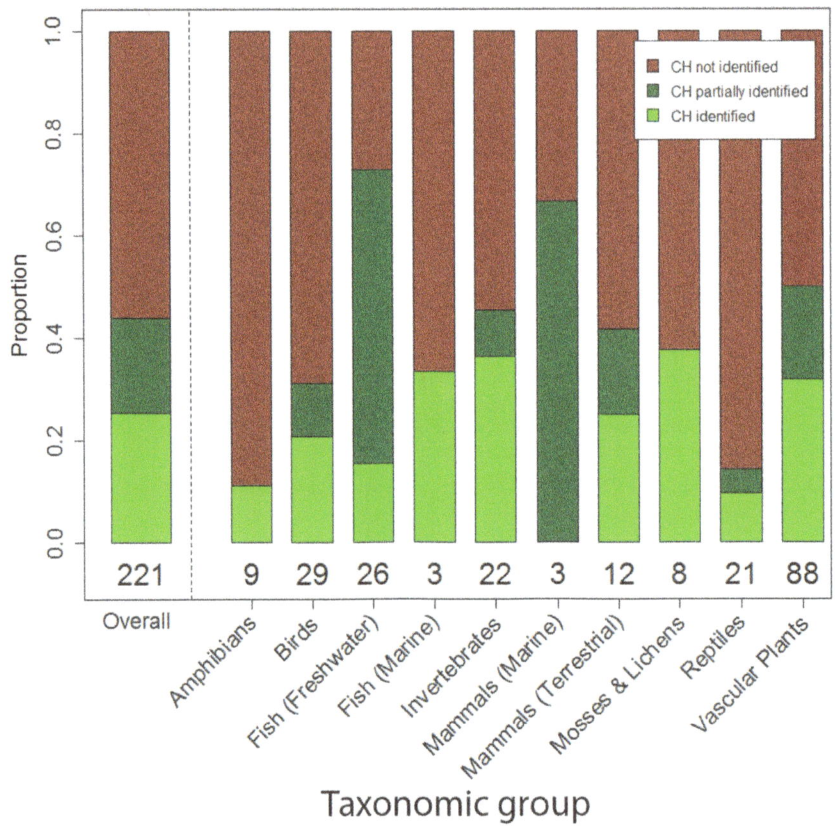

Figure 4. Proportion of SARA-listed species (listed as extirpated, endangered, or threatened) that have critical habitat (CH) fully identified (light green), partially identified (dark green), or not at all identified (dark red), overall and by taxonomic group. Values under bars indicate the number of species in each taxonomic category.

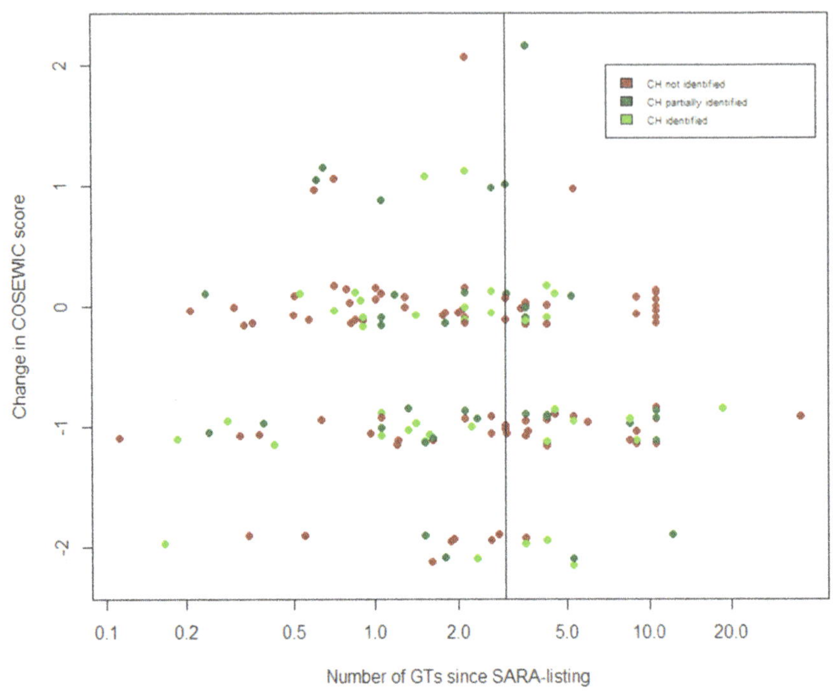

Figure 5. Change in COSEWIC assessment statuses versus the number of GT that have passed since initial SARA listing for all schedule 1 species listed as threatened, endangered, or extirpated. Dot colour indicates whether CH has been fully identified (green), partially identified (dark green), or not identified (red). The black vertical line indicates three GT.

imperiled species are usually threatened by loss of habitat [45], recent weakening of federal laws that protect habitat in Canada [25,46,47,48] will be unhelpful in the long term, as it may result in additional species declining to the point where they receive an at-risk designation. Habitat protection should not be limited to critical habitat – it should be managed appropriately for all habitat such that protection does not become critical. The experience in the United States demonstrates that protection of critical habitat is associated with improved conservation outcomes relative to species without such protection [44,47]. In addition, since we found that the most common outcome for a species assessed as special concern was to deteriorate, an assessment of special concern does not currently result in sufficient protection to promote recovery.

The single most common outcome across taxonomic groups was for species to remain at the same status across assessments, and the number of species that declined outnumbered those that recovered by a ratio of approximately 2:1. These findings are alarming for three reasons. First, by definition, a species that has been assessed as anything other than 'not at risk' is at elevated risk of extinction or extirpation given current conditions, and therefore maintaining a species at a threatened status should not be interpreted as a conservation victory. Second, it takes a substantial decline in population size or range size to trigger a change in assessment status, so real declines (or increases) could still occur within species held at the same COSEWIC threat level across assessments. Third, a 'not at risk' designation only means that the species is not at elevated risk of extinction or extirpation – it does not imply that the population has recovered to historical levels. Even species that are classified 'not at risk' can be heavily depleted and unable to serve their historic roles in ecosystem structure or function. This has implications for managers in the United States as well as Canada, because many at-risk species have ranges that extend into both countries. At the very least, a successful species at risk program should demonstrate species recovering to a point where they are not at risk of extinction or extirpation given current conditions. Currently, this goal is not being achieved in Canada for the overwhelming majority of species.

Recommendations

Our results lead us to make three core recommendations for at-risk species in Canada. First, given that it was much more common for species of special concern to deteriorate than to improve, we should recognize that a special concern listing warns of a coming deterioration, and we therefore suggest that the protections associated with this listing should be strengthened. Second, given the poor outcomes of at-risk species in Canada, it should be a policy priority to prevent species from becoming at-risk in the first place. The importance of critical habitat indicates that future legislation should be underpinned by a strong mandate to conserve habitat and we recommend that any legislative changes that may reduce habitat protection (e.g. the Fisheries Act [48]) should be reconsidered. Third, to experience conservation benefits from SARA, this law must be fully implemented. Implementation requires that critical habitat be fully identified and subsequently protected for SARA-listed species. The federal government should also be prepared to enact its 'safety net' provision, in the event that species in these regions are not receiving adequate protection to enable recovery (as it did with greater sage-grouse, *Centrocercus urophasianus* [49]).

Finally, even if these recommendations were accepted and put into effect, recovery takes time. Effective management requires that conservation measures be sustained over the long term, even if positive outcomes are not immediately observed.

Supporting Information

Table S1 Summary counts of species across taxonomic groups, for all species assessed more than once by COSEWIC. Values for species listed for >3 generation times (GT) refer to the subset of species for which GTs were reported.

Acknowledgments

We thank Environment Canada for providing us with dates that species were listed under SARA. The idea for this paper originated within a Research Derby conducted at the University of Victoria [50]. The 2013 UVic Research Derby author group consisted of the authors of this paper, as well as A.F. Martin, K. Pawluk, D. Roberts, and J. Robinson, and can be contacted through B. Favaro (brett.favaro@mi.mun.ca). We thank S. Pinkus and three anonymous reviewers for providing useful comments.

Author Contributions

Conceived and designed the experiments: BF DCC CHF CF JJH. Performed the experiments: BF DCC CHF CF JJH. Analyzed the data: BF DCC CHF CF JJH AR. Wrote the paper: BF DCC CHF CF JJH AR.

References

1. Pimm SL, Russell GJ, Gittleman JL, Brooks TM (1995) The future of biodiversity. Science 269: 347–350.
2. Chapin III FS, Zavaleta ES, Eviner VT, Naylor RL, Vitousek PM, et al. (2000) Consequences of changing biodiversity. Nature 6783: 234–242.
3. Rands MRW, Adams WM, Bennun L, Butchart SHM, Clements A, et al. (2010) Biodiversity conservation: Challenges beyond 2010. Science 329: 1298–1303.
4. Arponen A (2012) Prioritizing species for conservation planning. Biodivers Conserv 21: 875–893.
5. Hassan RM, Scholes R, Ash N, editors. (2005) Ecosystems and human well-being: Current state and trends. Findings of the condition and trends working group millennium ecosystem assessment series. Washington DC: Island Press.
6. Hooper DU, Chapin III FS, Ewel JJ, Hector A, Inchausti P, et al. (2005) Effects of biodiversity on ecosystem functioning: A consensus of current knowledge. Ecol Monogr 75: 3–35.
7. Barton J, Pretty J (2010) What is the best dose of nature and green exercise for improving mental health? A multi-study analysis. Environ Sci Tech 44: 3947–3955.
8. Waples RS, Nammack M, Cochrane JF, Hutchings JA (2013) A tale of two acts: Endangered species listing practices in Canada and the United States. Bioscience 63: 723–734.
9. Brooks TM, Mittermeier RA, Mittermeier CG, Da Fonseca GAB, Rylands AB, et al. (2002) Habitat loss and extinction in the hotspots of biodiversity. Conserv Biol 16: 909–923.
10. Kareiva P, Wennergren U (1995) Connecting landscape patterns to ecosystem and population processes. Nature 373: 299–302.
11. Wilcove DS, Rothstein D, Jason Dubow, Phillips A, Losos E (1998) Quantifying threats to imperiled species in the United States. Bioscience 48: 607–615.
12. Homan RN, Windmiller BS, Reed JM (2004) Critical thresholds associated with habitat loss for two vernal pool-breeding amphibians. Ecol App 14: 1547–1553.
13. Taylor MFJ, Sattler PS, Evans M, Fuller RA, Watson JEM, et al. (2011) What works for threatened species recovery? An empirical evaluation for Australia. Biodivers Conserv 20: 767–777.
14. "Secretariat of the Convention on Biological Diversity." (2010) Global biodiversity outlook 3. Montreal. 94 p.
15. COSEWIC (2013) Terms of reference: Committee on the status of endangered wildlife in Canada (COSEWIC), revised and approved by COSEWIC November 2011.
16. *SARA (Species at Risk Act)* (2002) An act respecting the protection of wildlife species at risk in Canada. SC 2002, c 29, s 15. Available: http://Laws-lois.justice.gc.ca/PDF/S-15.3.pdf. Accessed 2014 August 11.

17. *SARA (Species at Risk Act)* (2002) An act respecting the protection of wildlife species at risk in Canada. SC 2002, c 29, s 27. Available: http://Laws-lois.justice.gc.ca/PDF/S-15.3.pdf. Accessed 2014 August 11.
18. *SARA (Species at Risk Act)* (2002) An act respecting the protection of wildlife species at risk in Canada. SC 2002, c 29, ss 32–36. Available: http://Laws-lois.justice.gc.ca/PDF/S-15.3.pdf. Accessed 2014 August 11.
19. *SARA (Species at Risk Act)* (2002) An act respecting the protection of wildlife species at risk in Canada. SC 2002, c 29, ss 34(1)-35(1,2). Available: http://Laws-lois.justice.gc.ca/PDF/S-15.3.pdf. Accessed 2014 August 11.
20. *SARA (Species at Risk Act)* (2002) An act respecting the protection of wildlife species at risk in Canada. SC 2002, c 29, ss 3435. Available: http://Laws-lois.justice.gc.ca/PDF/S-15.3.pdf. Accessed 2014 August 11.
21. *SARA (Species at Risk Act)* (2002) An act respecting the protection of wildlife species at risk in Canada. SC 2002, c 29, ss 41–45. Available: http://Laws-lois.justice.gc.ca/PDF/S-15.3.pdf. Accessed 2014 August 11.
22. Auditor General of Canada (2013) Chapter 6: Recovery planning for species at risk. In: Anonymous Report of the commissioner of the environment and sustainable development. Office of the Auditor General of Canada. pp. 1–22. Available: http://www.oag-bvg.gc.ca/internet/English/parl_cesd_201311_06_e_38676.html. Accessed 2014 August 11.
23. *SARA (Species at Risk Act)* (2002) An act respecting the protection of wildlife species at risk in Canada. SC 2002, c 29, ss 65–72. Available: http://Laws-lois.justice.gc.ca/PDF/S-15.3.pdf. Accessed 2014 August 11.
24. Findlay C, Elgie S, Giles B, Burr L (2009) Species listing under Canada's Species at Risk Act. Conserv Biol 23: 1609–1617.
25. Mooers AO, Doak DF, Findlay CS, Green DM, Grouios C (2010) Science, policy, and species at risk in Canada. Bioscience 60: 843–849.
26. Schultz JA, Darling ES, Côté IM (2013) What is an endangered species worth? Threshold costs for protecting imperilled fishes in Canada. Mar Policy 42: 125–132. http://dx.doi.org/10.1016/j.marpol.2013.01.021.
27. McCune JL, Harrower WL, Avery-Gomm S, Brogan JM, Csergő A, et al. (2013) Threats to Canadian species at risk: An analysis of finalized recovery strategies. Biol Conserv 166: 254–265. http://dx.doi.org/10.1016/j.biocon.2013.07.006.
28. *Western Canada Wilderness Committee, David Suzuki Foundation, Greenpeace Canada, Sierra Club of British Columbia Foundation, and Wildsite v. Minister of Fisheries and Oceans and Minister of the Environment* (2014) Canada Federal Court, 2014 FC 148, T-1777-12. Available: http://cas-ncr-nter03.cas-satj.gc.ca/rss/T-1777-12%20SARA%20decision%2014-02-2014%20ENG.pdf. Accessed 2014 August 11.
29. Lukey JR, Crawford SS (2009) Consistency of COSEWIC species at risk designations: Freshwater fishes as a case study. Can J Fish Aquat Sci 66: 959–971.
30. Lukey JR, Crawford SS, Gillis D (2010) Effect of information availability on assessment and designation of species at risk. Conserv Biol 24: 1398–1406.
31. Powles H (2011) Assessing risk of extinction of marine fishes in Canada - the COSEWIC experience. Fisheries 36: 231–246.
32. Kowalchuk KA, Kuhn RG (2012) Mammal distribution in Nunavut: Inuit harvest data and COSEWIC's species at risk assessment process. Ecol Soc 17: 4.
33. COSEWIC (2013) Database of wildlife species assessed by COSEWIC. Available: http://www.cosewic.gc.ca/eng/sct1/searchform_e.cfm. Accessed 2014 August 11.
34. Government of Canada (2014) Species at risk act public registry. Available: http://www.sararegistry.gc.ca/default_e.cfm. Accessed 2014 August 11.
35. COSEWIC (2011) COSEWIC's Assessment Process and Criteria. Available: http://www.cosewic.gc.ca/pdf/Assessment_process_and_criteria_e.pdf. Accessed 2014 August 11.
36. COSEWIC (2005) Original criteria and definitions used in the status assessment of species from October 1999 to May 2001. Available: http://www.cosewic.gc.ca/eng/sct0/original_criteria_e.cfm. Accessed 2014 August 11.
37. Redford KH, Amato G, Baillie J, Beldomenico P, Bennet EL, et al. (2011) What does It mean to successfully conserve a (vertebrate) species? BioScience 61:39–48.
38. Westwood A, Reuchlin-Hugenholtz E, Keith DM (2014) Re-defining recovery: A generalized framework for assessing species recovery. Biological Conservation 172:155–162.
39. IUCN Standards and Petitions Subcommittee (2014) Guidelines for using the IUCN Red List categories and criteria. Version 11. Available: http://www.iucnredlist.org/documents/RedListGuidelines.pdf. Accessed 2014 August 11.
40. R Core Team (2013) R: A language and environment for statistical computing. R Foundation for Statistical Computing, Vienne, Austria. http://www.R-project.org.
41. Christensen RHB (2013) Ordinal – Regression models for Ordinal Data. R package version 2013.9-30. Available: http://www.cran.r-project.org/package=ordinal/.
42. Tutz G, Hennevogl W (1996) Random effects in ordinal regression models. Comput Stat Data An 22: 537–557.
43. Christensen RHB (2013) A tutorial on fitting Cumulative Link Models with the ordinal package. Available: ftp://ftp.stat.math.ethz.ch/R-CRAN/web/packages/ordinal/vignettes/clm_tutorial.pdf.
44. Taylor MFJ, Suckling KF, Rachlinski JJ (2005) The effectiveness of the endangered species act: A quantitative analysis. Bioscience 55: 360–367.
45. Venter O, Brodeur NN, Nemiroff L, Belland B, Dolinsek IJ, et al. (2006) Threats to endangered species in Canada. Bioscience 56: 903–910.
46. Favaro B, Reynolds JD, Côté IM (2012) Canada's weakening aquatic protection. Science 337: 154.
47. Kirchhoff D, Tsuji LJS (2014) Reading between the lines of the 'Responsible resource development' rhetoric: The use of omnibus bills to 'streamline' Canadian environmental legislation. Impact Assessment and Project Appraisal: 1–13. DOI: 10.1080/14615517.2014.894673.
48. Hutchings JA, Post JR (2013) Gutting Canada's fisheries act: No fishery, no fish habitat protection. Fisheries 38: 497–501.
49. Canada Gazette (2013) Emergency Order for the Protection of the Greater Sage-Grouse. Species At Risk Act. Government of Canada. P.C. 2013–1245. Available: http://canadagazette.gc.ca/rp-pr/p2/2013/2013-12-04/html/sup-eng.php Accessed 2014 July 22.
50. Favaro B, Braun DC, Earth2Ocean Research Derby (2013) The 'research derby': a pressure cooker for creative and collaborative science. Ideas Ecol Evol 6:40–66.
51. Vanderzwaag DL, Hutchings JA (2005) Canada's marine species at risk: Science and law at the helm, but a sea of uncertainties. Ocean Dev Int Law 36: 219–259.
52. Mooers AO, Prugh LR, Festa-Bianchet M, Hutchings JA (2007) Biases in legal listing under Canadian endangered species legislation. Conserv Biol 21: 572–575.
53. Taylor EB, Pinkus S (2013) The effects of lead agency, nongovernmental organizations, and recovery team membership on the identification of critical habitat for species at risk: Insights from the Canadian experience. Environ Rev 21: 93–102.

PERMISSIONS

All chapters in this book were first published in PLOS ONE, by The Public Library of Science; hereby published with permission under the Creative Commons Attribution License or equivalent. Every chapter published in this book has been scrutinized by our experts. Their significance has been extensively debated. The topics covered herein carry significant findings which will fuel the growth of the discipline. They may even be implemented as practical applications or may be referred to as a beginning point for another development.

The contributors of this book come from diverse backgrounds, making this book a truly international effort. This book will bring forth new frontiers with its revolutionizing research information and detailed analysis of the nascent developments around the world.

We would like to thank all the contributing authors for lending their expertise to make the book truly unique. They have played a crucial role in the development of this book. Without their invaluable contributions this book wouldn't have been possible. They have made vital efforts to compile up to date information on the varied aspects of this subject to make this book a valuable addition to the collection of many professionals and students.

This book was conceptualized with the vision of imparting up-to-date information and advanced data in this field. To ensure the same, a matchless editorial board was set up. Every individual on the board went through rigorous rounds of assessment to prove their worth. After which they invested a large part of their time researching and compiling the most relevant data for our readers.

The editorial board has been involved in producing this book since its inception. They have spent rigorous hours researching and exploring the diverse topics which have resulted in the successful publishing of this book. They have passed on their knowledge of decades through this book. To expedite this challenging task, the publisher supported the team at every step. A small team of assistant editors was also appointed to further simplify the editing procedure and attain best results for the readers.

Apart from the editorial board, the designing team has also invested a significant amount of their time in understanding the subject and creating the most relevant covers. They scrutinized every image to scout for the most suitable representation of the subject and create an appropriate cover for the book.

The publishing team has been an ardent support to the editorial, designing and production team. Their endless efforts to recruit the best for this project, has resulted in the accomplishment of this book. They are a veteran in the field of academics and their pool of knowledge is as vast as their experience in printing. Their expertise and guidance has proved useful at every step. Their uncompromising quality standards have made this book an exceptional effort. Their encouragement from time to time has been an inspiration for everyone.

The publisher and the editorial board hope that this book will prove to be a valuable piece of knowledge for researchers, students, practitioners and scholars across the globe.

LIST OF CONTRIBUTORS

Jacquomo Monk, Daniel Ierodiaconou and Alex Rattray
School of Life and Environmental Sciences, Faculty of Science and Technology, Deakin University, Warrnambool, Victoria, Australia

Euan Harvey
School of Plant Biology (Oceans Institute M470), The University of Western Australia, Crawley, Western Australia, Australia

Vincent L. Versace
School of Information Systems, Faculty of Business and Law, Deakin University, Warrnambool, Victoria, Australia

Warsha Singh
Science Institute, University of Iceland, Reykjavik, Iceland

Erla B. Örnólfsdóttir
Holar University College, Holar, Iceland

Gunnar Stefansson
Science Institute, University of Iceland, Reykjavik, Iceland

Ali M. Al-Aidaroos, Gopikrishna Mantha and Sathianeson Satheesh
Department of Marine Biology, Faculty of Marine Sciences, King Abdulaziz University, Jeddah, Saudi Arabia

Mohsen M. O. El-Sherbiny
Department of Marine Biology, Faculty of Marine Sciences, King Abdulaziz University, Jeddah, Saudi Arabia
Marine Science Department, Faculty of Science, Suez Canal University, Ismailia, Egypt

Susana Agustī
The UWA Oceans Institute and School of Plant Biology, University of Western Australia, Crawley, WA, Australia
Department of Global Change Research and LINC Global, IMEDEA (CSIC-UIB) Instituto Mediterráneo de Estudios Avanzados, Miquel Marqués, Esporles, Spain

Beatriz Carreja
Department of Global Change Research and LINC Global, IMEDEA (CSIC-UIB) Instituto Mediterráneo de Estudios Avanzados, Miquel Marqué
s, Esporles, Spain

Carlos M. Duarte
Department of Marine Biology, Faculty of Marine Sciences, King Abdulaziz University, Jeddah, Saudi Arabia
The UWA Oceans Institute and School of Plant Biology, University of Western Australia, Crawley, WA, Australia
Department of Global Change Research and LINC Global, IMEDEA (CSIC-UIB) Instituto Mediterráneo de Estudios Avanzados, Miquel Marqués, Esporles, Spain

Amy A. Wallace, David J. Hollander and Ernst B. Peebles
College of Marine Science, University of South Florida, St. Petersburg, Florida, United States of America

Elisabeth Debusschere
Bio-environmental research group, Institute for Agricultural and Fisheries Research, Oostende, Belgium
Biology Department, Ghent University, Ghent, Belgium

Bert De Coensel and Dick Botteldooren
Department of Information Technology, Ghent University, Ghent, Belgium

Aline Bajek
Ecloserie Marine de Gravelines, Gravelines, France

Kris Hostens, Sofie Vandendriessche and Karl Van Ginderdeuren
Bio-environmental research group, Institute for Agricultural and Fisheries Research, Oostende, Belgium

Jan Vanaverbeke and Magda Vincx
Biology Department, Ghent University, Ghent, Belgium

Steven Degraer
Biology Department, Ghent University, Ghent, Belgium
Operational Directorate Natural Environment (OD Nature), Royal Belgian Institute of Natural Sciences (RBINS), Brussels, Belgium

Alan M. Friedlander
Fisheries Ecology Research Laboratory, Department of Biology, University of Hawaii, Honolulu, Hawaii, United States of America
Pristine Seas, National Geographic Society, Washington, DC, United States of America

Enric Ballesteros
Centre d'Estudis Avançats-CSIC, Blanes, Spain

Michael Fay
Pristine Seas, National Geographic Society, Washington, DC, United States of America
Wildlife Conservation Society, Bronx, New York, United States of America
Special Advisor, Presidence de la République, Libreville, République Gabonaise

Enric Sala
Pristine Seas, National Geographic
Society, Washington, DC, United States of America
Centre d'Estudis Avançats-CSIC, Blanes, Spain

Xie Zhenzhen, Xiao Ling, Wang Dengdong, Fang Chao, Liu Qiongyu, Li Zihao, Liu Xiaochun, Zhang Yong and Li Shuisheng
State Key Laboratory of Biocontrol, Institute of Aquatic Economic Animals, and the Guangdong Province Key Laboratory for Aquatic Economic Animals, School of Life Sciences, Sun Yat-Sen University, Guangzhou, China

Lin Haoran
State Key Laboratory of Biocontrol, Institute of Aquatic Economic Animals, and the Guangdong Province Key Laboratory for Aquatic Economic Animals, School of Life Sciences, Sun Yat-Sen University, Guangzhou, China
College of Ocean, Hainan University, Haikou, Hainan, China

William T. White and Peter R. Last
CSIRO Oceans & Atmosphere Flagship, Hobart, Tasmania, Australia

Dharmadi, Ria Faizah and Umi Chodrijah
Research Centre for Fisheries Management and Conservation, Agency for Marine and Fisheries Research and Development, Ministry of Marine Affairs and Fisheries, Jakarta, Indonesia

Rik C. Buckworth and Catherine M. Dichmont
CSIRO Oceans & Atmosphere Flagship, Dutton Park, Queensland, Australia

Martin Stevens, Alice E. Lown and Alexander M. Denton
Centre for Ecology and Conservation, College of Life and Environmental Sciences, University of Exeter, Penryn Campus, Penryn, Cornwall, United Kingdom

Sandie Millot, Marco Cerqueira, Maria-Filipa Castanheira and Catarina I. M. Martins
CCMAR-CIMAR L.A., Universidade do Algarve, Campus de Gambelas, Faro, Portugal

Øyvind Øverli
Norwegian University of Life Sciences, Department of Animal and Aquacultural Sciences, Ås, Norway

Rui F. Oliveira
ISPA Unidade de Investigação em Eco-Etologia Integrative Behavioural Biology Group, Lisboa, Portugal
Champalimaud Neuroscience Programme, Instituto Gulbenkian de Ciência, Oeiras, Portugal

Frederic Vandeperre, Jorge Fontes, Marco Santos, Ricardo Serrão Santos and Pedro Afonso
Centre of IMAR of the University of the Azores; Department of Oceanography and Fisheries, Horta, Portugal
LARSyS Associated Laboratory, Lisboa, Portugal

Alexandre Aires-da-Silva
Inter-American Tropical Tuna Commission, La Jolla, California, United States of America

Karen Lau, Keng Po Lai, Anna Tse and Rudolf Shiu Sun Wu
School of Biological Sciences, Kadoorie Biological Sciences Building, The University of Hong Kong, Hong Kong SAR, China
The State Key Laboratory in Marine Pollution, Hong Kong, Hong Kong SAR, China

Jessie Yun Juan Bao, Na Zhang and Si Lok
Genome Research Centre, The Hong Kong Jockey Club Building for Interdisciplinary Research, The University of Hong Kong, Pokfulam, Hong Kong SAR, China

Jing Woei Li
School of Life Sciences, Hong Kong Bioinformatics Centre, The Chinese University of Hong Kong, Hong Kong SAR, China

Richard Yuen Chong Kong
The State Key Laboratory in Marine Pollution, Hong Kong, Hong Kong SAR, China

Wing Yee Lui and Alice Wong
School of Biological Sciences, Kadoorie Biological Sciences Building, The University of Hong Kong, Hong Kong SAR, China

Tauna L. Rankin
Division of Marine Biology and Fisheries, Rosenstiel School of Marine and Atmospheric Science, University of Miami, Miami, Florida, United States of America
NOAA National Marine Fisheries Service, Office of Habitat Conservation, Coral Reef Conservation Program, Silver Spring, Maryland, United States of America

Su Sponaugle
Division of Marine Biology and Fisheries, Rosenstiel School of Marine and Atmospheric Science, University of Miami, Miami, Florida, United States of America
Department of Integrative Biology, Oregon State University, Hatfield Marine Science Center, Newport, Oregon, United States of America

Chiara Maggi, Serena Lomiri, Bianca Di Lorenzo, Marco d'Antona and Maria Teresa Berducci
ISPRA, Institute for Environmental Protection and Research, Roma, Italy

Diego Alvarez-Berastegui
Balearic Islands Coastal Observing and Forecasting System (SOCIB), Palma de Mallorca, Balearic Islands, Spain

Lorenzo Ciannelli
College of Earth, Ocean, and Atmospheric Sciences (CEOAS), Oregon State University, Corvallis, Oregon, United States of America

Alberto Aparicio-Gonzalez, Patricia Reglero, Manuel Hidalgo, Jose Luis López-Jurado and Francisco Alemany
Instituto Españ ol de Oceanografía - Centre Oceanogràfic de les Balears (COB-IEO), Palma de Mallorca, Balearic Islands, Spain

Joaquín Tintore
Balearic Islands Coastal Observing and Forecasting System (SOCIB), Palma de Mallorca, Balearic Islands, Spain
Institute for Advanced Studies (IMEDEA), Consejo Superior de Investigaciones Científicas (CSIC) and the University of the Balearic
Islands (UIB). Esporles, Spain

Nina Sandlund
Research group Disease and Pathogen transmission, Institute of Marine Research, Bergen, Norway

Britt Gjerset, Renate Johansen and Ingebjørg Modahl
Section of Virology, National Veterinary Institute, Oslo, Norway

Øivind Bergh
Research group Oceanography and climate, Institute of Marine Research, Bergen, Norway

Niels Jørgen Olesen
Section of Virology, Technical University of Denmark, Frederiksberg C, Denmark

Nicole A. Hill, Justin Hulls and Neville Barrett
Institute for Marine and Antarctic Studies, University of Tasmania, Hobart, Tasmania, Australia

Emma Lawrence
Digital Productivity Flagship, Commonwealth Scientific and industrial Research Organisation (CSIRO), Brisbane, QLD, Australia

Jeffrey M. Dambacher and Keith R. Hayes
Digital Productivity Flagship, Commonwealth Scientific and industrial Research Organisation (CSIRO), Hobart, Tasmania, Australia

Alan Williams
Oceans and Atmosphere Flagship, Commonwealth Scientific and industrial Research Organisation (CSIRO), Hobart, Tasmania, Australia

Scott Nichol
Geoscience Australia, Canberra, ACT, Australia

David W. Pond
Scottish Association for Marine Science, Oban, Argyll, United Kingdom

Geraint A. Tarling
British Antarctic Survey, Natural Environment Research Council, Cambridge, United Kingdom

Daniel J. Mayor
Institute of Biological and Environmental Sciences, Oceanlab, University of Aberdeen, Newburgh, Aberdeenshire, United Kingdom

Kimmo Kalevi Kahilainen
Kilpisjärvi Biological Station, University of Helsinki, Kilpisjärvi, Finland
Department of Environmental Sciences, University of Helsinki, Helsinki, Finland

William Paul Patterson
Department of Geological Sciences, Saskatchewan, Isotope Laboratory, University of Saskatchewan, Saskatoon, Saskatchewan, Canada

Eloni Sonninen
Finnish Museum of Natural History, University of Helsinki, Helsinki, Finland

Chris Harrod
School of Biological and Chemical Sciences, Queen Mary University of London, London, United Kingdom
Instituto de Ciencias Naturales
Alexander Von Humboldt, Universidad de Antofagasta, Antofagasta, Chile

Mikko Kiljunen
Department of Biological and Environmental Sciences, University of Jyväskylä, Jyväskylä, Finland

Sara Vieira, Samir Martins, M. Alexandra Teodósio
Centro de Ciências do Mar, Universidade do Algarve, Campus de Gambelas, Faro, Portugal

Lucy A. Hawkes
University of Exeter, College of Life and Environmental Sciences, Penryn
Campus, Cornwall, United Kingdom

Adolfo Marco
Estación Biológica de Doñ ana (CSIC), Américo Vespucio, Sevilla, Spain

Melanie J. Underwood and Arill Engås
Institute of Marine Research, Bergen, Norway
Department of Biology, University of Bergen, Bergen, Norway

Shale Rosen and Elena Eriksen
Institute of Marine Research, Bergen, Norway

John D. Neilson and Irene Andrushchenko
Fisheries and Oceans Canada, 531 Brandy Cove Road, St. Andrews, New Brunswick E5B 2L9 Canada

Josh Loefer
Marine Resources Research Institute, South Carolina Department of Natural Resources, PO Box 12559, Charleston, South Carolina, United States of America

Eric D. Prince
National Marine Fisheries Service, Southeast Fisheries Science Center, 75 Virginia
Beach Drive, Miami, Florida 33149, United States of America

François Royer
Datasio, 22 rue Mage, 31000 Toulouse, France

Beatriz Calmettes, Philippe Gaspar and Rémy Lopez
Collecte Localisation Satellites, Parc Technologique du Canal, 8–10 rue Hérmes, 31526 Ramonville, France

Brett Favaro
Department of Biology, University of Victoria, Victoria, Canada
Centre for Sustainable Aquatic Resources, Fisheries and Marine Institute of Memorial University of Newfoundland, St John's, Canada
Department of Ocean Sciences, Memorial University of Newfoundland, St. John's, Canada

Danielle C. Claar, Cameron Freshwater and Jessica J. Holden
Department of Biology, University of Victoria, Victoria, Canada

Caroline H. Fox
Department of Geography, University of Victoria, Victoria, Canada
Raincoast Conservation Foundation, Sidney, Canada

Allan Roberts
Bamfield Marine Sciences Centre, Bamfield East, Canada

UVic Research Derby
Department of Biology, University of Victoria, Victoria, Canada
Department of Geography, University of Victoria, Victoria, Canada
School of Environmental Studies, University of Victoria, Victoria, Canada

Index